Learning Factories

Eberhard Abele · Joachim Metternich
Michael Tisch

Learning Factories

Concepts, Guidelines, Best-Practice Examples

Eberhard Abele
Institute of Production Management,
 Technology and Machine Tools
Technical University of Darmstadt
Darmstadt, Germany

Michael Tisch
Institute of Production Management,
 Technology and Machine Tools
Technical University of Darmstadt
Darmstadt, Germany

Joachim Metternich
Institute of Production Management,
 Technology and Machine Tools
Technical University of Darmstadt
Darmstadt, Germany

ISBN 978-3-319-92260-7   ISBN 978-3-319-92261-4 (eBook)
https://doi.org/10.1007/978-3-319-92261-4

Library of Congress Control Number: 2018951698

© Springer Nature Switzerland AG 2019

This work is subject to copyright. All rights are reserved by the Publisher, whether the whole or part of the material is concerned, specifically the rights of translation, reprinting, reuse of illustrations, recitation, broadcasting, reproduction on microfilms or in any other physical way, and transmission or information storage and retrieval, electronic adaptation, computer software, or by similar or dissimilar methodology now known or hereafter developed.

The use of general descriptive names, registered names, trademarks, service marks, etc. in this publication does not imply, even in the absence of a specific statement, that such names are exempt from the relevant protective laws and regulations and therefore free for general use.

The publisher, the authors and the editors are safe to assume that the advice and information in this book are believed to be true and accurate at the date of publication. Neither the publisher nor the authors or the editors give a warranty, express or implied, with respect to the material contained herein or for any errors or omissions that may have been made. The publisher remains neutral with regard to jurisdictional claims in published maps and institutional affiliations.

This Springer imprint is published by the registered company Springer Nature Switzerland AG
The registered company address is: Gewerbestrasse 11, 6330 Cham, Switzerland

Preface

Today's Major Educational Challenges in the World

Economic growth and global welfare are challenged by a lack of well educated and suitably qualified workforce. The three major challenges encountered are:

- Insufficient availability of skilled and qualified workforces and engineers,
- A mismatch between qualification of the available workforce and the changes in industrial demands,
- Declining awareness of the importance of effective technical education over all hierarchical levels.

Production-related competencies and skills are and will remain an important compound of countries' economic development and growth as well as industries' long-term competitiveness. Good qualifications are also the decisive factor for prosperity and advancement on a personal level. Solutions must be quickly found for approaching the mismatch of institutional qualification programs and actual market requirements. On the one hand, the future workforce must be qualified for current employment opportunities, while on the other hand, today's generation needs to be enabled to keep up with competences required in the future.

Universities are often not fully aware of the challenges their students face in working life. Knowledge in basic engineering fields, like mechanics and thermodynamics, is still necessary but must be enhanced by process-oriented domains. In order to be optimally prepared for a career in industry, students

- have to understand the complexity of the systems and processes in a real workshop,
- must develop adopt the capability to improve the value stream with up-to-date methods,

- need to know the applications of digitization in the context of production and its improvement and be able to plan such systems, and
- have to develop more social and personal competencies like teambuilding and leadership.

Learning factories offer a promising environment to address the challenges mentioned above in education and training and also in research on process-oriented improvement in production. While numerous learning factories have been built in industry and academia in the last decade, a comprehensive scientific overview of the topic is still missing.

This book intends to close this gap by reviewing the current state of research and practice on the subject of learning factories. In addition, it gives the reader an overview of existing learning factories, their hardware, their didactic, and their operating concept.

We are convinced that learning factories can play an important role in the excellence of future generations of engineers and production employees.

Darmstadt, Germany

Eberhard Abele
Joachim Metternich
Michael Tisch

About This Book

"Learning Factories" according to Encyclopedia CIRP, see Abele, E. (2016). Learning Factory. *CIRP Encyclopedia of Production Engineering*.

"A Learning Factory in a narrow sense is a learning environment specified by

- **processes** that are *authentic*, *include multiple stations*, and comprise *technical* as well as *organizational* aspects,
- a **setting** that is *changeable* and resembles a *real value chain*,
- a *physical* **product** being manufactured, and
- a **didactical concept** that comprises *formal, informal and non-formal learning*, enabled by *own actions of the trainees* in an *on-site learning* approach.

Depending on the **purpose** of the Learning Factory, learning takes place through *teaching*, *training* and/or *research*. Consequently, learning outcomes may be *competency development* and/or *innovation*. An operating model ensuring the sustained operation of the Learning Factory is desirable.

In a broader sense, learning environments meeting the definition above but with

- a setting that resembles a *virtual* instead of a *physical value chain*, or
- a *service* product instead of a *physical* product, or
- a didactical concept based on *remote learning* instead of *on-site learning*

can also be considered as Learning Factories."

Contents

1	**Challenges for Future Production/Manufacturing**		1
	1.1 Globalization		4
	1.2 New Technologies, Digitalization, and Networking		6
	1.3 Dynamic Product Life Cycles		11
	1.4 Limited Natural Resources		11
	1.5 Knowledge Society		14
	1.6 Risk of Instability		15
	1.7 Demographic Change		16
	1.8 Wrap-up of This Chapter		17
	References		18
2	**Competencies for Future Production**		23
	2.1 Competencies, Qualification, and Knowledge		23
	2.1.1 Knowledge		24
	2.1.2 Qualification		25
	2.1.3 Competence/Competency		27
	2.2 Learning Goals and Learning Outcomes		28
	2.3 Addressed Competencies in Learning Factories		30
	2.4 Relevant Competencies for Industrie 4.0		34
	2.5 A Domain-Specific Competency Model for Lean 4.0		43
	2.6 Wrap-up of This Chapter		50
	References		52
3	**Learning in Production, Learning for Production**		59
	3.1 Definition of Basic Terms and Notions		59
	3.2 Historical Development of Work-Related Learning		63
	3.3 Forms of Work-Related Learning for Production		66
	3.4 Types of Perceived Learning Concepts in Production		68

	3.5	Need for Learning Factories	74
	3.6	Wrap-up of This Chapter	75
	References		75
4	**Historical Development, Terminology, and Definition of Learning Factories**		**81**
	4.1	Historical Development of the Learning Factory Concepts	81
	4.2	Terminology of Learning Factories	84
	4.3	Definition of Learning Factories	86
	4.4	Wrap-up of This Chapter	93
	References		94
5	**The Variety of Learning Factory Concepts**		**99**
	5.1	Learning Factory Morphology: Dimension 1 "Operational Model"	100
		5.1.1 Economic or Financial Sustainability of the Learning Factory Concept	101
		5.1.2 Contentual or Thematic Sustainability of the Learning Factory Concept	104
		5.1.3 Personal Sustainability of the Learning Factory Concept	105
	5.2	Learning Factory Morphology: Dimension 2 "Targets and Purpose"	106
	5.3	Learning Factory Morphology: Dimension 3 "Process"	108
	5.4	Learning Factory Morphology: Dimension 4 "Setting"	110
	5.5	Learning Factory Morphology: Dimension 5 "Product"	113
	5.6	Learning Factory Morphology: Dimension 6 "Didactics"	116
	5.7	Learning Factory Morphology: Dimension 7 "Learning Factory Metrics"	117
	5.8	Learning Factory Database	119
	5.9	Wrap-up of This Chapter	121
	References		122
6	**The Life Cycle of Learning Factories for Competency Development**		**127**
	6.1	Learning Factory Planning and Design	128
		6.1.1 Overview Planning and Design Approaches	128
		6.1.2 The Darmstadt Approach to Competency-Oriented Planning and Design	130
	6.2	Learning Factory Built-up, Sales, and Acquisition	146
		6.2.1 Design and Built-up of Customer-Individual Learning Factories	147

		6.2.2	Built-up of Standardized Turnkey Learning Factories 150
		6.2.3	Offer of Learning Factory Trainings for Industrial Companies 156
	6.3	Learning Factory Operation, Evaluation, and Improvement ...	158
		6.3.1	Training Management for Learning Factories in Operation 158
		6.3.2	Quality System for Learning Factories Based on a Maturity Model 159
		6.3.3	Evaluation of the Success of Learning Factories 164
	6.4	Remodeling Learning Factory Concepts 191	
	6.5	Wrap-Up of This Chapter 193	
	References 193		
7	**Overview on Existing Learning Factory Application Scenarios** 199		
	7.1	Learning Factories in Education 214	
		7.1.1	Active Learning in Learning Factories 216
		7.1.2	Action-Oriented Learning in Learning Factories 218
		7.1.3	Experiential Learning and Learning Factories 219
		7.1.4	Game-Based Learning in Learning Factories and Gamification 220
		7.1.5	Problem-Based Learning in Learning Factories 226
		7.1.6	Project-Based Learning in Learning Factories 227
		7.1.7	Research-Based Learning in Learning Factories 228
		7.1.8	Best Practice Examples for Education 230
		7.1.9	Example: Learning Factories for Industrie 4.0 Vocational Education in Baden-Württemberg 232
	7.2	Learning Factories in Training 234	
		7.2.1	Competency Development in Course of Trainings in Learning Factories 235
		7.2.2	Best Practice Examples for Training 237
		7.2.3	Success Factors of Learning Factories for Education and Training 239
		7.2.4	Learning Factory Trainings as a Part of Change Management Approaches...................... 243
		7.2.5	Technology and Innovation Transfer in Course of Learning Factory Trainings 244
	7.3	Learning Factories in Research 246	
	7.4	Wrap-up of This Chapter 254	
	References 254		

8	Overview on the Content of Existing Learning Factories	263
	8.1 Learning Factories for Lean Production	264
	8.2 Learning Factories for Industrie 4.0	267
	8.3 Learning Factories for Resource and Energy Efficiency	270
	8.4 Learning Factories for Industrial Engineering	272
	8.5 Learning Factories for Product Development	274
	8.6 Other Topics Addressed in Learning Factories	275
	8.6.1 Learning Factories for Additive Manufacturing	276
	8.6.2 Learning Factories for Automation	276
	8.6.3 Changeability	277
	8.6.4 Complete Product Creation Processes	278
	8.6.5 Global Production Networks	279
	8.6.6 Intralogistics and Logistics	279
	8.6.7 Sustainability	281
	8.6.8 Workers Participation	281
	8.7 Learning Factories for Specific Industry Branches or Products	282
	8.8 Wrap-up of This Chapter	282
	References	283
9	Overview on Potentials and Limitations of Existing Learning Factory Concept Variations	289
	9.1 Potentials of Learning Factories	290
	9.2 Limitations of Learning Factories	290
	9.3 Learning Factory Concept Variations of Learning Factories in the Narrow Sense—Advantages and Disadvantages	295
	9.3.1 The Learning Factory Core Concept	295
	9.3.2 Model Scale Learning Factories	296
	9.3.3 Physical Mobile Learning Factories	299
	9.3.4 Low-Cost Learning Factories	300
	9.3.5 Digitally and Virtually Supported Learning Factories	301
	9.3.6 Producing Learning Factories	304
	9.4 Learning Factory Concept Variations of Learning Factories in the Broader Sense—Advantages and Disadvantages	305
	9.4.1 Digital, Virtual, and Hybrid Learning Factories	306
	9.4.2 Teaching Factories and Remotely Accessible Learning Factories	313
	9.5 Wrap-up of This Chapter	316
	References	318
10	Projects and Groups Related to Learning Factories	323
	10.1 Initiative on European Learning Factories	325
	10.2 Conferences on Learning Factories	327

	10.3	Netzwerk Innovativer Lernfabriken	328
	10.4	CIRP Collaborative Working Group on Learning Factories	329
	10.5	International Association of Learning Factories (IALF)	331
	10.6	Wrap-up This Chapter	331
	References		334
11	**Best Practice Examples**		**335**
	11.1	Best Practice Example 1: AutFab at the Faculty of Electrical Engineering of the University of Applied Sciences Darmstadt, Germany	335
		11.1.1 Starting Phase and Purpose	336
		11.1.2 Learning Environment and Products	337
		11.1.3 Operation	338
	11.2	Best Practice Example 2: Demonstration Factory at WZL, RWTH Aachen, Germany	338
		11.2.1 Starting Phase and Purpose	339
		11.2.2 Learning Environment and Products	340
		11.2.3 Operation	342
	11.3	Best Practice Example 3: Die Lernfabrik at IWF, TU Braunschweig, Germany	342
		11.3.1 Starting Phase and Purpose	343
		11.3.2 Learning Environment and Products	344
		11.3.3 Operation	345
	11.4	Best Practice Example 4: E|Drive-Center at FAPS, Friedrich-Alexander University Erlangen-Nürnberg, Germany	347
		11.4.1 Starting Phase and Purpose	347
		11.4.2 Learning Environment and Products	348
		11.4.3 Operation	349
	11.5	Best Practice Example 5: ESB Logistics Learning Factory at ESB Business School at Reutlingen University, Germany	350
		11.5.1 Starting Phase and Purpose	350
		11.5.2 Learning Environment and Products	351
		11.5.3 Operation	352
		11.5.4 Outlook and Future Development	354
	11.6	Best Practice Example 6: ETA-Factory at PTW, TU Darmstadt, Germany	354
		11.6.1 Starting Phase and Research in the ETA-Factory	355
		11.6.2 Learning Environment and Products	356
		11.6.3 Operation	358
	11.7	Best Practice Example 7: Festo Didactic Learning Factories	358

		11.7.1	History and Categories	359
		11.7.2	Products	361
		11.7.3	Customers and Operation Models	362
		11.7.4	Didactical Environment	362
	11.8	Best Practice Example 8: Festo Learning Factory in Scharnhausen, Germany		362
		11.8.1	Starting Phase and Purpose	363
		11.8.2	Learning Environment and Products	364
		11.8.3	Operation	365
	11.9	Best Practice Example 9: IFactory at the Intelligent Manufacturing Systems (IMS) Center, University of Windsor, Canada		367
		11.9.1	Starting Phase and Purpose	367
		11.9.2	Learning Environment and Product	368
		11.9.3	Operation	370
	11.10	Best Practice Example 10: IFA-Learning Factory at IFA, Leibniz University Hannover, Germany		371
		11.10.1	Starting Phase and Purpose	371
		11.10.2	Learning Environment and Products	372
		11.10.3	Operation	373
	11.11	Best Practice Example 11: Integrated Learning Factory at LPE & LPS, Ruhr-University Bochum, Germany		374
		11.11.1	Starting Phase and Purpose	374
		11.11.2	Learning Environment and Products	375
		11.11.3	Operation	376
	11.12	Best Practice Example 12: LEAN-Factory for a Pharmaceutical Company in Berlin, Germany		377
		11.12.1	Starting Phase and Purpose	378
		11.12.2	Learning Environment and Products	379
		11.12.3	Operation	380
	11.13	Best Practice Example 13: Learning and Innovation Factory at IMW, IFT, and IKT, TU Wien, Austria		381
		11.13.1	Starting Phase and Purpose	381
		11.13.2	Learning Environment and Products of the Learning & Innovation Factory	382
		11.13.3	Operation	382
	11.14	Best Practice Example 14: Learning Factory AIE at IFF, University of Stuttgart, Germany		383
	11.15	Best Practice Example 15: Learning Factory for Electronics Production at FAPS, Friedrich-Alexander University Erlangen-Nürnberg, Germany		388
		11.15.1	Starting Phase and Purpose	389

		11.15.2	Learning Environment and Products	390
		11.15.3	Operation	391
		11.15.4	Education	391
		11.15.5	Research	392
	11.16	Best Practice Example 16: Learning Factory for Innovation, Manufacturing and Cooperation at the Faculty of Industrial and Process Engineering of Heilbronn University of Applied Sciences, Germany		392
		11.16.1	Starting Phase and Purpose	393
		11.16.2	Learning Environment and Products	394
		11.16.3	Operation	395
	11.17	Best Practice Example 17: Learning Factory for Global Production at Wbk, KIT Karlsruhe, Germany		396
		11.17.1	Starting Phase and Purpose	397
		11.17.2	Learning Environment and Products	399
		11.17.3	Operation	399
	11.18	Best Practice Example 18: "Lernfabrik für schlanke Produktion", Learning Factory for Lean Production at iwb, TU München, Germany		399
		11.18.1	Starting Phase and Purpose	401
		11.18.2	Learning Environment and Products	402
		11.18.3	Operation	402
	11.19	Best Practice Example 19: "Lernfabrik für vernetzte Produktion" at Fraunhofer IGCV, Augsburg, Germany		403
		11.19.1	Starting Phase and Purpose	405
		11.19.2	Learning Environment and Products	406
		11.19.3	Operation	407
	11.20	Best Practice Example 20: LMS Factory at the Laboratory for Manufacturing Systems & Automation (LMS), University Patras, Greece		407
		11.20.1	Starting Phase and Purpose	408
		11.20.2	Learning Environment and Products	410
		11.20.3	Operation	410
	11.21	Best Practice Example 21: LPS Learning Factory at LPS, Ruhr-Universität Bochum, Germany		412
		11.21.1	Starting Phase and Purpose	412
		11.21.2	Learning Environment and Products	413
		11.21.3	Operation	414
	11.22	Best Practice Example 22: MAN Learning Factory at MAN Diesel & Turbo SE in Berlin, Germany		416
		11.22.1	Starting Phase and Purpose	417
		11.22.2	Learning Environment and Products	418
		11.22.3	Operation	419

11.23	Best Practice Example 23: MPS Lernplattform at Daimler AG in Sindelfingen, Germany		420
	11.23.1	Starting Phase and Purpose	420
	11.23.2	Learning Environment and Products	421
	11.23.3	Operation	421
11.24	Best Practice Example 24: MTA SZTAKI Learning Factory at the Research Laboratory on Engineering and Management Intelligence, MTA SZTAKI in Győr, Hungary		422
	11.24.1	Starting Phase and Purpose	424
	11.24.2	Learning Environment and Products	425
	11.24.3	Operation	425
11.25	Best Practice Example 25: Pilot Factory Industrie 4.0 at IMW, IFT, and IKT, TU Wien, Austria		427
	11.25.1	Starting Phase and Purpose	427
	11.25.2	Learning Environment and Products TU Wien Pilot Factory Industrie 4.0 (PF)	428
	11.25.3	Operation	428
11.26	Best Practice Example 26: Process Learning Factory CiP at PTW, TU Darmstadt, Germany		430
	11.26.1	Starting Phase and Purpose	431
	11.26.2	Learning Environment and Products	432
	11.26.3	Operation	434
11.27	Best Practice Example 27: Smart Factory at the Research Laboratory on Engineering & Management, MTA SZTAKI in Budapest, Hungary		435
	11.27.1	Starting Phase and Purpose	435
	11.27.2	Learning Environment and Products	436
	11.27.3	Operation	438
11.28	Best Practice Example 28: Smart Factory at DFKI, TU Kaiserslautern, Germany		438
	11.28.1	The Aim of SmartFactory^{KL} Is to Pave the Way for the Intelligent Factory of Tomorrow	439
	11.28.2	SmartFactory^{KL} is Industrie 4.0 You Can Touch—And Take Part In	440
	11.28.3	A Plant that Writes History	441
11.29	Best Practice Example 29: Smart Mini-Factory at IEA, Free University of Bolzano, Italy		442
	11.29.1	Starting Phase and Purpose	443
	11.29.2	Learning Environment and Products	443
	11.29.3	Operation	445

	11.30	Best Practice Example 30: Teaching Factory: An Emerging Paradigm for Manufacturing Education. LMS, University of Patras, Greece	445
		11.30.1 Purpose	447
		11.30.2 Learning Environment	447
		11.30.3 Operation	449
	11.31	Best Practice Example 31: VPS Center of the Production Academy at BMW in Munich, Germany	450
		11.31.1 Starting Phase and Purpose	451
		11.31.2 Learning Environment and Products	452
		11.31.3 Operation	452
	List of Contributors		456
	References		458
12	**Conclusion and Outlook**		461
	Reference		464

About the Authors

Prof. Dr.-Ing. Eberhard Abele was born in 1953 and studied mechanical engineering at the *Technische Hochschule Stuttgart*, Germany, where he received his diploma in 1977. Afterward, he was a research assistant and department manager at the Fraunhofer Institute for Production Technology and Automation, Stuttgart, (IPA). From 1983 to 1999, he held a leading position in the automotive supply industry as main department manager of manufacturing technology and factory manager with stations in Spain and France. The focus of his industrial activities was on automation, increase in productivity, and speeding up production. Since July 2000 he is leading the Institute of Production Management, Technology and Machine Tools (PTW) and contributed to more than 200 publications in the field of manufacturing organization, machine tool technology, and manufacturing processes.

The Process Learning Factory (CIP), which he initiated, has shown a novel path in the long-term qualification of university graduates and employees from companies in teaching, but also in further education in the field of production technology and lean management. He led the initiative *Produktionsforschung 2020* for the *Bundesministerium für Bildung und Forschung* (BMBF). He is the founder of the Initiative on European Learning Factories and chairman of the magazine *Werkstatt und Betrieb* of the Carl Hanser publisher.

He initiated the research project *eta-Fabrik* (learning factory for energy efficiency) with 35 partners with a total project sum of 16 million euros.

Since 2012 the PTW is managed together with Prof. Metternich.

Prof. Dr.-Ing. Joachim Metternich studied industrial engineering at the Technical University of Darmstadt and received his doctorate in 2001. After his time as assistant of the CEO of TRUMPF Werkzeugmaschinen GmbH, he headed a production group at Bosch Diesel s.r.o. in the Czech Republic. He then took over the responsibility for the worldwide lean production system of Knorr-Bremse SfS GmbH, a manufacturer of braking systems for rail vehicles. Since 2012 he has been one of the two directors of the Institute of Production Management, Technology and Machine Tools (PTW).

Dr.-Ing. Michael Tisch studied industrial engineering with a technical specialization in mechanical engineering at the Technical University of Darmstadt. Until July 2018 he worked as chief engineer at the Institute of Production Management, Technology and Machine Tools at the Technical University of Darmstadt. In his research, he deals with the competency-oriented learning factory design in the field of lean.

Chapter 1
Challenges for Future Production/Manufacturing

Education and training have numerous positive effects on the individuals, and the companies these individuals are working for as well as on society. If we bear this in mind, it is obvious that more high-quality education and training are beneficial to everyone (Fig. 1.1).

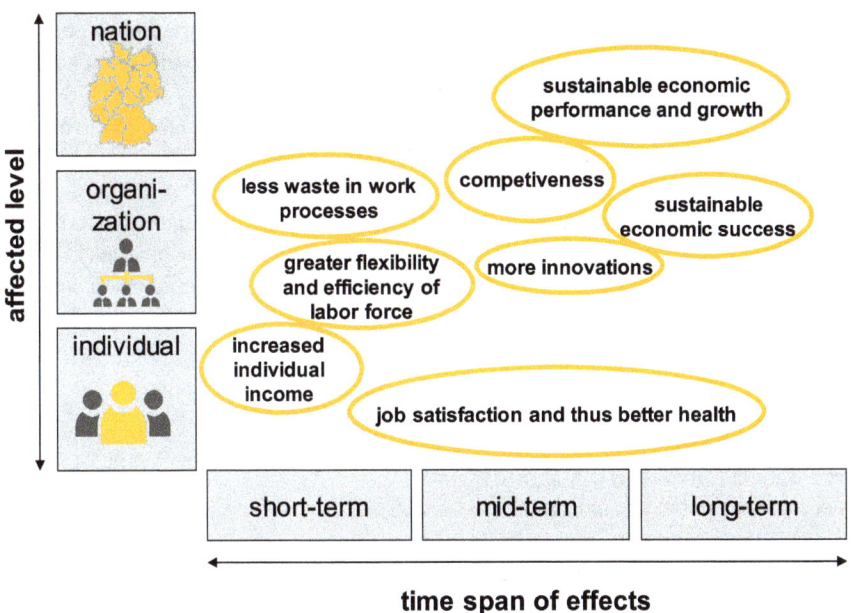

Fig. 1.1. Effects of education and training on national, organizational, and individual level

In economic terms, the general positive effects of educational quality on individual income, competitiveness of enterprises, and economic growth were shown.[1] It can be concluded that human capital is key to economic development.[2] On the personal level, incomes of individuals and cognitive skills are systematically related.[3]

Also for companies in the manufacturing sector in particular, the education and training play a crucial role. Today's and tomorrow's economic success of companies depend on the capability and the knowledge of engineers and managers.[4] It is mentioned that the lack of skills, e.g. entrepreneurial, managerial, or scientific management skills, greatly reduces the ability to innovate regarding fundamentally new products, process efficiency, productivity, and quality.[5] Studies forecast a significant shift in labor demand toward more knowledge- and competency-intensive jobs in the future.[6] For this reason, a larger number of specialists are needed, with gaps being expected in the future for certain professional functions and qualifications.[7] Positive effects of education for the recruitment of the new knowledge workers are also mentioned in this context.[8] In the literature of education controlling is stated that the return on training is almost always positive, can be very high, and can occur in many forms. Examples for the benefits induced by education are among others a greater level of value-adding activities, higher flexibility and a better innovation ability.[9]

In the long-term education and training are crucial for economic growth and the competitiveness of whole nations: "[…] a more skilled population – almost certainly including both a broadly educated population and a cadre of top performers – results in stronger economic performance for nations."[10]

"More and better education tends to shift comparative advantage away from primary production toward manufacturing and services, and thus to accelerate learning by doing and growth."[11]

Furthermore, for any nation, the industry sector is an important factor for the creation of wealth. For example, in Europe more than 26% of the value-added share in the non-financial business economy is accounted to the manufacturing sector.[12] Despite the widely propagated change from the industrial to the service and information society, the facts show production is still the backbone of the prosperity of

---

[1] See Gylfason (2001), Hanushek and Woessmann (2007), Barro (1996).
[2] Hanushek and Woessmann (2007).
[3] See Hanushek and Woessmann (2007).
[4] See O'Sullivan, Rolstadås, and Filos (2011).
[5] Tether, Mina, Consoli, and Gagliardi (2005).
[6] CEDEFOP (2010).
[7] See Vieweg (2011).
[8] See O'Sullivan et al. (2011).
[9] See Smith (2001).
[10] Hanushek and Woessmann (2007).
[11] Gylfason (2001).
[12] See Eurostat (2016).

the industrialized countries.[13] In Germany, over 8 million jobs are directly located in production areas.[14] In addition, approximately 6 million employees were allocated to the production-induced area of corporate services such as logistics and information technology.[15] This means 14 of in total 40 million employees in Germany are directly associated with the production sector. Some estimations conclude that in total 70% of jobs and 75% of the GDP in Europe are related to manufacturing.[16]

In order to obtain the importance of the sector in respective regions, excellent production processes are necessary to compete in the global race. Today, manufacturing is confronted with several megatrends and significant innovations, inter alia, regarding technologies, tools, and techniques. In this respect, we start with the question: What are the main drivers for the development of the production of the future? And based on this, continuing with the question: What particular competences do we need for the production of the future. This chapter therefore deals with the challenges for future production. In order to tackle those challenges properly, today's and future's engineers and blue-collar workers need the capability to learn and adjust to new situations—Chap. 2 deals with required competencies for future production. Chapter 3 addresses the ways that are available for competency development for production and concludes with the need for learning factories for manufacturing.

Learning factory design must pick up current and future developments in production. Those developments are accompanied by economic, ecologic, and social megatrends we currently recognize. Megatrends are seen as enormous economic, social, political, and technological changes with high probability that influence our lives for many years (7–10 or longer).[17] Although temporarily short-term developments may superimpose megatrends, in the longer term, they determine the direction of change regarding organizational, technological, and human-related issues.[18] These changes in production must be addressed with groundbreaking innovations regarding production processes, products, services, and technologies.[19] In Fig. 1.2 exemplarily an overview of the megatrends[20] can be found.

A loss of 20–25% of the production-related jobs is predicted in economies when companies do not adapt to these trends. These developments lead to a rapidly growing uncertainty and complexity in manufacturing companies, which will require new knowledge, skills, and competences. In addition, developed industries are under the pressure of an aging workforce, to secure their competitive advantage, the innovation of product and process must be supported by new approaches to develop

---

[13] Abele and Reinhart (2011).
[14] See DESTATIS (2016).
[15] See DESTATIS (2016).
[16] See O'Sullivan et al. (2011).
[17] Naisbitt (1982).
[18] See Abele and Reinhart (2011).
[19] See Grömling and Haß (2009).
[20] Identified by Abele and Reinhart (2011).

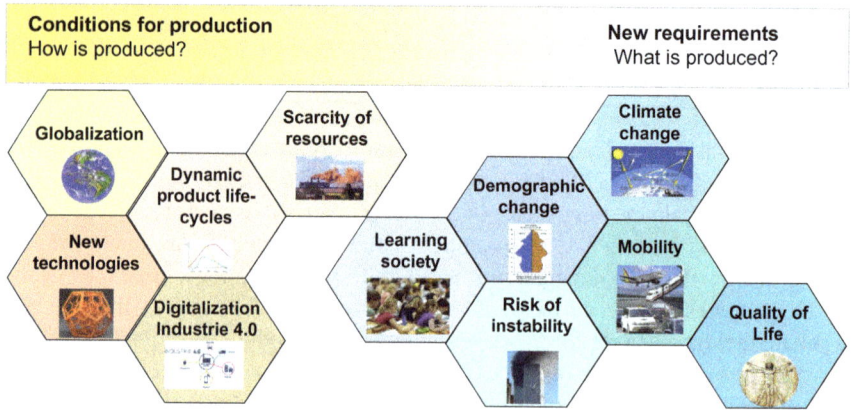

Fig. 1.2 Megatrends with crucial importance for production and products (Abele & Reinhart, 2011)

production-related competences at all hierarchical levels.[21] The recruitment and especially the development of competent employees are crucial competitive factors of companies that determine the success or failure. Figure 1.3 shows the result of a literature study looking at ten individual studies on the future of production.

In the overview, it is also shown whether the identified megatrends have an effect on the future design of the production processes or product characteristics.[22] The trends listed higher in the figure have a greater impact on the production, while the ones below have a corresponding effect on the product design. Globalization and thus an intensified competition, dynamic product life cycles, the emergence of new technologies, digitalization and networking, the scarcity of resources, the importance of knowledge, the risk of instability as well as demographic change are identified as the most important challenges for industrial production.[23] Figure 1.4 shows an overview of the structure of this chapter on the challenges for future production.

1.1 Globalization

The global integration of businesses, culture, politics, and other areas is referred to as globalization.[24] The main reasons for this development can be found in the progress of ICT, better quality of traffic technology, and the liberalization of world trade.[25]

---

[21] Abele and Reinhart (2011), Adolph, Tisch, and Metternich (2014).
[22] See also Abele and Reinhart (2011).
[23] See Adolph et al. (2014).
[24] Abele and Reinhart (2011).
[25] See Arndt (2008), Naisbitt (1982).

1.1 Globalization

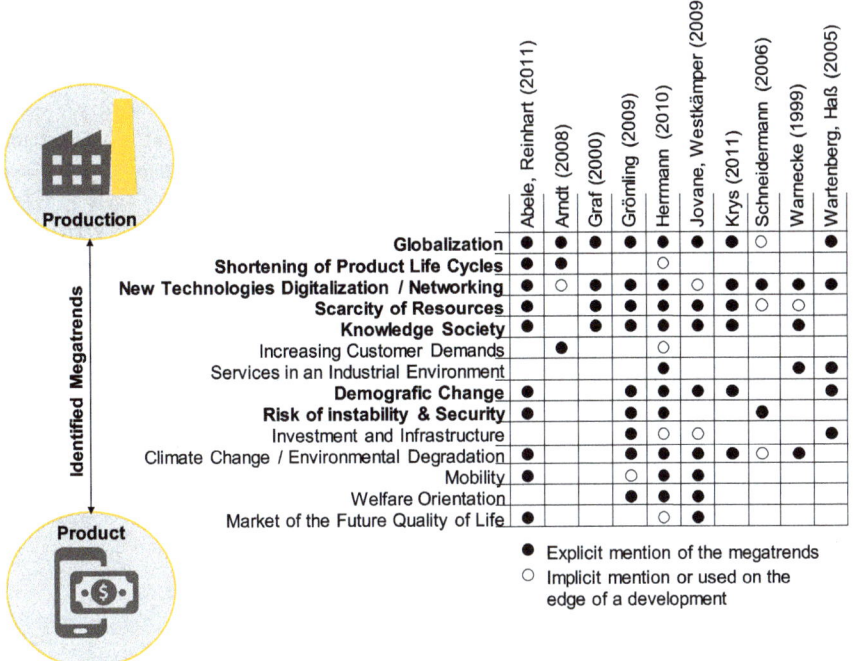

Fig. 1.3 Identified megatrends in literature shown in Adolph et al. (2014), based on Abele and Reinhart (2011), Arndt (2008), Graf (2000), Grömling and Haß (2009), Herrmann (2010), Jovane, Westkämper, and Williams (2009), Krys (2011), Warnecke (1999) und Wartenberg and Haß (2005)

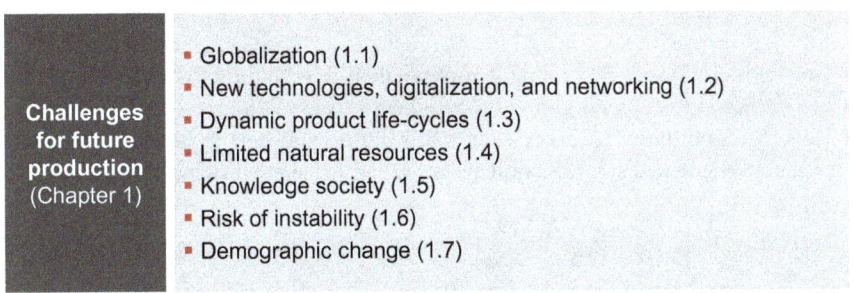

Fig. 1.4 Overview on the structure of this chapter

Although currently protectionist policies and a restriction of free trade seem to be gaining the upper hand, time will tell whether the long-term trends of globalization and liberalization are actually broken or will even be reversed, or whether only temporary phenomena can be observed. In the past, a successful export orientation enabled many European countries to benefit from the trend of global networks and thereby secure or even expand production and related jobs in their own country.[26] The example of German industry shows that production of high-quality goods in international networks has a positive effect on the national employment in industry, as long as the core production expertise will remain in the home country.[27] This preservation of jobs in the home country can only be realized with an educated and well-trained workforce. Furthermore, the globalization trend can also be perceived in international acquisitions of (mainly industrial) companies: The yearly number of acquisitions from China in Europe has risen by 48% from 2015 to 2016—and has multiplied more than sevenfold in the last ten years.[28] This phenomenon leads among others to the

- need for international cooperation,
- globally networked value chains that must be designed and managed, and
- a high demand for the worldwide standardization of production systems.

Figure 1.5 shows the number of acquisitions or investments of Chinese companies in Europe as well as a few prominent European companies acquisition China.

Consequently, in light of the megatrend globalization several challenges arise for industrial companies in high-wage countries:

- Achievement of leading productivity in international comparison,
- Availability of well-educated and excellent trained workforce; thinking globally, acting locally,
- Ensuring the highest quality of goods in the production network as a prerequisite,
- Achievement of high levels of changeability and flexibility of production systems.

In our discussion with Production and Human Resources Managers related to the megatrend globalization, it was often argued that future blue- and white-collar workers.

- Need possibilities to develop and improve their intercultural skills,
- Get earlier in touch with global procurement processes, and
- Have to opportunity to experience optimization and best practice examples of production processes, production systems, and value creation networks.

1.2 New Technologies, Digitalization, and Networking

Innovations are enabled by an increasing cooperation and integration of disciplines, since innovations often originate at interfaces of disciplines. For example, with the integration of first mechanics and electronics and subsequently with informatics,

---

[26] See Grömling and Haß (2009).
[27] Scherrer, Simons, and Westermann (1998).
[28] See Sun and Kron (2017).

1.2 New Technologies, Digitalization, and Networking

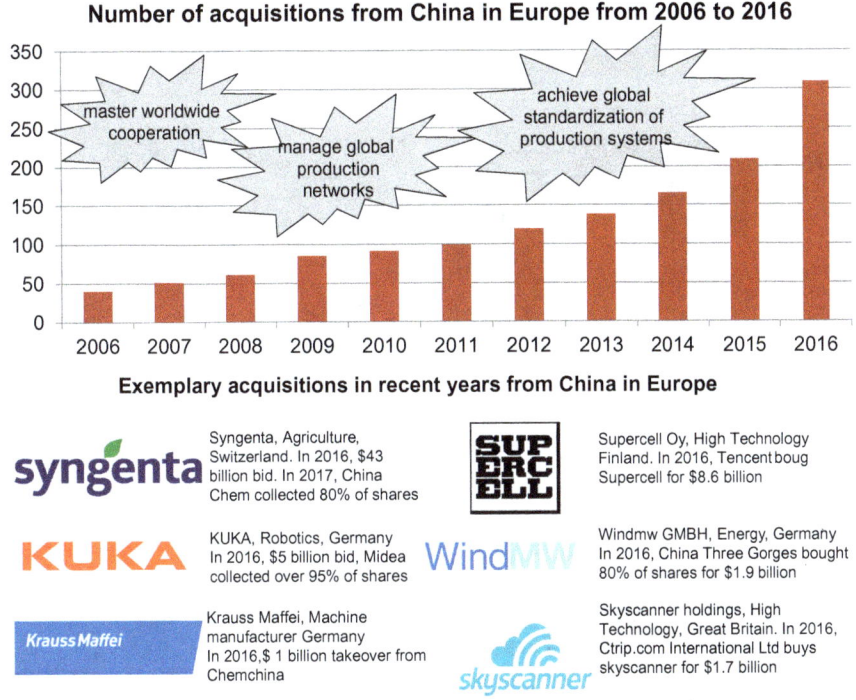

Fig. 1.5 Chinese acquisitions in Europe in recent years (will) lead to various challenges for production, data from Sun and Kron (2017)

innovative products and production processes are facilitated.[29] Most recently, the buzzword Industrie 4.0 (or also industrial Internet) falls into this category and has decisive influence on the way in which production processes are to be designed in the future. The Industrie 4.0 "project" envisions a factory with networked equipment, in which every product knows or even finds its way to finalization.[30] Consequently, the role of humans in production systems may shift if they are relieved of routine activities and on a broad data basis optimal decision making is enabled.

---

[29] Abele and Reinhart (2011).

[30] See Promotorengruppe Kommunikation der Forschungsunion Wirtschaft - Wissenschaft (2013).

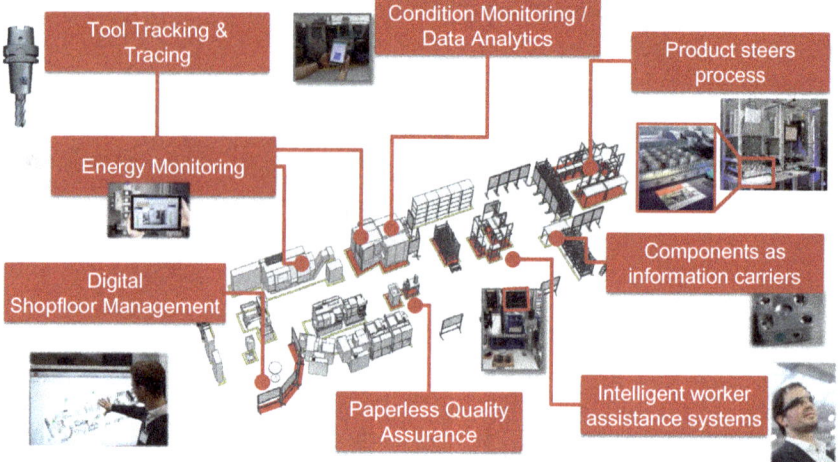

Fig. 1.6 Exemplary Industrie 4.0 concepts implemented in the Process Learning Factory CiP in Darmstadt

Figure 1.6 exemplarily shows some implemented Industrie 4.0 concepts and use cases in the Process Learning Factory CiP in Darmstadt:[31]

- Components as information carrier: In order to achieve efficient and future-oriented production in the sense of Industrie 4.0, the collection and the processing of the data that is generated during the value creation process are particularly important. In addition to the integration of necessary sensors into the production process, communication between all the systems and equipment involved is also necessary for the implementation of a media-free, digital and typically automatic data acquisition.
- Tool tracking and tracing: By integrating innovative sensor technology into the tool holder, the tool can be monitored and the entire tool circuit can be networked. The track and trace system on the control level makes it possible to optimize route planning, inventory management, procurement, storage location, and storage size actively.
- Condition and energy monitoring: With condition and energy monitoring, data from manufacturing machines can be used to access a real-time image of the quality or the energy consumption of the production process. The quality of the processing state here includes the control of product state, process state, and machine condition.
- Product steers process: The product variant is defined with a product configurator by the customer; the information is stored directly on the component. Before assembly, the component uses RFID to call a type-specific nonlinear assistant system for the respective operator, which allows the desired motor configuration

---

[31] A detailed description of the implemented use cases can be found in Abele et al. (2015) and PTW, TU Darmstadt (2017a).

1.2 New Technologies, Digitalization, and Networking

to be built. Data generated during processing such as assembly and screw protocols is stored cloud-based and can be accessed by the RFID information. In this way, the data remains permanently available.

- Digital shop floor management: In the context of the networked factory, employees face complex IT systems and autonomously operating machines. Employees in this environment have to simultaneously act as a flexible and creative problem solver. An instrument for supporting employees in this process is shop floor management enabled by now available real-time data. This serves as a central communication and collaboration platform for the employees at the shop floor in their daily tasks.
- Digital twin: In order to obtain all relevant information about the process in real time at all times, a digital value stream image is used. In this, all relevant information flows are networked across the entire value stream. The user-friendly visualization and linking of these previously separately collected and used data provides the basis for the rapid detection of potentials for improvement.
- Paperless quality assurance: A paperless, reliable, and automated quality assurance system is demonstrated in the manual assembly of the pneumatic cylinder. An electronic screw station is not activated until the upstream quality control releases the component. The screw station selects the corresponding screw program based on the present variant, and a work instruction is displayed to the worker. During the assembly of the cylinder, additional process characteristics for quality detection, e.g. torque or yield points are assigned to the identification number of the pneumatic cylinder currently being processed. The continuity of the documentation accompanying the process of the product quality as well as the test results enables a holistic traceability on the product level.
- Intelligent worker assistance systems: Assembly information is created and made available interactively from the 3D-CAD system for the assembly of small batch sizes. Parts of the implementation are intelligent networking of all components of the assembly workplace as well as systems for visual support and control in the assembly process. A bidirectional information flow between the system and the employee is enabled.

Another important key technology is seen in the additive manufacturing. Using the additive manufacturing with metallic materials, new shapes and geometrical features can be fabricated on-demand and customized.[32] But those additive manufactured parts will not only affect the possibilities to design products but also the possibilities and requirements of respective manufacturing processes. Accordingly, processes starting from CAD data creation over preprocessing, the actual additive manufacturing process, and post-processing have to be developed and designed which requires technological innovations and a broad variety of new additive manufacturing competences in companies in order to fully integrate and use the potential of additive manufacturing. Figure 1.7 illustrates a typical additive manufacturing process chain that can be part of a learning factory for additive manufacturing.

---

[32] See Vayre, Vignat, and Villeneuve (2012), Huang, Liu, Mokasdar, and Hou (2013).

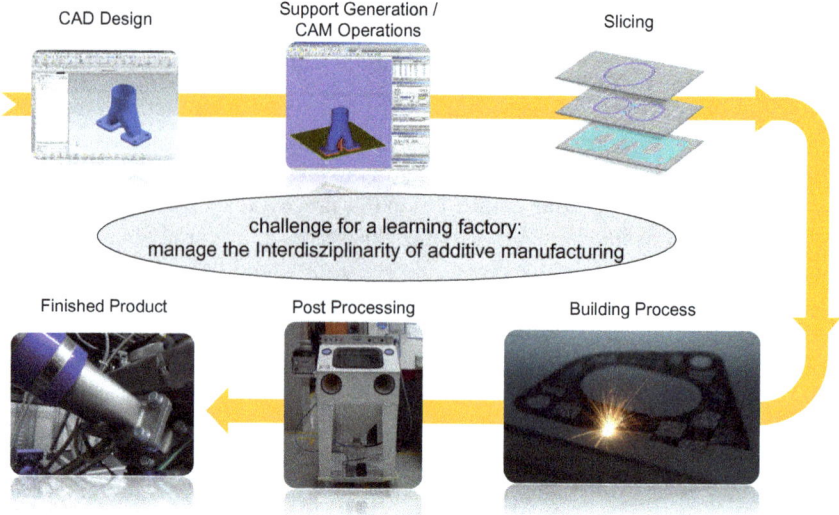

Fig. 1.7 Additive process chain changes the possibilities and requirements of manufacturing processes

Based on the mentioned new technologies for future production, the following challenges can be identified:

- Integration of innovative key technologies in production environments, e.g. additive production technologies,
- Digitalization and networking of existing production environments,
- Creation of simplified supply chains when powder or granulate replaces supplied parts,
- Consideration of adapted visions, principles, and methods in production systems.

Consequently, those challenges are related with a changing competency profile of today's and tomorrow's production workers and associated changing learning requirements:

- New competences and expertise for key technology-induced innovative products, processes, and production systems applying key technologies,
- Sound knowledge of visions, goals, and principles of production systems in general in order to overcome pure technology orientation and realistically assess the possibilities of new technologies in a benefit-oriented manner,
- Learning environments are needed in which blue- and white-collar production workers access and experience those key technologies like additive manufacturing and digitalization.

1.3 Dynamic Product Life Cycles

In the last decades, an increasing demand for customized products can be noticed which leads to shorter product life cycles reinforced by the technical advancements and the attempt to sell more products through more innovations in saturated markets. This means the time between two successive product generations is becoming shorter and shorter; average life cycles in the automotive industry in 1980 lasted approximately eight years, twenty years later the duration of life cycles is halved at four years. These developments lead to increasing demands with regard to changeability and adaptability of companies, their production systems, and their employees.[33] For the flexibility and the changeability of processes and organizations, it is argued that the investment in the lifelong competence development of the personnel is crucial[34].

In summary, by shorter cycles, it is possible to generate additional sales in mature markets with variations of products. A shorter time when a product is offered and declining sales of individual models are the results, i.e. individual models are produced in ever-smaller quantities. Consequently, the challenges and needs related to the capability of the workforce are:

- the risk of investments increases as a shorter amount of time is available to earn back invested capital.
- high demands on the ability to change and adapt of the whole company, the factories, and the employees.
- complex products raise the demands on the cognitive abilities of employees.
- more frequent production ramp-ups have to be mastered in a shorter amount of time; staff has to be prepared for those non-routine situations.
- higher demands on the flexibility of production plants and employees due to the individualization of products.
- suiting learning environments are needed that are able to map dynamic production systems.
- cost reduction effects resulting from learning curves based on repetitive activities are mitigated in these environments. Work-integrated learning methods to accelerate the learning curves are needed.

1.4 Limited Natural Resources

A shortage of resources is predicted in the coming years because of increased living standards, world population growth, and the partially irresponsible consumption of resources.[35] The publication of "The Limits of Growth" in 1972 launched a major controversy over how economies should grow and how they can grow sustainably in

---

[33] See Abele and Reinhart (2011), Westkämper and Zahn (2008), Arndt (2013).
[34] See Wagner, Heinen, Regber, and Nyhuis (2010), Adolph et al. (2014).
[35] Abele and Reinhart (2011).

the end.[36] The study, commissioned by the Club of Rome, predicts insurmountable and inevitable problems with unchanged economic development within few decades. From today's perspective, those prognoses have to be questioned knowing that the development of mining technology kept pace with the consumption of static resource range.[37] But to rely only on a technological advancement of the extraction, technology seems risky. In order not to impose a high raw material prices in the long term, manufacturing companies need not only innovations in mining technology, but above all resource-efficient production processes and the use of alternative materials for the produced goods.[38] Here an interdisciplinary cooperation between materials science and production technological research is required.[39] A future-oriented curriculum as well as a learning factory has to address four main questions:

- Which natural resources (energy, materials) have to be replaced in the coming years?
- Which alternative solutions are already today available?
- What are the challenges in product development and production engineering for the shift of resources?
- How can the new solutions be justified (economically, ecologically)?

In addition, over the past years, the issue regarding energy efficiency gained major interest from society, politics, and economy. Especially, interdisciplinary energy efficiency aspects have not yet been considered in industry, research, and education. Furthermore, until today, the energy efficiency is not yet integrated into engineering education. For example, aim of the research project ETA-Factory[40] was to construct a model factory, which integrates various interdisciplinary approaches reducing energy consumption and CO_2 emissions of production processes in industry, see Best Practice Example 6. Figure 1.8 shows an overview of the targets and the solution approaches to more energy-efficient production processes in course of the project.

Furthermore, in light of the energy transition and the associated challenge of a high proportion of wind and solar power, the generation and consumption of electrical energy must be timely coordinated with each other. This can be done on the one hand with innovative power storages, or on the other hand, with a so-called demand-side management (DSM), i.e. more flexible power consumption. For this, innovative and adapted technologies are needed for future industrial processes.[41]

In general, the efficient and flexible use of energy and non-energy raw materials in along with the complete product life cycle must be taken into consideration for economic and ecological reasons.[42]

---

[36] Meadows (1972).

[37] See Frondel (2008).

[38] See Abele and Reinhart (2011), Herrmann (2010).

[39] Abele and Reinhart (2011).

[40] See PTW, TU Darmstadt (2017b).

[41] See BMBF (2018).

[42] See Herrmann (2010), Bullinger, Jürgens, Eversheim, and Haasis (2013).

1.4 Limited Natural Resources

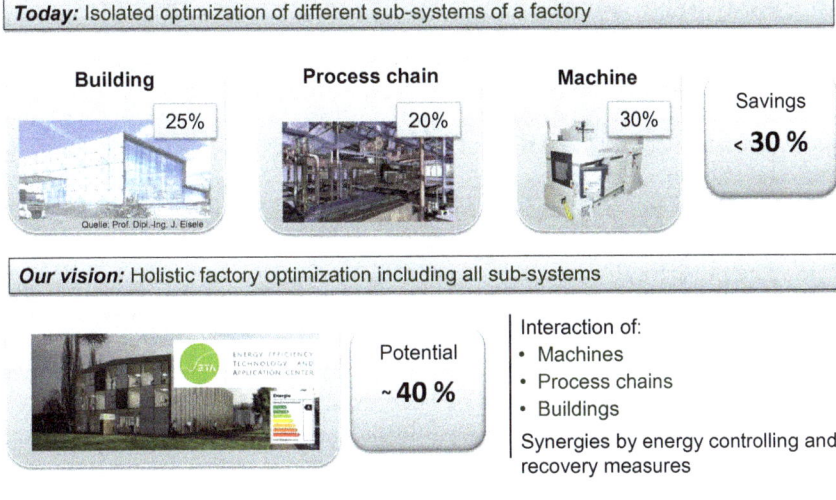

Fig. 1.8 Aim of the research project ETA-Factory (PTW, TU Darmstadt, 2017b)

The following challenges can be identified for future production:

- Energy and non-energy resource efficiency along the complete product life cycle,
- Innovative power storages and flexibility in the energy consumption of industrial processes,
- Alternative materials and production processes for innovative products.

Interviewed experts identify among others the following necessities for education, training, and research in the field of limited natural resources:

- Research environments are needed in which innovative technologies and processes for the efficient and flexible use of energy can be developed, tested, and transferred to industry.
- Sensitization for the topics resource and energy efficiency via integration of those topics in the curricula.
- Learning environments that are able to make energy-efficient production tangible, which is a challenge since energy flows are mostly not observable by the eye.
- Interdisciplinary education and training efforts in dedicated energy and resource-oriented programs.

1.5 Knowledge Society

In the twenty-first century, in many branches of Western industry the activities of workers are knowledge-based largely. Work processes are in those cases no longer highly dependent on manual skills, but on the knowledge of individuals and the organizational knowledge of companies.[43] Consequently, knowledge and education are the key resources for social and economic advancement.[44] Particularly in countries with no or few mineral resources, knowledge and the resulting innovation factors are decisive for prosperity. It is not surprising in this context that the countries with a low raw material-dependent share of GDP are investing heavily in education—and, on this basis show the significantly higher GDP growth rates.[45] Better education is usually associated with wage increases, which in turn must be justified by productivity gains, which in turn are again based on better education and training of employees who design and execute processes. In particular, three drivers can be identified that increase the importance of production-related knowledge and education:[46]

- Production technologies, that are continuously improved, tend to get more complex at the same time. On a higher level, this leads to significantly more complex production systems.
- As shown in Sect. 1.3, product life cycles are getting shorter. Product- and technology-related knowledge is correspondingly faster outdated.
- The length of employees staying in one position or department is diminishing. Exemplarily, from 1980 to 2010 the average stay of a production planner in one position went down from eight to just four years (Fig. 1.9).

Under these new conditions, production-related competencies in various domains must be developed more quickly. The knowledge associated with these competencies must be constantly identified, internalized, and transferred. As the production systems become more complex, the development of competencies becomes more difficult and has to be done more quickly.[47] In the future, innovative learning ways as well as methods and tools for the management of the rapidly generated knowledge are necessary.[48] Blue-collar workers and engineers require innovative lifelong ways of learning to keep up with described dynamics.[49] In summary, the following challenges can be identified:

---

[43] See Bullinger, Spath, Warnecke, and Westkämper (2009).
[44] See Abele and Reinhart (2011).
[45] See Gylfason (2001).
[46] See Abele and Reinhart (2011), Chryssolouris, Mavrikios, and Mourtzis (2013).
[47] See Adolph et al. (2014).
[48] See Abele and Reinhart (2011).
[49] See Chryssolouris et al. (2013).

1.5 Knowledge Society

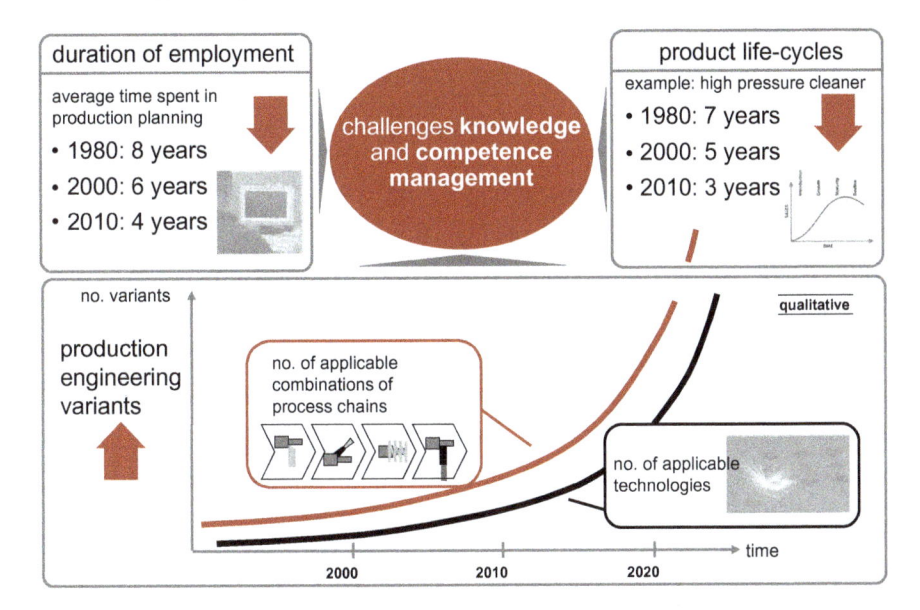

Fig. 1.9 Current challenges in production technology require efficient forms of knowledge and competence management, with slight changes according to Abele and Reinhart (2011)

- Establishment of learning organizations and work-integrated forms of learning in order to adapt on a corporate level to new situations and to preserve competitiveness,
- New ways of learning systems and methods used for education and training,
- Constant integration of research, industry, and education in order to have a two-way transfer of knowledge between academia and industry,
- Environments are needed that are able to integrate education, research, and industry,
- Ensure knowledge transfer related with innovation activities of organizations.

1.6 Risk of Instability

For company leaders, the growing market and economy dynamics make it increasingly difficult to foresee future relevant developments and adjust the company accordingly. These various instabilities, such as market breaks, resource bottlenecks, terrorist attacks, difficult-to-calculate policies, embargoes, refugee crises make long-term and stable planning of global production difficult or even impossible. Consequently, factories must be designed flexibly and in a versatile manner, which cannot always be achieved in highly automated environments. The following challenges related to industry's personnel can be derived from these developments:

- Anticipation of opportunities and threats because of possible changes in business environments,
- Robust and resilient reactions to changes in the corporate environment,
- Flexibility, in order to adapt quickly to potential, foreseeable changes,
- Changeability, in order to adapt to unforeseen changes in the business environment as quickly as possible without major efforts.

1.7 Demographic Change

In industrialized countries, the population's age structure will profoundly shift to an increased proportion of older people in the coming years because of an increase in the average life expectancy combined with a reduced birth rate.[50] Often, companies are trying to cope with the aging of the workforce with the preferential recruitment of younger employees in combination with the early retirement of older employees.[51] This reduces the average age in those companies, but at the same time the experience acquired over years is lost. A more sustainable approach to this would be to keep older employees in constant working capacity throughout the period of employment and to deploy them appropriately in relation to their skills and experiences.

It is also predicted that in the near future there will not be enough skilled workers—until 2030 at least a 15% decrease in the number of potential employees is predicted, projections until 2050 show a fall in the number of available workers by over 30%.[52] So along the way employees of all ages are needed, as a result compared to the status more older employees will work in production.

A further influence on the structure of the population[53] can be found in migration movements which have grown significantly in recent years. Most people migrating to Europe these days are in the age between 20 and 50 years old, Fig. 1.10 compares the age structure of European nationals to the non-nationals coming into the countries. To enable the integration of migrants into society and to open up opportunities for industry, immigrants must be prepared for the labor markets. For this purpose, targeted training and further education formats are required, which prepare refugees and immigrants for work in local industrial environments.

---

[50] See Abele and Reinhart (2011), Schmid (2013).
[51] See Roth, Wegge, and Schmidt (2007).
[52] See Fuchs and Dörfler (2005), Fuchs and Kubis (2016).
[53] And the potential labor force, see Fuchs and Kubis (2016).

1.7 Demographic Change

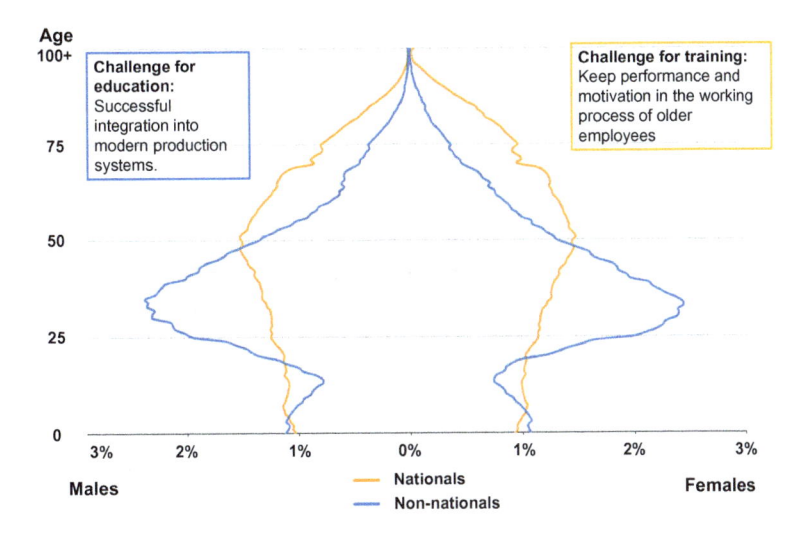

Fig. 1.10 Age structure of the national and non-national populations in EU-28, January 2016 (in %) (Eurostat, 2017)

Competency development and knowledge transfer have to adapt to these trends,[54] which implies the following:

- New forms of active learning that are based on experiences and real problem situations.
- Knowledge and competency management systems in industries need to be adapted to the new challenges.
- New methods for an efficient knowledge transfer inside and outside the work process.

1.8 Wrap-up of This Chapter

This chapter starts with the positive effects of education and training on individuals, organizations and society as a whole as a motivation. In particular, the importance of the production sector for the national economies is emphasized. Subsequently, long-term observable developments that have a decisive influence on industry and society are listed and the implications for the further development of organizations and employees at different levels of the hierarchy are presented. It is no longer enough

---

[54]See Abele and Reinhart (2011).

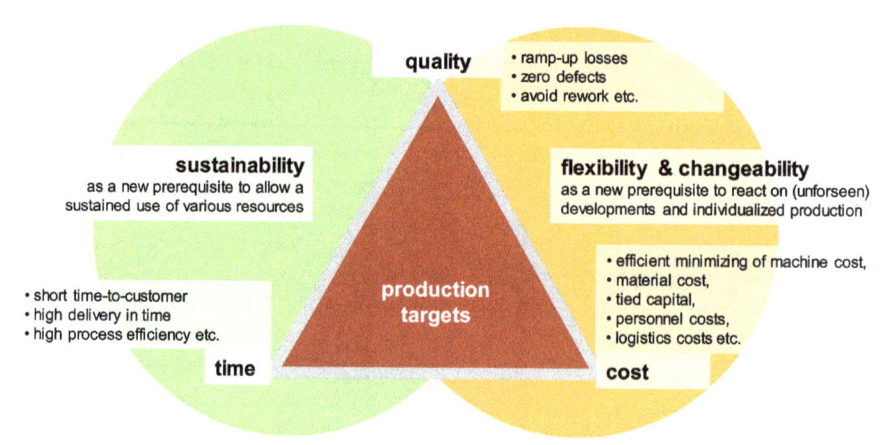

Fig. 1.11 Extended production targets

to only develop manufacturing processes, equipment, and machinery according to traditional production targets—cost, quality, and time. In light of the megatrends, industrial production targets have to be supplemented beyond the magic triangle of production by additional general conditions. Climate change, a growing scarcity of resources and demographic change mean that production targets have to be extended by the general condition of ecologic, economic, and social sustainability.[55] Furthermore, most of the presented megatrends lead to the strong need to be able to adjust quickly to changed conditions. This means that industry needs the capacity to quickly adapt to ever-changing market demands, shorter product life cycles, new technologies, the risk of instability, and the like. Consequently, the adaptability, the flexibility, and the changeability of production systems are required in respond to environmental changes as a second new general condition. Figure 1.11 visualizes the extended production targets.[56]

This chapter forms the basis and starting point for all further efforts in the field of training and further education for production with learning factories.

References

Abele, E., Anderl, R., Metternich, J., Wank, A., Anokhin, O., Arndt, A., et al. (2015). Effiziente Fabrik 4.0. *Zeitschrift für wirtschaftlichen Fabrikbetrieb (ZWF)*, *110*(3), 150–153. https://doi.org/10.3139/104.111293.

Abele, E., Metternich, J., Tenberg, R., Tisch, M., Abel, M., Hertle, C., et al. (2015). *Innovative Lernmodule und -fabriken: Validierung und Weiterentwicklung einer neuartigen Wissensplattform für die Produktionsexzellenz von morgen*. Darmstadt: Tuprints.

---

[55] See Abele and Reinhart (2011).

[56] According to Abele et al. (2015) based on Gienke, Kämpf, and Aldinger (2007), Kletti and Schumacher (2011), Abele and Reinhart (2011).

References

Abele, E., & Reinhart, G. (2011). *Zukunft der Produktion: Herausforderungen, Forschungsfelder, Chancen*. Munich: Hanser.

Adolph, S., Tisch, M., & Metternich, J. (2014). Challenges and approaches to competency development for future production. *Journal of International Scientific Publications – Educational Alternatives, 12*, 1001–1010.

Arndt, H. (2013). *Supply chain management: Optimierung logistischer Prozesse*: Gabler Verlag.

Arndt, H. (2008). *Supply chain management: Optimierung logistischer Prozesse* (4., aktualisierte und überarb. Aufl.). *Gabler Lehrbuch*. Wiesbaden: Gabler.

Barro, R. J. (1996). *Determinants of economic growth: A cross-country empirical study. NBER working paper series: Vol. 5698*. Cambridge, MA: National Bureau of Economic Research.

BMBF. (2018). *Kopernikus-Projekt SynErgie*. Retrieved from https://www.kopernikus-projekte.de/projekte/industrieprozesse.

Bullinger, H.-J., Jürgens, G., Eversheim, W., & Haasis, H. D. (2013). *Auftragsabwicklung optimieren nach Umwelt- und Kostenzielen: OPUS — Organisationsmodelle und Informationssysteme für einen produktionsintegrierten Umweltschutz*: Springer Berlin Heidelberg.

Bullinger, H.-J., Spath, D., Warnecke, H.-J., & Westkämper, E. (2009). *Handbuch Unternehmensorganisation: Strategien, Planung, Umsetzung* (3. neubearb. Aufl.). *VDI-Buch*. s.l.: Springer-Verlag.

CEDEFOP. (2010). *Skills supply and demand in Europe: Medium-term forecast up to 2020*. Luxembourg: Publications Office of the European Union.

Chryssolouris, G., Mavrikios, D., & Mourtzis, D. (2013). Manufacturing systems - skills & competencies for the future. In *46th CIRP Conference on Manufacturing Systems. Procedia CIRP, 7*, 17–24.

DESTATIS. (2016). 43,7 Millionen Erwerbstätige im 3. Quartal 2016: Pressemitteilung Nr. 407 vom 17.11.2016. Retrieved from https://www.destatis.de/DE/PresseService/Presse/Pressemitteilungen/2016/11/PD16_407_13321.html.

Eurostat. (2016). Manufacturing statistics - NACE Rev. 2: Relative importance of manufacturing (NACE Section C), 2013 (% share of value added and employment in the non-financial business economy total). Retrieved from http://ec.europa.eu/eurostat/statistics-explained/index.php/Manufacturing_statistics_-_NACE_Rev._2.

Eurostat. (2017). Migration and migrant population statistics. Retrieved from http://ec.europa.eu/eurostat/statistics-explained/index.php/Migration_and_migrant_population_statistics.

Frondel, M. (2008). Wettbewerb um Ressourcen – Rohstoffe als Trend oder Megatrend? In D. Bierbaum (Ed.), *So investiert die Welt: Globale Trends in der Vermögensanlage* (pp. 61–75). Wiesbaden: Gabler Verlag/Springer Fachmedien Wiesbaden GmbH Wiesbaden. https://doi.org/10.1007/978-3-8349-8969-7_5.

Fuchs, J., & Dörfler, K. (2005). *Projektion des Arbeitsangebots bis 2050: Demografische Effekte sind nicht mehr zu bremsen*. Retrieved from http://hdl.handle.net/10419/158191.

Fuchs, J., & Kubis, A. (2016). *Zuwanderungsbedarf und Arbeitskräfteangebot bis 2050*. Retrieved from https://www.destatis.de/DE/Publikationen/WirtschaftStatistik/2016/07_Sonderheft/ZuwanderungsbedarfBis2050_072016.pdf?__blob=publicationFile.

Gienke, H., Kämpf, R., & Aldinger, L. (Eds.). (2007). *Handbuch Produktion: Innovatives Produktionsmanagement: Organisation, Konzepte, Controlling*. München: Hanser.

Graf, H. G. (2000). *Globale Szenarien: Megatrends im weltweiten Kräftespiel*. Zürich: Verlag Neue Zürcher Zeitung.

Grömling, M., & Haß, H.-J. (2009). *Globale Megatrends und Perspektiven der deutschen Industrie. Forschungsberichte aus dem Institut der deutschen Wirtschaft Köln: Vol. 47*: Deutscher Instituts-Verlag.

Gylfason, T. (2001). Natural resources, education, and economic development. *European Economic Review: EER, 45,* 847–859.

Hanushek, E. A., & Woessmann, L. (2007). *The role of education quality in economic growth. Policy research working paper: Vol. 4122.* Washington, DC: World Bank Human Development Network Education Team.

Herrmann, C. (2010). *Ganzheitliches life cycle management: Nachhaltigkeit und Lebenszyklusorientierung in Unternehmen. VDI-Buch.* Berlin, Heidelberg: Springer.

Huang, S. H., Liu, P., Mokasdar, A., & Hou, L. (2013). Additive manufacturing and its societal impact: A literature review. *The International Journal of Advanced Manufacturing Technology, 67*(5), 1191–1203. https://doi.org/10.1007/s00170-012-4558-5.

Jovane, F., Westkämper, E., & Williams, D. J. (2009). *The ManuFuture Road: Towards Competitive and Sustainable High-Adding-Value Manufacturing.* Berlin: Springer.

Kletti, J., & Schumacher, J. (2011). *Die perfekte Produktion: Manufacturing excellence durch short interval technology (SIT).* Berlin: Springer.

Krys, C. (2011). Ausblick – Megatrends und ihre Implikationen auf Geschäftsmodelle. In T. Bieger (Ed.), *Academic Network, Innovative Geschäftsmodelle* (pp. 369–384). Berlin, Heidelberg: Springer. https://doi.org/10.1007/978-3-642-18068-2_19.

Meadows, D. H. (1972). *The limits to growth: A report for the Club of Rome's project on the predicament of mankind* (8. print). New York, NY: Universe Books.

Naisbitt, J. (1982). *Megatrends: Ten new directions transforming our lives (Di 1 ban).* New York: A Warner Communications Company.

O'Sullivan, D., Rolstadås, A., & Filos, E. (2011). Global education in manufacturing strategy. *Journal of Intelligent Manufacturing, 22*(5), 663–674. https://doi.org/10.1007/s10845-009-0326-2.

Promotorengruppe Kommunikation der Forschungsunion Wirtschaft - Wissenschaft. (2013). *Umsetzungsempfehlungen für das Zukunftsprojekt Industrie 4.0: Deutschlands Zukunft als Produktionsstandort sichern.* Abschlussbericht des Arbeitskreises Industrie 4.0.

PTW, TU Darmstadt. (2017a). Übersicht Use Cases. Retrieved from http://www.prozesslernfabrik.de/index.php/uebersicht-use-cases.

PTW, TU Darmstadt. (2017b). Welcome to ETA- Factory: The energy efficient model factory of the future. Retrieved from http://www.eta-fabrik.tu-darmstadt.de/eta/index.en.jsp.

Roth, C., Wegge, J., & Schmidt, K.-H. (2007). Konsequenzen des demographischen Wandels für das Management von Humanressourcen. *Zeitschrift für Personalpsychologie, 6*(3), 99–116. https://doi.org/10.1026/1617-6391.6.3.99.

Scherrer, P., Simons, R., & Westermann, K. (1998). *Von den Nachbarn lernen: Wirtschafts- und Beschäftigungspolitik in Europa. Standortdebatte.* Marburg: Schüren.

Schmid, S. (2013). Bevölkerungsentwicklung in Deutschland und weltweit. *Aus Politik und Zeitgeschichte, 4–5,* 46–52.

Smith, A. (2001). *Return on investment in training: Research readings.* Leabrook, S. Aust.: NCVER.

Sun, Y., & Kron, A. (2017). Chinesische Unternehmenskäufe in Europa: Eine Analyse von M&A-Deals 2006–2016. January 2017. Retrieved from http://www.ey.com/Publication/vwLUAssets/EY-ma-chinesische-investoren-januar-2017/$FILE/EY-ma-chinesische-investoren-januar-2017.pdf.

Tether, B., Mina, A., Consoli, D., & Gagliardi, D. (2005). *A literature review on skills and innovation: How does successful innovation impact on the demand for skills and how do skills drive innovation? A CRIC report for the department of trade and industry.* Manchester, England: ESRC Centre for Research on Innovation and Competition.

References

Vayre, B., Vignat, F., & Villeneuve, F. (2012). Designing for Additive Manufacturing. In *45th CIRP Conference on Manufacturing Systems. Procedia CIRP, 3*, 632–637.

Vieweg, H.-G. (2011). *Study on the competitiveness of the EU mechanical engineering industry.* Munich: Ifo Institute.

Wagner, C., Heinen, T., Regber, H., & Nyhuis, P. (2010). Fit for change – Der Mensch als Wandlungsbefähiger. *Zeitschrift für wirtschaftlichen Fabrikbetrieb (ZWF), 100*(9), 722–727.

Warnecke, H.-J. (Ed.). (1999). *Projekt Zukunft: Die Megatrends in Wissenschaft und Technik*. Köln: vgs.

Wartenberg, L.-G., & Haß, H.-J. (2005). *Investition in die Zukunft: Wie Deutschland den Anschluss an die globalisierte Welt findet* (1. Aufl.). Weinheim: Wiley-VCH.

Westkämper, E., & Zahn, E. (2008). *Wandlungsfähige Produktionsunternehmen: Das Stuttgarter Unternehmensmodell*: Springer.

Chapter 2
Competencies for Future Production

Following from the trends shown in the previous chapter, there is a need for a great diversity of competencies, which are based on knowledge and qualifications from different fields of activity that has to be developed at all hierarchy levels and along the complete value chain; see Fig. 2.1.

Figure 2.2 shows an overview of the topics addressed in this chapter regarding the competencies for future production.

2.1 Competencies, Qualification, and Knowledge

In everyday language, knowledge, qualifications, and competences are often used almost synonymously. Knowledge in the narrower sense, skills, and qualifications are most often not the goal of (further) education, but are necessary prerequisites.[1] The ability to act in complex situations is enabled through comprehensive competency development that includes more than just transferred knowledge.[2]

Knowledge (in the narrow sense) is a foundational element of the concept of competence.[3] Additionally, competencies are based on appropriate rules, values, and norms of individual persons or entire groups. However, these rules, values, and norms influence their own actions only when they have been internalized; when they have become a part of the personality, are no longer constantly questioned, and affect the actions of the individuals. Non-internalized rules, norms, or values, on the other hand, are not action-relevant. With increasing experience, competences are consolidated.[4] Figure 2.3 depicts the relation between knowledge, skills, qualification, and competency.

---

[1] See Kuhlmann and Sauter (2008).
[2] See Erpenbeck and Rosenstiel (2007).
[3] According to Heyse and Erpenbeck (2009).
[4] See Heyse and Erpenbeck (2009).

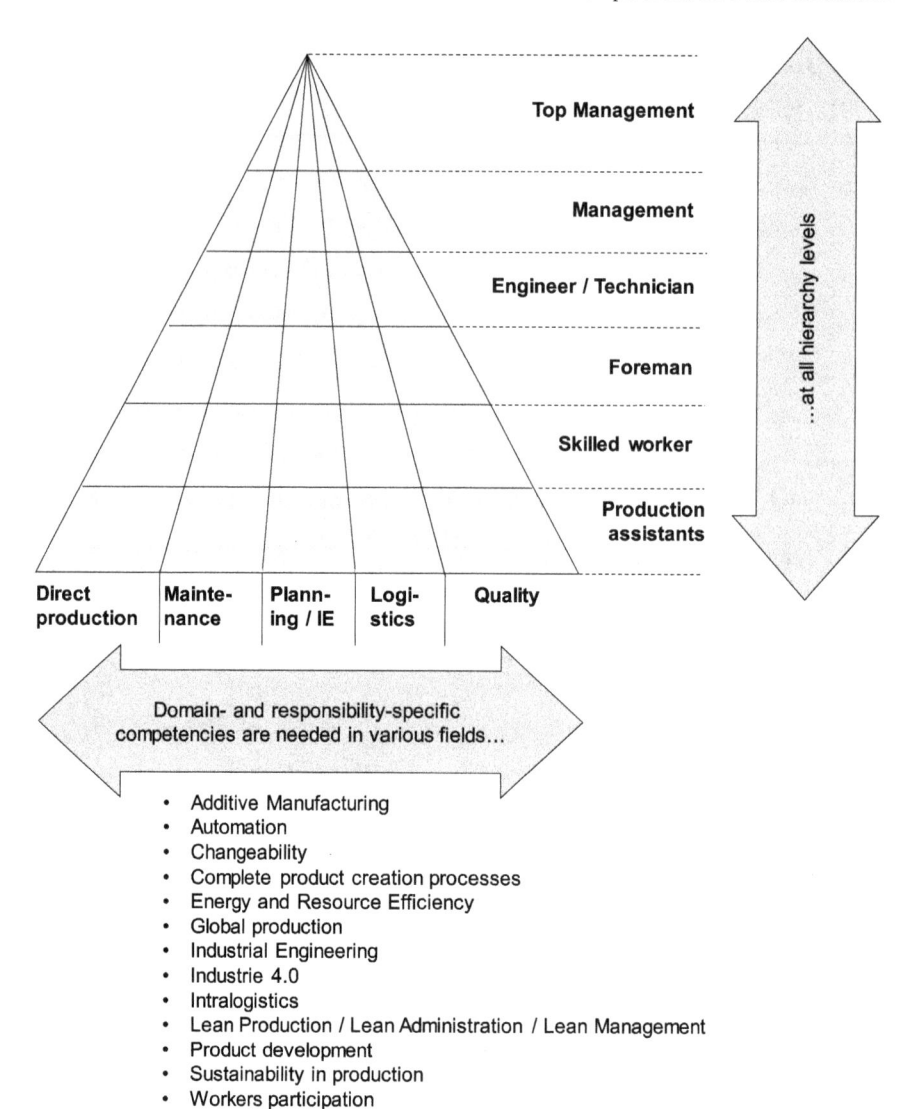

Fig. 2.1 New competencies are needed along the value chain at all hierarchy levels

2.1.1 Knowledge

The concept of knowledge is not clearly defined in the literature. Depending on whether business scientists, educators, politicians, philosophers, or managers define the term, given explanations are sometimes contradictory.[5] An intermixing of the

---
[5]Kuhlmann and Sauter (2008).

2.1 Competencies, Qualification, and Knowledge

Competencies for future production (Chapter 2)
- Competencies, Qualification, and knowledge (2.1)
- Learning goals and learning outcomes (2.2)
- Addressed competencies in learning factories (2.3)
- Relevant competencies for Industrie 4.0 (2.4)
- A domain-specific competency model for Lean 4.0 at the Process Learning Factory CiP (2.5)

Fig. 2.2 Overview of the structure of this chapter

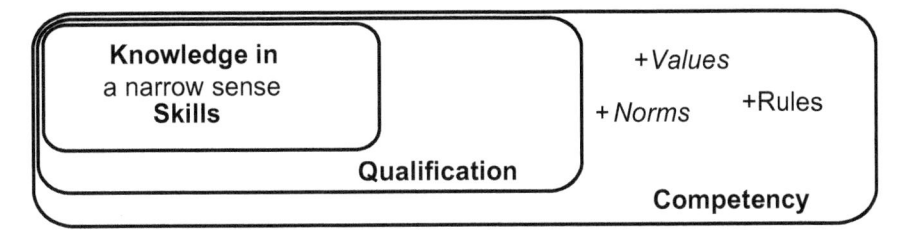

Fig. 2.3 Relation of competency, qualification, skills, and knowledge (Heyse & Erpenbeck, 2009)

different understandings in the discussion is more than a hindrance, especially for determining a training objective.[6] Knowledge arises through the networking of information. Accordingly, the European Commission describes knowledge as an "outcome of the assimilation of information through learning. Knowledge is the body of facts, principles, theories, and practices that is related to a field of work or study."[7] In the field of knowledge management, the term is defined in a similar fashion, and Fig. 2.4 exemplarily shows what is known as the "knowledge stair,"[8] which leads to the concept of knowledge using the depicted relationships of signs and data. The knowledge stair is continued up to the terms competence and competitiveness. This upper part of the knowledge stair is, however, not completely compatible with the general understanding of competence in this context.

2.1.2 Qualification

Not knowledge but adequate actions are in the center of the concept of qualification. In contrast to competencies, qualifications can be tested and verified without problem by means of a certification procedure, independently of the work process. Qualifications are an essential prerequisite for competencies.[9] In order to meet the

---
[6]See Heyse and Erpenbeck (2009).
[7]European Commission (2006).
[8]See North (2011).
[9]See Kuhlmann and Sauter (2008).

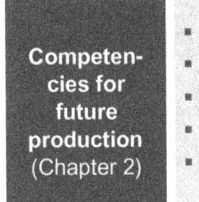

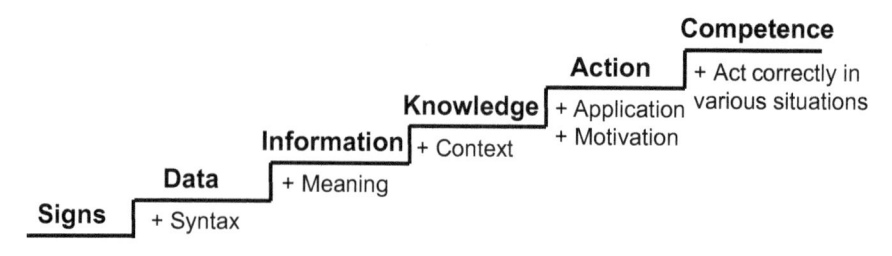

Competence
```
                                    Action     + Act correctly in
                         Knowledge  + Application  various situations
             Information + Context  + Motivation
      Data   + Meaning
Signs + Syntax
```

Fig. 2.4 Knowledge stair according to North (2011)

	workspace 1	workspace 2	workspace 3	workspace 4	workspace 5	workspace 6
Worker 1	4	2	2	3	3	1
Worker 2	1	1	3	3	1	2
Worker 3	3	3	3	2	4	2
Worker 4	2	4	4	4	3	3

4: Instructor
3: Can handle task and solve problems
2: Can work independently
1: Still needs training

Fig. 2.5 Exemplary qualification matrix for deployment and development of production staff

specific requirements of a job profile, personnel needs certain qualifications, which consist of knowledge in a narrow sense, abilities, and skills. Abilities are solidified systems of generalized psychophysical processes of action, which require the psychological and personal conditions.[10] Today, qualifications are often the central element for controlling deployment and development of employees in production. In most companies, for example, there are qualification matrices that describe the qualification of the individual employees for different activities and tasks, such as the operation of machines or workplaces. The qualification matrix can therefore be used for the deployment of workers and the further qualification planning inside the team. Figure 2.5 shows such a qualification matrix.

---

[10] See Hacker (1973).

2.1.3 Competence/Competency

The concept of competency (or also competence[11]) frequently leads to misinterpretation and confusion.[12] For clarification, three normative competence concept approaches are classified:[13]

- A behavioral, performance-oriented approach, that sees competencies as specific job-related and measurable behaviors blanking out underlying attributes,
- A generic approach, which includes underlying attributes like knowledge in the competence concept and ignores specific application contexts, and
- A holistic approach, integrating both above-mentioned approaches, i.e. competencies are conceptualized based on knowledge, attitudes, skills, performances, in explicit application contexts of professional life.

The roots of the competence concept are found in the linguistics concepts by Noam Chomsky.[14] Here, language competence is described as a person's ability to construct and understand self-organized an infinite number of unheard and unspoken sentences on the basis of finite basic elements in combination with rules for combination.[15] The close connection between competence and motivation was provided by the motivational psychologist Robert W. White. He described competences as abilities to act, formed by self-motivated interaction with the environment.[16] Accordingly, competences are seen as context-specific dispositions that built on the foundation of knowledge and skills,[17] enabling actions in open, unknown, and complex situations in a self-organized, creative manner.[18] In this context, the meaning of the similar terms "competence" and "competency" is not congruent but comparable[19]: While competence refers to a function, the term competency refers to a behavior.[20] Today, the term competency is used in an extended sense compared to its behavioral origins[21,22]; now, in addition the concept refers also to underpinning basic attributes like knowledge, skills.[23] The term "competency" is thus in the following used in the sense of the holistic approach, which regards both the underlying attributes and the

---

[11] For a terminological discussion of the two notions, see Le Delamare Deist and Winterton (2005), Rowe (1995) and Teodorescu (2006) or the differentiation below.
[12] See, for example, Short (1984), McMullan et al. (2003).
[13] See Short (1984).
[14] See Chomsky (1962).
[15] Chomsky (1962).
[16] White (1959).
[17] See European Commission (2006).
[18] See Erpenbeck and Rosenstiel (2007).
[19] See also, for example, Le Delamare Deist and Winterton (2005), Rowe (1995) and Teodorescu (2006).
[20] See Le Delamare Deist and Winterton (2005).
[21] McClelland (1976).
[22] See also Le Delamare Deist and Winterton (2005).
[23] See Spencer and Spencer (1993), Le Delamare Deist and Winterton (2005).

explicit professional application context. In contrast, the term competence is used in the sense of the above-mentioned generic approach, when a specific application context is ignored.

A distinction can be made between different classes of competences. In literature, numerous classification schemes for competences can be found. Exemplarily, the following classes are distinguished[24]:

- **Socio-communicative competences** entail the ability to communicative and cooperative self-organized action.
- **Technical and methodological competences** entail the ability to mental and physical self-organized action of technical problems.
- **Personal competences** entail abilities to act reflexively self-organized.
- **Activity and action competences** entail the ability to holistic, self-organized action. This includes the use of one's own motivations, emotions, experiences, and abilities as well as all other competences for the realization of successful actions.

Specific competences can now be assigned to the four competence classes described, although the four classes are closely linked. Exemplarily, Fig. 2.6 shows an overview of the competence classes and important competences allocated to the classes in the competence atlas.[25]

2.2 Learning Goals and Learning Outcomes

Learning goals can be defined on all aggregation levels: for universities and schools, educational programs, or single educational courses. Typically learning goals and learning objectives define the intention of an educational activity.[26] Often the comprehensive intention of an educational program is defined with goals, while objectives detail those goals more specifically.[27] The term "learning goal" describes a seen target and "learning objective" a target that is aimed at.[28] Generally, in the most famous taxonomies three domains of learning goals are differentiated[29]:

- Cognitive domain, i.e. the recognition of knowledge,[30]
- Affective domain, i.e. interest or attitudes of the learner,[31]

---

[24] See Erpenbeck and Rosenstiel (2007).

[25] See Heyse and Erpenbeck (2009).

[26] See Allan (1996).

[27] See Allan (1996), Tyler (1971).

[28] See Barkley and Major (2016).

[29] See Bloom, Engelhart, Furst, Hill, and Krathwohl (1956).

[30] The most important taxonomies in this field are delivered by Bloom et al. (1956) and Anderson, Krathwohl, and Airasian (2001).

[31] The most important taxonomy in this domain is by Krathwohl, Bloom, and Masia (1964).

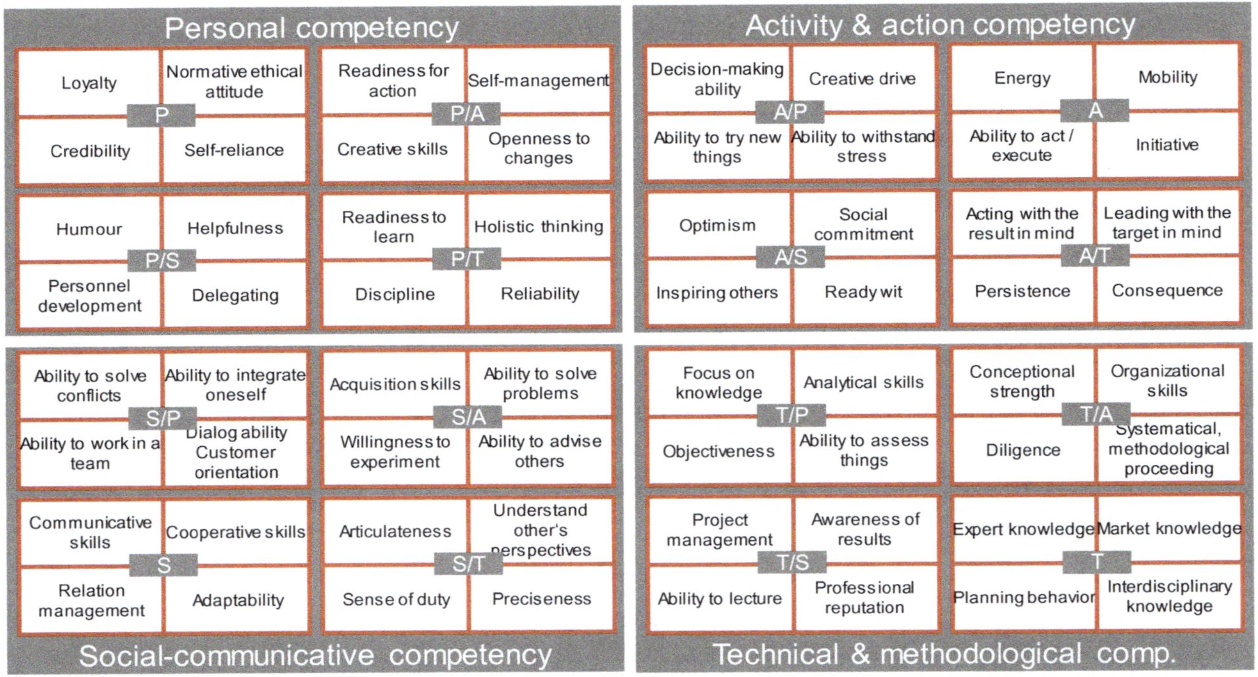

2.2 Learning Goals and Learning Outcomes

Personal competency

Loyalty	Normative ethical attitude	Readiness for action	Self-management
Credibility	Self-reliance	Creative skills	Openness to changes
	P		P/A	
Humour	Helpfulness	Readiness to learn	Holistic thinking
Personnel development	Delegating	Discipline	Reliability
	P/S		P/T	

Activity & action competency

Decision-making ability	Creative drive	Energy	Mobility
Ability to try new things	Ability to withstand stress	Ability to act/execute	Initiative
	A/P		A	
Optimism	Social commitment	Acting with the result in mind	Leading with the target in mind
Inspiring others	Ready wit	Persistence	Consequence
	A/S		A/T	

Social-communicative competency

Ability to solve conflicts	Ability to integrate oneself	Acquisition skills	Ability to solve problems
Ability to work in a team	Dialog ability Customer orientation	Willingness to experiment	Ability to advise others
	S/P		S/A	
Communicative skills	Cooperative skills	Articulateness	Understand other's perspectives
Relation management	Adaptability	Sense of duty	Preciseness
	S		S/T	

Technical & methodological comp.

Focus on knowledge	Analytical skills	Conceptional strength	Organizational skills
Objectiveness	Ability to assess things	Diligence	Systematical, methodological proceeding
	T/P		T/A	
Project management	Awareness of results	Expert knowledge	Market knowledge
Ability to lecture	Professional reputation	Planning behavior	Interdisciplinary knowledge
	T/S		T	

P: Personal competencies; P/A: Personal competencies with link to activity & action competency class; P/S: Personal competencies with link to social-communicative competency class; P/T: Personal competencies with link to technical & methodological competency class. A: Activity & action competencies; A/P: Activity & action competencies with link to personal competency class; A/S: Activity & action competencies with link to social-communicative competency class; A/T: Activity & action competencies with link to technical & methodological competency class. S: Social-communicative competencies; S/P: Social-communicative competencies with link to personal competency class; S/A: Social-communicative competencies with link to activity/ action competency class; S/T: Social-communicative competencies with link to technical & methodological competency class. T: Technical & methodological competencies; T/P: Technical & methodological competencies with link to personal competency class; T/A: Technical & methodological competencies with link to activity & action competency class; T/P: Technical & methodological competencies with link to social-communicative competency class.

Fig. 2.6 Competence atlas according to Heyse and Erpenbeck (2009)

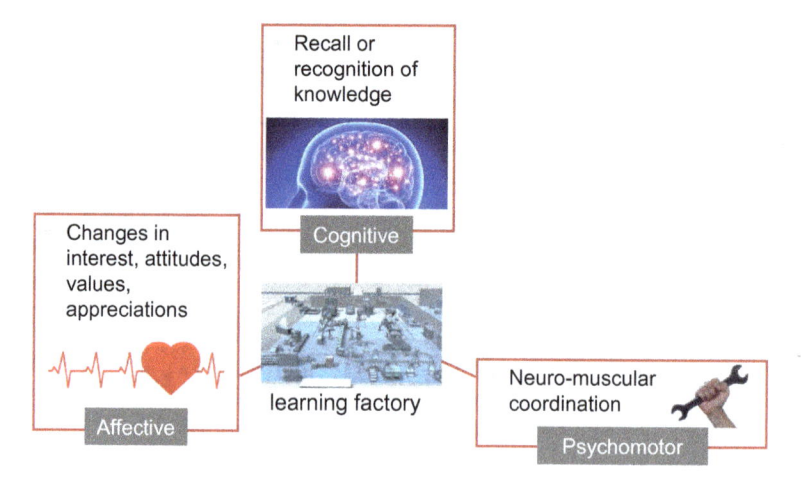

Fig. 2.7 Cognitive, affective, and psychomotor domain can be addressed in learning factories

- Psychomotor domain, i.e. task that entails neuromuscular coordination.[32]

Table 2.1 shows an overview of the most recognized learning goals taxonomies.[33]

Additionally to that, the term "learning outcome" describes the actual learning of your learners instead of the intention ("learning goal").[34] Learning outcomes that may relate to learner, subject, and teacher[35] are statements about expected effects at the end of an educational period.[36] "Defining the learning outcomes enables both the teacher and student to see what a student is expected to have achieved, and what progress he/she made with regard to his/her qualification goal".[37]

2.3 Addressed Competencies in Learning Factories

The learning factory concept offers the potential for competency development in all human performance areas (cognitive, affective, and psychomotor) and regarding all competency classes (Fig. 2.7):

---

[32] The most important taxonomy in this domain is provided by Dave (1970).

[33] The taxonomies are according to Bloom et al. (1956), Anderson et al. (2001), Krathwohl et al. (1964), and Dave (1970).

[34] See Barkley and Major (2016).

[35] See Eisner (1979).

[36] See Gosling and Moon (2001).

[37] Seliger, Reise, and Farland (2009).

2.3 Addressed Competencies in Learning Factories

Table 2.1 Taxonomies of cognitive, affective, and psychomotor learning goals according to Abele et al. (2017) based on Bloom et al. (1956), Anderson et al. (2001), Krathwohl et al. (1964), and Dave (1970)

Domains and learning goals		Description
Cognitive	Remember	Retrieve relevant knowledge from long-term memory
	Understand	Construct meaning from instructional messages, including oral, written, and graphic communication
	Apply	Carry out or use a procedure in a given situation
	Analyze	Break material into its constituent parts and determine how the parts relate to one another and to an overall structure or purpose
	Evaluate	Make judgments based on criteria and standards
	Create	Put elements together to form a coherent or functional whole; reorganize elements into a new pattern or structure
Affective	Receiving	Being aware of or attending to something in the environment
	Responding	Showing some new behavior as a result of experience
	Valuing	Showing some definite involvement or commitment
	Organization	Integrating a new value into one's general set of values, giving it some ranking among one's general priorities
	Characterization	Acting consistently with the new value
Psychomotor	Observing	Active mental attending of a physical event
	Imitating	Attempted copying of a physical behavior
	Practicing	Trying a specific physical activity over and over
	Adapting	Fine-tuning. Making minor adjustments in the physical activity in order to perfect it

- Professional and methodological competencies,[38]
- Socio-communicative competencies,[39]
- Personal competencies,[40] and
- Activity and action competencies.[41]

Although learning factories today are mostly used in the cognitive performance area, and primarily for the development of professional and methodological competencies, the learning environment is generally suitable to enable comprehensive competency development.[42] Other competency classes, like socio-communicative competencies, are not yet targeted primarily, but rather are to be regarded as by-catches, for example on the basis of the learning forms used. Examples might be the common use of group work or presentations of work results during learning factory courses, which does not necessarily mean that the "working in team" or the "presentation" competency is an intended outcome of all of those courses. Additionally, it must also be pointed out that competence development in these areas does not take place automatically if the learners are placed in a corresponding situation. Furthermore, required reflection is not necessarily initiated.[43]

The results of a literature research on the competences developed using learning factories[44] show that professional and methodological, social communicative as well as personal competences can be identified in the literature. An adapted and expanded literature research on the addressed competences in learning factories are shown in Table 2.2 (professional and methodological competences), Table 2.3 (socio-communicative competences), Table 2.4 (personal competences), and Table 2.5 (activity and action competences).

Additionally to the various more or less context-independent competences that are identified in the literature and shown in Table 2.2, an even greater variety of domain- and context-specific competencies can be identified. Those context- and domain-specific competencies are, in general, the primary goals of learning factory courses and very often the reason to develop and build up a learning factory in the first place. The variety of context- and domain-specific orientations of learning factories will be presented in Chap. 6 of this book in detail. For now, Table 2.6 shows an overview of the five competency domains or clusters of domains identified in the literature. In alphabetical order, they are:

- Energy and Resource Efficiency,

---

[38] See, e.g., Abele, Tenberg, Wennemer, and Cachay (2010), Jäger, Mayrhofer, Kuhlang, Matyas, and Sihn (2013), Steffen, Frye, and Deuse (2013), Tisch et al. (2013), Faller and Feldmüller (2015), Wagner, Prinz, Wannöffel, and Kreimeier (2015).

[39] See, e.g., Wagner, Heinen, Regber, and Nyhuis (2010), Steffen, May, and Deuse (2012), Jäger, Mayrhofer, Kuhlang, Matyas, and Sihn (2012), Helm, Reise, and Rößle (2014), Wagner et al. (2015).

[40] See, e.g., Steffen et al. (2012), Jäger et al. (2012).

[41] See, e.g., Blume et al. (2015), Goerke, Schmidt, Busch, and Nyhuis (2015), Abele et al. (2017), Müller-Frommeyer, Aymans, Bargmann, Kauffeld, and Herrmann (2017).

[42] Abele et al. (2015).

[43] See Steffen et al. (2012).

[44] See Müller-Frommeyer et al. (2017).

2.3 Addressed Competencies in Learning Factories

Table 2.2 Literature research on professional and methodological competences developed in learning factories, based on the results from Müller-Frommeyer et al. (2017), adapted and expanded

Competency classes	Competence	References
Professional and methodological	(Application of) professional knowledge	Abele et al. (2010), Balve and Albert (2015), Blume et al. (2015), Cachay, Wennemer, Abele, and Tenberg (2012), Goerke et al. (2015), Jäger et al. (2013), Jorgensen, Lamancusa, Zayas-Castro, and Ratner (1995), Kreimeier et al. (2014), Müller-Frommeyer et al. (2017), Steffen et al. (2012), Tisch et al. (2013), Wagner et al. (2010)
	Interdisciplinary knowledge and understanding	Jäger et al. (2012), Jorgensen et al. (1995), Lamancusa and Simpson (2004)
	Project management	Blume et al. (2015)
	Presentation skills	Müller-Frommeyer et al. (2017), Nöhring, Rieger, Erohin, Deuse, and Kuhlenkötter, (2015)
	Analytical thinking	Müller-Frommeyer et al. (2017)
	Domain-specific competencies	See Tables 2.6 and 2.7

- Industrial Engineering,
- Industrie 4.0,
- Lean production/lean administration/lean management, and
- Product development.

Beyond those domains, further specific foci of competency domains or combinations of competency domains can be identified. However, since those specific foci are part of or come in combination with at least one of the clusters mentioned above, those foci are subordinated here and therefore do not form a separate domain. The specific foci identified are in alphabetical order.

- Additive manufacturing,[45]
- Automation,[46]
- Changeability,[47]
- Complete product creation processes,[48]

---

[45] Often in combination with product development or Industrial Engineering.
[46] Combined with either global production or Industrie 4.0.
[47] In combination with either Industrial Engineering or lean production.
[48] As a combination of several topics such as product development and industrial engineering.

Table 2.3 Literature research on socio-communicative competences developed in learning factories, based on the results from Müller-Frommeyer et al. (2017), adapted and expanded

Competency classes	Competence	References
Socio-communicative	Adaptability	Wagner, AlGeddawy, ElMaraghy, and Müller (2012)
	Capability to work in teams	Blume et al. (2015), Goerke et al. (2015), Gräßler, Taplick, and Yang (2016), Jäger et al. (2013), Jorgensen et al. (1995), Müller-Frommeyer et al. (2017), Nöhring et al. (2015), Steffen et al. (2012), Tietze, Czumanski, Braasch, and Lödding (2013), Wagner et al. (2010), Wagner et al. (2012)
	Communication skills	Blume et al. (2015), Gräßler et al. (2016), Jorgensen et al. (1995), Müller-Frommeyer et al. (2017), Tietze et al. (2013), Veza, Gjeldum, and Mladineo (2015), Wagner et al. (2010)
	Problem-solving capability	Blume et al. (2015), Cachay and Abele (2012), Jäger et al. (2013), Micheu and Kleindienst (2014), Steffen et al. (2012), Tietze et al. (2013), Wagner et al. (2010)
	Leadership	Helm et al. (2014)

- Global production,[49]
- Intralogistics,[50]
- Sustainability in production,[51] and
- Workers' participation.[52]

2.4 Relevant Competencies for Industrie 4.0

In the broad-based study by Fraunhofer, IAO 21 Industrie 4.0 (I 4.0) experts and 661 practitioners from a wide range of industries were asked about the future of produc-

---

[49] In combination with lean production or Industrial Engineering.
[50] Often combined with either energy efficiency, Industrie 4.0, or lean production.
[51] In combination with lean production or energy and resource efficiency.
[52] In combination with either lean production or Industrie 4.0.

2.4 Relevant Competencies for Industrie 4.0

Table 2.4 Literature research on personal competences developed in learning factories, based on the results from Müller-Frommeyer et al. (2017), adapted and expanded

Competence classes	Competence	References
Personal	Creativity	Abele et al. (2015), Blume et al. (2015)
	Motivation	Blume et al. (2015), Dinkelmann, Siegert, and Bauernhansl (2014), Gräßler et al. (2016), Jorgensen et al. (1995), Müller-Frommeyer et al. (2017), Steffen et al. (2012), Tisch et al. (2013), Wagner et al. (2010)
	Personal responsibility	Balve and Albert (2015), Blume et al. (2015), Müller-Frommeyer et al. (2017), Tietze et al. (2013), Tisch et al. (2013), Tisch, Hertle, Abele, Metternich and Tenberg (2015)
	System thinking capability	Blume et al. (2015), Goerke et al., (2015), Helm et al. (2014), Jäger et al. (2013), Kreimeier et al., (2014), Steffen et al. (2012), Tietze et al. (2013)
	Result-oriented action	Blume et al. (2015), Micheu and Kleindienst (2014)
	Reflexion capability	Blume et al. (2015), Gräßler et al. (2016), Jäger et al. (2013), Steffen et al. (2012)
	Technology affinity	Müller-Frommeyer et al. (2017)
	Openness	Müller-Frommeyer et al. (2017)

Table 2.5 Literature research on activity and action competences developed in learning factories, based on the results from Müller-Frommeyer et al. (2017), adapted and expanded

Competence classes	Competence	References
Activity and action	Innovative capability	Balve and Albert (2015), Blume et al. (2015), Helm et al. (2014), Jäger et al. (2013)
	Decision-making	Blume et al. (2015), Goerke et al. (2015)
	Planning and realization capability	Nöhring et al. (2015)

Table 2.6 Domain-specific competencies in learning factories in the literature

Competency domain	Examples of domain-specific competencies	References with domain-specific competencies defined related with learning factory approaches
Energy and resource efficiency	Designing energy-efficient production systems; energetic optimization of machine tools; analysis of energy flows; eco-efficient design of products	Abele, Bauerdick, Strobel and Panten (2016), Blume et al. (2015), Kreitlein, Höft, Schwender and Franke (2015) Krückhans et al., (2015), Micheu and Kleindienst (2014), Plorin, Poller and Müller (2013); Focus intralogistics: Scholz et al. (2016)
Industrial engineering	Apply design for manufacturability/assembly; plan technology and production process; analysis and design of (ergonomic) workplaces; creating changeable production environments	Dinkelmann, Riffelmacher and Westkämper (2011), Jäger et al. (2013), Jorgensen et al. (1995), Micheu and Kleindienst (2014); Focus assembly systems: Nöhring et al. (2015), Schreiber, Funke and Tracht (2016), Steffen et al. (2012); Focus changeability: Bauernhansl, Dinkelmann and Siegert (2012), Matt, Rauch and Dallasega (2014), Wagner et al. (2010); Focus MTM: Morlock, Kreggenfeld, Louw, Kreimeier and Kuhlenkötter (2017)
Industrie 4.0	Use of innovative technologies (e.g. RFID); use ICT for material and worker tracking; application of data analytics methods	Blöchl and Schneider (2016), Erol, Jäger, Hold, Ott and Sihn (2016), Faller and Feldmüller (2015), Größler, Pöhler and Pottebaum (2016), Gronau, Ullrich and Teichmann (2017), Löhr-Zeidler, Hörner and Heer (2016), Merkel et al. (2017), Thiede, Juraschek and Herrmann (2016); Focus textile industry: Küsters, Praß and Gloy (2017); Focus workers participation: Reuter et al. (2017); Focus intralogistics: Hummel, Hyra, Ranz and Schuhmacher (2015); Focus automation: Simons, Abé, and Neser (2017)

(continued)

2.4 Relevant Competencies for Industrie 4.0

Table 2.6 (continued)

Competency domain	Examples of domain-specific competencies	References with domain-specific competencies defined related with learning factory approaches
Lean production/lean administration/lean management	Ability to perform systematic problem solving; ability to map and design value streams; ability to implement flow lines in production systems; material supply design	Abele, Eichhorn and Kuhn (2007), Block, Bertagnolli and Herrmann (2011), Cachay et al. (2012), Goerke et al. (2015), Kreimeier et al. (2014), Müller, Menn and Seliger (2017), Tietze et al. (2013), Tisch et al. (2015); With specific focus on the pharmaceutical context: Rybski and Jochem (2016); Focus changeability and flexibility: Matt et al. (2014); Focus workers participation: Wagner et al. (2015); Focus intralogistics: Blöchl and Schneider (2016), Micheu and Kleindienst (2014); Focus sustainability: Helm et al. (2014); Focus global production: Lanza, Moser, Stoll, and Haefner (2015), Lanza, Minges, Stoll, Moser, and Haefner (2016)
Product development	(Eco-efficient) product design (of components); coordinating product development and production; management of change requests	Bender, Kreimeier, Herzog and Wienbruch (2015), ElMaraghy and ElMaraghy, (2015), Schützer, Rodrigues, Bertazzi, Durão and Zancul (2017); With specific focus on additive manufacturing: Yoo et al. (2016)

tion in light of I 4.0. 96.9% of the interviewees indicated that human work in terms of planning, control, execution, and monitoring will be important or very important for production in five years.[53] The study also examines changes for the production environments of the future. The flexible response to customer requirements already is of great value. This value of flexible reactions on market volatility, which is mainly generated by competent employees, will continue to rise in the future. 98.6% of respondents consider the rapid response to customer requirements as important and very important for the future.[54] So it can be assumed that the efforts for I 4.0 will

---

[53] See Spath (2013).
[54] Spath (2013).

Table 2.7 Specific competency foci in learning factories in the literature

Competency foci	Examples of domain-specific competencies	References with domain-specific competencies defined related with learning factory approaches
Additive manufacturing	Designing products for additive manufacturing; developing value chains for additive manufactured products	Yoo et al. (2016)
Automation	Economic design automation systems; manage automation technology	Simons et al. (2017), Buergin, Minguillon, Wehrle, Haefner and Lanza, (2017)
Changeability	Creation and improvement of changeable production systems	Bauernhansl et al. (2012), Matt et al. (2014), Wagner et al. (2010)
Complete product creation processes	Analysis, design, and improvement of product creation processes; design, manufacturing, post-processing, quality control, and recycling in additive manufacturing	Gräßler, Taplick et al. (2016), Jäger et al. (2013), Jorgensen et al. (1995)
Global production	Analysis of global production networks; design location-specific production; planning of production networks	Lanza et al. (2015), Lanza et al. (2016); Liebrecht et al. (2017); with specific focus on automation: Buergin et al. (2017)
Intralogistics	Analyze, design, and improve intralogistics systems	Hummel et al. (2015), Scholz et al. (2016), Blöchl and Schneider (2016), Micheu and Kleindienst (2014)
Sustainability in production	Creation of socially, economically, and ecologically sustainable working environments	Helm et al. (2014), Seliger et al. (2009)
Workers' participation	Creation of mechanisms for workers' participation and cooperation in production	Wagner et al., (2015), Reuter et al. (2017)

change production, maybe even to a high degree, production employees will need an expanded and adapted competence profile.[55] The experts from industry and academia agree that the necessary competence development should in particular address the area of knowledge work and higher education.[56]

---

[55] See Acatech (2016), Kagermann, Wahlster, and Helbig (2013).

[56] See Richter, Heinrich, Stocker, and Unzeitig (2015), Erol, Jäger, Hold, Ott, & Sihn, (Erol et al. 2016), Prifti, Knigge, Kienegger, and Krcmar (2017).

2.4 Relevant Competencies for Industrie 4.0

In order to be able to make use of learning factories in the context of I 4.0 in a meaningful and goal-oriented manner, the relevant required competencies for employees in production-related areas must be defined as a basis for the learning factory and learning module design. The irreconcilable problem with the necessary competences for I 4.0 is, however, that the topic of I 4.0—in contrast to the topics lean production, energy efficiency or product development for example—is still largely in the theoretical debate about how the production of the future, be it in 5, 10, or 20 years, will look like.

One way of approaching the problem is the evaluation of expert opinions from academia and industry that provide information about which competencies are expected to become more important in the age of I 4.0 than others.[57] The second way of approaching the relevant competencies for Industrie 4.0 is to define a theoretical future scenario for "I 4.0 production" including associated technological, organizational, and social trends and consequently derive necessary competencies of employees in order to be able to produce and work in the predefined hypothetical scenario successfully.[58] Of course, both approaches are not offering exact predictions, but they still provide a way to get an impression of how the competency profile of industrial employees could change. In the literature, there are several publications that define necessary competencies for "I 4.0" expert- and/or scenario-based from different point of views.[59] An exemplary overview of selected competency models for I 4.0 (which can be seen as a compilation or a structure of necessary competencies for I 4.0) identified in the literature is provided in the following.

In one of the first publications addressing competences for I 4.0, Dombrowski et al. (2014) derive requirements for employees through I 4.0 based on identified technical, organizational, and human-oriented changes. As a result of changing working tasks, higher demands on professional, social, methodological, and personal competences in general are derived from this. For the four competence classes, no specific competences are named. As requirements for production in times of I 4.0, a higher level of problem-solving activities, flexibility, and exact communication is predicted. In a more detailed competency study, Erol et al. (2016) derive various personal, social/interpersonal, action-related, and domain-related competencies from a presented vision of I 4.0 and future work processes. Specific relevant competencies are highlighted in each competency class. Furthermore, the competency model is used as a fundament for a scenario-based learning factory approach showing that the identified relevant I 4.0 competencies can be addressed in experiential learning setups using a learning factory. Irrespective of the application in learning factories, Hecklau et al. (2016) develop a comprehensive competence model for I 4.0. First, based on the definition of I 4.0 upcoming economic, social, technical, environmental, political, and legal challenges are identified. As a next step of the approach, competences are derived from the identified challenges, and finally, the derived competences

---

[57] See, e.g., Acatech (2016).
[58] See, e.g., Vernim, Reinhart, and Bengler (2017) and Erol et al. (2016).
[59] See, e.g., Dombrowski, Riechel, and Evers (2014), Erol et al. (2016), Hecklau, Galeitzke, Flachs, and Kohl (2016), Acatech (2016), Gronau et al. (2017), Prifti et al. (2017), Vernim et al. (2017).

are aggregated and categorized to the four competence classes: technical, methodological, social, and personal competences. Based on this, rated actual competences and a minimum level of required competences can be compared. In the evaluation especially competence gaps are identified for personal (motivation to learn and sustainable mindset), social (intercultural skills, leadership skills, ability to transfer knowledge, cooperation, teamwork, and language skills), and methodological competencies (creativity, entrepreneurial thinking, analytical skills). Since the presented minimum requirements as well as the employee profile are designated as only exemplary, no general conclusions about competence development programs for Industrie 4.0 can be drawn.[60]

Another perspective at competences for I 4.0 gives the competence development study for I 4.0.[61] The study is based on an online survey with 345 participating German companies. Additionally, 38 guided interviews with experts from academia and industry were conducted. Results are reported in terms of key company competences[62] and the relevant competences of employees. Furthermore, a distinction is made between large companies and SMEs. The results of the study are visualized in Fig. 2.8. From this study, two interesting conclusions can be drawn regarding the qualification and the competence development for Industrie 4.0:

- First, it is apparent that nearly all staff competence requirements tend to be rated with an average value by the SMEs—in contrast to the ratings of the larger companies. In other words: It can be assumed that while the large companies (seem to) know which competences will be important for future employees, the SMEs are largely in the dark. Consequently, SMEs rate all competences as equally important. This suggests that, in particular, SMEs need support in the identification and definition of success-critical competences in the I 4.0 age.
- Second, taking a look at the highest rated competence requirements for employees (interdisciplinary thinking and acting, better process know-how, competence to lead, working in innovation processes, problem-solving and optimization competence), it becomes clear that these competences can be excellently addressed by means of learning factories. Learning factories support the interdisciplinary thinking and acting; they sharpen the eye for value-added processes in experiential learning setups, enable the risk-free testing of management routines, are a playground for innovations and innovation processes in the field of research, and are extremely well suited to develop problem-solving and optimization competences in problem- and scenario-based learning forms. At this point, it should be emphasized that, on the one hand, it is possible to address these competences in learning factories and that, on the other hand, few to no other available learning possibilities of the work processes can be identified which are capable of effectively developing these required competences.

---

[60] See Hecklau et al. (2016).
[61] See Acatech (2016).
[62] That is, competences on company-level.

2.4 Relevant Competencies for Industrie 4.0

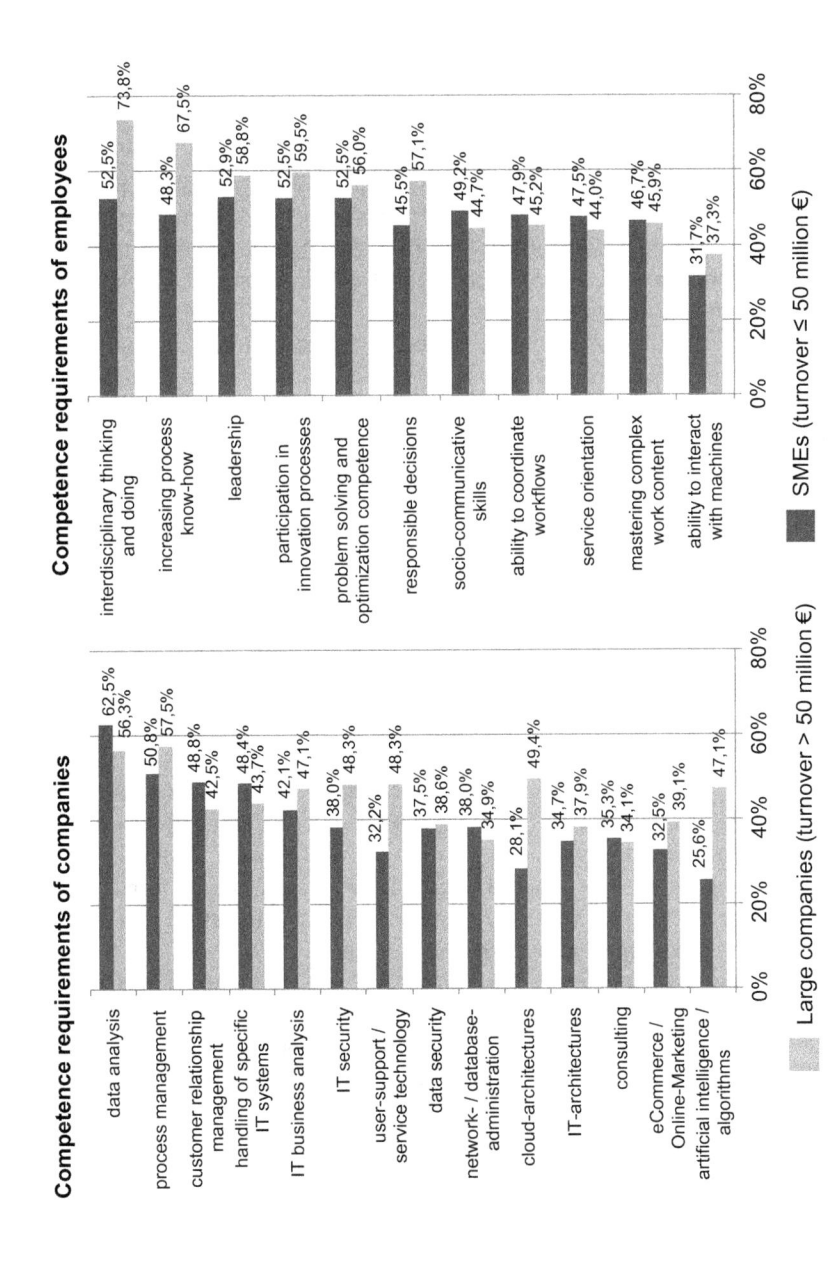

Fig. 2.8 Competence requirements of companies and the need for competences of employees according to Acatech (2016), literally translated

Gronau et al. (2017) derive "vocational action competences" from the two identified main drivers of Industrie 4.0: cyber-physical systems and the Internet of Things. Furthermore, the identified action competences are decomposed to "competence facets" which represent a mixture of often recognized competence classes and generic meta-competences. The described competences are further addressed in a learning factory approach.

Vernim et al. (2017) derive from the changes of working processes in production by Industrie 4.0 the profile of the production employee of the future. A distinction is made here between Industrie 4.0 (flexibility, IT comprehension, etc.) and employee-specific aspects (better work-life balance, lifelong further development, etc.) in the changes of the employee profile. In order to meet these changes in the profile, comprehensive qualification requirements are defined, which are assigned to the clusters of interdisciplinary, data, and standards.[63]

Prifti et al. (2017) develop a competence model that extends the Universal Competency Framework[64] on the basis of a comprehensive literature research on the topics of competences for Industrie 4.0, Internet of Things, and digitization. The literature study is followed by focus groups with academic experts. A total of 3.363 articles were found in relevant databases, screened, and reduced to 17 articles describing explicitly required competences for Industrie 4.0. Competences are defined related to information systems, computer science, and engineering. The identified generic competences (and issues related to Industrie 4.0) are clustered in eight groups "leading and deciding," "supporting and cooperating," "interacting and presenting," "creating and conceptualizing," "organizing and executing," "adapting and coping," "enterprising and performing." Most often, in over 70% of the selected papers the communication competence is identified as relevant for Industrie 4.0, followed by IT/technology affinity, big data, and problem solving identified each in nearly half of the publications. Figure 2.9 shows the relative frequency of the identified relevant competences in the selected publications. In this much aggregated and unspecified form of competence naming, it is difficult to specifically define the role of the system learning factory in these areas. It is, however, obvious that each of these competences and issues can be linked with the use of learning factories through education and training, research, and technology transfer. The learning factory can therefore play a central role in enabling and implementing Industrie 4.0 in practice.

In these analyzes, competences, for which a growing importance is predicted, are defined generically, i.e. as meta-competences.[65] To a considerable extent, in addition, domain-specific competencies need to be identified for individual job profiles, specific fields of activity, or areas of responsibility. These domain-specific competencies will be just as decisive for the professional capacity for action of all parties involved as the expected increase in the meaning of vague meta-competence. Erol et al. (2016)

---

[63]For a comprehensive overview on these qualification clusters, see Vernim et al. (2017).
[64]See Bartram (2013).
[65]Meta-competences are competences that are defined more or less without a context, for further explanations on the concept of meta-competence, as well as the problems associated with the concept see for example Rogalla (2013).

2.4 Relevant Competencies for Industrie 4.0

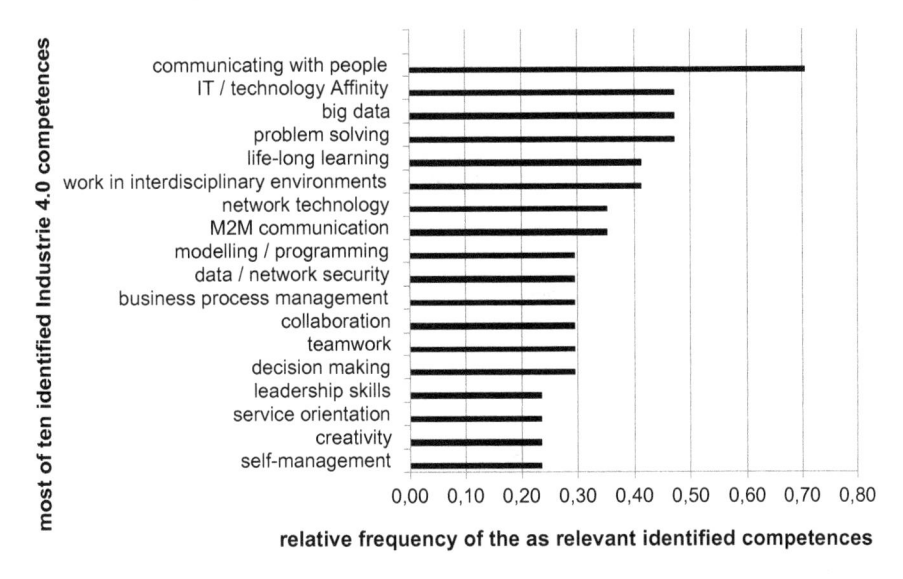

Fig. 2.9 Relative frequency of as relevant identified competences for Industrie 4.0 according to the literature analysis of Prifti et al. (2017)

name the domain-related competencies as a separate, important competency class which is not easy to model. As examples for the domain-related competencies for one scenario in the learning factory, the "application of lean thinking and methods in manufacturing," the "application of conceptual modeling methods," and the "application of information and communication technology for material tracking and worker tracking" are given.[66,67]

2.5 A Domain-Specific Competency Model for Lean 4.0

In order to be able to define domain-specific competencies meaningfully, it is helpful to address the following questions as completely as possible:[68]

1. What are the characteristics of the industrial environment, the target organization or the target sector in which the competencies defined should unfold?
2. What are the specific production technologies, social, and organizational goals of the organization(s)? And how can these goals be achieved? Which fields of actions are crucial?

---

[66] See Erol et al. (2016).

[67] For the field of vocational education, these ideas on necessary competencies in the future working world are carried out by Tenberg and Pittich (2017).

[68] According to the first didactic transformation described in Tisch et al. (2013, 2015).

44   2 Competencies for Future Production

1 **Charcteristics of industrial environment**	2 **Goals and strategies to tackle the challenges of industry**	3 **Target group which is involved to pursue goals and strategies**
• Mass or one-off production? • Discrete or continuous production? • Focus on machining or assembly? • Current trends influencing the industry? • …	• Develop innovative products? • Boost productivity of production processes? • Resource and energy efficient production? • Pursuing a learning organization? • …	• Production workers? • Executives? • Indirect functions? • Production technology experts? • IT experts? • Process and value creation experts? • …

Domain- and context-specific competency model

Fig. 2.10 Questions to be addressed when defining a domain- and context-specific competency model

3. Which employees are addressed specifically? What are the tasks and activities of these employees?

Figure 2.10 shows an overview of the questions that have to be addressed when defining a domain- and context-specific competency model. These questions are also part of the first didactic transformation in learning factory design.[69]

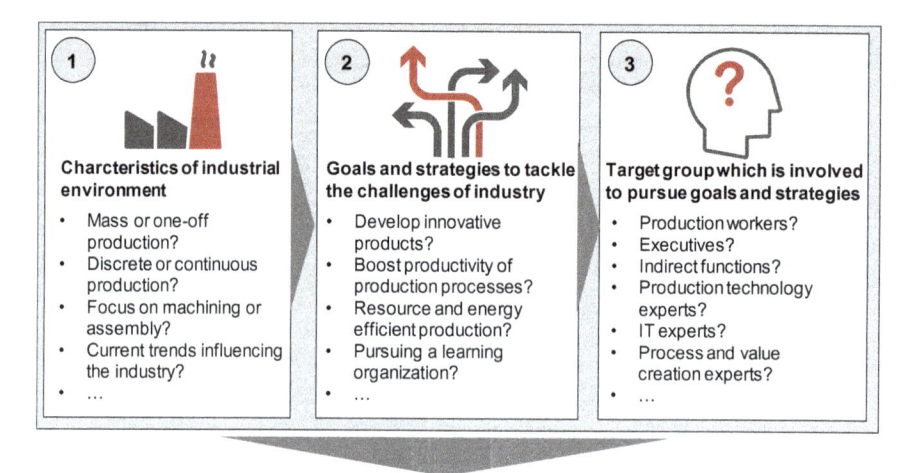

In this section, a competency model for one of the numerous possible domain constellations is derived. The description of this domain- and context-specific competency model is only to be seen as an example; it fits the identified framework conditions of the Process Learning Factory CiP and its industry partners and should not be understood as an universal or referential competency model.

Regarding the characteristics of the industrial environment (first requirement field), the aim is to create a competency model, which is as specific as possible on the one hand, but can also be applied to as many cases as possible on the other. A good compromise is given with the focus on direct areas of manufacturing companies, mainly from discrete production, which is containing in general manufacturing as well as assembly and logistics tasks.

Regarding the second requirement field, goals of the industrial companies using the training program are defined. Lean production systems have been established in industry as starting point for operational excellence. Among other things, the challenge here is always to provide the employees with the necessary competencies and motivations. Furthermore, in times of Industrie 4.0, the companies with lean processes and lean production system are wondering what the I 4.0-trend could

---
[69] See also Sect. 6.1.

2.5 A Domain-Specific Competency Model for Lean 4.0

The ideal of digitalization
- Wait & search times for information = 0 (real-time)
- Complete mobile transparency
- Minimal information complexity / perfect decision support
- Complete collection and use of product data throughout the entire life cycle
- 100% IT security

supports

Toyota's True North
- 0 defects
- 100% value added
- One piece flow (in sequence, on customer demand)
- Respect for employees

Toyota's True North

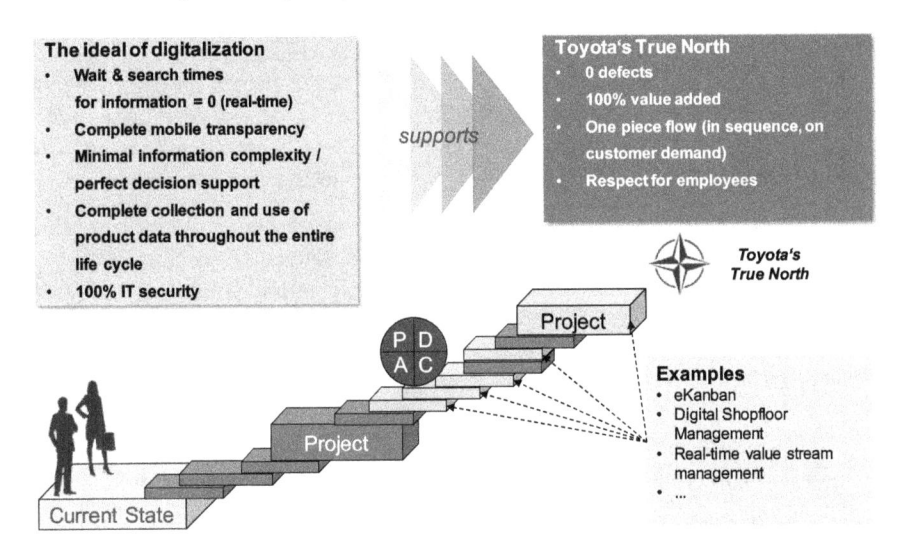

Project

PDCA

Examples
- eKanban
- Digital Shopfloor Management
- Real-time value stream management
- ...

Project

Current State

Fig. 2.11 Ideal of digitalization and Toyota's True North according to (VDMA & PTW, TU Darmstadt, 2018)

mean to them. These companies are increasingly aware of the fact that Industrie 4.0, lacking value-orientation, design guidelines and visions, cannot replace the existing lean production system but can merely supplement it.[70] The on-site analyses of PTW on the potentials of digitalization show that I 4.0 implementation is not useful, where processes are unstable, no standards exist, and basic performance indicators are missing. Only when a team has understood the lean thinking and applied their instruments sovereignly, digitalization will ensure the next productivity boost.[71] Goal is to use Industrie 4.0 and digitalization in order to expand the possibilities to create more efficient processes and to leverage lean production systems to the next level. This means, the vision of Toyota's True North (0 defects, 100% value added, one piece flow, respect employees) remains the same, while now digital tools might help to approach this vision.[72] The digitalization therefore additionally approaches the supporting ideal of digitalization; see Fig. 2.11. In order to achieve this goal, Lean 4.0 experts that transfer the identified potential to the companies' value-adding processes are needed. In summary, goals and strategies of those companies are needed to sustainably boost productivity of the production system by integrating Industrie 4.0 in existing lean production system.

With regard to the third requirement field, it is generally possible to differentiate between various roles in the context of Industrie 4.0, each of which must possess different competencies. First of all, there are the workers employed in machining and assembly areas as well as in supporting functions like maintenance or logistics,

---

[70] Metternich, Müller, Meudt, and Schaede (2017).
[71] See Metternich (2017).
[72] See Meudt, Leipoldt, and Metternich (2016).

who need the skills to make use of new digital tools and react flexibly on challenges of individualized production. In addition, competency demands on executives on different hierarchy levels, who must orchestrate the digital transformation, will also change. Furthermore, in times of Industrie 4.0 several additional roles can be identified, which relate equally to informatics, computer sciences, and engineering:

- I 4.0 system designer, who is responsible for the development of cyber-physical production systems in order to provide specific predefined system functions,
- I 4.0 system integrator, who is responsible for the integration of developed cyber-physical systems in running production processes,
- I 4.0 system administrator, who is responsible for the administration, maintenance and updating of cyber-physical production systems in order to ensure the stability of the systems,
- I 4.0 Process Expert, who is responsible for the design of material and information flows throughout the whole value creating system in order to boost productivity and the conceptualization of work organization, leadership routines, and others.

The competency model described in the following is focused on the I 4.0 Process Expert who is integrating Industrie 4.0 solutions in lean production system in order to develop production processes and people. This includes responsibility for material and information flow design throughout the whole value chain, conceptualization of work organization, leadership routines, and many others to improve the overall value creating system. Based on the three requirement fields

- intended competencies, that are necessary for the above-mentioned I 4.0 Process Expert, are derived and outlined methods based on interviews and expert discussions,
- interconnections between these competencies are discussed and related competencies are bundled in competence clusters, and
- a timeline for the competency development along the competency clusters is proposed.

In the following those three steps (based on the three requirement fields) are described in detail, see also A, B, and C in Fig. 2.12

A. Method-based definition of intended competencies for the I 4.0 Process Expert

In the last years, various "Industrie 4.0 demonstrators" are established in learning factories in order to exemplify the benefits of digital solutions for value creation. Following this technology-oriented approach, the actual learning focusses on isolated technical solutions in specific processes. An exact transfer of the exemplary Industrie 4.0 solutions demonstrated to other companies' environments does not make sense most often. Alternatively to this, structuring and defining learning goals, i.e. intended competencies, in a method-based approach allow a flexible deployment of Lean 4.0 methods and principles in varying industrial environments. In this context, the technological solutions are rather used as waste reduction tools that follow the principles of existing lean production systems. The term Lean 4.0 describes the use of digital possibilities in existing lean production systems.

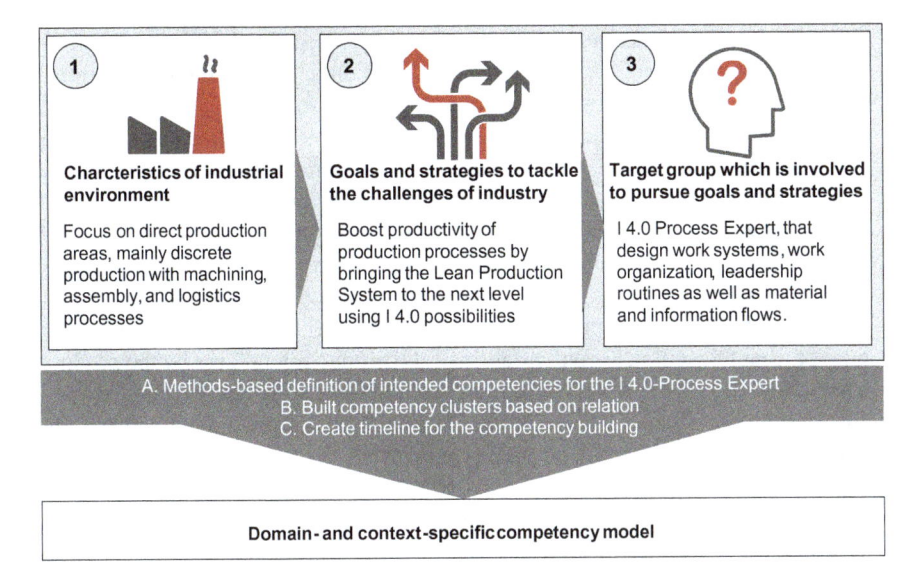

2.5 A Domain-Specific Competency Model for Lean 4.0

Fig. 2.12 Steps to the creation of the competency model for the I 4.0 Process Expert

This method-based approach is used to adapt and expand the existing training program of the Process Learning Factory CiP. In the last years, the training program addressed technical and methodological competencies in the lean production field in order to further develop the training program with regard to the specific requirements of the I 4.0 Process Expert summarized in Fig. 2.12 various experts of lean production and Industrie 4.0 got involved in the creation of the new Lean 4.0 curriculum. In the first step, the intented competencies' part of this curriculum is defined, and in the following, they are listed in alphabetical order:

- Analysis 4.0: The trainee is able to analyze the material and information flow in multiple production environments in terms of the classic seven types of waste as well as the new digital waste types. The learners visually identify and visualize the weaknesses in found material and information flows.
- Continuous improvement processes 4.0: The trainee is able to design and implement a leadership system for continuous improvement that describes the path for process improvement accompanied by a coach from an actual state via obstacles to a target state in the direction of the vision for production processes. Furthermore, the trainee is able to create a concept for a digital continuous improvement process support for the implemented system.
- Design 4.0: Based on the previous analysis, the trainee is able to design and gradually implement a short- to medium-term target image for material and information flows taking into account the possibilities of Industrie 4.0 in the direction of the Toyota and the digital true north.

- Digital customized machining: The trainee is able to design and improve the multi-variant order processing at the interface between engineering and manufacturing implementing and using digital support systems.
- Flexible manpower systems: Trainees have the ability to flexibilize manufacturing systems with regard to various customer demands. Digital tools are conceptualized and implemented as needed.
- Just-in-time 4.0: The trainee is able to analyze, implement, and improve pull systems and flow lines including suiting standard in process stocks. On this basis, the trainee is able to decide when it makes sense, to implement product individual aspects of production processes (work instructions, machine program, material, etc.) or even the complete configuration of the production process by the product. The learner is able to create, implement, and evaluate corresponding concepts.
- Just in time for machining: The learner is able to conceptualize, implement, and standardize flow in the machine-intensive areas.
- Predictive maintenance: The trainee is able to decide if predictive maintenance strategies (identification of necessity and implementation planning) with the help of which Industrie 4.0 possibilities (sensors, smart glasses, etc.) make sense economically and production wise. The trainee is able to plan, design, and implement the predictive maintenance strategies on existing machine tools.
- Problem-solving 4.0: Learner is able to design, use, implement, and evaluate problem-solving processes using when needed digital tools and assistance systems to support root cause discovery and structured problem solving.
- Production control 4.0: The trainee can design and implement concepts of production control. He or she is able to select suiting methods and smooth the production plan at the interface between the customer and production. Digital resources are designed and implemented if necessary.
- Quality 4.0: The learner is able to analyze existing production systems with regard to quality requirements taking into account the Jidoka principle, to identify weak points, to develop digital and non-digital alternative solutions, to select adequate solutions systematically, to implement the selected solutions, and to evaluate the solutions.
- Shop floor management digitally supported: The learner is able to design holistic context- and company-specific shop floor management (SFM) systems. In addition, the trainee is able to create a concept for implementation and use in a digital application based on the designed analog SFM system.
- SMED: Learners are able to decide when to use the SMED method, to apply the method, and to analyze the results.
- Stabilized, standardized, and individualized production: The trainee is able to evaluate the meaningfulness of individual employee and product-related standard work, e.g. in the form of digital standard operating instructions, digitally supported balancing and employee assistance systems, to design and implement the systems. Furthermore, the learner is able to arrange the collection of data in existing production environments, to design information flows consistently, and to analyze data and draw appropriate conclusions for the production system that is under consideration.

2.5 A Domain-Specific Competency Model for Lean 4.0

- Statistical quality techniques: The learner is able to select statistical quality techniques adequately, to improve statistical process quality, to implement the methods in existing production environments, and to analyze the results.
- Value creation excellence in the indirect areas: The trainee is able to analyze work processes of the indirect areas, to recognize weak points, and to adapt them to the new Lean 4.0 production processes. The trainee is also able to weigh and design the use of possible Industrie 4.0 tools in indirect areas.

B. Definition of competency clusters

Based on those defined intended competencies for the I 4.0 Process Expert listed above, the learning targets are clustered in expert discussions with regard to the similarity of thematic foci. Finally, the following clusters are identified (in alphabetical order):

- Lean 4.0 Basics
  - Analysis 4.0,
  - Design 4.0,
  - Stabilized, standardized, and individualized production,
  - Value creation excellence in the indirect areas.
- Lean 4.0 Machining
  - Digital customized machining,
  - Just in time for machining,
  - Predictive maintenance,
  - SMED.
- Lean 4.0 Material and Information Flow
  - Flexible manpower systems,
  - Just-in-time 4.0,
  - Production control 4.0.
- Lean 4.0 Quality
  - Problem-Solving 4.0,
  - Quality 4.0,
  - Statistical quality techniques.
- Lean 4.0 Thinking
  - Continuous improvement processes 4.0,
  - Shop floor management digitally supported.

C. Timeline for competency building

As a final step, the clusters are placed in an order in which they are reasonably completed by the learners. The training program is finally arranged in the three phases "Lean 4.0 Understanding," "Lean 4.0 Core Elements," and "Lean 4.0 Culture," while the second phase "Lean 4.0 Core Elements" contains the clusters "Lean 4.0 Machining," "Lean 4.0 Material and Information Flow," and "Lean 4.0 Quality." The three phases contain in total 17 learning modules—each for every intended competency—with one to three days duration. The competency model can be as well be applied in training to further develop today's production experts to I 4.0 Process Experts as also in education to educate tomorrows production experts properly. Figure 2.13 visualizes the training program for the I 4.0 Process Experts for the industry partners of the Process Learning Factory CiP.

2.6 Wrap-up of This Chapter

This chapter lays the foundations for important terms in the field of learning such as competences, qualifications, and knowledge as well as learning objectives and learning outcomes. Within the framework of a literature research, it is analyzed and explained which professional and overarching competencies are addressed in learning factories. A hot topic here is the identification of the necessary competencies in the context of Industrie 4.0. There are already various studies regarding necessary competencies for Industrie 4.0 that mostly identify vague meta-competencies. The challenge here is that these meta-competencies cannot serve as a basis for the development of learning factory training due to their vagueness and lack of context. As an example, it is shown how competencies can be defined for a sub-area of industrie 4.0 (the combination of Industrie 4.0 and Lean Production) that supports the learning factory developer in the goal-oriented development of learning factory offers.

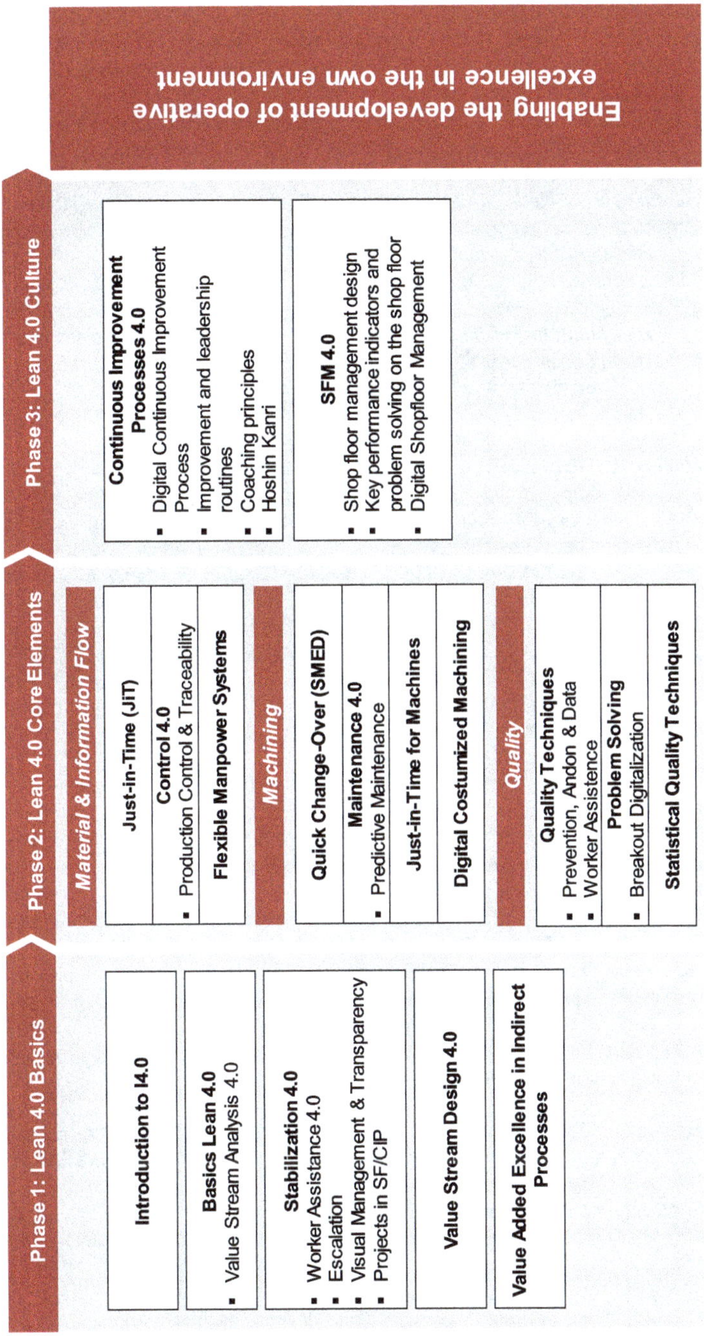

2.6 Wrap-up of This Chapter

Enabling the development of operative excellence in the own environment

Phase 1: Lean 4.0 Basics
- Introduction to I4.0
- Basics Lean 4.0
  - Value Stream Analysis 4.0
- Stabilization 4.0
  - Worker Assistance 4.0
  - Escalation
  - Visual Management & Transparency
  - Projects in SF/CIP
- Value Stream Design 4.0
- Value Added Excellence in Indirect Processes

Phase 2: Lean 4.0 Core Elements

Material & Information Flow
- Just-in-Time (JIT)
- Control 4.0
  - Production Control & Traceability
- Flexible Manpower Systems

Machining
- Quick Change-Over (SMED)
- Maintenance 4.0
  - Predictive Maintenance
- Just-in-Time for Machines
- Digital Costumized Machining

Quality
- Quality Techniques
  - Prevention, Andon & Data
  - Worker Assistence
- Problem Solving
  - Breakout Digitalization
- Statistical Quality Techniques

Phase 3: Lean 4.0 Culture
- Continuous Improvement Processes 4.0
  - Digital Continuous Improvement Process
  - Improvement and leadership routines
  - Coaching principles
  - Hoshin Kanri
- SFM 4.0
  - Shop floor management design
  - Key performance indicators and problem solving on the shop floor
  - Digital Shopfloor Management

Fig. 2.13 Visualization of the Lean 4.0 curriculum of the Process Learning Factory CiP

References

Abele, E., Bauerdick, C., Strobel, N., & Panten, N. (2016). ETA learning factory: A holistic concept for teaching energy efficiency in production. *6th CIRP-sponsored Conference on Leanring Factories. Procedia CIRP, 54*, 83–88.

Abele, E., Chryssolouris, G., Sihn, W., Metternich, J., ElMaraghy, H. A., Seliger, G., et al. (2017). Learning Factories for future oriented research and education in manufacturing. *CIRP Annals - Manufacturing Technology, 66*(2), 803–826.

Abele, E., Eichhorn, N., & Kuhn, S. (2007). Increase of productivity based on capability building in a learning factory. *Computer Integrated Manufacturing and High Speed Machining: 11th International Conference on Production Engineering, Zagreb*, pp. 37–41.

Abele, E., Metternich, J., Tenberg, R., Tisch, M., Abel, M., Hertle, C., Eißler, S., Enke, J., & Faatz, L. (2015). *Innovative Lernmodule und -fabriken: Validierung und Weiterentwicklung einer neuartigen Wissensplattform für die Produktionsexzellenz von morgen*. Darmstadt: tuprints.

Abele, E., Metternich, J., Tisch, M., Chryssolouris, G., Sihn, W., ElMaraghy, H. A., Hummel, V., & Ranz, F. (2015). Learning factories for research, education, and training. *5th CIRP-sponsored Conference on Learning Factories. Procedia CIRP, 32*, 1–6. https://doi.org/10.1016/j.procir.2015.02.187.

Abele, E., Tenberg, R., Wennemer, J., & Cachay, J. (2010). Kompetenzentwicklung in Lernfabriken für die Produktion. *Zeitschrift für wirtschaftlichen Fabrikbetrieb (ZWF), 105*(10), 909–913.

Acatech. (2016). *Kompetenzen für Industrie 4.0. Qualifizierungsbedarfe und Lösungsansätze.* acatech POSITION. Munich: Herbert Utz Verlag.

Allan, J. (1996). Learning outcomes in higher education. *Studies in Higher Education, 21*(1), 93–108.

Anderson, L. W., Krathwohl, D. R., & Airasian, P. W. (Eds.). (2001). *A taxonomy for learning, teaching, and assessing: A revision of Bloom's taxonomy of educational objectives* (Complete ed.). New York: Longman.

Balve, P., & Albert, M. (2015). Project-based learning in production engineering at the Heilbronn learning factory. *5th CIRP-sponsored Conference on Learning Factories. Procedia CIRP, 32*, 104–108. https://doi.org/10.1016/j.procir.2015.02.215.

Barkley, E. F., & Major, C. H. (2016). *Learning assessment techniques: A handbook for college faculty* (1st ed.). San Francisco, CA: Jossey-Bass, A Wiley Brand.

Bartram, D. (2013). *The universal competency framework*. Retrieved from https://www.cebglobal.com/content/dam/cebglobal/us/EN/talent-management/talent-assessment/pdfs/Universal-competency-framework-white-paper.pdf.

Bauernhansl, T., Dinkelmann, M., & Siegert, J. (2012). Lernfabrik advanced industrial Engineering Teil 1: Lernkonzepte und Struktur. *Werkstattstechnik Online: WT, 102*(3), 80–83.

Bender, B., Kreimeier, D., Herzog, M., & Wienbruch, T. (2015). Learning Factory 2.0 – Integrated view of product development and production. *5th CIRP-sponsored Conference on Learning Factories. Procedia CIRP, 32*, 98–103. https://doi.org/10.1016/j.procir.2015.02.226.

Blöchl, S. J., & Schneider, M. (2016). Simulation game for intelligent production logistics – The PuLL® learning factory. *Procedia CIRP, 54,* 130–135. https://doi.org/10.1016/j.procir.2016.04.100.

Block, M., Bertagnolli, F., & Herrmann, K. (2011). Lernplattform - Eine neue Dimension des Lernens von schlanken Abläufen/Learning plattform - a new dimension to learn about lean processes. *Productivity Management, 4,* 52–55.

Bloom, B. S., Engelhart, M. D., Furst, E. J., Hill, W. H., & Krathwohl, D. R. (1956). *Taxonomy of educational objectives: The classification of educational goals (Book I: Cognitive domain)*. New York: McKay.

Blume, S., Madanchi, N., Böhme, S., Posselt, G., Thiede, S., & Herrmann, C. (2015). Die Lernfabrik – Research-based learning for sustainable production engineering. *5th CIRP-sponsored Conference on Learning Factories. Procedia CIRP, 32*, 126–131. https://doi.org/10.1016/j.procir.2015.02.113.

References

Buergin, J., Minguillon, F. E., Wehrle, F., Haefner, B., & Lanza, G. (2017). Demonstration of a concept for scalable automation of assembly systems in a learning factory. *Procedia Manufacturing, 9*, 33–40. https://doi.org/10.1016/j.promfg.2017.04.026.

Cachay, J., & Abele, E. (2012). Developing competencies for continuous improvement processes on the shop floor through learning factories–conceptual design and empirical validation. *45th CIRP Conference on Manufacturing Systems. Procedia CIRP, 3*(3), 638–643.

Cachay, J., Wennemer, J., Abele, E., & Tenberg, R. (2012). Study on action-oriented learning with a learning factory approach. *Procedia - Social and Behavioral Sciences., 55,* 1144–1153.

Chomsky, N. (1962). Explanatory models in linguistics. In E. Nagel, P. Suppes, & A. Tarski (Eds.), *Logic, methodology and philosophy of science* (pp. 528–550). Stanford: Stanford University Press.

Dave, R. H. (1970). Psychomotor Levels. In R. J. Armstrong (Ed.), *Developing and writing behavioral objectives* (pp. 21–22). Tucson, Arizona: Educational Innovators Press.

Dinkelmann, M., Riffelmacher, P., & Westkämper, E. (2011). Training Concept and Structure of the Learning Factory advanced Industrial Engineering. *Enabling Manufacturing Competitiveness and Economic Sustainability*, 623–629.

Dinkelmann, M., Siegert, J., & Bauernhansl, T. (2014). Change management through learning factories. In M. F. Zäh (Ed.), *Enabling manufacturing competitiveness and economic sustainability* (pp. 395–399). Berlin: Springer. https://doi.org/10.1007/978-3-319-02054-9_67.

Dombrowski, U., Riechel, C., & Evers, M. (2014). Die Rolle des Menschen in der 4. Industriellen Revolution. In W. Kersten, H. Koller, & H. Lödding (Eds.), *Schriftenreihe der Hochschulgruppe für Arbeits- und Betriebsorganisation e.V. (HAB). Industrie 4.0: Wie intelligente Vernetzung und kognitive Systeme unsere Arbeit verändern* (pp. 129–153). Berlin: GITO.

Eisner, E. W. (1979). *The educational imagination: On the design and evaluation of school programs*. New York: Macmillan.

ElMaraghy, H. A., & ElMaraghy, W. (2015). Learning integrated product and manufacturing systems. *5th CIRP-sponsored Conference on Learning Factories. Procedia CIRP, 32,* 19–24.

Erol, S., Jäger, A., Hold, P., Ott, K., & Sihn, W. (2016). Tangible industry 4.0: A scenario-based approach to learning for the future of production. *6th CIRP-sponsored Conference on Leanring Factories. Procedia CIRP, 54,* 13–18.

Erpenbeck, J., & von Rosenstiel, L. (2007). *Handbuch Kompetenzmessung: Erkennen, verstehen und bewerten von Kompetenzen in der betrieblichen, pädagogischen und psychologischen Praxis (2., überarb. und erw. Aufl.)*. Stuttgart: Schäffer-Poeschel.

European Commission. (2006). *Implementing the community lisbon programme: Proposal for a RECOMMENDATION OF THE EUROPEAN PARLIAMENT AND OF THE COUNCIL on the establishment of the European Qualifications Framework for Lifelong Learning. COM(2006) 479 final*. Brussels.

Faller, C., & Feldmüller, D. (2015). Industry 4.0 learning factory for regional SMEs. *5th CIRP-sponsored Conference on Learning Factories. Procedia CIRP, 32,* 88–91. https://doi.org/10.1016/j.procir.2015.02.117.

Goerke, M., Schmidt, M., Busch, J., & Nyhuis, P. (2015). Holistic approach of lean thinking in learning factories. *5th CIRP-sponsored Conference on Learning Factories. Procedia CIRP, 32,* 138–143. https://doi.org/10.1016/j.procir.2015.02.221.

Gosling, D., & Moon, J. A. (2001). *How to use learning outcomes and assessment criteria. How to series*. London: SEEC.

Gräßler, I., Pöhler, A., & Pottebaum, J. (2016). Creation of a learning factory for cyber physical production systems. *6th CIRP-sponsored Conference on Leanring Factories. Procedia CIRP, 54,* 107–112. https://doi.org/10.1016/j.procir.2016.05.063.

Gräßler, I., Taplick, P., & Yang, X. (2016b). Educational learning factory of a holistic product creation process. *Procedia CIRP, 54,* 141–146. https://doi.org/10.1016/j.procir.2016.05.103.

Gronau, N., Ullrich, A., & Teichmann, M. (2017). Development of the industrial IoT competences in the areas of organization, process, and interaction based on the learning factory concept. *Procedia Manufacturing, 9,* 254–261. https://doi.org/10.1016/j.promfg.2017.04.029.

Hacker, W. (1973). *Allgemeine Arbeits- und Ingenieurpsychologie: Psychische Struktur und Regulation von Arbeitstätigkeiten*. Berlin: Deutscher Verlag der Wissenschaften.

Hecklau, F., Galeitzke, M., Flachs, S., & Kohl, H. (2016). Holistic approach for human resource management in industry 4.0. *Procedia CIRP, 54*, 1–6. https://doi.org/10.1016/j.procir.2016.05.102.

Helm, R., Reise, C., & Rößle, D. (2014). Learning factories for sustainable manufacturing - A generic design approach. *Advanced Materials Research, 1018*, 517–524.

Heyse, V., & Erpenbeck, J. (2009). *Kompetenztraining: 64 modulare Informations- und Trainingsprogramme für die betriebliche pädagogische und psychologische Praxis (2., überarb. und erw. Aufl.)*. Stuttgart: Schäffer-Poeschel.

Hummel, V., Hyra, K., Ranz, F., & Schuhmacher, J. (2015). Competence development for the holistic design of collaborative work systems in the logistics learning factory. *5th CIRP-sponsored Conference on Learning Factories. Procedia CIRP, 32*, 76–81. https://doi.org/10.1016/j.procir.2015.02.111.

Jäger, A., Mayrhofer, W., Kuhlang, P., Matyas, K., & Sihn, W. (2012). The "Learning Factory": An immersive learning environment for comprehensive and lasting education in industrial engineering. *16th World Multi-Conference on Systemics, Cybernetics and Informatics, 16*(2), 237–242.

Jäger, A., Mayrhofer, W., Kuhlang, P., Matyas, K., & Sihn, W. (2013). Total immersion: Hands and heads-on training in a learning factory for comprehensive industrial engineering education. *International Journal of Engineering Education, 29*(1), 23–32.

Jorgensen, J. E., Lamancusa, J. S., Zayas-Castro, J. L., & Ratner, J. (1995). The learning factory: Curriculum integration of design and manufacturing. *4th World Conference on Engineering Education*, pp. 1–7.

Kagermann, H., Wahlster, W., & Helbig, J. (2013). Recommendations for implementing the strategic initiative INDUSTRIE 4.0: Securing the future of German manufacturing industry. Final report of the Industrie 4.0 Working Group. Retrieved from http://www.acatech.de/fileadmin/user_upload/Baumstruktur_nach_Website/Acatech/root/de/Material_fuer_Sonderseiten/Industrie_4.0/Final_report__Industrie_4.0_accessible.pdf.

Krathwohl, D. R., Bloom, B. S., & Masia, B. (1964). *Taxonomy of educational objectives*. London: Longman.

Kreimeier, D., Morlock, F., Prinz, C., Krückhans, B., Bakir, D. C., & Meier, H. (2014). Holistic learning factories - A concept to train lean management, resource efficiency as well as management and organization improvement skills. *47th CIRP Conference on Manufacturing Systems. Procedia CIRP, 17*, 184–188.

Kreitlein, S., Höft, A., Schwender, S., & Franke, J. (2015). Green factories Bavaria: A network of distributed learning factories for energy efficient production. *5th CIRP-sponsored Conference on Learning Factories. Procedia CIRP, 32*, 58–63. https://doi.org/10.1016/j.procir.2015.02.219.

Krückhans, B., Wienbruch, T., Freith, S., Oberc, H., Kreimeier, D., & Kuhlenkötter, B. (2015). Learning factories and their enhancements - A comprehensive training concept to increase resource efficiency. *5th CIRP-sponsored Conference on Learning Factories. Procedia CIRP, 32*, 47–52. https://doi.org/10.1016/j.procir.2015.02.224.

Kuhlmann, A. M., & Sauter, W. (2008). *Innovative Lernsysteme: Kompetenzentwicklung mit blended learning und social Software*. X.media.press. Berlin and Heidelberg: Springer.

Küsters, D., Praß, N., & Gloy, Y.-S. (2017). Textile learning factory 4.0 – Preparing Germany's textile industry for the digital future. *Procedia Manufacturing, 9*, 214–221. https://doi.org/10.1016/j.promfg.2017.04.035.

Lamancusa, J. S., & Simpson, T. (2004). The learning factory: 10 years of impact at penn state. *Proceedings International Conference on Engineering Education*, pp. 1–8.

Lanza, G., Minges, S., Stoll, J., Moser, E., & Haefner, B. (2016). Integrated and modular didactic and methodological concept for a learning factory. *6th CIRP-sponsored Conference on Leanring Factories. Procedia CIRP, 54*, 136–140. https://doi.org/10.1016/j.procir.2016.06.107.

Lanza, G., Moser, E., Stoll, J., & Haefner, B. (2015). Learning Factory on Global Production. *5th CIRP-sponsored Conference on Learning Factories. Procedia CIRP, 32*, 120–125. https://doi.org/10.1016/j.procir.2015.02.081.

References

Le Delamare Deist, F., & Winterton, J. (2005). What is competence? *Human Resource Development International, 8*(1), 27–46. https://doi.org/10.1080/1367886042000338227.

Liebrecht, C., Hochdörffer, J., Treber, S., Moser, E., Erbacher, T., Gidion, G., et al. (2017). Concept development for the verification of the didactic competence promotion for the learning factory on global production. *Procedia Manufacturing, 9,* 315–322. https://doi.org/10.1016/j.promfg.2017.04.019.

Löhr-Zeidler, B., Hörner, R., & Heer, J. (2016). Handlungsempfehlungen Industrie 4.0 – Umsetzungshilfen für Lehrerinnen und Lehrer der beruflichen Schulen. *Berufsbildung, 70*(159), 11–14.

Matt, D. T., Rauch, E., & Dallasega, P. (2014). Mini-Factory – A learning factory concept for students and small and medium sized enterprises. *47th CIRP Conference on Manufacturing Systems. Procedia CIRP, 17,* 178–183.

McClelland, D. (1976). *A guide to job competency assessment*. Boston: McBer.

McMullan, M., Endacott, R., Gray, M. A., Jasper, M., Miller, C. M. L., Scholes, J., et al. (2003). Portfolios and assessment of competence: A review of the literature. *Journal of Advanced Nursing, 41*(3), 283–294. https://doi.org/10.1046/j.1365-2648.2003.02528.x.

Merkel, L., Atug, J., Merhar, L., Schultz, C., Braunreuther, S., & Reinhart, G. (2017). Teaching smart production: An insight into the learning factory for cyber-physical production systems (LVP). *Procedia Manufacturing, 9,* 269–274. https://doi.org/10.1016/j.promfg.2017.04.034.

Metternich, J. (2017). Digitalisierung im Maschinenbau: Chancen nutzen, Risiken erkennen und vermeiden. Retrieved from http://mav.industrie.de/allgemein/digitalisierung-im-maschinenbau/#slider-intro-1.

Metternich, J., Müller, M., Meudt, T., & Schaede, C. (2017). Lean 4.0 – zwischen Widerspruch und Vision. *Zeitschrift für wirtschaftlichen Fabrikbetrieb (ZWF), 112*(5), 346–348. https://doi.org/10.3139/104.111717.

Meudt, T., Leipoldt, C., & Metternich, J. (2016). Der neue Blick auf Verschwendungen im Kontext von Industrie 4.0. *Zeitschrift für wirtschaftlichen Fabrikbetrieb (ZWF), 111*(11), 754–758. https://doi.org/10.3139/104.111617.

Micheu, H.-J., & Kleindienst, M. (2014). Lernfabrik zur praxisorientierten Wissensvermittlung: Moderne Ausbildung im Bereich Maschinenbau und Wirtschaftswissenschaften. *Zeitschrift für wirtschaftlichen Fabrikbetrieb (ZWF), 109*(6), 403–407.

Morlock, F., Kreggenfeld, N., Louw, L., Kreimeier, D., & Kuhlenkötter, B. (2017). Teaching methods-time measurement (MTM) for workplace design in learning factories. *Procedia Manufacturing, 9,* 369–375. https://doi.org/10.1016/j.promfg.2017.04.033.

Müller, B. C., Menn, J. P., & Seliger, G. (2017). Procedure for experiential learning to conduct material flow simulation projects, enabled by learning factories. *Procedia Manufacturing, 9,* 283–290. https://doi.org/10.1016/j.promfg.2017.04.047.

Müller-Frommeyer, L. C., Aymans, S. C., Bargmann, C., Kauffeld, S., & Herrmann, C. (2017). Introducing competency models as a tool for holistic competency development in learning factories: Challenges, example and future application. *Procedia Manufacturing, 9,* 307–314. https://doi.org/10.1016/j.promfg.2017.04.015.

Nöhring, F., Rieger, M., Erohin, O., Deuse, J., & Kuhlenkötter, B. (2015). An interdisciplinary and hands-on learning approach for industrial assembly systems. *5th CIRP-sponsored Conference on Learning Factories. Procedia CIRP, 32,* 109–114. https://doi.org/10.1016/j.procir.2015.02.112.

North, K. (2011). *Wissensorientierte Unternehmensführung: Wertschöpfung durch Wissen* (5., aktualisierte und erweiterte Auflage). *Lehrbuch*. Wiesbaden: Gabler.

Plorin, D., Poller, R., & Müller, E. (2013). Advanced Learning Factory (aLF): Integratives Lernfabrikkonzept zur praxisnahen Kompetenzentwicklung am Beispiel der Energieeffizienz. *Werkstattstechnik Online: WT, 103*(3), 226–232.

Prifti, L., Knigge, M., Kienegger, H., & Krcmar, H. (2017). A Competency model for "Industrie 4.0" employees. In J. M. Leimeister & W. Brenner (Eds.), *Proceedings der 13. Internationalen Tagung Wirtschaftsinformatik (WI 2017)* (pp. 46–60). St. Gallen.

Reuter, M., Oberc, H., Wannöffel, M., Kreimeier, D., Klippert, J., Pawlicki, P., et al. (2017). Learning Factories' Trainings as an enabler of proactive workers' participation regarding Industrie 4.0. *Procedia Manufacturing, 9,* 354–360. https://doi.org/10.1016/j.promfg.2017.04.020.

Richter, A., Heinrich, P., Stocker, A., & Unzeitig, W. (2015). Der Mensch im Mittelpunkt der Fabrik von morgen. *HMD Praxis der Wirtschaftsinformatik, 52*(5), 690–712. https://doi.org/10.1365/s4 0702-015-0173-x.

Rogalla, I. (2013). *Metakompetenzen - Die neuen Schlüsselqualifikationen? Ein Plädoyer für einen gehaltvollen Kompetenzbegriff* (1., Auflage). *Berufliche Handlungskompetenz: Vol. 4.* Berlin: R & W Verlag der Editionen.

Rowe, C. (1995). Clarifying the use of competence and competency models in recruitment, assessment and staff development. *Industrial and Commercial Training, 27*(11), 12–17.

Rybski, C., & Jochem, R. (2016). Benefits of a learning factory in the context of lean management for the pharmaceutical industry. *Procedia CIRP, 54,* 31–34. https://doi.org/10.1016/j.procir.201 6.05.106.

Scholz, M., Kreitlein, S., Lehmann, C., Böhner, J., Franke, J., & Steinhilper, R. (2016). Integrating intralogistics into resource efficiency oriented learning factories. *Procedia CIRP, 54,* 239–244. https://doi.org/10.1016/j.procir.2016.05.067.

Schreiber, S., Funke, L., & Tracht, K. (2016). BERTHA - A flexible learning factory for manual assembly. *Procedia CIRP, 54,* 119–123. https://doi.org/10.1016/j.procir.2016.03.163.

Schützer, K., Rodrigues, L. F., Bertazzi, J. A., Durão, L. F. C. S., & Zancul, E. (2017). Learning environment to support the product development process. *Procedia Manufacturing, 9,* 347–353. https://doi.org/10.1016/j.promfg.2017.04.018.

Seliger, G., Reise, C., & Farland, R.M. (2009). Outcome-oriented learning environment for sustainable engineering education. *7th Global Conference on Sustainable Manufacturing: Sustainable Product Development and Life Cycle Engineering,* pp. 91–98.

Short, E. C. (1984). Competence reexamined. *Educational Theory, 34*(3), 201–207. https://doi.or g/10.1111/j.1741-5446.1984.50001.x.

Simons, S., Abé, P., & Neser, S. (2017). Learning in the AutFab – The fully automated Industrie 4.0 learning factory of the university of applied sciences darmstadt. *Procedia Manufacturing, 9,* 81–88. https://doi.org/10.1016/j.promfg.2017.04.023.

Spath, D. (Ed.). (2013). *Produktionsarbeit der Zukunft - Industrie 4.0.* Stuttgart: Fraunhofer Verlag.

Spencer, L. M., & Spencer, S. M. (1993). *Competence at work: Models for superior performance.* New York: Wiley.

Steffen, M., Frye, S., & Deuse, J. (2013). The only source of Knowledge is experience: Didaktische Konzeption und methodische Gestaltung von Lehr-Lern-Prozessen in Lernfabriken zur Aus- und Weiterbildung im Industrial Engineering. *TeachING LearnING.EU. Innovationen für die Zukunft der Lehre in den Ingenieurwissenschaften,* pp. 117–129.

Steffen, M., May, D., & Deuse, J. (2012). The industrial engineering laboratory: Problem based learning in industrial engineering education at TU Dortmund University. *Global Engineering Education Conference (EDUCON), IEEE, - Collaborative Learning & New Pedagogic Approaches in Engineering Education, Marrakesch, Marokko; 17.-20. 04.2012,* pp. 1–10.

Tenberg, R., & Pittich, D. (2017). Ausbildung 4.0 oder nur 1.2? Analyse eines technisch-betrieblichen Wandels und dessen Implikationen für die technische Berufsausbildung. *Journal of Technical Education (JOTED), 5*(1).

Teodorescu, T. (2006). Competence versus competency: What is the difference? *Performance Improvement, 45*(10), 27–30.

Thiede, S., Juraschek, M., & Herrmann, C. (2016). Implementing cyber-physical production systems in learning factories. *6th CIRP-sponsored Conference on Leanring Factories. Procedia CIRP, 54,* 7–12. https://doi.org/10.1016/j.procir.2016.04.098.

Tietze, F., Czumanski, T., Braasch, M., & Lödding, H. (2013). Problembasiertes Lernen in Lernfabriken. *Werkstattstechnik online: wt, 103*(3), 246–251.

Tisch, M., Hertle, C., Abele, E., Metternich, J., & Tenberg, R. (2015). Learning factory design: A competency-oriented approach integrating three design levels. *International Journal of Com-*

References

puter Integrated Manufacturing, 29(12), 1355–1375. https://doi.org/10.1080/0951192X.2015.1033017.

Tisch, M., Hertle, C., Cachay, J., Abele, E., Metternich, J., & Tenberg, R. (2013). A systematic approach on developing action-oriented, competency-based learning factories. *46th CIRP Conference on Manufacturing Systems. Procedia CIRP, 7*, 580–585.

Tyler, R. W. (1971). *Basic principles of curriculum and instruction. Set books/Open University.* Chicago, London: University of Chicago Press.

VDMA, & PTW, TU Darmstadt. (2018). *Leitfaden Industrie 4.0 trifft Lean: Wertschöpfung ganzheitlich steigern.* Frankfurt am Main: VDMA Verlag. Retrieved from https://industrie40.vdma.org/documents/4214230/26095707/Leitfaden_I40_Lean_1524489604061.pdf/39b9d595-fc44-8212-d7ec-c5ff420647dd.

Vernim, S., Reinhart, G., & Bengler, K. (2017). Qualifizierung des Produktionsmitarbeiters in der Industrie 4.0. In G. Reinhart (Ed.), *Handbuch Industrie 4.0: Geschäftsmodelle, Prozesse, Technik* (pp. 60–66). Carl Hanser.

Veza, I., Gjeldum, N., & Mladineo, M. (2015). Lean learning factory at FESB – University of Split. *5th CIRP-sponsored Conference on Learning Factories. Procedia CIRP, 32*, 132–137. https://doi.org/10.1016/j.procir.2015.02.223.

Wagner, C., Heinen, T., Regber, H., & Nyhuis, P. (2010). Fit for change – Der Mensch als Wandlungsbefähiger. *Zeitschrift für wirtschaftlichen Fabrikbetrieb (ZWF), 100*(9), 722–727.

Wagner, P., Prinz, C., Wannöffel, M., & Kreimeier, D. (2015). Learning factory for management, organization and workers' participation. *5th CIRP-sponsored Conference on Learning Factories. Procedia CIRP, 32*, 115–119. https://doi.org/10.1016/j.procir.2015.02.118.

Wagner, U., AlGeddawy, T., ElMaraghy, H. A., & Müller, E. (2012). The state-of-the-art and prospects of learning factories. *45th CIRP Conference on Manufacturing Systems. Procedia CIRP, 3*, 109–114.

White, R. (1959). Motivation reconsidered: The concept of competence. *Psychological Review, 66*, 297–333.

Yoo, I. S., Braun, T., Kaestle, C., Spahr, M., Franke, J., Kestel, P., et al. (2016). Model factory for additive manufacturing of mechatronic products: Interconnecting world-class technology partnerships with leading AM players. *Procedia CIRP, 54*, 210–214. https://doi.org/10.1016/j.procir.2016.03.113.

Chapter 3
Learning in Production, Learning for Production

In this chapter, the forms and variations of work-related learning in and for production, the historic development of work-related learning as well as the importance and relevance of work-related learning in form of education and training are explained. In the following section, the basic terms and notions in the field of learning are defined in order to show the relevance and importance of the learning factory approach to competency development and research in production-related areas (Fig. 3.1).

3.1 Definition of Basic Terms and Notions

Learning is regularly seen as the acquisition and modification of

- Knowledge,
- Behavior,
- Beliefs,
- Attitudes,
- Skills,

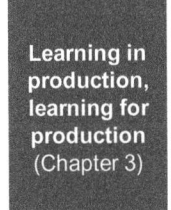

Learning in production, learning for production (Chapter 3)	• Definition of basic terms and notions (3.1) • Historical development of work-related learning (3.2) • Forms of work-related learning for production (3.3) • Types of perceived learning concepts in production (3.4) • Need for learning factories (3.5)

Fig. 3.1 Overview on the structure of this chapter on the learning in production and the learning for production

- Strategies,
- Values,
- and others.[1]

Acquisitions and modifications can take manifold forms and may concern linguistic, cognitive, motor, and social areas.[2] Usually, achievement areas are distinguished in a cognitive, a psychomotor, and an affective dimension.[3] Learning builds and makes connections to and between existing experiences and knowledge. Learning-induced changes might have deep and lasting impact; therefore, effects of learning can be understood as relatively permanent.[4]

Often, learning is classified into formal and informal learning. Formal learning is planned and regularly based on related targets. Informal learning is mostly incidental and occurs during experiential and implicit learning.[5] Between the two forms of formal and informal learning, the non-formal learning concept is positioned. Non-formal learning describes various loosely organized learning situations.[6] The learning factory concept can be located in the non-formal learning field; it possesses characteristics from both formal and informal learning; see also Fig. 3.2.

A variety of learning theories can be identified in the literature. The theories are defined as conceptual frameworks on how to see and interpret learning processes. The

formal learning
- planned
- individual learning
- vertical (hierarchical) knowledge
- educational institutions

informal learning
- incidental
- learning through everyday embodied practices
- horizontal (segmented) knowledge
- non-educational settings

the learning factory concept

non-formal learning
- planned with room for incidental learning
- learning in social context / in groups
- Knowledge is generated vertically and horizontally
- no traditional education setting

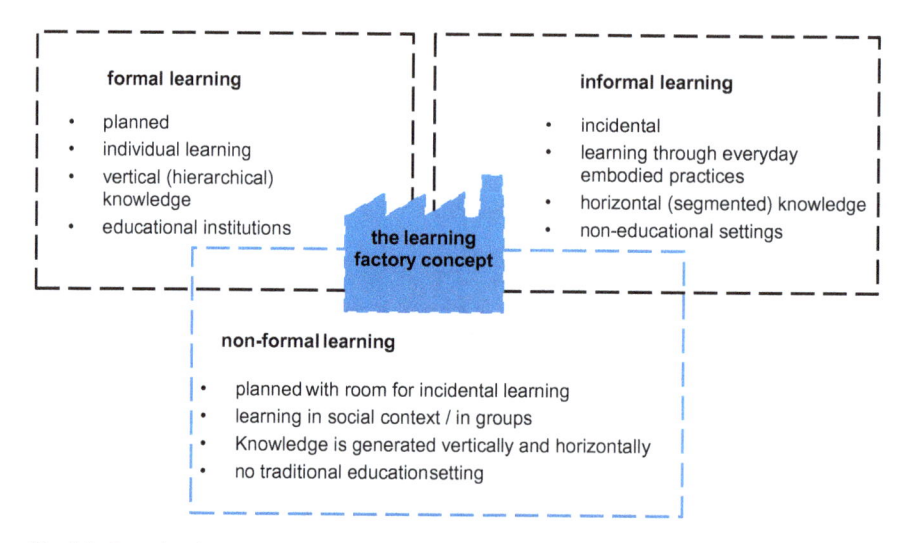

Fig. 3.2 Learning factory concept between formal and informal learning

---

[1]See, for example, Schunk (1996) and Kirkpatrick (1996).
[2]See Schunk (1996).
[3]Based on Bloom, Engelhart, Furst, Hill, and Krathwohl (1956), for a descriptions and taxonomies of the three domains, see also Sect. 2.2.
[4]See Schacter, Gilbert, and Wegner (2009).
[5]See Dehnbostel (2007).
[6]See Eraut (2000).

3.1 Definition of Basic Terms and Notions

most famous theories of behaviorism, cognitivism, and constructivism are explained briefly in the following.

Behaviorism is a learning psychological theory that explains behavior or the change of behavior of the learner by the expected consequences rather than the result of cognitive processes.[7] Within the behaviorism, three key forms of learning can be identified:

- Reactions (behaviors) are modified through the use of associated stimuli (classical conditioning).[8]
- Behaviors are modified through effected reward or punishment (operant conditioning).[9]
- Behaviors are modified by observation and imitation of other people's actions (modeling).[10]

Shortly after this and opposed to the behavioristic learning theory, the cognitivist learning theory focuses on the investigation of cognitive processes inside the learners that were treated by the behaviorists as a black box. In this theory, which is also referred to as cognitivism, thinking processes are the starting point for the explanation of specific behavioral changes.[11,12]

Both learning theories, behaviorism and cognitivism, can be attributed to objectivism. Objectivist learning theories are based on the understanding that in course of learning processes the learner is processing learner-external realities. This understanding is confronted by the learning theory called constructivism. In contrast to objectivism, in constructivism, the individual experiences of the learners determine learner-individual reality.[13] This differentiation leads to variations in the perception and the design of learning and the definition of respective learning goals. Table 3.1 summarizes and contrasts the convictions of the three described learning theories regarding different characteristics of learning like learning processes, learning results, the didactic design.

As in the context of other learning systems, both objectivist and constructivist learning theories can be helpful for learning factories.[14] Depending on which goals are pursued with the respective learning factory, which target group is active, and which learning objectives and learning contents are to be conveyed, a different view of the learning process can be useful.

---

[7] See Skinner (1976).
[8] See Skinner (1976), Pavlov (2003).
[9] See Skinner (1976), Thorndike (1965).
[10] See Bandura (1974).
[11] See Neisser (1967).
[12] A detailed overview of this approach and the related developments can, for example, be found in Hartley (1985), Cooper (1993), Tenberg (2011), and Kerres (2012).
[13] See Jonassen (1991).
[14] See Footnote 13.

Table 3.1 Perspectives of the learning theories taken from Tisch (2018) according to Kerres (2012) with additions according to Ertmer and Newby (1993), Schunk (1996), and Jonassen (1991)

	Behaviorism	Cognitivism	Constructivism
Learning process	Response to environment	Construction of cognitive structures	(Re-)construction of knowledge, participation in cultural practice
Learning result	Stimulus-response connection	Abstract, generalizable knowledge	Contextualized, situationally applicable knowledge
Demand for didactic design	Structuring learning contents into small learning units	Adaptation of the learning material to learning requirements or learning progress	Integration into application contexts, authenticity, learning material, location
Didactic methods	Sequentially prepared exposure (performance); repeated demonstration and execution of activities	Exposure and exploration (discovery); active integration of the learner, content structuring for easier processing	Exploration, project method, cooperation
Learning process control	External control	Learning progress-dependent external/self-control	Self-control
Learning transfer	Through generalization	Through understanding how knowledge is to be applied specifically in the context	By integration into authentic tasks in meaningful contexts
Learning forms explained by theory	Simple forms of learning: memory of facts, definition of terms, use of explanatory patterns, implementation of specific procedures	Complex forms of learning: problem solving, information processing, logical thinking, etc.	No context-independent forms identifiable in constructivism; applicable to the content of advanced knowledge-based knowledge
Learning success control	Mandatory adaption of the learning offer with every learning step	After a meaningful learning unit, embedded as much as possible in learning tasks	For self-diagnosis, application-oriented tasks, ensure transfer
Role of media	Control and regulation of the learning process	Presentation of knowledge, interactivity, adaptability	Offers for construction activities

3.2 Historical Development of Work-Related Learning

In medieval times, learning a craft in a workshop can be seen as a classic example of workplace learning. Apprentices gained technical skills and knowledge working and learning in the workplace under and supported by master craftsmen.[15] With the industrialization, the workplace was not the only place for learning anymore. New learning places and formal forms of learning more and more replaced the workplace as learning place.[16] Due to the comprehensive qualification requirements, first production-independent vocational training workshops emerged in the late eighteenth century. Teaching and learning processes were decoupled from the production area and the pressure to produce (see for example Fig. 3.3).[17] With this, a gap between the learning and the specific workplace requirements was created. In addition, in this context, several negative accompanying phenomena could be identified: Learning and motivation problems, a reduced level of self-initiative and self-organized learning, increased costs of learning and prolonged training periods have been identified.[18]

Fig. 3.3 Vocational training workshop of German company AEG in Mühlheim-Saarn, approximately taken in the year 1956, picture taken from Bundesarchiv, B 145 Bild-P048595/CC-BY-SA 3.0

---

[15] See Schaper (2000).
[16] See Münch (1977).
[17] See Dehnbostel, Holz, Novak, and Arndt (1996).
[18] See Footnote 15.

After the Second World War, Japan established systematic quality systems. With the help of American experts, these quality systems were developed. In addition to quality management, quality circles (problem-solving groups, invented by Kaoru Ishikawa) were established in 1962. By the end of the 1960s, there were no imitators in other countries, while in Japan, quality circles were spreading rapidly. In the West, the concept became of interest much later (in the second half of the 1980s).[19]

Independently from the quality circles and before they emerged in the West, in Germany the learning in small groups was established in the beginning of the 1970s with the concept of the so-called Lernstatt [a composition of the German words "lernen" (to learn) and "Werkstatt" (factory/workshop)]. The starting point for the Lernstatt concept was the attempt to eliminate misunderstandings of foreign employees in German companies by means of a practice-oriented, operational learning form. The concept was developed by social scientists (namely, Helga Cloyd and Waldemar Kasprzik) and first implemented at BMW. Today, the concept and its history are still inseparably linked to BMW.[20] Further learning approaches in Germany were built on this. The development and use of the Lernstatt can be divided into two phases[21]:

- In the first phase, it was used as conceptualized for teaching language skills with a special form of work-related communication for foreign employees.
- The origin of the second phase was the fact that German workers were almost excluded from the traditional training courses and that they also wanted a possibility of exchange of experiences and practical training in the company. Consequently, the Lernstatt was transformed from a language-learning approach to a concept for company training regarding specialist knowledge. Thus, the Lernstatt became the first concept of industrial employee training.

In the 1980s, another shift in bringing the learning processes closer to the working processes can be recognized (Fig. 3.4). The importance of work-integrated learning was recognized due to an increasingly clear gap between the training scenarios compared to the company's real situations as well as more complex production environments and tasks. This renaissance of the learning at the workplace was facilitated by the emergence and spread of new information and communication technologies as well as new work and organizational concepts;[22] which is an interesting parallel to the present situation in production.[23] In the end of the 1980s, Mercedes-Benz, at Gaggenau plant, started a project funded by the German Federal Institute for vocational training with the intention to develop a requirement-oriented form of learning for modern automotive production. In this project, learning bays have been developed, mainly for the direct production process, in which trainees independently handle com-

---

[19] See Deppe (1992).

[20] See Footnote 19.

[21] Deppe (2013).

[22] See Footnote 15.

[23] See Schuh, Gartzen, Rodenhauser, and Marks (2015), Dehnbostel and Schröder (2017).

Fig. 3.4 Coaching in industrial environments as a form of work-integrated learning

plex production orders. Those learning bays were subsequently also implemented, tested, and used by many SMEs.[24]

At the beginning of the 1990s, new learning theoretical approaches were developed with regard to application-oriented learning. Learning should be related to context and situation, furthermore included learning tasks and solved problems in course of the learning process should be realistic and authentic.[25] This is the time when first learning factories are developed, among others for vocational education (Festo Didactic) or as training platform for Computer-Integrated Manufacturing.[26] In the following years and decades, approaches were developed on this basis, while still new approaches are following. Examples for this kind of learning approaches are the "Lernwerkstatt" concept (literally translated from German: learning workshop), the learning factory, and the teaching factory concept. Furthermore, new problem-solving-related learning processes (e.g. work and learning tasks or coaching-supported approaches) are established. Today, additional efforts can be noticed in digitization-supported, work-integrated learning processes. Those digitally supported learning approaches are in an early idea stage, especially for the effective development of professional competencies at the workplace (in contrast to knowledge supply at the workplace) innovative concepts are needed. All those current learning approaches for and in production are further described and classi-

---

[24] See Footnote 17.
[25] See Footnote 15.
[26] See Reith (1988).

fied in the Sects. 3.3 and 3.4. Fig 3.5 illustrates the development of work-related learning beginning from the medieval age until today. Dividing the approaches in work-integrated/-connected (on the upper part of the figure) and work-oriented (on the lower part of the figure) forms, a pendulum movement between the two forms can be noticed. For the description of work-integrated, work-connected, and work-oriented approaches, see Sect. 3.3.

3.3 Forms of Work-Related Learning for Production

Forms of work-related learning are getting more important in recent years, especially with the developments around Industrie 4.0 and digitalization, the topic is getting a new boost.[27] In the field of work-related learning manifold, variants can be differentiated. Distinguished variations of the semantical broad "work-related learning" field according to the relation between the places of learning and working are

- Work-integrated,
- Work-connected, and
- Work-oriented learning.[28]

The concept of work-integrated (WI) learning is defined by a learning location that is identical to the workplace.[29] This means that the learning and the working processes are interwoven with one another. Examples in the field of production are the job rotation method,[30] traditional apprenticeship,[31] or job instruction often based on the Training Within Industry (TWI) program.[32] Today, work-integrated forms of learning focus very much on psychomotor competencies necessary to carry out the work process. In the future, support will be available in those learning processes (e.g. exoskeleton), which can provide more detailed and accurate feedback on the execution of the work process. In addition to the skills for work process execution, also interesting and necessary methods, like coaching approaches, that include competences needed for continuous improvement of work processes and structured solution of problems are mainly implemented in bigger companies.[33] There are also so far hardly approaches that address social-communicative competencies.

Work-connected (WC) learning uses separated, but spatially and organizationally connected, learning and working locations.[34] In production-related areas, work-

---

[27] See Dehnbostel and Schröder (2017).
[28] According to Dehnbostel (2009).
[29] See Footnote 27.
[30] See Ortega (2001).
[31] See Ryan (2012).
[32] See Robinson and Schroeder (1993).
[33] See Footnote 27.
[34] See Footnote 27.

3.3 Forms of Work-Related Learning for Production   67

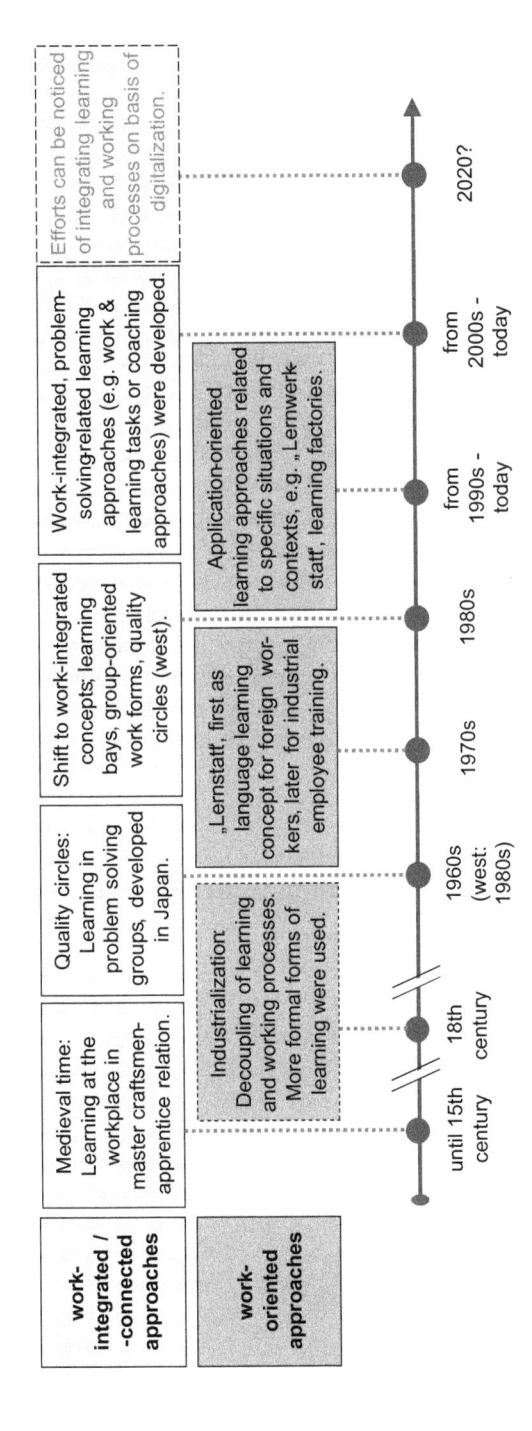

Fig. 3.5 Historical development of work-related learning approaches

connected learning can be implemented, for example, with quality circles[35] and learning and training stations close to the production area.[36] In the context of work-connected learning, the possibilities of digitalization allow dynamic learning stations, which can be adapted to the learning situation and the target group, in the immediate vicinity of the work process in order to develop professional and methodical competencies of employees.

In work-oriented (WO) learning concepts, learning takes place in locations separated from the workplace, like training centers, vocational schools, and universities. Learning environments contain some kind of simulation of work processes and tasks in order to make it oriented toward work, while a most authentic replication of the workplace characteristics is aimed at.[37] Examples in the field of production are ranging from a low degree of authenticity with abstract LEGO games[38] toward a high authenticity degree in learning factories.[39] Learning factories can compensate for a major drawback of work-oriented learning approaches, a tendency to low motivation and acceptance, with a high degree of realism and learner activity. A great advantage of these learning environments is seen in the opportunity to test and develop approaches in a risk-free (with regard to the consequences of incorrect decisions) but highly contextualized environment.[40] For future innovative learning—work approaches, development potentials were identified in the area of the necessary resource use for the development and operation of the learning systems, the mobility, the scalability, and the effectiveness of the approaches as well as the possibility of the mapping ability of individual problem situations[41]—for a further discussion of this current limitation of the learning factory concept, see also Sect. 9.2. Table 3.2 gives an overview on established and innovative work-related approaches to competency development for production.

3.4 Types of Perceived Learning Concepts in Production

Within the approaches mentioned in the previous section, in today's production environments, a broad variety of different types of learning concepts can be identified. Five major types of learning concepts are identified with small changes in connection with production environments[42]:

---

[35] See Lloyd, Rehg, National Center for Research in Vocational Education, USA. Office of Vocational and Adult Education (1983).
[36] See Cachay (2013).
[37] See Footnote 27.
[38] See, for example, Stier (2003).
[39] See, for example, Abele (2016).
[40] See Abele et al. (2015).
[41] See also Tisch and Metternich (2017).
[42] In reference to Dehnbostel and Schröder (2017).

3.4 Types of Perceived Learning Concepts in Production

Table 3.2 Work-related learning approaches for production (Adolph, Tisch, & Metternich, 2014)

	Work-based learning	Work-connected learning	Work-bound learning
Established approaches	Training workshops Training centers Practice firms Learning factory Self-learning programs	Quality circle Workshop circle Learning station	Training station Guiding text method Instruction Informal learning by doing in a real work process
Innovative approaches (exemplary)	Location-independent (remote) learning factories using ICT equipment and blended learning Virtual learning factories with dynamic adaptation to target group needs …	Process-oriented, virtual learning stations for demand-based methodical support …	Ad hoc skills development during the work process Learning tools (e.g. exoskeleton) …

- Type 1: Learning by doing approaches in real production processes: Learning while actively participating in production processes can be seen as the most common type of learning.[43] The coupling of serious work processes with the learning process is particularly well suited to gather valuable experience, to strengthen the motivation, and to learn to act in the social environment. This type of work-related learning can also be assigned to the widely used job rotation method.[44] An advantage of this type is that the learning process takes place directly in real working and social situations and consequently a transfer of the learned is facilitated or even omitted for this reason.[45] Most often this type of learning is identified as a (team-driven) on-the-job training.[46]
- Type 2: Learning in enriched working processes: Discussed in literature and implemented in individual companies are work-related learning approaches that somehow enrich working environments and processes with enhanced possibilities for learning (Fig. 3.6). For example, the learning bay which was first implemented in a couple of Daimler plants is such a concept.[47] Learning bays are defined as company-dependent real workstations enriched with learning equipment that provides good possibilities for learning.[48] Learning bays are supervised and supported by skilled workers, and additionally group work is implemented. Another example

---

[43] See also Dehnbostel and Schröder (2017).
[44] See Footnote 30.
[45] See Chalofsky (2014).
[46] See, for example, Walter (2001).
[47] See Millward (2005).
[48] Dehnbostel (2002).

Fig. 3.6 Learning approaches using enriched working processes

of this type is working and learning tasks that can be used for targeted competence development.[49]

- Type 3: Instruction and training approaches (Fig. 3.7): A systematic introduction is usually part of the training program for new employees. Here, the most simple and common form of instruction is based on imitation of the perceived.[50] For example, the traditional four-step method is carried out in the steps: (1) preparation of the worker, (2) presentation of the operation, (3) tryout performance (imitation), and (4) follow-up (training and testing).[51] Similar sequences can be found in other methods inside the TWI framework derived from the four-step method.[52] The effect of these methods on competence development is limited due to missing links to related knowledge and the strict external control of the learning process.[53]
- Type 4: Coaching and problem-solving approaches: In recent years, approaches that embed their learning processes in continuous improvement, problem-solving, and coaching processes are broadly spreading in industry.[54] Here, the traditional

---

[49] For further information on this concept, see Schröder (2009) and Dehnbostel and Molzberger (2006).

[50] See Footnote 27.

[51] See, for example, Graupp and Wrona (2015).

[52] See Dinero (2005).

[53] See Footnote 27.

[54] See, for example, Bessant, Caffyn, and Gallagher (2001), Rother (2009), Cachay and Abele (2012), and Hertle, Siedelhofer, Metternich, and Abele (2015).

3.4 Types of Perceived Learning Concepts in Production

Fig. 3.7 Job instruction approach in production

quality circles[55] as well as related forms such as the problem-solving circle[56] can be assigned to this type. Success factor of those approaches can be seen in a combination of formal and informal learning,[57] while in contrast to traditional schooling and education, here lies a risk that the formal learning side is neglected; blind operationalism should be prevented by discovering, providing, and reflecting about necessary related justification knowledge. This is seen as a part of the coach's task.[58]

- Type 5: Learning factories, laboratories, and approaches using abstract simulated processes: Learning factories include a quasi-real factory environment, which includes complete value streams that are used for the complex development of competences both in the university and in the industrial sector. Learning factories are based on didactic concepts and are found in a broad variety.[59] Similar learning environments with a lower degree of reality exist also.[60] Additionally, laboratories in manufacturing[61] are similar to learning factories but focus less on the

---

[55] See Lloyd and Rehg (1983).

[56] See Aderhold, Rosenberger, and Wetzel (2015).

[57] See Schröder (2009), Dehnbostel and Schröder (2017).

[58] See Footnote 36.

[59] See, for example, Abele et al. (2015). A closer look at the learning factory concept is presented in Chaps. 4 and 5.

[60] See, for example, Stier (2003).

[61] See Tekkaya et al. (2016).

Fig. 3.8 Learning approaches using simulated production processes

production processes or industrial training and more on production technologies and engineering education. For the research-based learning in those environments, it is often not mandatory to show complete value streams; in many cases, single workstations and machines are sufficient for the purposes (Fig. 3.8).[62]

The mentioned types of work-related learning should not be considered complete. Furthermore, there are other types of work-related learning mentioned like (accompanied) communities of practice, learning facilitation, and internship programs for (university) students. Used work-related learning concepts in production will evolve further, new concepts will come up over the time, and maybe other concepts will completely lose their relevance. Consequently, the mentioned types have to be seen as today practicable approaches to work-related competency development. Table 3.3 gives an overview on the mentioned perceivable types of work-related learning in production, assigns the respective form of learning, names exemplary concepts and addressable dimensions (affective, cognitive, psychomotor) for each type, classifies the achievement of complex competency development, and compares speed and predictability of competency development.

A big advantage of learning factories compared to other approaches from other types of work-related learning is a valuable combination of characteristics of the learning approach:

---

[62]Tekkaya et al. (2016) give an extensive overview on the laboratory in engineering education.

3.4 Types of Perceived Learning Concepts in Production

Table 3.3 Types of work-related learning in production in the overview

	Form of learning	Exemplary concepts	Addressable dimensions (affective, cognitive, psychomotor)	Achievable competence complexity	Speed and predictability of competency development
Type 1: learning in real production processes	WI	Learning on the job, job rotation, apprenticeship, etc.	All dimensions can be addressed alike	Not limited in advance due to very open competency development approach	Due to the informal learning process, learning is not predictable and may take some time
Type 2: learning in enriched working processes	WI/WC	Working and learning tasks, learning bay, etc.	All dimensions can be addressed alike	Well suited for complex competency development	Predictability and speed are improved compared to Type 1 due to the integrated learning concept
Type 3: instruction and training	WI/WC	traditional four-step instruction, etc.	Focus on the psychomotor dimension	No development of complex competencies	Speed and predictability very high
Type 4: coaching and problem-solving approaches	WI/WC	Coaching-KATA, problem-solving cycles, etc.	Especially good opportunities for cognitive and affective dimension	Well suited for complex competency development when informal and formal learning are integrated	Speed and predictability depend to a large extent on the organizational framework of the concept
Type 5: learning factories and laboratories	WO	Learning factories, laboratories, simulation games, etc.	Especially for the cognitive and to a lesser extent also for the affective domain. Psychomotor dimension also addressable	Well suited for complex competency development	Speed and predictability tend to be higher than in most WI and WC approaches

- Learning factories are especially suitable for a complex competency development but at the same time offers.
- a high learning speed and predictability.

This is a combination of characteristics that are not found in other approaches. While simple issues can be addressed in most work-related learning approaches, complex competency development can be achieved in some approaches of Type 2 "Learning in enriched working processes" and Type 4 "Coaching and Problem-Solving Approaches," although those approaches cannot regard with the learning factory with regard to learning predictability or efficiency. Furthermore, approaches from Type 3 "Instruction and Training" have a good predictability and a high learning speed, but are only suited for a low complexity of the competency development. At this point, it has to be stated that those are only the advantages of learning factories in the field of competency development compared to other work-related learning approaches. Beyond that, there are numerous good reasons to develop and build up learning factories; some of those reasons in fields like education, training, research, innovation transfer, or business creation are mentioned in the next section. No learning factory resembles the other so there are various reasons to implement the concept.

3.5 Need for Learning Factories

In order to address the challenges mentioned previously, competence development methods are necessary that enable a target-oriented, effective, and efficient training of today's and tomorrow's engineers.[63] Regarding education and training for production systems, learning factories will contribute significantly to the continuous supply of well-prepared young manufacturing engineers and to the constant update and modernization of intellectual capital in industry in the context of the challenges identified earlier. But also in a lot of other fields like research, innovation transfer, or business creation, learning factories contribute largely to a healthy connection between industry, research, and education. Various positive effects of the learning factory concepts are described in the literature which cannot be achieved with other learning approaches from Type 5 or the learning approaches of Type 1–4:

- Enhancement of the quality of education and training in general,[64]
- Creation of possibilities for research, innovation, and technology transfer,[65]
- Catalyst for business creation,[66]

---

[63] See Adolph et al. (2014).

[64] See Jäger, Mayrhofer, Kuhlang, Matyas, and Sihn (2013), Abele, Eichhorn, and Kuhn (2007), Cachay, Wennemer, Abele, and Tenberg (2012), Steffen, May, and Deuse (2012), Plorin and Müller (2014).

[65] See Entingh, Andrews, Kenkeremath, Mock, and Janis (1987), Çambel and Mock (1995), Rentzos, Doukas, Mavrikios, Mourtzis, and Chryssolouris (2014), Hamid, Mohd Halimudin Mohd Isa, Masrom, and Salim (2014).

[66] See Dorer (2018).

3.5 Need for Learning Factories

- Development of soft skills and interdisciplinary competences,[67]
- Strengthening of attitude and work philosophy,[68]
- Motivation to learn and motivation and confidence to apply,[69]
- Overcoming problems of traditional teaching methods,[70]
- Strengthening of the link between industry and academia,[71]
- Strengthening the link between education and research,[72]
- Improving the prestige of industry[73] and manufacturing and design education.[74]

3.6 Wrap-up of This Chapter

In this chapter, the types of learning for and in production are systematized, the historical development is shown, and application contexts of the different work-oriented, work-connected, and work-integrated forms of learning are discussed. From this systematization of different learning forms for and in production, the necessity for and the benefit of the further spreading of the learning factory concept can be derived.

References

Abele, E. (2016). Learning factory. *CIRP Encyclopedia of Production Engineering*.
Abele, E., Eichhorn, N., & Kuhn, S. (2007). Increase of productivity based on capability building in a learning factory. In *Computer Integrated Manufacturing and High Speed Machining: 11th International Conference on Production Engineering*, Zagreb (pp. 37–41).
Abele, E., Metternich, J., Tisch, M., Chryssolouris, G., Sihn, W., ElMaraghy, H. A., Hummel, V., & Ranz, F. (2015). Learning factories for research, education, and training. In *5th CIRP-Sponsored Conference on Learning Factories. Procedia CIRP, 32,* 1–6. https://doi.org/10.1016/j.procir.2015.02.187.
Aderhold, J., Rosenberger, M., & Wetzel, R. (2015). *Modernes Netzwerkmanagement: Anforderungen—Methoden—Anwendungsfelder*. Gabler Verlag.

---

[67] See Jäger, Mayrhofer, Kuhlang, Matyas, and Sihn (2012), Kreimeier et al. (2014), Wagner, Heinen, Regber, and Nyhuis (2010), Nöhring, Rieger, Erohin, Deuse, and Kuhlenkötter (2015), Hamid, Masrom, and Salim (2014), Martawijaya (2012), Badurdeen, Marksberry, Hall, and Gregory (2010).

[68] See Badurdeen et al. (2010).

[69] See Stier (2003), Dinkelmann, Siegert, and Bauernhansl (2014), Tisch, Hertle, Abele, Metternich, and Tenberg (2015), Elbadawi, McWilliams, and Tetteh (2010), Plorin and Müller (2014).

[70] See Jorgensen, Lamancusa, Zayas-Castro, and Ratner (1995), Tisch et al. (2013), Cachay et al. (2012).

[71] See Jorgensen et al. (1995), Tietze, Czumanski, Braasch, and Lödding (2013).

[72] See Blume et al. (2015).

[73] See Chryssolouris, Mavrikios, and Mourtzis (2013), Abele et al. (2015).

[74] See Jorgensen et al. (1995).

Adolph, S., Tisch, M., & Metternich, J. (2014). Challenges and approaches to competency development for future production. *Journal of International Scientific Publications—Educational Alternatives, 12*, 1001–1010.

Badurdeen, F., Marksberry, P., Hall, A., & Gregory, B. (2010). Teaching lean manufacturing with simulations and games: A survey and future directions. *Simulation & Gaming, 41*(4), 465–486.

Bandura, A. (Ed.). (1974). *Psychological modeling: Conflicting theories*. New York: Lieber-Atherton.

Bessant, J., Caffyn, S., & Gallagher, M. (2001). An evolutionary model of continuous improvement behaviour. *Technovation, 21*(2), 67–77. https://doi.org/10.1016/S0166-4972(00)00023-7.

Bloom, B. S., Engelhart, M. D., Furst, E. J., Hill, W. H., & Krathwohl, D. R. (1956). *Taxonomy of educational objectives: The classification of educational goals (book I: Cognitive domain)*. New York: McKay.

Blume, S., Madanchi, N., Böhme, S., Posselt, G., Thiede, S., & Herrmann, C. (2015). Die Lernfabrik—Research-based learning for sustainable production engineering. In *5th CIRP-Sponsored Conference on Learning Factories. Procedia CIRP, 32*, 126–131. https://doi.org/10.1016/j.procir.2015.02.113.

Cachay, J. (2013). *Methode zur kompetenzorientierten Gestaltung und nachhaltigen Verankerung von proaktiven Verbesserungsprozessen in der Produktion. Schriftenreihe des PTW: "Innovation Fertigungstechnik"*. Herzogenrath: Shaker.

Cachay, J., & Abele, E. (2012). Developing competencies for continuous improvement processes on the shop floor through learning factories–conceptual design and empirical validation. In *45th CIRP Conference on Manufacturing Systems. Procedia CIRP, 3*(3), 638–643.

Cachay, J., Wennemer, J., Abele, E., & Tenberg, R. (2012). Study on action-oriented learning with a learning factory approach. *Procedia—Social and Behavioral Sciences, 55*, 1144–1153.

Çambel, A. B., & Mock, J. E. (1995). Expediting technology transfer with multimedia. *Technological Forecasting and Social Change: An International Journal, 48*(1), 1–5.

Chalofsky, N. F. (2014). *Handbook of human resource development*. Wiley.

Chryssolouris, G., Mavrikios, D., & Mourtzis, D. (2013). Manufacturing systems—Skills and competencies for the future. In *46th CIRP Conference on Manufacturing Systems. Procedia CIRP, 7*, 17–24.

Cooper, P. A. (1993). Paradigm shifts in designed instruction: From behaviorism to cognitivism to constructivism. *Educational Technology, 33*(5), 12–19.

Dehnbostel, P. (2002). Bringing work-related-learning back to authentic work contexts. In P. Kämäräinen, G. Attwell, & A. Brown (Eds.), *CEDEFOP Reference series: Vol. 37. Transformation of learning in education and training: Key qualifications revisited* (pp. 190–202). Luxembourg.

Dehnbostel, P. (2007). *Lernen im Prozess der Arbeit. Studienreihe Bildungs- und Wissenschaftsmanagement: Bd. 7*. Münster, New York, Munich, Berlin: Waxmann.

Dehnbostel, P. (2009). New learning strategies and learning cultures in companies. In R. Maclean & D. Wilson (Eds.), *International handbook of education for the changing world of work: Bridging academic and vocational learning* (1st ed., pp. 2629–2645). Dordrecht: Springer Netherlands. https://doi.org/10.1007/978-1-4020-5281-1_173.

Dehnbostel, P., Holz, H., Novak, H., & Arndt, H. (Eds.). (1996). *Neue Lernorte und Lernortkombinationen—Erfahrungen und Erkenntnisse aus dezentralen Berufsbildungskonzepten. Berichte zur beruflichen Bildung* (Vol. 195). Bielefeld: Bertelsmann.

Dehnbostel, P., & Molzberger, G. (2006). Combination of formal learning and learning by experience in industrial enterprises. In J. N. Streumer (Ed.), *Work-related learning* (pp. 181–194). Netherlands: Springer.

Dehnbostel, P., & Schröder, T. (2017). Work-based and work-related learning—Models and learning concepts. *bwp@*. (9), 1–16.

Deppe, J. (1992). *Quality circle und lernstatt: Ein integrativer ansatz* (1992nd ed.). *Bochumer Beiträge zur Unternehmungsführung und Unternehmensforschung: Vol. 35*. Wiesbaden: Gabler.

Deppe, J. (2013). *Quality Circle und Lernstatt: Ein integrativer Ansatz*. Gabler Verlag.

References

Dinero, D. (2005). *Training within industry: The foundation of lean*. Taylor & Francis.

Dinkelmann, M., Siegert, J., & Bauernhansl, T. (2014). Change Management through learning factories. In M. F. Zäh (Ed.), *Enabling manufacturing competitiveness and economic sustainability* (pp. 395–399). Springer. https://doi.org/10.1007/978-3-319-02054-9_67.

Dorer, C. (2018). Ecodesign circle: Learning factory ecodesign. Retrieved from https://www.ecodesigncircle.eu/resources-for-you/learning-factory-ecodesign.

Elbadawi, I., McWilliams, D. L., & Tetteh, E. G. (2010). Enhancing lean manufacturing learning experience through hands-on simulation. *Simulation and Gaming, 41*(4), 537–552. https://doi.org/10.1177/1046878109334333.

Entingh, D. J., Andrews, C. J., Kenkeremath, D. C., Mock, J. E., & Janis, F. T. (1987). *Guide book for technology transfer managers: Moving public R & D to the marketplace*. Washington, D.C., Oak Ridge, Tenn.: United States Department of Energy; Distributed by the Office of Scientific and Technical Information, U.S. Department of Energy.

Eraut, M. (2000). Non-formal learning and tacit knowledge in professional work. *British Journal of Educational Psychology, 70*, 113–136.

Ertmer, P. A., & Newby, T. J. (1993). Behaviorism, cognitivism, constructivism: Comparing critical features from an instructional design perspective. *Performance Improvement Quarterly, 6*(4), 50–72. https://doi.org/10.1111/j.1937-8327.1993.tb00605.x.

Graupp, P., & Wrona, R. J. (2015). *The TWI workbook: Essential skills for supervisors* (2nd ed.). CRC Press.

Hamid, M. H. M. I., Masrom, M., & Salim, K. R. (2014). Review of learning models for production based education training in technical education. In *International Conference on Teaching and Learning in Computing and Engineering, IEEE* (pp. 206–211).

Hartley, J. R. (1985). Some psychological aspects of computer-assisted learning and technology. *Programmed Learning and Educational Technology, 22*(2), 140–149.

Hertle, C., Siedelhofer, C., Metternich, J., & Abele, E. (2015). The next generation shop floor management—How to continuously develop competencies in manufacturing environments. In *The 23rd International Conference on Production Research, 03.08.2015*, Manila, Philippines.

Jäger, A., Mayrhofer, W., Kuhlang, P., Matyas, K., & Sihn, W. (2012). The "learning factory": An immersive learning environment for comprehensive and lasting education in industrial engineering. *16th World Multi-Conference on Systemics, Cybernetics and Informatics, 16*(2), 237–242.

Jäger, A., Mayrhofer, W., Kuhlang, P., Matyas, K., & Sihn, W. (2013). Total immersion: Hands and heads-on training in a learning factory for comprehensive industrial engineering education. *International Journal of Engineering Education, 29*(1), 23–32.

Jonassen, D. H. (1991). Objectivism versus constructivism: Do we need a new philosophical paradigm? *Educational Technology Research and Development, 39*(3), 5–14. https://doi.org/10.2307/30219973.

Jorgensen, J. E., Lamancusa, J. S., Zayas-Castro, J. L., & Ratner, J. (1995). The learning factory: Curriculum integration of design and manufacturing. *4th World Conference on Engineering Education*, 1–7.

Kerres, M. (2012). *Mediendidaktik: Konzeption und Entwicklung mediengestützter Lernangebote* (3., vollst. überarb. Aufl.). *Informatik 10-2012*. München: Oldenbourg.

Kirkpatrick, D. L. (1996). Great ideas revisited: Revisiting kirkpatrick's four level model. *Training and Development., 1*, 54–59.

Kreimeier, D., Morlock, F., Prinz, C., Krückhans, B., Bakir, D. C., & Meier, H. (2014). Holistic learning factories—A concept to train lean management, resource efficiency as well as management and organization improvement skills. In *47th CIRP Conference on Manufacturing Systems. Procedia CIRP, 17*, 184–188.

Lloyd, R. F., Rehg, V. R. (1983). National center for research in vocational education, and United States. Office of Vocational and Adult Education. In *Quality circles: Applications in vocational education*. National Center for Research in Vocational Education, Ohio State University.

Martawijaya, D. H. (2012). Developing a teaching factory learning model to improve production competencies among mechanical engineering students in a vocational senior high school. *Journal of Technical Education and Training, 4*(2), 45–56.

Millward, L. (2005). *Understanding Occupational and Organizational Psychology*. SAGE Publications.

Münch, J. (Ed.). (1977). *Lernen, aber wo? Der Lernort als pädagogisches und lernorientiertes Problem* (1st ed.). Trier: Spee.

Neisser, U. (1967). *Cognitive psychology. Century psychology series*. Englewood Cliffs NJ: Prentice Hall.

Nöhring, F., Rieger, M., Erohin, O., Deuse, J., & Kuhlenkötter, B. (2015). An Interdisciplinary and hands-on learning approach for industrial assembly systems. In *5th CIRP-Sponsored Conference on Learning Factories. Procedia CIRP, 32*, 109–114. https://doi.org/10.1016/j.procir.20 15.02.112.

Ortega, J. (2001). Job rotation as a learning mechanism. *Management Science, 47*(10), 1361–1370. https://doi.org/10.1287/mnsc.47.10.1361.10257.

Pavlov, I. P. (2003). *Conditioned reflexes (Republ)*. Mineola, NY: Dover Publ.

Plorin, D., & Müller, E. (2014). Developing an ambient assisted living environment applying the advanced learning factory (aLF): A conceptual approach for the practical use in the research project A^{2}LICE. *ISAGA, 2013,* 69–76.

Reith, S. (1988). *Außerbetriebliche CIM-Schulung in der „Lernfabrik"*. Berlin, Heidelberg: Springer. Retrieved from http://link.springer.com/content/pdf/10.1007%2F978-3-662-01109-6_24.pdf.

Rentzos, L., Doukas, M., Mavrikios, D., Mourtzis, D., & Chryssolouris, G. (2014). Integrating manufacturing education with industrial practice using teaching factory paradigm: A construction equipment application. In *47th CIRP Conference on Manufacturing Systems. Procedia CIRP, 17*, 189–194.

Robinson, A. G., & Schroeder, D. M. (1993). Training, continuous improvement, and human relations: The U.S. TWI programs and the Japanese management style. *California Management Review, 35*(2), 35–57.

Rother, M. (2009). *Toyota Kata: Managing People for Improvement*. Adaptiveness and Superior Results: McGraw-Hill Education.

Ryan, P. (2012). Apprenticeship: Between theory and practice, school and workplace. In M. Pilz (Ed.), *SpringerLink Bücher. The future of vocational education and training in a changing world* (pp. 402–432). Wiesbaden: Springer VS. https://doi.org/10.1007/978-3-531-18757-0_23.

Schacter, D. L., Gilbert, D. T., & Wegner, D. M. (2009). *Psychology*. New York: Worth Publishers.

Schaper, N. (2000). *Gestaltung und Evaluation arbeitsbezogener Lernumgebungen*. Heidelberg: Habilitationsschrift, Ruprecht-Karls-Universität Heidelberg.

Schröder, T. (2009). *Arbeits- und lernaufgaben für die weiterbildung: Eine lernform für das lernen im prozess der arbeit. Berufsbildung, Arbeit und Innovation—Dissertationen und Habilitationen—Band 15*. Bielefeld [Germany]: W. Bertelsmann Verlag.

Schuh, G., Gartzen, T., Rodenhauser, T., & Marks, A. (2015). Promoting work-based learning through industry 4.0. In *5th CIRP-Sponsored Conference on Learning Factories. Procedia CIRP, 32*, 82–87. https://doi.org/10.1016/j.procir.2015.02.213.

Skinner, B. F. (1976). *About behaviorism*. New York: Vintage Books.

Schunk, D. H. (1996). *Learning theories: An educational perspective* (2nd ed.). Englewood Cliffs, New Jersey: Prentice Hall.

Steffen, M., May, D., & Deuse, J. (2012). The industrial engineering laboratory: Problem based learning in industrial engineering education at TU Dortmund University. In *Global Engineering Education Conference (EDUCON), IEEE—Collaborative Learning and New Pedagogic Approaches in Engineering Education*, Marrakesch, Marokko, April, 17–20 2012 (pp. 1–10).

Stier, K. W. (2003). Teaching lean manufacturing concepts through project-based learning and simulation. *Journal of Industrial Technology, 19*(4), 1–6.

References

Tekkaya, A. E., Wilkesmann, U., Terkowsky, C., Pleul, C., Radtke, M., & Maevus, F. (Eds.). (2016). *Das Labor in der ingenieurwissenschaftlichen Ausbildung: Zukunftsorientierte Ansätze aus dem Projekt IngLab: acatech Studie. acatech Studie*. München: Herbert Utz Verlag.

Tenberg, R. (2011). *Vermittlung fachlicher und überfachlicher Kompetenzen in technischen Berufen: Theorie und Praxis der Technikdidaktik*. Stuttgart: Steiner.

Thorndike, E. L. (1965). *Animal intelligence: Experimental studies*. New York: Hafner Pub. Co.

Tietze, F., Czumanski, T., Braasch, M., & Lödding, H. (2013). Problembasiertes Lernen in Lernfabriken. *Werkstattstechnik online: wt, 103*(3), 246–251.

Tisch, M. (2018). *Modellbasierte Methodik zur kompetenzorientierten Gestaltung von Lernfabriken für die schlanke Produktion*. Dissertation, Darmstadt. Aachen: Shaker.

Tisch, M., Hertle, C., Abele, E., Metternich, J., & Tenberg, R. (2015). Learning factory design: A competency-oriented approach integrating three design levels. *International Journal of Computer Integrated Manufacturing, 29*(12), 1355–1375. https://doi.org/10.1080/0951192X.2015.1033017.

Tisch, M., Hertle, C., Cachay, J., Abele, E., Metternich, J., & Tenberg, R. (2013). A systematic approach on developing action-oriented, competency-based learning factories. *46th CIRP Conference on Manufacturing Systems. Procedia CIRP, 7*, 580–585.

Tisch, M., & Metternich, J. (2017). Potentials and limits of learning factories in research, innovation transfer, education, and training. *7th CIRP-Sponsored Conference on Learning Factories. Procedia Manufacturing* (In Press).

Wagner, C., Heinen, T., Regber, H., & Nyhuis, P. (2010). Fit for Change—Der Mensch als Wandlungsbefähiger. *Zeitschrift für wirtschaftlichen Fabrikbetrieb (ZWF), 100*(9), 722–727.

Walter, D. (2001). *Training on the job: A new team-driven approach that empowers employees*. Is Quick to Implement, Gets Bottom-Line Results: ASTD.

Chapter 4
Historical Development, Terminology, and Definition of Learning Factories

Learning factory approaches have been intensely discussed and used in recent years. But where and how did the learning factory concept emerge? Moreover, what are learning factories precisely and how can they be differentiated from similar approaches in education and research? This chapter gives an overview (Fig. 4.1).

4.1 Historical Development of the Learning Factory Concepts

The historical development of different learning factory concepts can be understood in three consecutive phases:

- Phase 1: First local learning factories are established, predominantly in the USA.
- Phase 2: Second wave of learning factories, mainly in Europe, takes place.
- Phase 3: Formation of learning factory networks and scientific consideration of the learning factory.

Historical Development, Terminology, and Definition of Learning Factories (Chapter 4)	▪ Historical development of the learning factory concepts (4.1) ▪ Terminology of learning factories (4.2) ▪ Definition of learning factories (4.3)

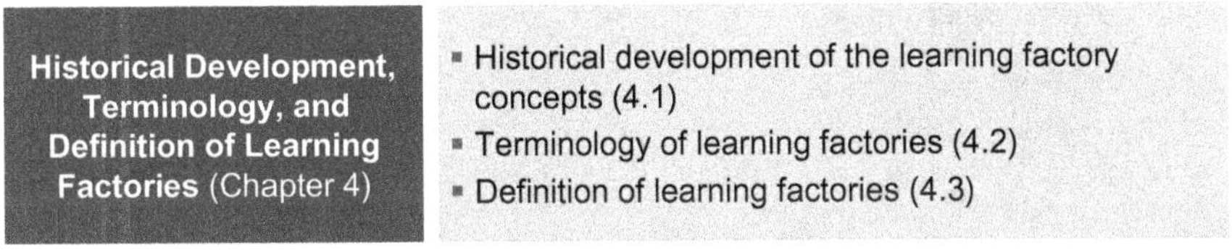

Fig. 4.1 Overview of the structure of this chapter regarding the historical development, terminology, and definition of learning factories

Phase 1

The most prominent example of the learning factories built-up in phase 1 is the Bernard M. Gordon Learning Factory at the Penn State University. In the year 1994, a consortium led by the Penn State University received a grant by the National Science Foundation (NSF) to develop a "learning factory". At this time, the term "learning factory" was coined and became known. Basically, the concept devised here refers to industry-partnered engineering design projects with interdisciplinary and hands-on focus. Since 1995, numerous design projects sponsored by industry partners were carried out in a 2000 sqm facility that is equipped with appropriate machines, materials, and tools.[1] In 2006, the learning factory won the Bernard M. Gordon Prize of the National Academy of Engineering for "Innovations in Engineering Education." This early implementation of a learning factory accentuates the practical experience through the application of knowledge learned in the course of engineering studies. Thus, real problems of the industry can be solved and products can be designed to meet the identified needs.[2] Additionally to this approach at Penn State, which is focused on the use of the learning factory in education, also more industry-directed approaches can be identified. For example, in the late 1980s a today comparatively less famous learning factory was established at the IAO in Stuttgart, Germany called the "Lernfabrik."[3] This facility intended to qualify industrial personnel regarding Computer-Integrated Manufacturing (CIM).[4] Another concept appears shortly after the turn of the millennium, with the Teaching Factory concept which has attracted interest primarily in the USA, leading to a series of pilot activities addressing educational as well as business purposes.[5] Although this concept is called "teaching factory" and not "learning factory," the two approaches are very similar. The concept of the Teaching Factory has its roots in the medical field; more specifically, the Teaching Hospitals are seen as a model—here medical schools are working side by side with a hospital that enables real hospital experience and training for the students. The Teaching Factory draws the parallel for engineering: Real industrial practice (in real factories) is integrated with manufacturing training and education.[6]

Phase 2

Approximately one decade ago, predominantly in Europe more and more learning factory implementations appeared.[7] Here, the learning factory concept was implemented in large variety, in order to improve the learning experiences in a wide range of

---

[1] See Penn State University (2017).

[2] See Jorgensen, Lamancusa, Zayas-Castro, and Ratner (1995), Lamancusa, Zayas, Soyster, Morell, and Jorgensen (2008).

[3] German translation of "learning factory".

[4] See Reith (1988).

[5] See Alptekin, Pouraghabagher, McQuaid, and Waldorf (2001), Dizon (2000).

[6] For examples of teaching factory activities see Alptekin et al. (2001), Dizon (2000) as well as Best Practice Example 20 and Best Practice Example 30 in Chap. 11.

[7] See Wagner, AlGeddawy, ElMaraghy, and Müller (2012), Abele et al. (2015), Micheu and Kleindienst (2014).

applications, industries, and target groups.[8] One of the first learning factories of this wave was the Process Learning Factory CiP (Center for industrial Productivity) inaugurated in 2007 at the Institute of Production Management, Technology and Machine Tools (PTW), TU Darmstadt.[9] The multistage production processes within the Process Learning Factory include machining, manual, and semi-automatic assembly as well as integrated functions of logistics and quality assurance. Lean manufacturing as well as lately Industrie 4.0 can be experienced and learned in this environment, ensuring a sustainable transfer of knowledge.[10] In recent years, many learning factories with other learning content focus and physical manifestations have been established.

Phase 3

In 2011, the "1st Conference on Learning Factories"[11] took place in Darmstadt and at this occasion a group of European academic learning factory operators founded the Initiative on European Learning Factories (IELF) with the aim of starting joint research projects, making the learning factory concept known worldwide and improving it together. President of the initiative in the first years (from 2011 to 2016) was Prof. Eberhard Abele from PTW, TU Darmstadt. Beginning from summer 2016 Prof. Joachim Metternich, also PTW, TU Darmstadt, followed as president of the Initiative on European Learning Factories, which was renamed due to an ongoing internationalization of the learning factory topic to "International Association of Learning Factories."[12]

One important project initiated by the members of the IELF under the lead of ESB Reutlingen was the Network of Innovative Learning Factories (NIL) funded by the German Academic Exchange Service (DAAD) and the Federal Ministry of Education and Research (BMBF). The goal of the project is described by the project partners as:

> The execution of the joint idea as a funded research project is a consequent step to pursue the ambitioned targets of the Initiative on European Learning Factories. The partners of the strategic network NIL with their scientists of high reputation have settled the strategic goal to significantly contribute to an internationally recognized standard of the "Learning Factory", to support international mobility, to establish innovative educational programs and to enhance the quality of existing and future Learning Factories. (ESB Reutlingen, 2017)

As a result of these activities, a joint Europe-wide collaboration was established. The following conferences on Learning Factories in Vienna (2012), Munich (2013), Stockholm (2014), Bochum (2015), Gjovik (2016), and Darmstadt (2017) are growing in popularity and internationality. At the 2017 conference in Darmstadt, 150 participants from 18 countries participated in the two-day conference. Since 2015,

---

[8] See Abele et al. (2015).
[9] See Abele, Eichhorn, and Kuhn (2007).
[10] A detailed description of the Process Leaning Factory CiP can be found in the Best Practice Example 26 in Chap. 11.
[11] See Abele, Cachay, Heb, and Scheibner (2011) for the proceedings.
[12] See also Chap. 10 for a detailed description.

the conference has been CIRP-sponsored, which can be seen as an indication of the growing importance of learning factories in manufacturing research.

In addition, in 2014, on the fundament of the IELF with additional learning factory partners inside CIRP,[13] a CIRP Collaborative Working Group on "Learning factories for future-oriented research and education in manufacturing" (short: CIRP CWG on learning factories) was initiated. The CIRP CWG started with the intention to organize research related to learning factories globally, form a common understanding of used terms in the field, collect global state-of-the-art knowledge about learning factories, reinforce the connection between academia and industry regarding this topic and publish the results of the CWG in a keynote paper about "Learning Factories for future-oriented research and education in manufacturing."[14] The CIRP CWG on learning factories was successfully finished with the completion of the 2017 STC-O keynote paper.[15]

In connection with the end of the Collaborative Working Group on learning factories inside CIRP, in the year 2017, the members of the Initiative on European Learning Factories decided to open up the initiative for worldwide members and to change the name in this process to "International Association of Learning Factories" (IALF). The goals, current members, running projects, and the like of the currently founded International Association of Learning Factories are presented in detail in Chap. 8.

The growth of the learning factory community in the last decades can also be seen looking at the number of publications addressing the topic. Figure 4.2 illustrates this by showing the number on Google Scholar indexed documents for the search of the terms "learning factory"/"Lernfabrik" and "teaching factory" plus corresponding plural forms each year for the last thirty years. Search results in which the terms were used in a different context were excluded accordingly from the results. Those are mainly "learning factory" in context of critique for over-formalized schooling, "learning factory" in the sense of the learning organization and "learning factory" in regard to machine learning.

4.2 Terminology of Learning Factories

The term "learning factory" consists of the words "learning" and "factory." It can be concluded that the term is used for systems that address both parts of the term.[16] Learning factories are therefore simulated, authentic factory environments that are used for educational, training and/or research purposes—the learning factory concept is defined in more detail in Sect. 4.3. In literature, the term "learning factory" is also

---

[13] The International Academy for Production Engineering.
[14] See Abele, Chryssolouris, Sihn, and Seifermann (2014).
[15] Abele et al. (2017).
[16] See Wagner et al. (2012).

4.2 Terminology of Learning Factories

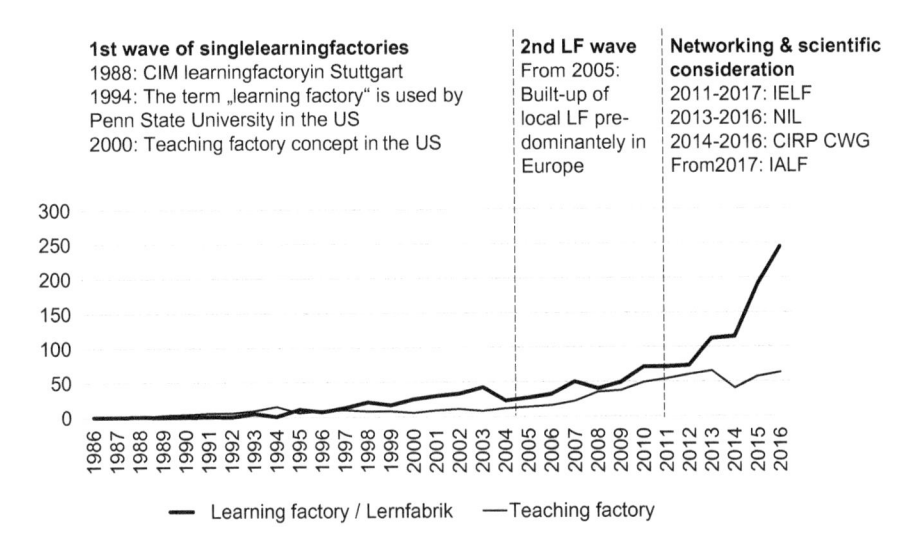

1st wave of singlelearningfactories	2nd LF wave	Networking & scientific
1988: CIM learningfactoryin Stuttgart	From 2005:	consideration
1994: The term „learning factory" is used by	Built-up of	2011-2017: IELF
Penn State University in the US	local LF pre-	2013-2016: NIL
2000: Teaching factory concept in the US	dominantely in	2014-2016: CIRP CWG
	Europe	From 2017: IALF

Fig. 4.2 Historical development of learning factory approaches and the number of indexed documents on Google Scholar regarding learning and teaching factories (Tisch & Metternich, 2017)

used in two other contexts, while sometimes those different uses are getting mixed up. In order to avoid confusion, these uses are explained and delimited in the following[17]:

- The term "learning factory" (also "knowledge factory"[18] or "learning laboratory"[19]) is used to describe a factory that succeeds in integrating internal and external knowledge, problem-solving, experimentation, and innovation into the daily work routine.[20] This concept strives to shape the factories of the future at all levels as a place of continuous improvement.[21] The term "learning factory" in this context describes the transfer of the concept of the "learning organization"[22] by integrating workplace and learning location on the operational factory level.[23] In particular, this "Learning Factory" model includes the elements of teamwork, the involvement of all employees and training.[24] This element of the planned training can be addressed with a learning factory in the sense of this book.
- Furthermore, in some cases, the terms "learning factory" as well as the German term "Lernfabrik" are used with a negative connotation to describe schools and universities, which, according to their respective authors, are characterized by

---

[17] Based on Tisch, Hertle, Abele, Metternich, and Tenberg (2015).
[18] Roth, Marucheck, Kemp, and Trimbl (1993).
[19] Leonard-Barton (1992).
[20] See Leonard-Barton (1992), Barton and Delbridge (2001), Roth et al. (1993).
[21] See Leonard-Barton (1992).
[22] For the learning organization see, for example, (Senge, 1990, Argyris & Schön, 1999).
[23] See Barton and Delbridge (2001).
[24] See Barton and Delbridge (2001).

pedagogically wrong concepts and values, over-formalized schooling or too large class sizes.[25] In this metaphor, the learner is the "product" of the "factories" school and university.

Furthermore, other terms are used more or less synonymously. In the following, those terms are described and linked to the learning factory concept: (Tisch et al., 2015)

- The "teaching factory" terminologically and conceptually approaches the American "teaching hospitals."[26] In the teaching factory concept, the industrial production environment is transmitted into the classroom using video streams and modern information and communication technologies. This way, students can learn about and on the basis of real industrial problems. Compared to the learning factory concept, in this manner, a less experiential but still problem-based, less investment-intensive learning is enabled. The notion of learning factory compared to teaching factory emphasizes the importance of the experimental and experiential learning and linked to this the own actions of learners in the learning factory.[27]
- The term "model factory"[28] describes the factory environment as an image of a real factory—the didactic concept, which is linked to the learning factory, is terminologically not considered in the concept of the model factory.

In the following section, the learning factory concept is defined.

4.3 Definition of Learning Factories

Starting with the appearance of learning factories, various definitions were implicitly and explicitly formulated. While early definitions are to a big extent based on learning factory use case descriptions, in recent years a lively scientific exchange of thoughts has taken place.

In a definition related to one of the early learning factory approaches that are used in engineering education, the authors describe a learning factory as "[…] an activity-based facility which is designed to be used across the curriculum."[29] The concept contains an active experience of process and product realization, which is dynamic in the sense that the students' projects define structure and use of the environment. The learning factory definitions of more recent implementations are largely based on the same assumptions but broaden the application to education and

---

[25]For this use of the term learning factory see, for example, Bachmann (1986), Beattie (1988), van der Zee (2006), Bögel (2011), McLaren and Jandric (2015), Bichow (2015).

[26]See Mavrikios, Papakostas, Mourtzis, and Chryssolouris (2013), Alptekin et al. (2001).

[27]See Abele et al. (2015).

[28]See, for example, McKinsey & Company (2017).

[29]See Jorgensen et al. (1995).

4.3 Definition of Learning Factories

training and underline the importance of authentic factory environments and close-to-reality-processes.[30] In addition, in descriptions of the teaching factory concept[31] real factories are bidirectionally interlinked with classrooms using advanced ICT.[32] An investigation with more than 25 learning factories emphasizes the postulation of learning factories' changeability and authenticity.[33] Likewise, the suitability of the learning factory approach for different targets groups as well as the testing of theoretical knowledge and the transfer of industry is claimed. Additionally, in connection with the systematic design of learning factories, reference is made to action-oriented learning processes and the competency orientation in the learning factory concept.[34] In general, two overarching perspectives are differentiated[35]:

- a didactic perspective according to which learning factories are seen as highly complex learning environments for topics in the fields of production technology and production management, and
- an operational perspective according to which the learning factory can be seen as an idealized replica of (parts of) real manufacturing systems in which formal and informal learning can take place.

In 2013, the members of the Initiative on European Learning Factories agreed on a broad definition which is often referred to:

> A learning factory is a learning environment, where processes and technologies are based on a real industrial site which allows a direct approach to product creation process. Learning factories are based on a didactical concept emphasizing experiential and problem-based learning. The continuous improvement philosophy is facilitated by own actions and interactive involvement of the participants. (IELF, 2013)

Based on this definition, physical and virtual variations of the learning factory concept are distinguished.[36] There are further definitions that cover additional facets of the concepts.[37] The scope of respective learning factory definitions is sketched in Table 4.1.

Based on an analysis of the definitions described above, discussions within the international community of the CIRP CWG on learning factories lead to a comprehensive and generally accepted definition which was published in the CIRP Encyclopedia[38]:

---

[30] See Abele, Tenberg, Wennemer, and Cachay (2010).
[31] See Chryssolouris, Mavrikios, and Mourtzis (2013), Chryssolouris, Mavrikios, and Rentzos (2016), Metternich et al. (2014).
[32] Information and communication technology.
[33] See Wagner et al. (2012).
[34] See Tisch et al. (2013).
[35] See Tisch et al. (2015).
[36] See Sihn (2014).
[37] See, e.g., Kreimeier et al. (2014), Müller, Plorin, and Ackermann (2012) or Tracht (2014).
[38] See Abele (2016).

Table 4.1 Scope of learning factory definitions (Tisch, 2018)

Dimension	Aspect	Jorgensen et al. (1995)	Abele et al. (2010)	Müller et al. (2012)	Wagner et al. (2012)	Initiative on European Learning Factories (2013)	Chryssolouris (2014)	Kreimeier et al. (2014)	Sihn (2014)	Tracht (2014)	Abele (2016), Abele et al. (2017)
Purpose and targets	Education	■■■	■□□	■■■	■■■	■□□	■■■	■■■	■□□	■■■	■■■
Process	Training	■□□	■□□	■■■	■□□	■□□	■□□	■■■	■□□	■□□	■■■
	Research	■□□	■□□	■■■	■□□	■□□	■□□	■□□	■□□	■□□	■■■
	Authentic	■■■	■■■	■■■	■■■	■■■	■□□	■■■	■■■	■■■	■■■
	Multistaged	■■■	■□□	■■■	■□□	■□□	■□□	■□□	■■■	■■□	■■■
	Technological and organizational	■■■	■■■	■□□	■■■	■■■	■■■	■□□	■□□	■■■	■■■
Setting	Changeable	■□□	■■■	■□□	■□□	■□□	■□□	■■■	■□□	■□□	■■■
	Real setting	■□□	■■■	■■■	■■□	■□□	■■■	■□□	■■■	■■□	■■■
	Virtual setting	■□□	■□□	■■■	■□□	■□□	■□□	■□□	■■■	■□□	■■■
Product	Material product	■□□	■□□	■■■	■□□	■■■	■□□	■□□	■□□	■■■	■■■
	Service	■□□	■□□	■□□	■□□	■□□	■□□	■□□	■□□	■□□	■■■

(continued)

4.3 Definition of Learning Factories

Table 4.1 (continued)

Dimension	Aspect	Jorgensen et al. (1995)	Abele et al. (2010)	Müller et al. (2012)	Wagner et al. (2012)	Initiative on European Learning Factories (2013)	Chryssolouris (2014)	Kreimeier et al. (2014)	Sihn (2014)	Tracht (2014)	Abele (2016), Abele et al. (2017)
Didactics	Didactical concept	■■□	■□■	■□■	■□■	■■■	■□□	□□■	■■■	■■■	■■■
	Integration formal and informal learning	■■□	■■■	■□□	■□□	■■■	□□□	□□□	■■■	□□□	■■■
	Own actions of learners	■■■	■■■	■□■	■■■	■■■	□□□	□□□	■■■	■■■	■■■
	On-site learning in factory environment	■■■	■■■	■■■	■■■	■■■	□□□	□□■	■■■	■■■	■■■
	Remote learning	□□□	□□□	□□□	□□□	□□□	■■■	□□□	□□□	□□□	■■■

(continued)

Table 4.1 (continued)

Dimension	Aspect	Jorgensen et al. (1995)	Abele et al. (2010)	Müller et al. (2012)	Wagner et al. (2012)	Initiative on European Learning Factories (2013)	Chryssolouris (2014)	Kreimeier et al. (2014)	Sihn (2014)	Tracht (2014)	Abele (2016), Abele et al. (2017)
Operational Model	Economically sustainable	□□□	□□□	□□□	□□□	□□□	□□□	□□□	□□□	□□□	■■■
	Contentual sustainable	□□□	□□□	□□□	□□□	□□□	□□□	□□□	□□□	□□□	■■■
	Personnel-related sustainable	□□□	□□□	□□□	□□□	□□□	□□□	□□□	□□□	□□□	■■■
■■■	Aspect is explicitly mentioned in the definition
■■□	Aspect is addressed within the paper
■□□	Aspect is implicitly regarded
□□□	Aspect is not addressed

4.3 Definition of Learning Factories

A learning factory in a narrow sense is a learning environment specified by **processes** that are authentic, include multiple stations, and comprise technical as well as organizational aspects, a **setting** that is changeable and resembles a real value chain, a physical **product** being manufactured, and a **didactical concept** that comprises formal, informal and non-formal learning, enabled by own actions of the trainees in an on-site learning approach. Depending on the **purpose** of the learning factory, learning takes place through teaching, training and/or research. Consequently, learning outcomes may be competency development and/or innovation. An operating model ensuring the sustained operation of the learning factory is desirable. In a broader sense, learning environments meeting the definition above but with a setting that resembles a virtual instead of a physical value chain, or a service product instead of a physical product, or a didactical concept based on remote learning instead of on-site learning can also be considered as learning factories. (Abele, 2016)

It is a broad consensus in literature, that this "learning" in a "factory" environment includes training of industrial employees, academic education of students and similar forms of learning.[39] Furthermore, research (identification of research gaps, validation of results, etc.[40]) or the innovation transfer (implementation, dissemination, etc.[41]) can be understood as learning processes.[42] Research and transfer facilitated by simulated, quasi-realistic production environments are therefore also covered in the term, see Fig. 4.3.[43]

The production processes depicted within the value stream of a learning factory should be realistic.[44] The value stream must contain several workstations, since only single machines or individual workstations do merely characterize an authentic factory and exclude most of relevant topics for learning factory education and research.[45] Furthermore, a technical demonstrator, or in general an exclusively technical process, does not correspond to the learning environment of a learning factory. Organizational, process-oriented aspects must be considered additionally.

A great potential of the learning factory is the possibility to make learning processes experience-based.[46] In order to enable this kind of learning, the learning factory environment should be changeable in order to be open to the ideas of the learners.[47] Physical equipment and a physical setting[48] are part of the learning factory in the narrow sense, while virtual, computer-simulated factories, or other virtual representations of the value chain[49] are assigned to the learning factory in the broader

---

[39] See Abele et al. (2017).

[40] See, for example, Seifermann, Metternich, and Abele (2014).

[41] See, for example, Schuh, Gartzen, Rodenhauser, and Marks (2015).

[42] See Abele et al. (2017).

[43] Abele et al. (2015).

[44] See Abele et al. (2007), Jäger, Mayrhofer, Kuhlang, Matyas, and Sihn (2012), Tisch et al. (2013).

[45] See Abele et al. (2015), Tisch et al. (2015).

[46] See Hempen, Wischniewski, Maschek, Thomas, and Deuse (2010), Lamancusa et al. (2008), Plorin and Müller (2014), Hamid, Masrom, and Salim (2014), Tisch and Metternich (2017).

[47] See Tisch et al. (2013).

[48] See, for example, Abele et al. (2007), Jorgensen et al. (1995).

[49] See, for example, Plorin and Müller (2014), Riffelmacher (2013), Sivard, Eriksson, Florin, Shariatzadeh, and Lindberg (2016), Sivard and Lundholm (2013).

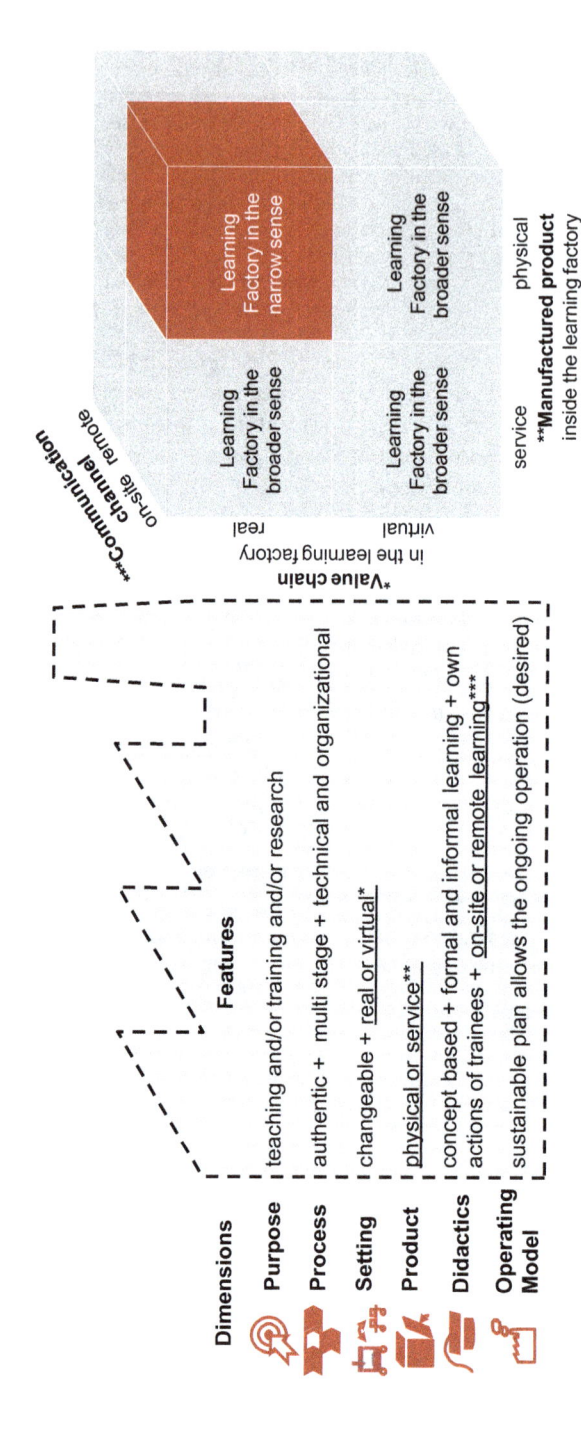

Fig. 4.3 Key characteristics of learning factories and learning factories in the narrow and in the broader sense (Abele et al., 2015)

4.3 Definition of Learning Factories

sense. Likewise, learning factories which represent the production of a physical product are assigned to learning factories in the narrow sense, while those which represent the provision of services[50] are assigned to the learning factory in a broad sense.

In terms of didactics, the learning factory integrates formal and informal learning. An institution that offers only space for self-directed learning,[51] for example, learning centers[52] or learning spaces,[53] is not yet a learning factory. The learning factory concept includes a didactic concept that defines who learns what and how,[54] or more comprehensively: "Who should learn what, from whom, when, with whom, where, how, with what and for which purpose".[55] Learners have the possibility to shape the factory environment on the basis of own considerations.[56] The learning factory in the narrow sense entails the physical presence of learners (on-site learning) while learning factories in the broad sense contain the possibility of remote learning via ICT.[57] For the differentiation of learning factories in the narrow and the broad sense, see also Fig. 4.3

The learning factory concept in the narrow sense offers the most suitable conditions for effective competency development. Deviations from the learning factory concept in the narrow sense threaten to diminish the effectiveness of the competency development approach because they show less of the identified success factors for competency development (see also Sect. 7.2.3) of the leaning factory concept:

- When using virtual instead of physical environments, competency development processes suffer from a tendency to less active learning, less authentic, and a less concrete learning environment.
- When learners are remotely attending the learning or factory environment, actions of the learners cannot be affected directly in the environment, and the learning process tends to be more passive and less immersive. An integration of thinking and doing in the learning process is impeded.
- Mapping a service, rather than the production of a physical product, has a different character but can be made less tangible.

4.4 Wrap-up of This Chapter

In this chapter, the historical development of the learning factory concept from individual local learning factory approaches from the end of the 1980s to European and

---

[50] See, for example, Hammer (2014).
[51] See Knowles (1975).
[52] See Piskurich (1993).
[53] See Jouault and Seta (2014).
[54] See Abele et al. (2015).
[55] According to Jank and Meyer (2002) in literal translation of Zierer and Seel (2012).
[56] See Hamid et al. (2014), Cachay, Wennemer, Abele, and Tenberg (2012).
[57] See Mavrikios et al. (2013), Rentzos, Mavrikios, and Chryssolouris (2015).

finally worldwide associations on the learning factory concept is presented. The term "learning factory" is regularly understood in a different way. This chapter creates the terminology and presents the now generally accepted definition of the learning factory system.

References

Abele, E. (2016). Learning factory. *CIRP Encyclopedia of Production Engineering*.
Abele, E., Cachay, J., Heb, A., & Scheibner, S. (Eds.). (2011). *1st Conference on Learning Factories, Darmstadt*. Darmstadt: Institute of Production Management, Technology and Machine Tools (PTW).
Abele, E., Chryssolouris, G., Sihn, W., Metternich, J., ElMaraghy, H. A., Seliger, G., et al. (2017). Learning factories for future oriented research and education in manufacturing. *CIRP Annals - Manufacturing Technology, 66*(2), 803–826.
Abele, E., Chryssolouris, G., Sihn, W., & Seifermann, S. (2014, January). *CIRP collaborative working group – Learning factories for future oriented research and education in manufacturing.* CIRP. CIRP Winter Meeting, Paris.
Abele, E., Eichhorn, N., & Kuhn, S. (2007). Increase of productivity based on capability building in a learning factory. In *Computer Integrated Manufacturing and High Speed Machining: 11th International Conference on Production Engineering, Zagreb*, 37–41.
Abele, E., Metternich, J., Tisch, M., Chryssolouris, G., Sihn, W., ElMaraghy, H. A., Hummel, V., & Ranz, F. (2015). Learning factories for research, education, and training. In *5th CIRP-Sponsored Conference on Learning Factories, Procedia CIRP, 32,* 1–6. https://doi.org/10.1016/j.procir.2015.02.187.
Abele, E., Tenberg, R., Wennemer, J., & Cachay, J. (2010). Kompetenzentwicklung in Lernfabriken für die Produktion. *Zeitschrift für wirtschaftlichen Fabrikbetrieb (ZWF), 105*(10), 909–913.
Alptekin, S. E., Pouraghabagher R., McQuaid, P., & Waldorf D. (2001). Teaching factory. *American Society for Engineering Education*, 1–8.
Argyris, Chris; Schön, Donald A. (1999). Die lernende organisation. Grundlagen, Methode, Praxis. Stuttgart: Klett-Cotta.
Bachmann, R. (1986). Ecology in the school environment. *Health Promotion International, 1*(3), 325–334. https://doi.org/10.1093/heapro/1.3.325.
Barton, H., & Delbridge, R. (2001). Development in the learning factory: Training human capital. *Journal of European Industrial Training, 25*(9), 465–472.
Beattie, C. (1988). Studying, taming, and exploiting the micro. *Journal of Curriculum Studies, 20*(2), 181–188.
Bichow, S. (2015). "Verfolgung und Ermordung der Universitätswürde 1968" - Die Studentenproteste an der Christian-Albrechts-Universität. *Christian-Albrechts-Universität zu Kiel*, 622–636.
Bögel, R. (2011, July 4). Keine Lernfabrik. *Stuttgarter Nachrichten*.
Cachay, J., Wennemer, J., Abele, E., & Tenberg, R. (2012). Study on action-oriented learning with a learning factory approach. *Procedia - Social and Behavioral Sciences., 55,* 1144–1153.
Chryssolouris, G. (2014). Definitions of a „Learning Factory". In E. Abele, G. Chryssolouris, W. Sihn, & S. Seifermann (Eds.), *CIRP collaborative working group - Learning factories for future oriented research and education in manufacturing* (Vol. 28.08.2014, p. 11). Nantes, France.
Chryssolouris, G., Mavrikios, D., & Mourtzis, D. (2013). Manufacturing systems - skills & competencies for the future. In *46th CIRP Conference on Manufacturing Systems. Procedia CIRP, 7,* 17–24.

References

Chryssolouris, G., Mavrikios, D., & Rentzos, L. (2016). The teaching factory: A manufacturing education paradigm. In *49th CIRP Conference on Manufacturing Systems. Procedia CIRP, 57*, 44–48.

Dizon, D. P. (2000). Making social sciences relevant to engineering students: The Greenfield Coalition experience. In *Proceedings International Conference on Engineering Education*, 1–4.

ESB Reutlingen. (2017). NIL network of innovative learning factories. Retrieved from http://www.esb-business-school.de/en/research/research-activities/nil-network-innovative-learning-factories/.

Hamid, M. H. M. I., Masrom, M., & Salim, K. R. (2014). Review of learning models for production based education training in technical education. In *International Conference on Teaching and Learning in Computing and Engineering, IEEE*, 206–211.

Hammer, M. (2014, August). *Making Operational Transformations Successful with Experiential Learning*. CIRP Collaborative Working Group – Learning Factories for future oriented research and education in manufacturing, CIRP General Assembly, Nantes, France.

Hempen, S., Wischniewski, S., Maschek, T, & Deuse, J. (2010). Experiential learning in academic education: A teaching concept for efficient work system design. In *14th Workshop of the Special Interest Group on Experimental Interactive Learning in Industrial Management of the IFIP Working Group 5.7*, 71–78.

IELF. (2013). *General assembly of the initiative on European learning factories*. München.

Initiative on European Learning Factories. (2013). *General assembly of the initiative on European learning factories*. Munich.

Jäger, A., Mayrhofer, W., Kuhlang, P., Matyas, K., & Sihn, W. (2012). The "Learning Factory": An immersive learning environment for comprehensive and lasting education in industrial engineering. In *16th World Multi-Conference on Systemics, Cybernetics and Informatics, 16*(2), 237–242.

Jank, W., & Meyer, H. (2002). *Didaktische Modelle (5., völlig überarb. Aufl.)*. Berlin: Cornelsen-Scriptor.

Jorgensen, J. E., Lamancusa, J. S., Zayas-Castro, J. L., & Ratner, J. (1995). The learning factory: Curriculum integration of design and manufacturing. In *4th World Conference on Engineering Education*, 1–7.

Jouault, C., & Seta, K. (2014). Content-dependent question generation for history learning in semantic open learning space. In S. Trausan-Matu, K. Boyer, M. Crosby, & K. Panourgia (Eds.), *Intelligent tutoring systems: 12th international conference, ITS 2014, Honolulu, HI, USA* (pp. 300–305). Berlin: Springer.

Knowles, M. S. (1975). *Self-directed learning: A guide for learners and teachers* (4th ed.). New York: Cambridge The Adult Education Comp.

Kreimeier, D., Morlock, F., Prinz, C., Krückhans, B., Bakir, D. C., & Meier, H. (2014). Holistic learning factories - A concept to train lean management, resource efficiency as well as management and organization improvement skills. In *47th CIRP Conference on Manufacturing Systems. Procedia CIRP, 17*, 184–188.

Lamancusa, J. S., Zayas, J. L., Soyster, A. L., Morell, L., & Jorgensen, J. E. (2008). The learning factory: Industry-partnered active learning - 2006 Bernard M. Gordon prize lecture. *Journal of Engineering Education, 97*(1), 5–11.

Leonard-Barton, D. (1992). The factory as a learning laboratory. *Sloan Management Review, 34*(1), 3–14.

Mavrikios, D., Papakostas, N., Mourtzis, D., & Chryssolouris, G. (2013). On industrial learning and training for the factories of the future: A conceptual, cognitive and technology framework. *Journal of Intelligent Manufacturing, 24*(3), 473–485. https://doi.org/10.1007/s10845-011-0590-9.

McKinsey & Company. (2017). Model factories and offices: Building operations excellence. Retrieved from https://capability-center.mckinsey.com/files/mccn/2017-03/emea_model_factories_brochure_1.pdf.

McLaren, P., & Jandric, P. (2015). The critical challenge of networked learning: Using information technologies in the service of humanity. In P. Jandric & D. Boras (Eds.), *Critical learning in digital networks* (pp. 199–226). Cham, Heidelberg, New York, Dordrecht, London: Springer.

Metternich, J., Abele, E., Chryssolouris, G., Sihn, W., ElMaraghy, H. A., Tracht, K., Tolio, T., Mavrikios, D., Mourtzis, D., Jäger, A., Tisch, M., & Seifermann, S. (2014, August). *WP 1: Definitions of "Learning Factories"*. CIRP. CIRP Collaborative Working Group – Learning Factories for future oriented research and education in manufacturing, CIRP General Assembly, Nantes, France.

Micheu, H.-J., & Kleindienst, M. (2014). Lernfabrik zur praxisorientierten Wissensvermittlung: Moderne Ausbildung im Bereich Maschinenbau und Wirtschaftswissenschaften. *Zeitschrift für wirtschaftlichen Fabrikbetrieb (ZWF), 109*(6), 403–407.

Müller, E., Plorin, D., & Ackermann, J. (2012). Fachkompetenzentwicklung in der advanced Learning Factory (aLF) als Antwort auf den demografischen Wandel. In E. Müller (Ed.), *Demographischer Wandel: Herausforderung für die Arbeits- und Betriebsorganisation der Zukunft* (pp. 3–26). Berlin: GITO.

Penn State University. (2017). *Bernard M. Gordon Learning Factory: We bring the real world into the classroom*. Retrieved from http://www.lf.psu.edu/.

Piskurich, G. M. (1993). *Self-directed learning: A practical guide to design, development, and implementation* (1st ed.). Hoboken: Jossey-Bass, A Wiley Brand.

Plorin, D., & Müller, E. (2014). Developing an ambient assisted living environment applying the advanced Learning Factory (aLF): A conceptual approach for the practical use in the research project A^{2}LICE. *ISAGA, 2013, 69*–76.

Reith, S. (1988). *Außerbetriebliche CIM-Schulung in der „Lernfabrik"*. Berlin, Heidelberg: Springer. Retrieved from http://link.springer.com/content/pdf/10.1007%2F978-3-662-01109-6_24.pdf.

Rentzos, L., Mavrikios, D., & Chryssolouris, G. (2015). A two-way knowledge interaction in manufacturing education: The teaching factory. In *5th CIRP-sponsored Conference on Learning Factories, Procedia CIRP, 32*, 31–35. https://doi.org/10.1016/j.procir.2015.02.082.

Riffelmacher, P. (2013). *Konzeption einer Lernfabrik für die variantenreiche Montage*. Dissertation, Stuttgart. Stuttgarter Beiträge zur Produktionsforschung: Vol. 15. Stuttgart: Fraunhofer Verlag.

Roth, A. V., Marucheck, A. S., Kemp, A., & Trimbl, D. (1993). The knowledge factory for accelerated learning practices. *Strategy & Leadership, 3,* 26–46.

Schuh, G., Gartzen, T., Rodenhauser, T., & Marks, A. (2015). Promoting work-based learning through Industry 4.0. In *5th CIRP-sponsored Conference on Learning Factories, Procedia CIRP, 32*, 82–87. https://doi.org/10.1016/j.procir.2015.02.213.

Seifermann, S., Metternich, J., & Abele, E. (2014, January). *Learning Factories – Benefits for Research and exemplary Results*. CIRP, CIRP January Meeting, STC-O Technical Presentation, Paris, France.

Senge, Peter M. (1990): The fifth discipline. The art and practice of the learning organization. 18th ed. New York, N.Y.: Doubleday/Currency.

Sihn, W. (2014). Definitions of a „Learning Factory". In E. Abele, G. Chryssolouris, W. Sihn, & S. Seifermann (Eds.), *CIRP collaborative working group - learning factories for future oriented research and education in manufacturing* (Vol. 28.08.2014, p. 12). Nantes, France.

Sivard, G., Eriksson, Y., Florin, U., Shariatzadeh, N., & Lindberg, L. (2016). Cross-disciplinary design based on the digital factory as a boundary object. *26th CIRP design conference procedia CIRP, 50,* 565–570.

Sivard, G., & Lundholm, T. (2013). XPRES - a digital learning factory for adaptive and sustainable manufacturing of future products. In G. Reinhart, P. Schnellbach, C. Hilgert, & S. L. Frank (Eds.), *3rd conference on learning factories, Munich: May 7th, 2013* (pp. 132–154). Augsburg.

Tisch, M. (2018). *Modellbasierte Methodik zur kompetenzorientierten Gestaltung von Lernfabriken für die schlanke Produktion*. Dissertation, Darmstadt. Aachen: Shaker.

Tisch, M., Hertle, C., Abele, E., Metternich, J., & Tenberg, R. (2015). Learning factory design: A competency-oriented approach integrating three design levels. *International Journal of Computer Integrated Manufacturing, 29*(12), 1355–1375. https://doi.org/10.1080/0951192X.2015.1033017.

References

Tisch, M., Hertle, C., Cachay, J., Abele, E., Metternich, J., & Tenberg, R. (2013). A systematic approach on developing action-oriented, competency-based learning factories. In *46th CIRP conference on manufacturing systems. Procedia CIRP, 7*, 580–585.

Tisch, M., & Metternich, J. (2017). Potentials and limits of learning factories in research, innovation transfer, education, and training. In *7th CIRP-sponsored Conference on Learning Factories. Procedia Manufacturing.* (In Press).

Tracht, K. (2014). Definitions of a „Learning Factory". In E. Abele, G. Chryssolouris, W. Sihn, & S. Seifermann (Eds.), *CIRP collaborative working group - learning factories for future oriented research and education in manufacturing* (Vol. 28.08.2014, p. 12). Nantes, France.

Van der Zee, H. (2006). The learning society. *International Journal of Lifelong Education, 10*(3), 213–230. https://doi.org/10.1080/0260137910100305.

Wagner, U., AlGeddawy, T., ElMaraghy, H. A., & Müller, E. (2012). The State-of-the-Art and Prospects of Learning Factories. In *45th CIRP Conference on Manufacturing Systems. Procedia CIRP, 3*, 109–114.

Zierer, K., & Seel, N. M. (2012). General didactics and instructional design: Eyes like twins; a transatlantic dialogue about similarities and differences, about the past and the future of two sciences of learning and teaching. *Springer, 1*(15), 1–22. https://doi.org/10.1186/2193-1801-1-15.

Chapter 5
The Variety of Learning Factory Concepts

A single learning factory can only represent sections of the complex reality of a production company. Building up a learning factory, it is decided which part of reality is mapped in the learning environment. But which different facets of the learning factory orientation can be distinguished in this context? In this chapter, the possible dimensions of a learning factory are systematically derived and the variety of learning factories is presented in morphological boxes (morphologies). Furthermore, a learning factory database is presented in Sect. 5.8 that allows the systematization of learning factories around the globe using the morphology. Finally, success factors of the broad variants of learning factories are discussed in Sect. 7.2.3.

The multi-dimensional description model was mutually developed and validated in course of the CIRP CWG on Learning Factories[1] in cooperation with the NIL project.[2] As the morphology was developed within the CIRP CWG, it represents an academic consensus on included features.[3] The learning factory morphology intends to

- provide orientation in course of the design of new learning factories,
- describe, distinguish, compare, and structure existing learning factory concepts, and
- identify dominant designs of learning factories that are more suiting than others in specific situations.

It should be pointed out that the presented morphology is merely a current and not a final state. New research results, new technologies, or current trends in manufacturing and in industry will require adjustments or extensions of the described dimensions. The actual morphology consists of seven dimensions that are operationalized with

---

[1]See also Sect. 10.4 for a detailed description.
[2]Network of Innovative Learning Factories, see also Sect. 10.3 for a detailed description.
[3]An overview over other models, morphologies, and typologies is provided in Abele et al. (2017a). The morphology of learning factories is described in detail in Tisch, Ranz, Abele, Metternich, and Hummel (2015) and Tisch (2018).

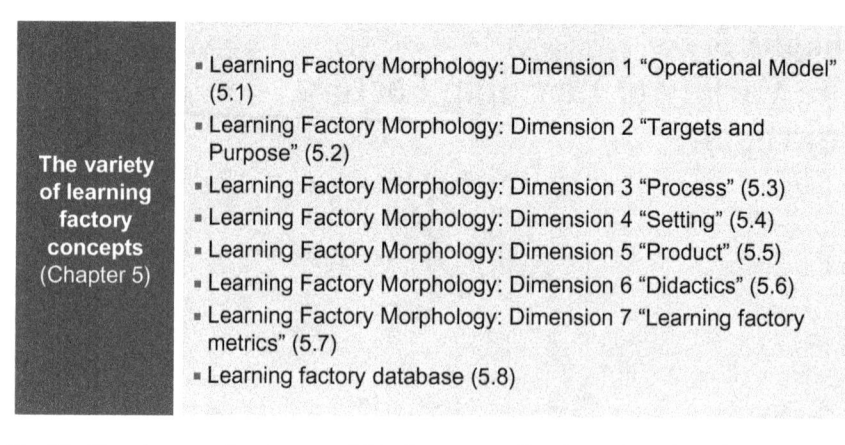

- Learning Factory Morphology: Dimension 1 "Operational Model" (5.1)
- Learning Factory Morphology: Dimension 2 "Targets and Purpose" (5.2)
- Learning Factory Morphology: Dimension 3 "Process" (5.3)
- Learning Factory Morphology: Dimension 4 "Setting" (5.4)
- Learning Factory Morphology: Dimension 5 "Product" (5.5)
- Learning Factory Morphology: Dimension 6 "Didactics" (5.6)
- Learning Factory Morphology: Dimension 7 "Learning factory metrics" (5.7)
- Learning factory database (5.8)

Fig. 5.1 Overview over the structure of this chapter regarding the variety of learning factory concepts

in total 59 characteristics and respective attributes. The features of learning factories are clustered in the seven dimensions:

- Operational model,
- Targets and purpose,
- Process,
- Setting,
- Product,
- Didactics, and
- Learning factory metrics.

In the following, those dimensions, assigned characteristics, and suitable learning factory examples are described in detail. Figure 5.1 shows the structure of this chapter.

5.1 Learning Factory Morphology: Dimension 1 "Operational Model"

The operational model of a learning factory describes how operating organizations succeed in ensuring the sustainable operation of the learning factory and thereby continuously achieving competence development across all hierarchical levels and innovation in technological and organizational areas.[4] Of course, the decision which forms the operational model of the learning factory is fundamentally influenced by the operating organization itself. Today, most learning factory implementations are operated by academic organizations; see, e.g. the proceedings of the yearly Con-

---

[4] See Tisch (2018), Tisch et al. (2015).

ference on Learning Factories.[5] Two studies conducted on the subject of learning factories confirm this[6] and lead to the conclusion that more than three quarters of existing learning factories are from the academic sector. It can be assumed, however, that these figures could be falsified due to the increased publication activity of academic operators.[7] A second important group of operational organizations are profit-oriented businesses in form of consulting firms[8] or companies mainly from the automotive industry.[9] Additionally, variations of the learning factory concept are operated in vocational schools.[10]

For the intended sustainable operation, it is not enough to possess required production equipment and a facility. In order to operate and adapt the learning factory concept continuously three dimensions of sustainability are identified:

- Economic or financial sustainability of the learning factory concept,
- Contentual or thematic sustainability of the learning factory concept, and
- Personal sustainability of the learning factory concept.

In the following sections, those dimensions are explained in detail; see also Fig. 5.2.

5.1.1 *Economic or Financial Sustainability of the Learning Factory Concept*

Financial factors play an important role in the complete life cycle of learning factory.[11] Initially, the learning factory equipment and facility, the development as well as the built-up must be financed. When the learning factory is in operation, a business model is needed that creates values for potential partners and revenues to finance all operating costs from salaries to consumed material—an overview over nonrecurring and operating costs associated with learning factories is given in. In Sect. 6.3.3.4 regarding the "economy of learning factories."

The most important nonrecurring costs of learning factories are

- Internal and external cost for the planning of the learning system,
- Acquisitions of machinery, tools, other equipment, or even lean or buildings,
- Costs for the construction of a learning factory building,

---

[5]See Abele, Cachay, Heb, and Scheibner (2011), Sihn and Jäger (2012), Reinhart, Schnellbach, Hilgert, and Frank (2013), KTH (2014), Kreimeier (2015), Martinsen (2016), Metternich and Glass (2017).
[6]See Plorin (2016), Micheu and Kleindienst (2014).
[7]Tisch (2018).
[8]See Hammer (2014).
[9]See Herrmann and Stäudel (2014), Werz (2012), Oberthuer (2013).
[10]See Zinn (2014) and Didactic (2015).
[11]See Tisch and Metternich (2017).

Economic sustainability	Personal sustainability	Contentual sustainability
▪ What is the value proposition for potential learning factory partners? ▪ How is the value created in which configuration? Which services are offered in which markets? ▪ What is the plan for the initial financiation (equipment, facility etc.)? ▪ How can on-going expenditures be financed? Which revenues are generated from which sources?	▪ How can suiting learning factory personnel be found and acquired? ▪ How can the relevant practical experience of the staff be ensured? ▪ How can learning factory personnel be developed? ▪ What are the requirements in the field of professional competencies? ▪ What are the requirements in the field of personal and social-communication skills?	▪ How can the professional expertise of the organization in the addressed area be ensured? ▪ How can important new issues be identified? ▪ How can new topics be developed? ▪ How can new relevant content be included in the learning factory program?

Fig. 5.2 Three sustainability dimensions of learning factory operational models

- Staff costs in order to be able to recruit and train the needed learning factory staff, and
- Related investments in neighboring sectors of the learning factory (e.g. HR, production).

Additionally, the most important operating costs of learning factories are

- Costs for consumed raw material, supplies, or energy in the learning factory,
- Personnel cost for the operation of the learning factory (e.g. trainers, support, production staff),
- Cost for external services in course of maintenance or remodeling issues,
- Several costs allocated to the learning factory like room cost, general works service, or IT, and
- Cost for debt service, interest, and depreciation.

In order to cover both nonrecurring and operating costs, financing by internal funds, public funds, or third-party funding is conceivable in different time horizons. In learning factory practice, the financing via payments for offered training programs is widespread, while the training programs come in open form (club model (subscription) or fees) as well as individually designed for single companies. A combination of the models is also observed.

For universities and other academic operators of learning factories, the most common types of financing of learning factory activities are

5.1 Learning Factory Morphology: Dimension 1 …

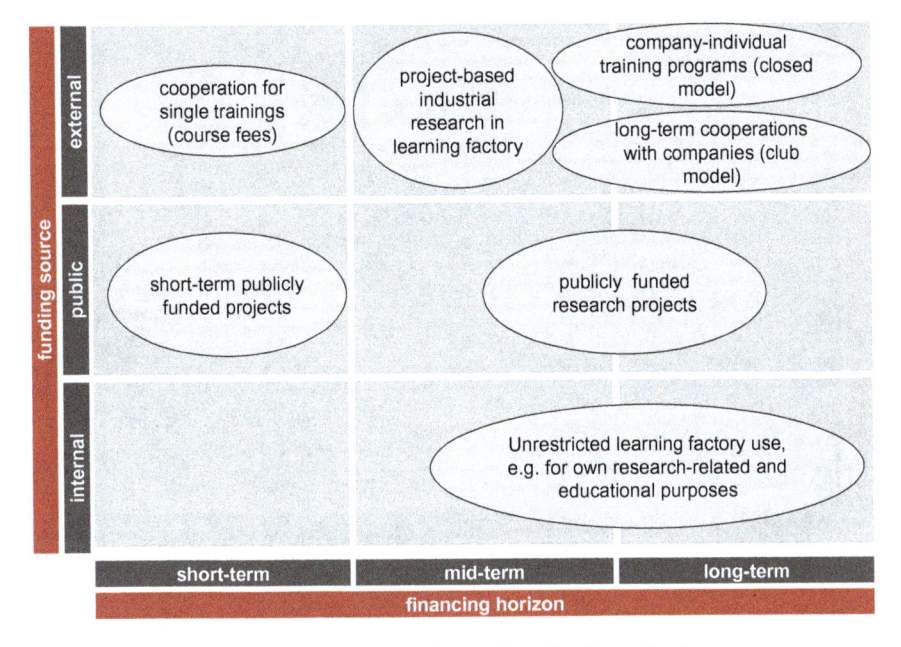

Fig. 5.3 Common types of ongoing learning factory financing for academic operators

- Internal funding for (unrestricted) use in research and education,
- Cooperation with industry for single trainings (course fees),
- Closed models with company individual training programs,
- Open club models with long-term cooperation with industry,
- Project-based research activities in learning factories for industry, and
- Publicly funded projects in relation to the learning factory.

Figure 5.3 shows the most common types of financing for academic learning factories.

For industrial operators of learning factories, the most common types of financing of learning factory activities are

- Company internal funding of learning factory activities,
- Course fees for trainings for other companies, and
- Publicly funded projects are possible, but not common in this case.

Common types of ongoing financing for industrial learning factory operators are shown in Fig. 5.4.

In industry, combinations of the types indicated in Fig. 5.4 can also be identified. For example, a manufacturer of brake systems for rail and commercial vehicles operates a learning factory on the company premises together with a consulting firm. The company provides the facility and equipment ("company internal funding of learning factory activities"). The consulting form operates the learning factory and offers trainings for employees of the company. Instead of receiving payments for the

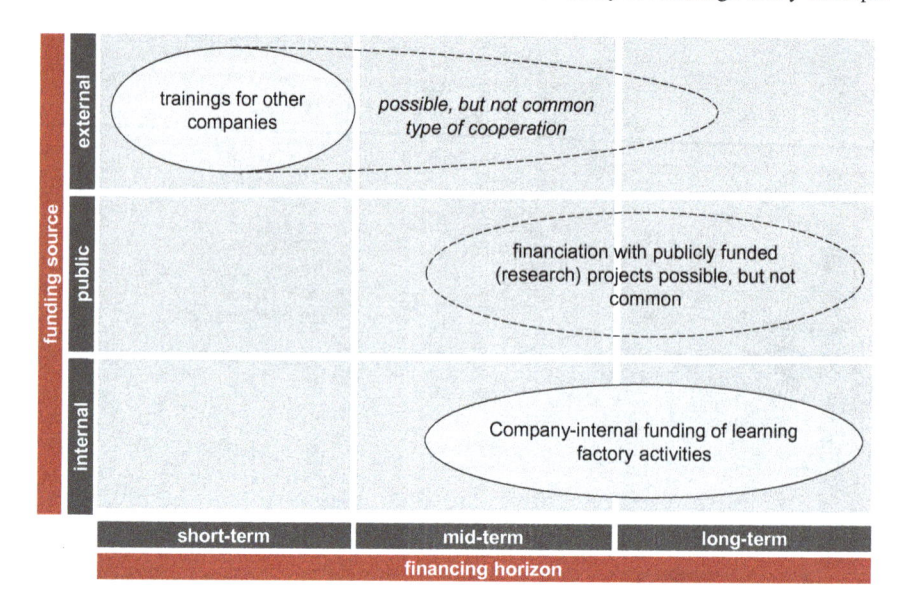

Fig. 5.4 Common types of ongoing learning factory financing for industrial operators

training, the consulting firm is allowed to train own customers in the learning factory ("trainings for other companies").

5.1.2 Contentual or Thematic Sustainability of the Learning Factory Concept

Learning factories have to address manufacturing industry- and education-relevant issues adequately. To achieve this in a sustainable manner, a learning factory should be organized in a way that it can integrate new and interesting content into the learning program. In this context, the following four fields of activity influence and enrich each other:

- Research in the learning factory,
- Education and training in the learning factory,
- Transfer from the learning factory to industry,
- Consulting, coaching, and other industrial projects, and
- New business creation.

Figure 5.5 illustrates the dependencies of the three fields of activity. Research as well as innovative industrial projects may serve as a basis or input for the initiation of new learning factory trainings and transfer measures, while on the other hand the experience gained in learning factory training and transfer activities provides

5.1 Learning Factory Morphology: Dimension 1 …

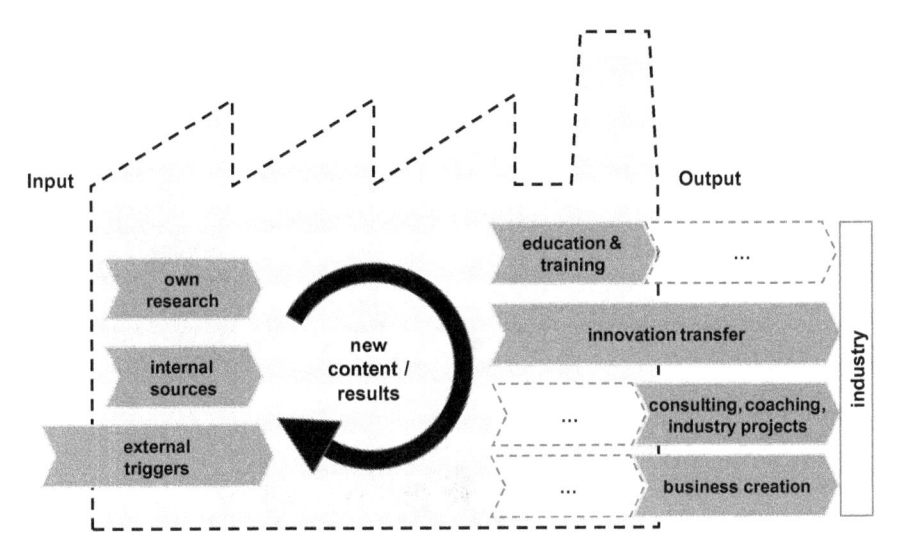

Fig. 5.5 Dependencies of research, transfer, education and training, industrial projects, and business creation in learning factories

valuable feedback for research and implementation in real-life factories. In general, upcoming issues and trends are first addressed in research and sometimes also at an early stage in industrial pilot projects. Naturally, learning factory trainings and transfer in general lag some time behind. Upcoming topics can either be combined with existing trainings or form a completely new training module and therefore enhance the learning factory program.

5.1.3 Personal Sustainability of the Learning Factory Concept

Personnel aspects play an important role in the successful and high-quality built-up and operation of a learning factory. In addition to the necessary professional expertise, the learning staff needs certain didactic skills for the moderation and coaching of the learners as well as practical experience in the addressed topics. Appropriate personnel, like research assistants or engineers depending on the type of operating organization, are to be not only selected and convinced of the task but also prepared and continually evolved regarding professional as well as soft skills. In order to meet these requirements of the personnel profile, the field of activity of the learning factory staff is frequently combined with other responsibilities in the fields of research or consulting services. The complete morphological description of the dimension 1 "operational model" is depicted in Fig. 5.6.

1.1	operator	academic institution (university, college etc.)	non-academic institution (vocational school, high school, chamber, union, industrial network etc.)		profit-oriented operator (consulting business, producing company etc.)	
1.2	trainer	researcher	student assistant	technical expert	manager	consultant	educationalist
1.3	development	own development	external assisted development		external development	
1.4	initial funding	internal funds	public funds		external funds	
1.5	ongoing funding	internal funds	public funds		external funds	
1.6	funding continuity	short term funding (e.g. single events)	mid term funding (projects and programs < 3 years)		long term funding (projects and programs > 3 years)	
1.7	business model for trainings	open models			closed models (training program only for single company)	
		club model	course fees			

Fig. 5.6 Learning factory morphology, dimension 1: operational model, according to Tisch et al. (2015) and Tisch (2018)

5.2 Learning Factory Morphology: Dimension 2 "Targets and Purpose"

As mentioned earlier, learning factories can be used for different purposes that all involve some kind of learning. Primary purposes of learning factories are defined as education of students, training of industrial personnel, and production-related research.[12,13] This means the learning inside a learning factory takes either place in the form of competency development, in the cases of education and training, or innovation, in case of the use in research.[14] Additionally to this, secondary purposes can be identified in the demonstration and the technology testing and transfer as well as the industrial production of goods and the creation of a positive public image.[15] In recent years, in particular demonstration and technology transfer in learning factories in connection with the topics digitization and industry 4.0 are of great importance. In those cases, often new Industrie 4.0 possibilities and process innovations are demonstrated to visitors from industry, with the goal that these visitors will transfer the ideas shown in their own factory.[16]

With regard to education and training in learning factories, various target groups can be addressed in varying setups.[17] Depending on the objective pursued, those

---

[12] See Fig. 5.8, 2.1.

[13] See also the learning factory definition in Sect. 4.3 as well as the overview on learning factories with examples from the different fields in Chaps. 7–9.

[14] See Abele et al. (2015).

[15] See Fig. 5.8, 2.2.

[16] Descriptions for such a use of a learning factory can, for example, be found in Wank et al. (2016), Erol, Jäger, Hold, Ott, and Sihn (2016), and Schuh, Gartzen, Rodenhauser, and Marks (2015).

[17] See Fig. 5.8, 2.3.

5.2 Learning Factory Morphology: Dimension 2 …

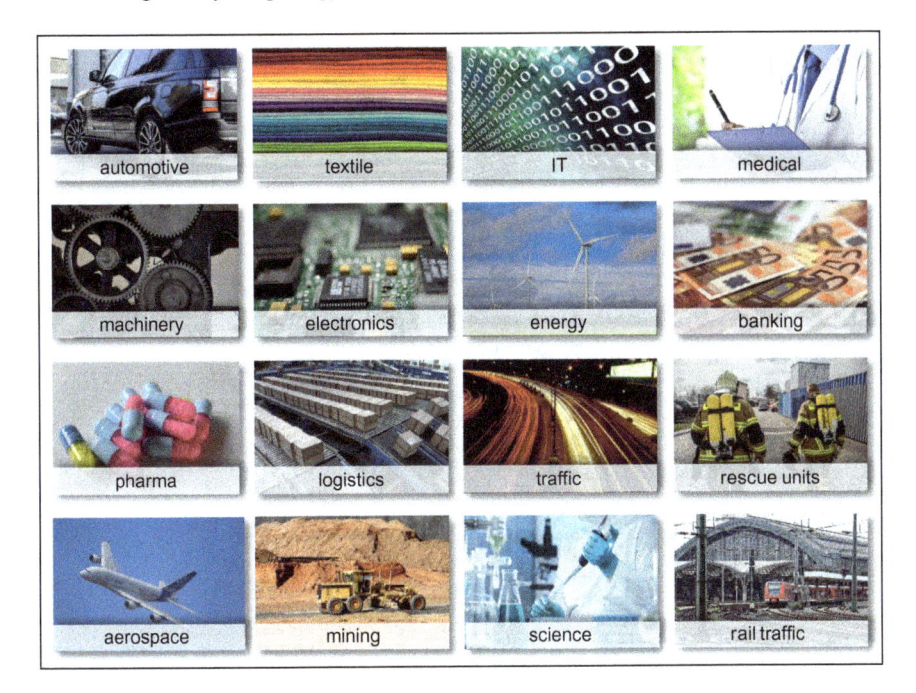

Fig. 5.7 Current and future application areas of learning factories in producing and non-producing sectors

target groups composed homogeneously or heterogeneously.[18] The concept of heterogeneity of the group compilations may refer to a variety of different parameters, which may include the prior knowledge, the hierarchical level, the company affiliation, or the role in the company.

In learning factories, different branches of industry can be specifically addressed;[19] examples focusing on very specific industrial branches are learning factories in the automotive industry,[20] for the pharmaceutical industry[21] or the textile industry.[22] Figure 5.7 gives an overview on potential industry branches where learning factories can be beneficial.

In addition, the objectives of the learning factories differ with respect to the learning content.[23,24] Similar to this, distinct research fields in learning factories

---

[18] See Fig. 5.8, 2.4.

[19] See Fig. 5.8, 2.5.

[20] See, for example, Oberthuer (2013), Herrmann and Stäudel (2014).

[21] See Rybski and Jochem (2016).

[22] See Küsters, Praß, and Gloy (2017).

[23] See Fig. 5.8, 2.6.

[24] For an extensive overview on learning factories addressing different learning content and industry branches see Chap. 6.

2.1	main purpose	education					vocational training				research		
2.2	secondary purpose	test environment / pilot environment					industrial production			innovation transfer	public image		
2.3	target groups for education & training	schoolchildren	students			employees					self-employed	unemployed	open public
			bachelor	master	phd students	apprentices	skilled workers	semi-skilled worker	unskilled	managers			
										lower mgmt	middle mgmt	top mgmt	
2.4	group constellation	homogenous				heterogenous (Knowledge level, hierarchy, students+employees, etc.)							
2.5	targeted industries	mechanical & plant eng.	automotive		logistics		transportation		FMCG		aerospace		
		chemical industry	electronics		construction		insurance / banking		textile industry		...		
2.6	subject-rel. learning contents	product creation processes	energy & resource efficiency		global production		industrial engineering		Industrie 4.0	lean mgmt	design	...	
2.7	role of LF for research	research object						research enabler					
2.8	research topics	product creation processes	energy & resource efficiency		global production		industrial engineering		Industrie 4.0	lean mgmt	design	didactics	...

Fig. 5.8 Learning factory morphology, dimension 2: targets and purpose, with small adaptations according to Tisch et al. (2015) and Tisch (2018)

can be identified,[25] while a distinction can be made between whether the learning factory serves as a research object or research enabler in the sense of a research laboratory.[26,27] The complete morphological description of dimension 2 "targets and purpose" is shown in Fig. 5.8.

5.3 Learning Factory Morphology: Dimension 3 "Process"

In the dimension 3 "process," the production processes depicted in the learning factory are specified more in detail. As a first step, system boundaries are defined. For this purpose, the four interrelated, production-relevant life cycles,[28] namely the product-, the factory-, the order-, and the technology life cycle, are identified.[29] The individual life cycles exceed the system boundary of a production system, but all of them intersect in production (production, assembly, logistics) and therefore influence the production system. Those life cycles can be used for describing the

---

[25] See Fig. 5.8, 2.8.
[26] See Fig. 5.8, 2.7.
[27] See also Sect. 7.3.
[28] See Bauernhansl et al. (2014).
[29] See Fig. 5.10, 3.1–3.4.

5.3 Learning Factory Morphology: Dimension 3 "Process"

Product life cycle describes the life phases of a product from the product idea to recycling or disposal. For detailed explanations on the product life cycle see also Westkämper (2006), Bauernhansl et al. (2014), Umeda et al. (2012), and Westkämper et al. (2006).

| product planning | product development | prototyping | manufacturing, assembly, logistics | service | recycling |

Factory life cycle describes the life phases of a factory from the initial planning to dismantling or recycling. For detailed explanations on the factory life cycle see also Westkämper et al. (2006), Schenk et al. (2014), and Grundig (2015).

| investment planning | factory concept & process planning | ramp-up | manufacturing, assembly, logistics | maintenance | recycling / dismantling |

Order life cycle describes the life phases of a customer order from the first coordination and order to the dispatch of the finished product. For detailed explanations see also Schuh (2006), Dürr (2013), and Bauernhansl et al. (2014).

| configuration / order | order sequencing | planning & scheduling | manufacturing, assembly, logistics | picking & packaging | shipping |

Technology life cycle describes the life phases of technologies from initial planning, over development, use in production to modernization.

| planning | development | virtual testing | manufacturing, assembly, logistics | maintenance | modernization |

Fig. 5.9 Life cycles of production (phases of production highlighted in red) according to Tisch (2018) based on Bauernhansl et al. (2014), Westkämper (2006), Westkämper, Constantinescu, and Hummel (2006), Umeda et al. (2012), Schenk, Wirth, and Müller (2014), Grundig (2015), Schuh (2006), Dürr (2013)

foci of production-related learning factories, although not all life cycles are relevant for all learning factories. For example, no dedicated learning factory activities are known at the present time with regard to the technology life cycle. However, it is possible that all life cycles are addressed or simulated at once in a learning factory. This is further proof of the versatility of the learning factory concept. In learning factories for production, all four life cycles meet in the production-related steps of the life-cycles assembly, machining, and logistics. The four life cycles are shown and described in Fig. 5.9.

Furthermore, indirect functions[30] inside the learning factory concept, which are only indirectly involved in the production process, are identified, such as purchasing or sales. Here, all value-added activities,[31] which are not assigned to the "operations" area, are named. The further elements describe the modeled production processes with regard to the material flow,[32] the process type,[33] the production organization,[34] the degree of automation of the production,[35] the production processes,[36] and tech-

---

[30] See Fig. 5.10, 3.5.
[31] According to Porter (2008).
[32] See Fig. 5.10, 3.6.
[33] See Fig. 5.10, 3.7.
[34] See Fig. 5.10, 3.8.
[35] See Fig. 5.10, 3.9.
[36] See Fig. 5.10, 3.10.

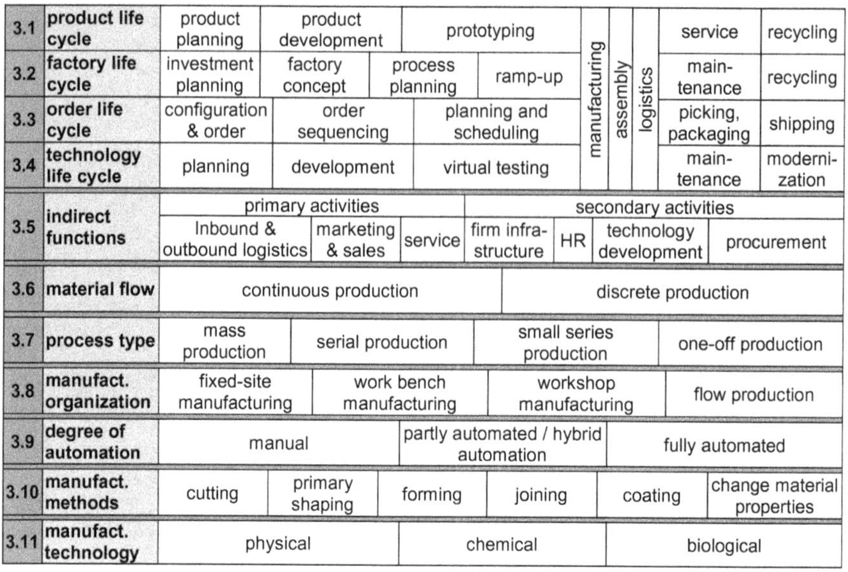

3.1	product life cycle	product planning	product development	prototyping	manufacturing	assembly	logistics	service	recycling
3.2	factory life cycle	investment planning	factory concept	process planning	ramp-up			maintenance	recycling
3.3	order life cycle	configuration & order	order sequencing	planning and scheduling				picking, packaging	shipping
3.4	technology life cycle	planning	development	virtual testing				maintenance	modernization
3.5	indirect functions	primary activities						secondary activities	
		Inbound & outbound logistics	marketing & sales	service	firm infrastructure	HR	technology development	procurement	
3.6	material flow	continuous production				discrete production				
3.7	process type	mass production	serial production		small series production		one-off production			
3.8	manufact. organization	fixed-site manufacturing	work bench manufacturing		workshop manufacturing		flow production			
3.9	degree of automation	manual		partly automated / hybrid automation			fully automated			
3.10	manufact. methods	cutting	primary shaping	forming	joining || coating	change material properties		
3.11	manufact. technology	physical			chemical			biological			

Fig. 5.10 Learning factory morphology, dimension 3: process, according to Tisch et al. (2015) and Tisch (2018)

nologies.[37] The complete morphological description of dimension 3 "process" is shown in Fig. 5.10.

5.4 Learning Factory Morphology: Dimension 4 "Setting"

The dimension 4 "setting" describes the learning environment and its characteristics. Learning factories in the narrow sense contain a physical factory environment[38] in which learners can discover and experiment. Alternatively, to this, other digital and virtual representations of the included factory environment can be recognized[39]:

- These physical learning factories can be supported by means of digital factory systems and tools (ERP, MES, etc.); most physical learning factories have some kind of digital systems implemented.
- The physical value streams can be expanded virtually.[40]

---

[37] See Fig. 5.10, 3.11.
[38] See also Sect. 4.3.
[39] See Fig. 5.12, 4.1.
[40] See, for example, Riffelmacher (2013).

5.4 Learning Factory Morphology: Dimension 4 "Setting"

- The factory environment can also be implemented as a purely virtual representation (learning factory in the broad sense).[41]

Additionally, the factory environment can make use of either life-size equipment that is also used in real factories[42] or scaled down factory equipment[43] that means smaller models conceptually inspired by life-size factory equipment.[44] An exemplary selection of physical, virtual, life-size, and scaled down learning factories is shown in Fig. 5.11.

By definition,[45] learning factories include more than just a single workplace or a single machine. Thus, the mapping of learning factories can range from single manufacturing or assembly cells, over entire factories, to even to factory networks.[46] The respective subordinated factory levels are also part of the learning factory setup.[47]

Furthermore, the flexibility and changeability of the learning environment are of great importance for learning factories. The use of the terms flexibility and changeability in the context of learning factories is in line with the generally accepted distinction between the concepts[48]:

- The flexibility allows a rapid planned conversion of the factory environment within trainings (in reference to a planned learning path),
- while the changeability describes the ability to adapt the factory environment to various unforeseen changes or unforeseen ideas of the learners.[49]

The dimensions in respect of which the factory environment has to be flexible and changeable[50] can be distinguished in product, process, organization, and layout.[51] In order to ensure the ability to change, analogous to real production systems, the learning factory environment needs special properties. These properties are referred to as change enablers.[52,53] The primary change enablers are mobility, modularity, compatibility, scalability, and universality.

---

[41] See, for example, Haghighi, Shariatzadeh, Sivard, Lundholm, and Eriksson (2014) and FBK (2015).
[42] See, e.g. Abele, Tenberg, Wennemer, and Cachay (2010).
[43] See Fig. 5.12, 4.2.
[44] See, for example, Didactic (2016) and Kaluza et al. (2015).
[45] See Sect. 4.3.
[46] See Fig. 5.12, 4.3.
[47] The factory levels used in the morphology are according to Wiendahl, Reichardt, and Nyhuis (2009).
[48] See, e.g. Nyhuis, Reinhart, and Abele (2008).
[49] See Tisch (2018).
[50] Changeability dimensions, see Fig. 5.12, 4.5.
[51] See Wiendahl et al. (2007).
[52] See Fig. 5.12, 4.4.
[53] According to Morales (2003).

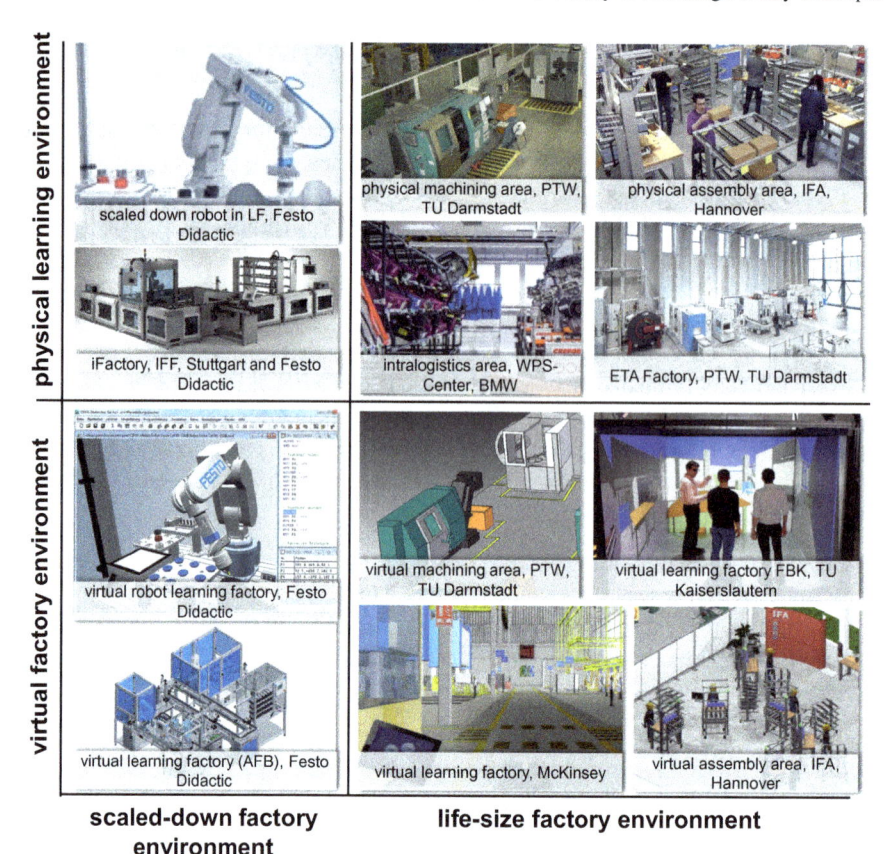

Fig. 5.11 Examples for physical, virtual, life-size, and scaled down learning environments in learning factories based on Tisch (2018), from left to right and top to bottom pictures, were taken from Didactic (2017c, 2017a, 2017b), PTW, TU Darmstadt (2017a), BMW (2015), Jäger, Sihn, Hummel, and Ranz (2015), Hammer (2014), IFA (2017), PTW, TU Darmstadt (2017b), FBK (2015), and Görke, Bellmann, Busch, and Nyhuis (2017)

With regard to IT support in learning factories, different IT systems can be distinguished depending on the relation to the production phase[54,55]:

- Before the start of production, there are systems like CAD and CAM.
- After the start of production, there are systems like ERP and MES.
- And after the production phase, systems like CRM and PLM are used.

The complete morphological description of dimension 4 "setting" is shown in Fig. 5.12.

---

[54] See Fig. 5.12, 4.6.

[55] See Tisch et al. (2015).

5.5 Learning Factory Morphology: Dimension 5 "Product" 113

4.1	learning environment	purely physical (planning + execution)	physical LF supported by digital factory (see line "IT-Integration")	physical value stream of LF extended virtually		purely virtual (planning + execution)
4.2	environment scale	scaled down			life-size	
4.3	work system levels	station	cell	system	segment	factory	network
4.4	enablers for changeability	mobility	modularity	compability	scaleability		universality
4.5	changeability dimensions	product		process	organization		building & layout
4.6	IT-integration	IT before SOP (CAD, CAM, simulation)	IT after SOP (PPS, ERP, MES)			IT after production (CRM, PLM...)

Fig. 5.12 Learning factory morphology, dimension 4: setting, according to Tisch et al. (2015) and Tisch (2018)

5.5 Learning Factory Morphology: Dimension 5 "Product"

The characteristics of the products, which are produced within the simulated factory environments, form an important dimension of the learning factory concept. The characteristics of the product and its variants must be adapted to the overarching concept of the learning factory. In addition, the choice of the product and the complexity of the product have influences on

- the degree of complexity of the learning factory scenarios and the duration required by the learners to get into the represented processes,
- the material and personnel cost for operation and maintenance of the learning factory, and
- the possibilities of modeled value-adding processes.[56]

The products used in learning factories are fundamentally in most cases based or oriented on the products of real factories. In most cases, material products are used,[57] whereby the use of intangible products (services) can also be observed.[58] Material products in general can be subdivided into general cargo, bulk goods, and flow products.[59,60] Figure 5.13 exemplarily shows the range of products used in learning factories today.

---

[56] See Tisch (2018).
[57] See learning factory in the narrow sense.
[58] See Fig. 5.15, 5.1.
[59] See Fig. 5.15, 5.2.
[60] See Schenk et al. (2014).

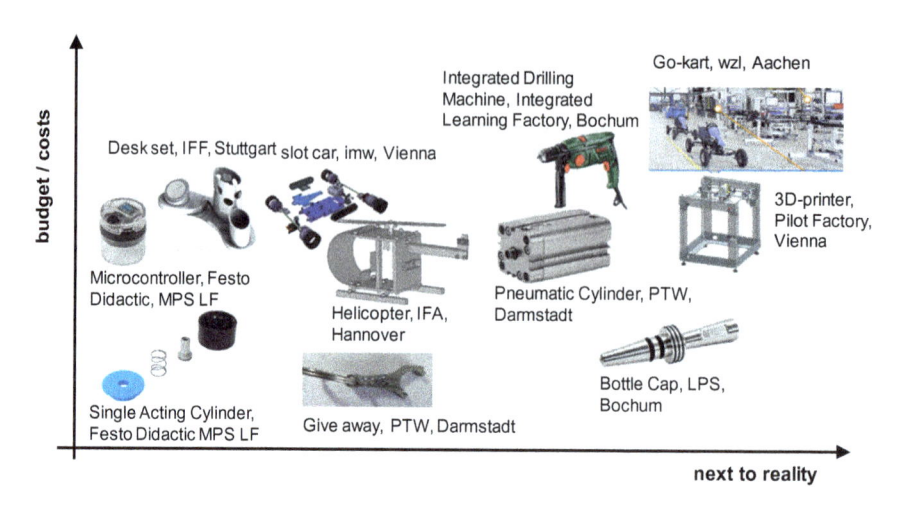

Fig. 5.13 Examples for learning factory products (Abele et al., 2017b)

In contrast to the usual product development process,[61] the product used in the learning factory is either selected specifically from the products available on the market[62] or even specially developed for use in the learning factory;[63] see also Fig. 5.14.[64]

In the case of products specifically developed for the learning factory, a distinction can be made between products which are available on the market but have been didactically simplified and products not available on the market.[65] Additionally, learning factory products can also be distinguished in terms of their functionality, from fully functional products over didactically adapted products with reduced functionality to products without any function.[66] The products especially developed for the learning factory can be subdivided into proprietary developments of the operator, externally assigned developments, or even developments by learners in course of the learning modules.[67] In most learning factories, dismountable products are used for reason of cost, so that single or as many as possible product components[68] can be reused. Few learning factory operators sell the produced goods after the learning module[69] [70]. The complete morphological description of dimension 5 "product" is shown in Fig. 5.15.

---

[61] See Fig. 5.14, (a).

[62] See Fig. 5.14, (c).

[63] See Fig. 5.14, (b).

[64] See Metternich, Abele, and Tisch (2013), Wagner, AlGeddawy, ElMaraghy, and Müller (2014, 2015), Tisch, Hertle, Abele, Metternich, and Tenberg (2015), Tisch (2018).

[65] See Fig. 5.15, 5.4.

[66] See Fig. 5.15, 5.5.

[67] See Fig. 5.15, 5.3.

[68] See Fig. 5.15, 5.8.

[69] See as a rare example Kreimeier et al. (2014).

[70] Other options for the further use of learning factory products are listed in Fig. 5.15, 5.9.

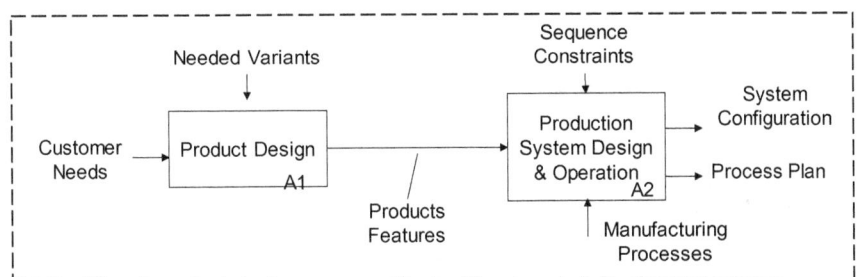

5.5 Learning Factory Morphology: Dimension 5 "Product"

(a) Traditional product design process: The traditional product development process shown is used for traditional product and production system design. The development of the product on the basis of customer requirements forms the starting point. Based on this, the production system and production operations are designed. For learning factories, this approach to product and production system development is not effective.

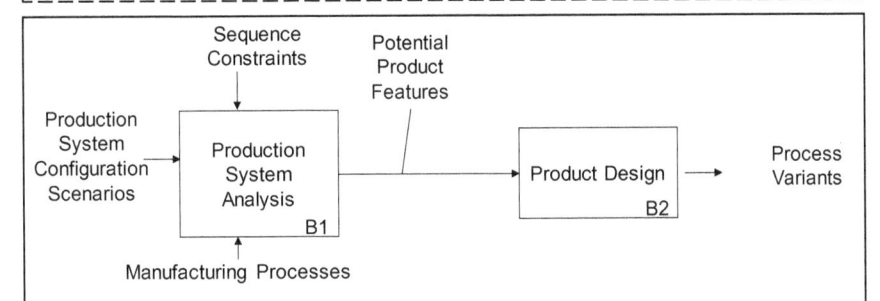

(b) Possible learning factory product design process 1: Realizable production systems are analyzed based on constrains, different configurations and included manufacturing processes. Based on the predefined production system potential product features and finally a suiting learning factory product is defined. This approach reverses the traditional product development process.

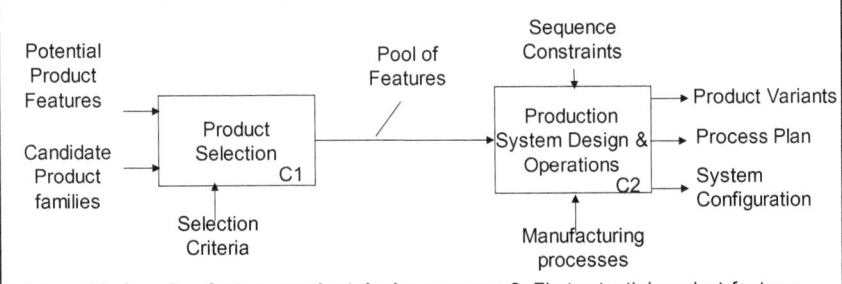

(c) Possible learning factory product design process 2: First potential product features and product families are collected and based on defined selection criteria a suiting product for the learning factory concept is selected. Based on this, similar to the traditional product design process the production system is designed.

Fig. 5.14 Comparison of traditional product design process (**a**) and the product design process for learning factories (**b, c**) with changes inspired by Wagner et al. (2014), similar also in Abele et al. (2017a)

5.1	materiality	material (physical product)		immaterial (service)		
5.2	form of product	general cargo		bulk goods		flow products
5.3	product origin	own development		development by participants		external development
5.4	marketability of product	available on the market		available on the market but didactically simplified		not available on the market
5.5	functionality of product	functional product		didactically adapted product with limited functionality		without function/ application, for demonstration only
5.6	no. of different products	1 product	2 products	3-4 products	> 4 products	flexible, developed by participants	acceptance of real orders
5.7	no. of variants	1 variant	2-4 variants	5-20 variants	...	flexible, depending on participants	determined by real orders
5.8	no. of components	1 comp.	2-5 comp.	6-20 comp.	21-50 comp.	51-100 comp.	> 100 comp.
5.9	further product use	re-use / re-cycling		exhibition / display	give-away	sale	disposal

Fig. 5.15 Learning factory morphology, dimension 5: product, according to Tisch et al. (2015) and Tisch (2018)

In the best practice examples shown in Chap. 11, several different concepts for the selection of a product can be identified; among other there are:

- An own developed fantasy product in the IFA Learning Factory, see Best Practice Example 10.
- Own developed products that are produced in the learning factory for the market in the Demonstration Factory at WZL in Aachen, see Best Practice Example 2.
- A product available on the market from an industrial company, for example, in the Process Learning Factory CiP, see Best Practice Example 26.

5.6 Learning Factory Morphology: Dimension 6 "Didactics"

The "didactics" are an integral part of learning factory concepts, which address one of the primary purposes of education and training.[71] Regarding this dimension, the following questions are important[72]:

- **What should be learned** in terms of competency classes[73] and different learning objectives[74]?

---

[71] See Abele et al. (2015).
[72] See Tisch (2018).
[73] See Fig. 5.16, 6.1.
[74] See Fig. 5.16, 6.2.

5.6 Learning Factory Morphology: Dimension 6 "Didactics"

- **How should be learned** regarding the learning scenario,[75] the degree of autonomy of the learners,[76] the format of the learning modules[77] and any standardization,[78] the embedding of the systematizing elements into the learning module[79] as well as the role of the trainer in this[80]?
- **Where should be learned** in relation to the type of learning environment[81] or the communication channel[82]?
- **How should we evaluate** based on the evaluation levels[83] and the type of evaluation instrument[84]?

The complete morphological description of dimension 6 "didactics" is shown in Fig. 5.16.

5.7 Learning Factory Morphology: Dimension 7 "Learning Factory Metrics"

This section presents quantitative characteristics of learning factory concepts, such as the number of participants per learning module,[85] the average duration of individual learning modules,[86] or the available learning area.[87] The different parameters provide a rough framework for the learning factory concepts related to selected quantitative figures. Individual parameters could also be assigned to the other dimensions of the morphology.[88] The complete morphological description of dimension 7 "learning factory metrics" is shown in Fig. 5.17.

---

[75] See Fig. 5.16, 6.3, from a rigid instruction to an open learning scenario.
[76] See Fig. 5.16, 6.6.
[77] See Fig. 5.16, 6.8.
[78] See Fig. 5.16, 6.9.
[79] See Fig. 5.16, 6.10.
[80] See Fig. 5.16, 6.7.
[81] See Fig. 5.16, 6.4.
[82] See Fig. 5.16, 6.5.
[83] See Fig. 5.16, 6.11.
[84] See Fig. 5.16, 6.12.
[85] See Fig. 5.17, 7.1.
[86] See Fig. 5.17, 7.3.
[87] See Fig. 5.17, 7.6.
[88] For example, the elements 7.1–7.3 in Fig. 5.17 are closely related to the dimension 6 "didactics" and the elements 7.4–7.7 could also be assigned to the dimension 1 "operator".

6.1	competence classes	technical and methodological competencies	social & communication competencies	personal competencies	activity and implementation oriented competencies
6.2	dimensions learn. targets	cognitive		affective	psychomotor
6.3	learn. scenario strategy	instruction	demonstration	closed scenario	open scenario
6.4	type of learn. environment	greenfield (development of factory environment)		brownfield (improvement of existing factory environment)	
6.5	communication channel	onsite learning (in the factory environment)		remote connection (to the factory environment)	
6.6	degree of autonomy	instructed	self-guided/ self-regulated		self-determined/ Self-organized
6.7	role of the trainer	presenter	moderator	coach	instructor
6.8	type of training	tutorial	practical lab course	seminar	workshop	project work
6.9	standardization of trainings	standardized trainings		customized trainings	
6.10	theoretical foundation	prerequisite	in advance (en bloc)	alternating with practical parts	based on demand	afterwards
6.11	evaluation levels	feedback of participants	learning of participants	transfer to the real factory	economic impact of trainings	return on trainings / ROI
6.12	learning success evaluation	knowledge test (written)	knowledge test (oral)	written report	oral presentation	practical exam	none

Fig. 5.16 Learning factory morphology, dimension 6: didactics, according to Tisch et al. (2015)

7.1	no. of participants per training	1-5 participants	6-10 participants	11-15 participants	16-30 participants	>30 participants
7.2	no. of standardized trainings	1 training	2-4 trainings	5-10 trainings		> 10 trainings
7.3	aver. duration of a single training	≤ 1 day	> 1 day until ≤ 2 days	> 2 days until ≤ 5 days	> 5 days until ≤ 10 days	> 10 days bis ≤ 20 days	> 20 days
7.4	participants per year	< 50 participants	50-200 participants	201-500 participants	501-1000 participants	> 1000 participants
7.5	capacity utilization	< 10%	> 10 until ≤ 20%	> 20% until ≤ 50%	> 50% until ≤ 75%	> 75%
7.6	size of LF	≤ 100 sqm	> 100sqm bis ≤ 300sqm	> 300sqm bis ≤ 500sqm	>500sqm bis ≤ 1000sqm	> 1000 sqm
7.7	FTE in LF	< 1	2-4	5-9	10-15	> 15

Fig. 5.17 Learning factory morphology, dimension 7: learning factory metrics, according to Tisch et al. (2015)

5.8 Learning Factory Database

Based on the dimensions and the morphology shown in the previous section, a database for the collection of different learning factory concepts has been established in course of the CIRP CWG on learning factories[89] in order to create an overview on existing learning factory concepts around the globe, classify existing approaches systematically, and maybe identify certain characteristics of different learning factory types. The structure of the database consists of three main entities: the user, the associated facility, and the application scenarios.[90] The three entities are described shortly in the following:

- User: The user can log in into the database, view existing entries, and create new entries for the own learning factory. Newly created facilities are then associated with the respective user.
- Associated Facility: The associated facility describes the structure and the use of the complete learning factory concept. Inside of a single learning factory concept, different application scenarios (scenarios of use of the learning factory) can be defined by the user. Application scenarios are associated with individual facilities. Furthermore, learning factory equipment can be defined and associated with the respective facility.
- Application Scenario: Application scenarios are taken into account since the structure and use of specific learning factories can vary widely in different (individually defined) application scenarios, such as research, teaching students in the bachelor's/master's, further education, educational projects, learning factory enriched lectures. For each application scenario created, the user of the respective learning factory defines the characteristics of the application scenario regarding the dimension and the features of the morphology presented in the previous section—this means each application scenario contains a set of eight associated entities: operational model, purpose and targets, process, setting, product, didactics, learning factory metrics, and additionally an entity that allows the possibility to store and retrieve videos that are also associated with specific application scenarios.

For a detailed description of the various learning factory dimensions or subentities in connection with the database, see the previous section.

The database is accessible via a Web application called "Learning Factories Morphologies application" that enables editing and visualizing data. The technical implementation of the Web application which is deploying to a Java servlet container and follows a three-tier architecture is described in detail in Mavrikios et al. (2017). The application can be used for storing new information of characteristics of a (own) specific learning factory as well as for browsing and viewing information on learning factories morphologically characterized in the database. The Learning Factory Morphology application is available under:

---

[89] See Mavrikios, Sipsas, Smparounis, Rentzos, and Chryssolouris (2017).
[90] An entity relationship diagram describing those three entities and its relations can be found in Mavrikios et al. (2017).

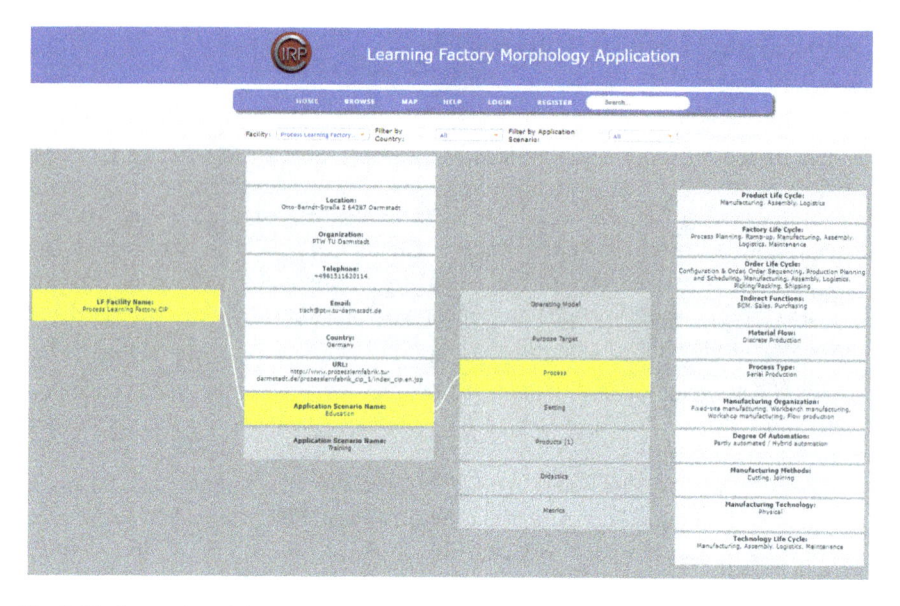

Fig. 5.18 Screenshot of the "learning factories morphology Web application" (LMS, 2015)

http://syrios.mech.upatras.gr/LF/ (LMS, 2015).

At the top of the Web site, the user can navigate through all the functions of the application; see also Fig. 5.18. The functions of the application (current state, 2017) are shortly described in the following:

- Home: In the home screen, information on already inserted learning factories is shown. The user can see the information on the specific learning factory by selecting the learning factory name directly in the drop-down menu on the left entitled "Facility": Furthermore, it is possible to apply filters by choosing the country of interest ("Filter by Country") and/or the application scenario of interest ("Filter by Application Scenario"). When the user has selected a specific learning factory facility, he or she can see general information of the learning factory (location, contact, country, etc.) as well as the characteristics of the defined application scenarios associated with the facility. Furthermore, it is possible to switch from the visualized graph to a printable version of the information, which shows all values of the selected learning factory in a table format.
- Browse: With the browse function of the application, the user is able to search for learning factory concepts with specific characteristics. For example, for a company which is looking for a learning factory near the company location that is able to train their employees regarding specific learning content, like lean production, energy efficiency, or Industrie 4.0, this functionality is useful.
- Map: The map function of the application gives a geographic overview on the locations of the learning factories registered in the database. A screenshot of the current status is shown in Fig. 5.19. The majority of characterized learning factories

5.8 Learning Factory Database

Fig. 5.19 Screenshot of the map function of the "learning factories morphology Web application" (LMS, 2015)

is located in Europe, while a good share of those is in Germany. By clicking on an icon on the map, the user is able to see all the inserted information on the corresponding learning factory facility.

- Help: The help button leads to the user's manual that gives general information on the application and shortly describes the functionalities of the Web application.
- Login: At the login screen, a registered user is able to log in in order to edit existing or create new learning factory facilities and learning factory application scenarios.
- Register: In order to register for the Learning Factory Morphology Application, the user has to fill in name, organization, email address, and a short description of the own learning factory.
- Search: Terms of interest can be typed in the "search" field. The application displays learning factories based on the search.

Currently, twenty learning factories from nine countries are registered in the database. Since the use of the application increases greatly with the increasing number of users and consequently more characterized facilities, all learning factory operators should be encouraged to register and characterize their current learning factory concepts using the Learning Factory Morphology Application: http://syrios.mech.upatras.gr/LF/register.

5.9 Wrap-up of This Chapter

In this chapter, the variety of the learning factory concept is illustrated by interspersing many practical examples from existing learning factories around the globe. The diversity of the existing learning factories is shown along seven dimensions of the learning factory system. Morphological models within the dimensions sharpen the systematic view on learning factory systems.

References

Abele, E., Cachay, J., Heb, A., & Scheibner, S. (Eds.). (2011). *1st conference on learning factories, Darmstadt*. Darmstadt: Institute of Production Management, Technology and Machine Tools (PTW).
Abele, E., Chryssolouris, G., Sihn, W., Metternich, J., ElMaraghy, H. A., Seliger, G., et al. (2017a). Learning factories for future oriented research and education in manufacturing. *CIRP Annals—Manufacturing Technology, 66*(2), 803–826.
Abele, E., Chryssolouris, G., Sihn, W., Metternich, J., ElMaraghy, H. A., Seliger, G., Sivard, G., ElMaraghy, W., Hummel, V., Tisch, M., & Seifermann, S. (2017b, August). *Learning factories for future oriented research and education in manufacturing*. Presentation CIRP STC-O Keynote-Paper, GA 2017. CIRP General Assebly 2017, Lugano, Switzerland.
Abele, E., Metternich, J., Tisch, M., Chryssolouris, G., Sihn, W., ElMaraghy, H. A., Hummel, V., & Ranz, F. (2015). Learning factories for research, education, and training. In *5th CIRP-Sponsored Conference on Learning Factories. Procedia CIRP, 32*, 1–6. https://doi.org/10.1016/j.procir.2015.02.187.
Abele, E., Tenberg, R., Wennemer, J., & Cachay, J. (2010). Kompetenzentwicklung in Lernfabriken für die Produktion. *Zeitschrift für wirtschaftlichen Fabrikbetrieb (ZWF), 105*(10), 909–913.
Bauernhansl, T., Siegert, J., Groß, E., Dinkelmann, M., Abele, E., Metternich, J., et al. (2014). Kompetenzbildung in der Wertschöpfung. *Werkstattstechnik online: wt, 104*(11/12), 776–780.
BMW. (2015). *WPS-center der BMW group: Lernwerkstatt auf 1500 m^2 Fläche*. Retrieved from https://www.press.bmwgroup.com/deutschland/pressDetail.html?title=wps-center-der-bmw-group-lernwerkstatt-auf-1500-m%C2%B2-fl%C3%A4che&outputChannelId=7&id=T0214544DE&left_menu_item=node__5247.
Dürr, P. (2013). *Modell zur Bewertung der Effizienz der IT-Unterstützung im Auftragsabwicklungsprozess von produzierenden KMU*. Univ., Diss.–Stuttgart, 2013. *Stuttgarter Beiträge zur Produktionsforschung* (Vol. 16). Stuttgart: Fraunhofer Verlag.
Erol, S., Jäger, A., Hold, P., Ott, K., & Sihn, W. (2016). Tangible industry 4.0: A scenario-based approach to learning for the future of production. In *6th CIRP-Sponsored Conference on Leanring Factories. Procedia CIRP, 54*, 13–18.
FBK, T. U. K. (2015). *Virtuelle Lernfabrik*. Retrieved from http://www.wgp.de/fileadmin/Produktionsakademie/wgp-KL2.pdf.
Didactic, F. (2015). *Festo: Schulterschluss für die Fachkräfte der Zukunft: Kooperation mit der Gewerblichen Schule Göppingen—Lernfabrik eingeweiht*. Esslingen. Retrieved from https://www.festo.com/net/SupportPortal/Files/354477/CC_S_1_15_Kooperation.rtf.
Didactic, F. (2016). *MPS® the modular production system: Das Konzept im detail*. Retrieved from http://www.festo-didactic.com/de-de/lernsysteme/lernfabriken,cim-fms-systeme/cp-factory/mps-transfer-factory-das-konzept-im-detail.htm?fbid=ZGUuZGUuNTQ0LjEzLjE4LjEyOTMuNzY0Mw.
Didactic, F. (2017a). *iFactory: Innovative training factory: For advanced industrial engineering (aIE)*. Retrieved from http://www.festo-didactic.com/int-en/news/ifactory-innovative-training-factory.htm?fbid=aW50LmVuLjU1Ny4xNy4xNi4yOTUy.
Didactic, F. (2017b). *Individuelle Lösungen: AFB factory Hybride Produktion*. Retrieved from http://www.festo-didactic.com/de-de/lernsysteme/lernfabriken,cim-fms-systeme/afb-factory-hybride-produktion/individuelle-loesungen-afb-factory-hybride-produktion.htm?fbid=ZGUuZGUuNTQ0LjEzLjE4Ljk5Ny43Nzc4.

References

Didactic, F. (2017c). *Robot vision cell: Trends der Robotik im Fokus*. Retrieved from http://www.festo-didactic.com/de-de/lernsysteme/lernfabriken,cim-fms-systeme/robot-vision-cell/robot-vision-cell-trends-der-robotik-im-fokus.htm?fbid=ZGUuZGUuNTQ0LjEzLjE4LjEyNzEuNzYzMw.

Görke, M., Bellmann, V., Busch, J., & Nyhuis, P. (2017). Employee qualification by digital learning games. *Procedia Manufacturing, 9*, 229–237. https://doi.org/10.1016/j.promfg.2017.04.040.

Grundig, C.-G. (2015). *Fabrikplanung [Elektronische Ressource]: Planungssystematik—Methoden—Anwendungen Claus-Gerold Grundig*. München: Carl Hanser.

Haghighi, A., Shariatzadeh, N., Sivard, G., Lundholm, T., & Eriksson, Y. (2014). Digital learning factories: Conceptualization, review and discussion. In *The 6th Swedish Production Symposium (SPS14)*. Retrieved from http://conferences.chalmers.se/index.php/SPS/SPS14/paper/viewFile/1729/401.

Hammer, M. (2014, August). Making operational transformations successful with experiential learning. In *CIRP collaborative working group—Learning factories for future oriented research and education in manufacturing*. Nantes, France: CIRP General Assembly.

Morales, R. H. (2003). *Systematik und Wandlungsfähigkeit in der Fabrikplanung* (Als Ms. gedr). *Fortschritt-Berichte/VDI. Reihe 16, Technik und Wirtschaft. Nr. 149*. Düsseldorf: VDI-Verl.

Herrmann, S., & Stäudel, T. (2014). Learn and experience VPS in the BMW learning factory. In *4th Conference on Learning Factories*. Stockholm, Sweden (pp. 1–18).

IFA. (2017). *IFA Lernfabrik*. Retrieved from http://www.ifa-lernfabrik.de/.

Jäger, A., Sihn, W., Hummel, V., & Ranz, F. (2015, August). Physical and digital learning factories—Differentiation and collaboration. In *CIRP CWG "Learning factories for future oriented research and education in manufacturing"*. Kapstadt, Südafrika.

Kaluza, A., Juraschek, M., Neef, B., Pittschellis, R., Posselt, G., Thiede, S., & Herrmann, C. (2015). Designing learning environments for energy efficiency through model scale production processes. In *5th CIRP-Sponsored Conference on Learning Factories. Procedia CIRP, 32*, 41–46. https://doi.org/10.1016/j.procir.2015.02.114.

Kreimeier, D. (Ed.). (2015). 5th conference on learning factories. *Procedia CIRP, 32*.

Kreimeier, D., Morlock, F., Prinz, C., Krückhans, B., Bakir, D. C., & Meier, H. (2014). Holistic learning factories—A concept to train lean management, resource efficiency as well as management and organization improvement skills. In *47th CIRP Conference on Manufacturing Systems. Procedia CIRP, 17*, 184–188.

KTH. (2014). *4th conference on learning factories*. Retrieved from https://www.kth.se/en/itm/inst/iip/4clf/presentations/presentations-1.459320.

Küsters, D., Praß, N., & Gloy, Y.-S. (2017). Textile learning factory 4.0—Preparing Germany's textile industry for the digital future. *Procedia Manufacturing, 9*, 214–221. https://doi.org/10.1016/j.promfg.2017.04.035.

LMS. (2015). *Learning factory morphology application*. Retrieved from http://syrios.mech.upatras.gr/LF/.

Martinsen, K. (Ed.). (2016). 6th CIRP conference on learning factories. *Procedia CIR, 54*.

Mavrikios, D., Sipsas, K., Smparounis, K., Rentzos, L., & Chryssolouris, G. (2017). A web-based application for classifying teaching and learning factories. In *7th CIRP-Sponsored Conference on Learning Factories. Procedia Manufacturing* (In Print).

Metternich, J., Abele, E., & Tisch, M. (2013). Current activities and future challenges of the process learning factory CiP. In G. Reinhart, P. Schnellbach, C. Hilgert, & S. L. Frank (Eds.), *3rd Conference on Learning Factories*, Munich, May 7th, 2013 (pp. 94–107). Augsburg.

Metternich, J., & Glass, R. (Eds.). (2017). 7th conference on learning factories, CLF 2017. *Procedia Manufacturing, 9*.

Micheu, H.-J., & Kleindienst, M. (2014). Lernfabrik zur praxisorientierten Wissensvermittlung: Moderne Ausbildung im Bereich Maschinenbau und Wirtschaftswissenschaften. *Zeitschrift für wirtschaftlichen Fabrikbetrieb (ZWF), 109*(6), 403–407.

Nyhuis, P., Reinhart, G., & Abele, E. (Eds.). (2008). *Wandlungsfähige Produktionssysteme: Heute die Industrie von morgen gestalten*. Garbsen: PZH Produktionstechnisches Zentrum.

Oberthuer, C. (2013). Integration of process simulations into the CIP of energy efficiency at daimler trucks. In G. Reinhart, P. Schnellbach, C. Hilgert, & S. L. Frank (Eds.), *3rd Conference on Learning Factories*, Munich, May 7th, 2013 (pp. 38–47). Augsburg.

Plorin, D. (2016). *Gestaltung und Evaluation eines Referenzmodells zur Realisierung von Lernfabriken im Objektbereich der Fabrikplanung und des Fabrikbetriebes*. Dissertation, Chemnitz. *Wissenschaftliche Schriftenreihe des Instituts für Betriebswissenschaften und Fabriksysteme: Heft 120*. Chemnitz: Techn. Univ. Inst. für Betriebswiss. und Fabriksysteme.

Porter, M. E. (2008). *Competitive advantage: Creating and sustaining superior performance* (2nd ed.). Riverside: Free Press.

PTW, TU Darmstadt. (2017a). *Prozesslernfabrik CiP: Der Weg zur operativen Exzellenz*. Retrieved from http://www.prozesslernfabrik.de/.

PTW, TU Darmstadt. (2017b). *Welcome to ETA-factory: The energy efficient model factory of the future*. Retrieved from http://www.eta-fabrik.tu-darmstadt.de/eta/index.en.jsp.

Reinhart, G., Schnellbach, P., Hilgert, C., & Frank, S. L. (Eds.). (2013, May 7th). *3rd conference on learning factories*. Munich. Augsburg.

Riffelmacher, P. (2013). Konzeption einer Lernfabrik für die variantenreiche Montage. *Dissertation, Stuttgart. Stuttgarter Beiträge zur Produktionsforschung* (Vol. 15). Stuttgart: Fraunhofer Verlag.

Rybski, C., & Jochem, R. (2016). Benefits of a learning factory in the context of lean management for the pharmaceutical industry. *Procedia CIRP, 54*, 31–34. https://doi.org/10.1016/j.procir.2016.05.106.

Schenk, M., Wirth, S., & Müller, E. (2014). *Fabrikplanung und Fabrikbetrieb: Methoden für die wandlungsfähige, vernetzte und ressourceneffiziente Fabrik* (2., vollst. überarb. und erw. Aufl.). *VDI-Buch*. Berlin: Springer-Vieweg.

Schuh, G. (Ed.). (2006). *Produktionsplanung und -steuerung: Grundlagen, Gestaltung und Konzepte* (3., völlig neu bearbeitete Auflage). *VDI-Buch*. Berlin, Heidelberg: Springer.

Schuh, G., Gartzen, T., Rodenhauser, T., & Marks, A. (2015, July). *Promoting work-based learning through industry 4.0*. RU Bochum. In *5th Conference on Learning Factories*, Bochum, Germany.

Sihn, W., & Jäger, A. (Eds.). (2012). *2nd conference on learning factories—Competitive production in Europe through education and training*.

Tisch, M. (2018). Modellbasierte Methodik zur kompetenzorientierten Gestaltung von Lernfabriken für die schlanke Produktion. *Dissertation, Darmstadt*. Aachen: Shaker.

Tisch, M., Hertle, C., Abele, E., Metternich, J., & Tenberg, R. (2015a). Learning factory design: A competency-oriented approach integrating three design levels. *International Journal of Computer Integrated Manufacturing, 29*(12), 1355–1375. https://doi.org/10.1080/0951192X.2015.1033017.

Tisch, M., & Metternich, J. (2017). Potentials and limits of learning factories in research, innovation transfer, education, and training. In *7th CIRP-Sponsored Conference on Learning Factories. Procedia Manufacturing* (In Press).

Tisch, M., Ranz, F., Abele, E., Metternich, J., & Hummel, V. (2015). Learning factory morphology: Study on form and structure of an innovative learning approach in the manufacturing domain. *TOJET, July 2015* (Special Issue 2 for International Conference on New Horizons in Education 2015), 356–363.

Umeda, Y., Takata, S., Kimura, F., Tomiyama, T., Sutherland, J. W., Kara, S., et al. (2012). Toward integrated product and process life cycle planning—An environmental perspective. *CIRP Annals—Manufacturing Technology, 61*, 681–702.

Wagner, U., AlGeddawy, T., ElMaraghy, H. A., & Müller, E. (2014). Product family design for changeable learning factories. In *47th CIRP Conference on Manufacturing Systems. Procedia CIRP, 17*, 195–200. https://doi.org/10.1016/j.procir.2014.01.119.

Wagner, U., AlGeddawy, T., ElMaraghy, H. A., & Müller, E. (2015). Developing products for changeable learning factories. *CIRP Journal of Manufacturing Science and Technology, 9*, 146–158.

Wank, A., Adolph, S., Anokhin, O., Arndt, A., Anderl, R., & Metternich, J. (2016). Using a learning factory approach to transfer industrie 4.0 approaches to small- and medium-sized enterprises. In

References

6th CIRP-Sponsored Conference on Leanring Factories. *Procedia CIRP, 54*, 89–94. https://doi.org/10.1016/j.procir.2016.05.068.

Werz, F. (2012). Excellent qualified and trained employees: The key for successful implementation of lean production. In W. Sihn & A. Jäger (Eds.), *2nd Conference on Learning Factories—Competitive Production in Europe Through Education and Training* (pp. 106–123).

Westkämper, E. (2006). *Einführung in die Organisation der Produktion. Springer-Lehrbuch*. Berlin, Heidelberg: Springer.

Westkämper, E., Constantinescu, C., & Hummel, V. (2006). New paradigm in manufacturing engineering: Factory life cycle. *Production Engineering, 13*(1), 143–146.

Wiendahl, H.-P., ElMaraghy, H. A., Nyhuis, P., Zäh, M. F., Wiendahl, H.-H., Duffie, N., et al. (2007). Changeable manufacturing—Classification, design and operation. *CIRP Annals—Manufacturing Technology, 56*(2), 783–809. https://doi.org/10.1016/j.cirp.2007.10.003.

Wiendahl, H.-P., Reichardt, J., & Nyhuis, P. (2009). *Handbuch Fabrikplanung: Konzept, Gestaltung und Umsetzung wandlungsfähiger Produktionsstätten*. München, Wien: Carl Hanser.

Zinn, B. (2014). Lernen in aufwendigen technischen Real-Lernumgebungen. *Die berufsbildende Schule, 66*(1), 23–26.

Chapter 6
The Life Cycle of Learning Factories for Competency Development

In recent years, a variety of models and other approaches in the field of learning factories has been developed and published. Existing approaches to the design, operation, and implementation of the learning factory concept are broken down into the following categories[1]:

- Description of specific learning factories or learning modules (use cases),
- Description models of the learning factory system,[2]
- Methods for the design of learning factories (general and specific),[3]
- Methods for the design of learning modules,[4]
- Didactic-medial design of learning arrangements,[5]
- Success measurements as well as methods for success measurement and evaluation.[6]

This chapter deals with planning, designing, evaluating, and improving competency development in learning factories referring among others to the approaches and models named above. The structure of the chapter can be explained along the learning factory life cycle; see also Fig. 6.1.

---

[1] See Tisch (2018).
[2] See, e.g. Steffen, Frye, and Deuse (2013b), Tisch, Ranz, Abele, Metternich, and Hummel (2015).
[3] See, e.g., Reiner (2009), Tisch, Hertle, Abele, Metternich, and Tenberg (2015), Kaluza et al. (2015).
[4] See, e.g., Enke, Kraft, and Metternich (2015), Abele et al. (2015), Plorin (2016).
[5] See, e.g., Pittschellis (2015), Tvenge, Martinsen, and Kolla (2016).
[6] See, e.g., Cachay, Wennemer, Abele, and Tenberg (2012), Tisch, Hertle, Metternich, and Abele (2014), Tisch, Hertle, Metternich, and Abele (2015), Enke, Glass, and Metternich (2017).

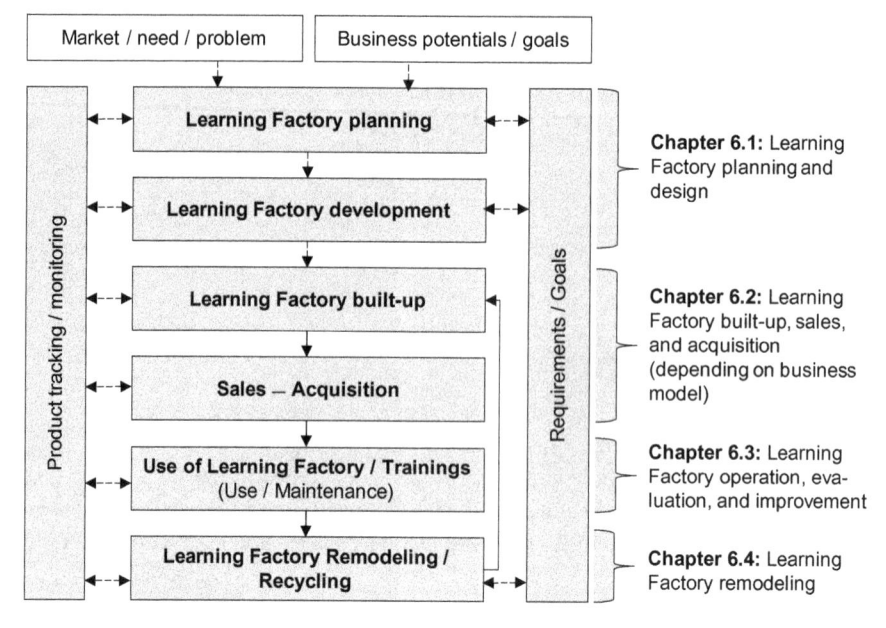

Fig. 6.1 Learning factory life cycle (Tisch & Metternich, 2017)

6.1 Learning Factory Planning and Design

6.1.1 Overview Planning and Design Approaches

In the design-oriented approaches, learning factory systems (or subsystems) are developed from two perspectives, while a combination of the two perspectives is also recognized:

1. From a factory perspective: Learning factories can be interpreted as an idealized representation of real production environments.[7] A variety of design approaches takes up this perspective.[8] Only the smaller part of these approaches focus exclusively on this perspective.[9]

---

[7] Tisch, Hertle, Abele et al. (2015).

[8] see, e.g., Tisch, Hertle, Abele et al. (2015), Plorin (2016), Dinkelmann (2016), Kaluza et al. (2015), or Rentzos, Doukas, Mavrikios, Mourtzis, and Chryssolouris (2014).

[9] as representatives of this Kaluza et al. (2015), Gebbe et al. (2015), and Rentzos et al. (2014) can be named.

6.1 Learning Factory Planning and Design

2. From a learning perspective: Learning factories can be interpreted as a complex learning environment.[10] A variety of approaches can be identified that design learning factories as complex learning environments.[11]
3. From the integrated learning and factory perspective: In the combination of the perspectives 1 and 2, learning factories are seen as learning environments targeting issues of socio-technical systems while integrating a model of a socio-technical system. For the design and configuration of the entire learning factory concept (including the factory environment, the learning modules, and the learning situations), didactical, social, and technological implications have to be kept in mind.[12]

In recent years, several linear, sequential approaches have been proposed, which are similar to product life phases of well-known VDI guidelines.[13] Reiner (2009) describes the integration of the learning factory concept into a lean transformation and uses a generic three-step approach for learning factory design: "requirements," "development and setup," and "operation and use." Likewise, Doch, Merkler, Straube, and Roy (2015) use similar phases: "Analysis of requirements," "conceptualization," and "design and implementation." Riffelmacher (2013) describes a learning factory concept for the high-mix assembly using a physical as well as a virtual learning environment, while based on this specific development, no general design process for learning factories is derived. Dinkelmann (2016) describes a method for the participation of employees in the change management of multi-variant series production using learning factories. Within this change management approach, problems that are actually occurring in industrial companies are mapped in learning factories, where they are resolved by employees with the aim of transferring motivation and solution ideas into the daily work environment. Plorin (2016) describes a design approach based on a reference model, consisting of structure, design, integration, and quality model, for the realization of learning factories for factory planning and factory operation. The comprehensive approach addresses a broad field of applications, which particularly describes the didactic-medial arrangement of the learning arrangements and the design of individual learning modules. With regard to the design of the learning environment, the model particularly addresses a learning module-induced adaptation of existing environments.

In the approach followed in Darmstadt, the intended competencies crucially influence the learning factory design process, since domain-specific competency development is in general seen as key objective of learning factories.[14] The competency-

---

[10] Tisch, Hertle, Abele et al. (2015).

[11] See, e.g., Reiner (2009), Riffelmacher (2013), Tisch et al. (2013), Tisch, Hertle, Abele et al. (2015), Abele et al. (2015), Plorin (2016).

[12] Tisch, Hertle, Abele et al. (2015).

[13] See, e.g., VDI (1993).

[14] See Tisch, Hertle, Abele et al. (2015), Reiner (2009), Abele, Tenberg, Wennemer, and Cachay (2010), Abel, Czajkowski, Faatz, Metternich, and Tenberg (2013), Tisch et al. (2013), Steffen, Frye, and Deuse (2013a), Jäger, Mayrhofer, Kuhlang, Matyas, and Sihn (2013), Matt, Rauch, and Dallasega (2014).

oriented learning factory design follows a holistic approach in two major transformations on three design levels.[15] The Darmstadt learning factory design approach is presented in Sect. 6.1.2 in detail.

Other approaches focus on parts of the learning factory design. Wagner, AlGeddawy, ElMaraghy, and Müller (2014) and Wagner, AlGeddawy, ElMaraghy, and Müller (2015) describe a product development approach for learning factories, arguing that, compared to traditional product development, learning factory products must be designed to complement the learning factory concept. In general, two ways can be identified in the literature to the appropriate learning factory product[16]:

- Industrial products that are available on the market are selected (and optionally didactically modified) with the goal of completing the learning factory configuration or
- Learning factory products that are specifically developed for fitting into the learning factory concept.

In the context of the selection of learning factory products, it is important to consider that decisions on the mapped processes and the manufactured product are interwoven with each other. A decision regarding the product always affects needed or possible production processes and vice versa. The interactions between the factory and the product life cycle visualizes these interconnections; see Fig. 6.2.

Kaluza et al. (2015) describe a procedure for creating scaled-down learning environments using the example of energy efficiency. The focus is on the exact mapping of the technical system to ensure that the scaled-down environment has the same behavior or characteristics of the real factory environment.

Gebbe et al. (2015) mention that existing design approaches focus on learning factories used for education and training, while a systematic procedure for the design of demonstration and research factories is not available. Tvenge et al. (2016) propose opening up of the learning factory approach to a broader modern workplace learning framework[17] that enables continuous, autonomous social learning on demand and on-the-go. For an overview see Table 6.1.

6.1.2 The Darmstadt Approach to Competency-Oriented Planning and Design

The Darmstadt approach to learning factory design should address in particular the weak spots of previous learning factory design processes; namely, they are[18]:

---

[15] See, e.g., Tisch, Hertle, Abele et al. (2015).
[16] Metternich, Abele, and Tisch (2013), Wagner et al. (2014, 2015), Tisch, Hertle, Abele et al. (2015).
[17] See Hart (2015).
[18] See Tisch et al. (2013), Tisch (2018).

6.1 Learning Factory Planning and Design

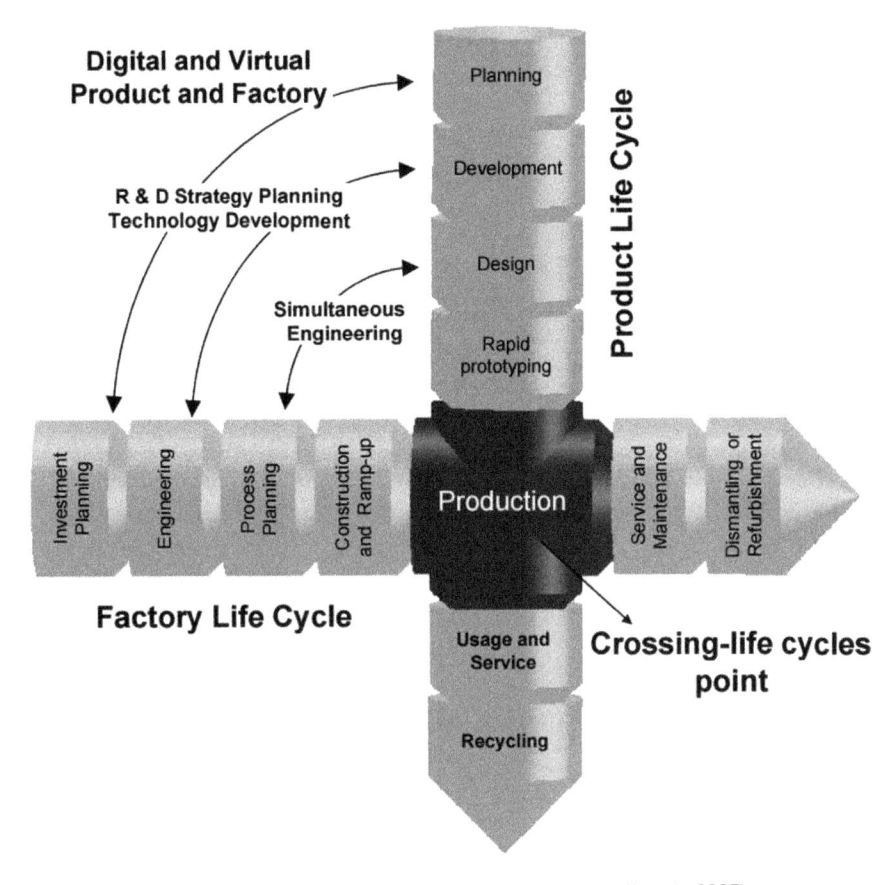

Fig. 6.2 Interconnection of product and factory life cycle (Pedrazzoli et al., 2007)

1. Learning factories are usually developed with a predominantly engineering background and neglect the didactic principles for designing learning systems.
2. So far, no detailed procedure model for the structured design of learning factories is followed, which means that every learning factory development is accompanied by large, sometimes unnecessary, redesign efforts.
3. Learning modules in learning factories are rarely designed to be goal- or competence-oriented, although competence development is generally recognized as the goal of learning factory activities.
4. In addition to the often undefined objectives, there are no practicable methods and procedures for checking the achievement of goals.
5. In many cases, the transfer of learning from learning factories to the real factory environment is hampered by a lack of user or target group orientation.

To tackle current problems of learning factory design, a competency-oriented approach was developed that designs learning factory systems on three design levels:

Table 6.1 Overview on learning factory design approaches (Abele et al., 2017)

Approach	Design object	Focus	Description
Reiner (2009)	Learning factory	Training	Generic three-step approach
Riffelmacher (2013)	Digital and physical learning environment	Training	Development of a learning environment for high-variant assembly systems. No design approach derived
Tisch et al. (2013), Tisch, Hertle, Abele et al. (2015), Tisch (2018)	Learning factory, learning modules, learning situations	Training, education	Holistic learning factory design approach on three conceptual levels
Wagner et al. (2014), (2015)	Learning factory product	Research, education	Product design for learning factories
Doch et al. (2015)	Learning factory	Training	Generic four-step approach
Kaluza et al. (2015)	Scaled-down learning environment	Equipment	Approach to develop scaled-down learning equipment for resource efficiency
Plorin, Jentsch, Hopf, and Müller (2015), Plorin (2016)	Adjustment of learning environment, learning modules	Training	Iterative approach for the adjustment of learning environments

the macro level, the meso level and the micro level; see Fig. 6.3. The design levels are based on structuring of other learning offers[19] and are used in order to structure the complex system learning factory in an appropriate and clear way. Following this, in the learning factory design at the design levels, different design objects are addressed[20]:

- The **macro level** focuses in general on the design of complete educational programs. Transferred to the learning factory concept, this level involves significantly more than traditional or new media-aided learning programs, as the "educational program" of a learning factory also manifests itself physically, particularly through a factory environment. The macro level of the learning factory model also includes the socio-technical learning factory infrastructure (including the quasi-real production environment and the production processes, the product and the employees), the content of the learning factory program to be designed, and a basic didactic concept. As part of the design process of the learning factory program, the overarching learning goals, target groups (or learners) as well as other stakeholders (trainers, operators) are to be included.

---
[19] See Seufert and Euler (2005).
[20] See Tisch (2018).

6.1 Learning Factory Planning and Design

```
1 – Macro level (Learning Factory): Design of the Learning Factory infrastructure including the production environment as well as fundamental parts of intended learning processes

2 – Meso level (teaching module): Design of teaching modules including the explication of specific sub-competencies and the definition of general teaching-learning sequences

3 – Micro level (learning situation): Design of specific teaching-learning situations
```

1st didactic transformation → 2nd didactic transformation →

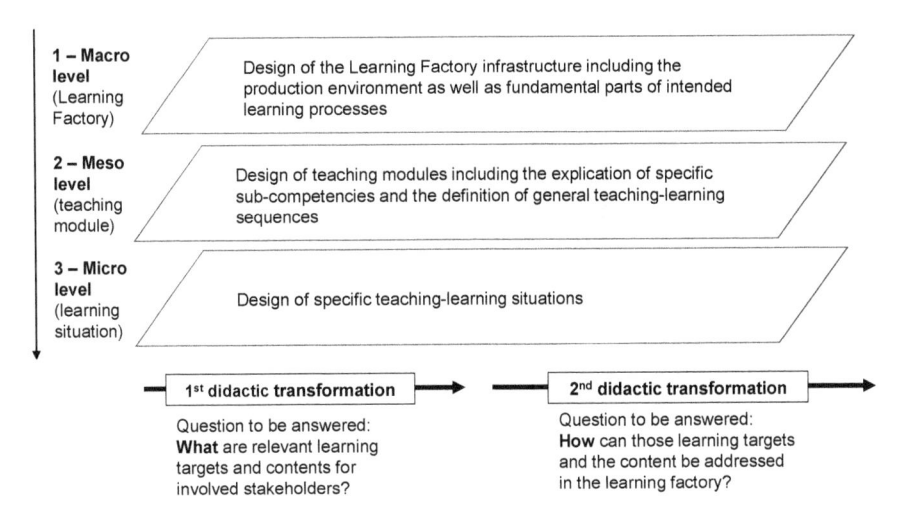

Question to be answered: **What** are relevant learning targets and contents for involved stakeholders?

Question to be answered: **How** can those learning targets and the content be addressed in the learning factory?

Fig. 6.3 Levels of learning factory design (Tisch, Hertle, Abele et al., 2015)

- In general, the **meso level** focuses on designing a teaching event or course module. Transferred to the terminology of learning factories, individual learning modules are designed on the meso level. At the learning module level, specific learning modules are located, which are part of the learning factory program (macro) and use the infrastructure defined there. Within a learning environment (macro), a large number of learning modules are usually located. However, the learning modules may not necessarily be interdependent. The design of a learning module includes not only the sequencing of learning processes but also the planning of the changeability of the socio-technical factory infrastructure.
- At the **micro level**, in general learning scenarios and learning resources are designed. The micro level of learning factories accordingly consists of specific learning situations that are part of a learning module. Learning situations can have exploratory or experimental as well as systematizing or reflective character.[21] Learning situations within a learning module have strong dependencies. Thus, they have to be adjusted to the learning objectives of the learning module level as well as to other situations within the learning module. The design of the learning situations includes both the didactic-methodical design of the individual phases and the preparation of the factory environment used as a learning support, learning material, learning medium, and learning product in one.[22]

The procedure for designing the learning factories can be subdivided into two didactic transformations within the individual levels. The first transformation relates to the derivation of curricula, the second transformation to their conceptual implementation. The first didactic transformation includes the determination of compe-

---

[21] See Abele et al. (2015).
[22] See Footnote 21.

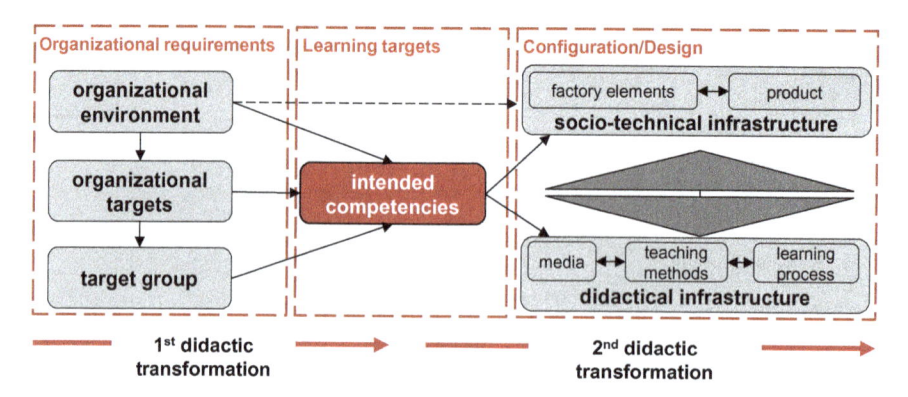

Fig. 6.4 Conceptual relationships in the two didactic transformations in learning factory design (Tisch, Hertle, Abele et al., 2015)

tency requirements from an overall operational and technical situation. Those competency requirements are highly individualized with regard to the target industry, the organizational targets, and target groups. These competency requirements must be formulated as learning goals, so that they can then be conceptually implemented in the second transformation. In anticipation of specific learning processes, learning methods are established which accentuate the relevant products and production processes in such a way that a purposeful alternation of action and understanding is triggered. The conceptual relationships inside the two didactic transformations are visualized in Fig. 6.4. In the following sections, the approach on the individual design levels is sketched.[23]

6.1.2.1 Learning Factory Design on the Macro Level

On the macro level, the general didactic-methodological concept and the socio-technical learning environment as well as the overall learning factory program are created. Figure 6.5 gives an overview over the design steps on macro level.[24]

In course of the **general target and framework definition**, requirements, targets, and conditions are defined. First, learning factory-specific framework conditions, company goals, and organizational requirements are identified in connection with the learning system. In the following steps, further relevant requirements are analyzed. The other non-competence-related requirement sources are the target work system, the target group, and the learning content.

The starting point for the definition of learning factory-specific framework conditions is the morphological description of the learning factory system presented in

---

[23] A detailed process model in BPMN 2.0 and a description of the design process on the three design levels can be found in Tisch (2018).

[24] A detailed description is given in Tisch (2018).

6.1 Learning Factory Planning and Design

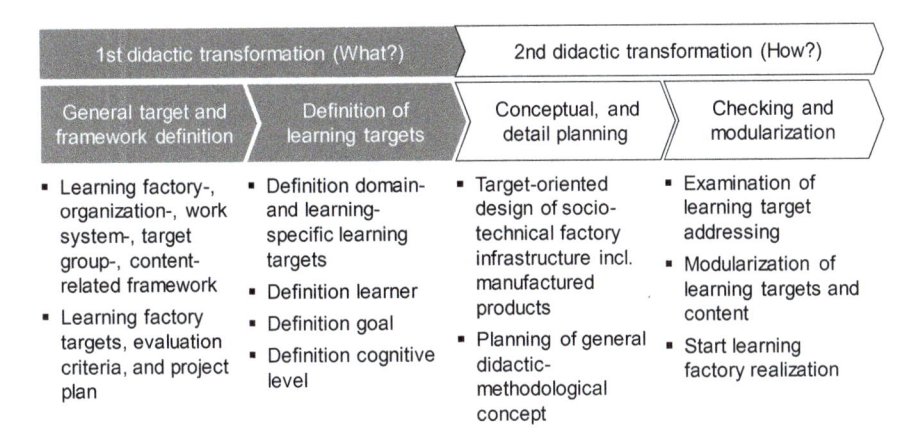

Fig. 6.5 Simplified learning factory design process on macro level according to Tisch (2018)

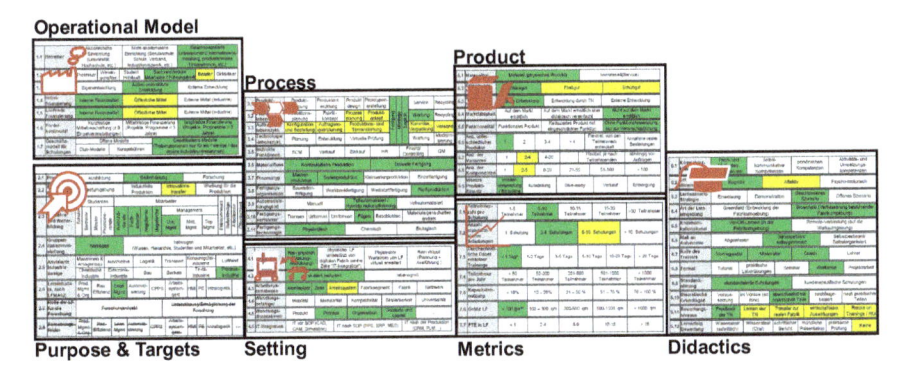

Fig. 6.6 Exemplary filled in the form of the definition of learning factory-specific requirements (Tisch, 2018)

Secs. 5.1–5.7. For this purpose, the morphology is converted into a standardized sheet that can be used in consultation with the relevant stakeholders to structure the organizational goals, requirements, and framework conditions for the learning factory design. In this context, it is discussed and defined which characteristic of individual design elements should be part of the system. In this case, a distinction is made between mandatory (green) and optional requirements (yellow) based on the practice in the preparation of specifications (Fig. 6.6).[25]

In the context of the **definition of learning targets**, the overarching definition of intended competences takes place. The learning objectives are closely related to the results of the previous section. The aim of this step is to identify and formulate the dispositions that are necessary for successful coping with relevant problem situations. The identification and formulation of the competences are strongly domain-

---
[25] See Tisch (2018).

	Addressing	Cognitive level	Content
Description	Reference to the learner and the targeted goal (often as an introductory phrase)	Description of the addressed cognitive level of the learning objective	Specific description of the learning target related content
Example	After participating in the learning module "Design of Shopfloor Management-Systems" the experts of lean production are able to ...	...to analyze, design and evaluate...	...shop floor management systems including related hardware systems as well as the underlying processes and routines.

Fig. 6.7 Components of the competency formulation (e.g. design of shop floor management systems)

and content-specific (especially technical and methodical ones). Components of the competency formulation include a reference to the learner and the intended goal, a specific description of the learning objective-related content as well as a description of the cognitive level of the learning goal; see Fig. 6.7. To assist in the description and definition of the cognitive level of the learning objective, the relevant key terms from relevant learning taxonomies[26] can be used.[27]

In the **conceptual and detail planning phase**, based on the in the previous step identified competencies, the general didactic-methodological concept as well as the socio-technical infrastructure is designed. Here, all modeled manufacturing processes, the products used in the concept, as well as the general didactic-methodological framework are geared to the predefined competency targets. First, the structural planning determines the breadth of the life-cycle phases (horizontal) as well as the socio-technical scope on the individual factory levels (vertical); see Fig. 6.8. The typical scopes of today's learning factories are highlighted with a bold line.

Based on this, design alternatives of the different learning factory elements are generated, evaluated, and put together to one concept. The solution generation follows the top-down structural planning approach with a bottom-up approach. First of all, a variety of alternative solutions are derived for employee and station, then at cell, system, segment, and factory levels. Higher levels contain the subordinate levels, while only the factory levels defined as relevant in advance are considered. Solutions on subordinate levels are aggregated into coherent, fundamentally different (not only minor variations of concepts) alternative concepts. These aggregated alternatives include an understanding of the socio-technical-didactic system with respect to

- the different factory levels,
- their ability to change,

---

[26] Bloom, Engelhart, Furst, Hill, and Krathwohl (1956), Anderson, Krathwohl, and Airasian (2001).
[27] See inter alia Schermutzki (2007).

6.1 Learning Factory Planning and Design

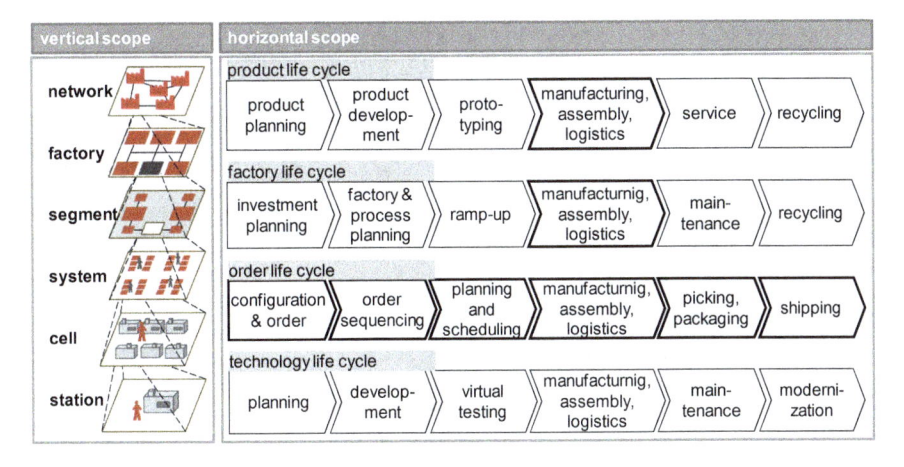

Fig. 6.8 Vertical and horizontal scope in the structural planning

- the product to be manufactured as well,
- the employees and teams acting in it.

At the respective factory levels, in particular the learning environment-related flexibility and changeability with regard to the desired contents and learning objectives are to be considered, so that at least one initial and one target state is created; see Fig. 6.9. The need to change the learning environment arises from the coupling of the socio-technical and didactic systems. Flexibility and changeability are two concepts serving the planning or partial anticipation of learning processes in the learning environment. Due to the flexible design of the factory environment, the learners can vary the environment within a short time; this has a high relevance especially in connection with the use of closed learning scenarios. Beyond that, the changeability of the environment enables the implementation of solutions that have not been planned and prepared in advance giving the learners a much wider possibilities for interference; this form of changeability is used accordingly in (partially) open scenarios. In the context of the anticipated changes, the state-based design of the factory levels should not lead to obvious or even trivial problems. Here, the use of ill-structured problems can help. Those are the predominant type of problems in industrial practice.[28]

A detailed procedure, which defines furthermore (a) whether the product or the process or both should be defined at the same time and (b) how alternatives are evaluated and selected is shown in Tisch (2018).

After selecting the alternative, in the **checking and modularization** phase it should be critically examined whether the chosen alternative is able to address the defined intended competences and whether it meets the underlying requirements and objectives. If the test comes to a positive result, the preparation and the monitoring of the learning factory realization follow. If an addressing of defined targets is not

---

[28] See Tisch (2018).

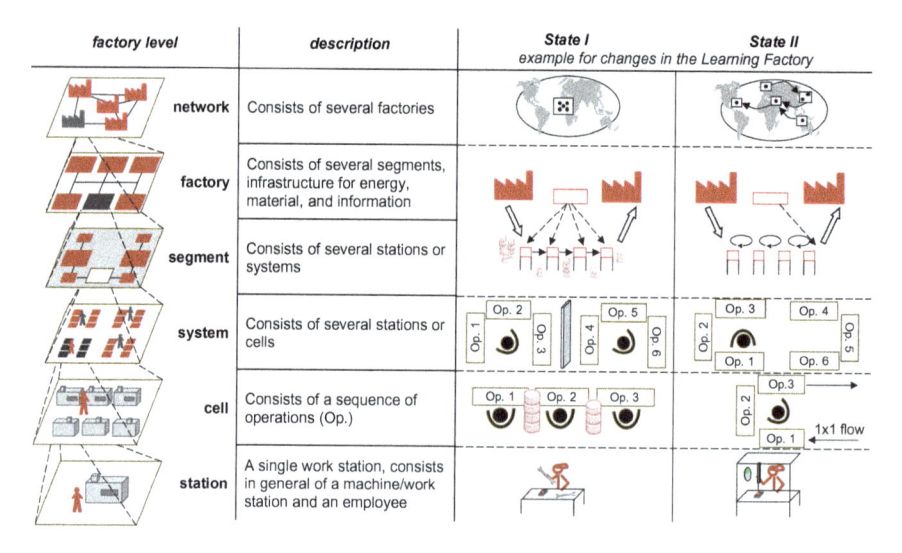

factory level	description	State I	State II
		example for changes in the Learning Factory	
network	Consists of several factories		
factory	Consists of several segments, infrastructure for energy, material, and information		
segment	Consists of several stations or systems		
system	Consists of several stations or cells	Op. 1, Op. 2, Op. 3, Op. 4, Op. 5, Op. 6	Op. 1, Op. 2, Op. 3, Op. 4, Op. 5, Op. 6
cell	Consists of a sequence of operations (Op.)	Op. 1, Op. 2, Op. 3	Op. 1, Op. 2, Op.3 — 1x1 flow
station	A single work station, consists in general of a machine/work station and an employee		

Fig. 6.9 Use of initial and target states on different factory levels in course of the learning factory design (Tisch, Hertle, Abele et al., 2015)

possible adequately, an extension or adaptation of the learning factory concept follows. In the context of the modularization, the higher-level learning objectives are grouped into modules in a way that

- individual modules can be meaningfully addressed in the course of separate, time-limited events, and
- the modules are posed together in an explainable (content-related or action-related) relationship.

The sum of all modules forms the learning factory program, which is not to be regarded as rigid but continuously expandable.

6.1.2.2 Learning Factory Design on the Meso Level

On the meso level, for the modularized learning targets and content respective learning modules are created. For this purpose, in the first didactic transformation on meso level, it is specified what should be learned in the module, and in the second didactic transformation, it is defined how the sequence and the learning environment of the learning module look like (Fig. 6.10).

The procedure at meso level is performed once per learning module. In the first step, the **requirements and the learning module framework** are defined. On the one hand, general conditions and requirements of the learning module are defined and mutually dependent, and on the other hand, the roles addressed in the learning module are identified and described. The procedure for identifying and describing the roles is content-specific. Exemplarily, roles in the field of the shop floor management

6.1 Learning Factory Planning and Design

1st didactic transformation (What?)		2nd didactic transformation (How?)
Requirements analysis and learning module framework	Operationalization of learning targets	Technical-methodical design of learning module
• Identification and description of relevant roles and activities • Definition of learning module framework conditions	• Derivation of sub-competencies • Performance- and knowledge-oriented operationalization of learning targets via competency transformation	• Planning of learning sequences of the learning module • Detailing, adapting and expanding the socio-technical infrastructure

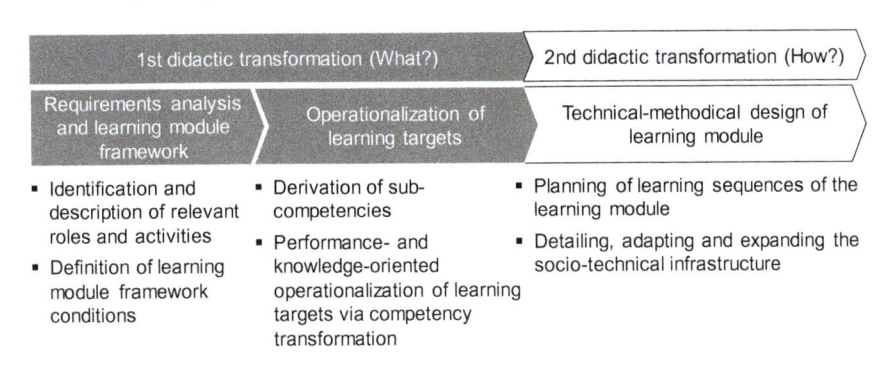

Fig. 6.10 Simplified learning factory design process on meso level according to Tisch (2018)

Checklist for the learning module framework

- What is the title of the planned learning module?
- What is the content of the learning module? (Short description: 2-3 sentences)
- Who are the planned trainers of the module to be developed? (see also learning factory morphology 2.3)
- How many coaches are available?
- Who is the target group of the learning module to be developed? (see also morphology 2.3)
- Is this target group homogeneous or heterogeneous? With regard to which features? (if heterogeneous, see also morphology 2.4)
- For how many participants should the training be designed? (see also morphology 7.1)
- What is the duration of the learning module to be developed? (see also morphology 7.3)
- Other special features (for the learning environment, setting, etc.) which must be specified after defining the framework conditions at the macro level. (Description in bullet-points)

Fig. 6.11 Checklist for the learning module framework definition (Tisch, 2018)

topic as well as a role description for the "shop floor management expert" are shown in Fig. 6.12. In this description, tasks and activities with particular connection to the learning content of the respective roles are described. For the general definition of the framework conditions at the learning module level, a checklist for learning module framework definition presented in Fig. 6.11 can be used.

For the **operationalization of learning targets**, a competency transformation table can be used to operationalize identified competencies based on the requirements identified in the previous step using related knowledge elements as well as competency-related performances; see Fig. 6.13. Within this competence transformation, the overarching competencies are divided in subcompetencies in order to

Description

The shop floor management expert supports the introduction of shop floor management. The expert is involved in designing the instruments within the shop-floor-management-system.

Activities in the context of shop floor management

The shop floor management experte...

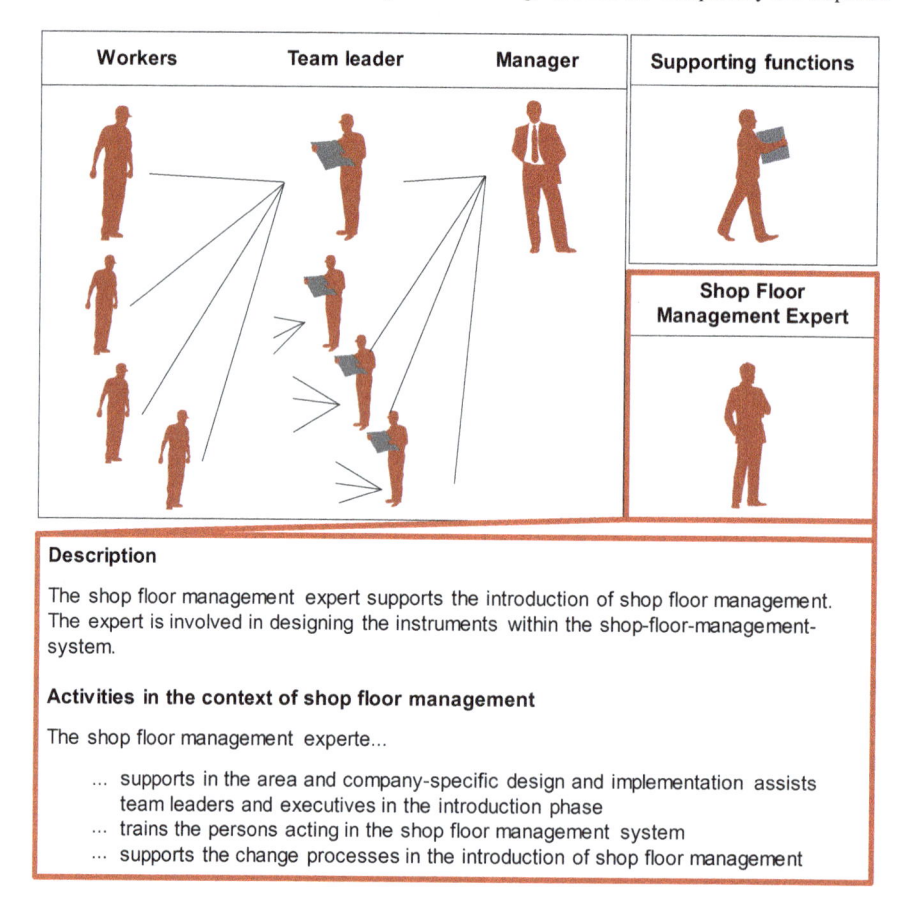

- ... supports in the area and company-specific design and implementation assists team leaders and executives in the introduction phase
- ... trains the persons acting in the shop floor management system
- ... supports the change processes in the introduction of shop floor management

Fig. 6.12 Roles in the shop floor management; tasks and activities of a shop floor management expert

structure the learning targets for the learning module. Correspondingly, competencies and subcompetencies are formulated and defined in this step. Furthermore, corresponding performances of the learners as well as required knowledge elements are related to the listed learning targets. In general, knowledge elements can be divided in professional knowledge[29] and conceptual knowledge.[30] Both knowledge types have to be considered.

Depending on whether this instrument is used for the design of learning modules or for the analysis in course of a redesign of existing learning modules, a different approach is proposed:

- For the design of new learning modules, the intended competences are leading the approach. Based on the competencies, it is concluded which performances within

---

[29] Knowledge that addresses mainly questions regarding the "what?" and the "how?".

[30] Knowledge that entails answers to "why?" questions.

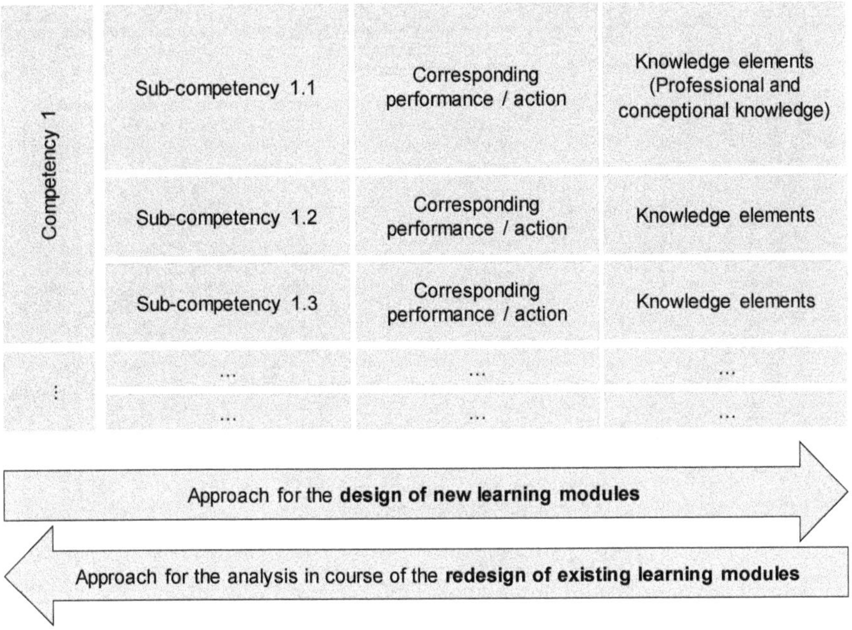

6.1 Learning Factory Planning and Design

Fig. 6.13 Structure of the competency transformation for the design and redesign of learning modules according to Tisch et al. (2013)

the learning module the participants carry out; subsequently, required knowledge aspects are derived. In the generation of the competency transformation, subactions are related to differentiated knowledge elements in order to describe the intended competences.

- In order to redesign existing learning modules, firstly, relevant content is identified through the analysis of learning materials by assigning the contents to the various knowledge aspects. Subsequently, specific performances are linked to the respective professional knowledge and the corresponding conceptual knowledge elements. Based on those assignments, competency-oriented learning targets can be derived. Based on the resulting current state of the learning module, the competency transformation can be reworked.

Using the competency transformation for existing learning modules, various problems or improvement potentials of intuitively designed learning modules can be discovered and recorded.[31] Problems and potentials of existing learning modules are mainly in the following areas:

- Missing knowledge aspects can be supplemented.
- Recorded knowledge aspects to which no performance is assigned can be identified as inert knowledge.

---

[31] A use case for the use of the competency transformation for existing learning modules is given in Abele et al. (2015).

Competency	Sub-competencies	Action	Knowledge base
The participants have the ability to explain the methods and tools for the implementation of Jidoka* and for the solution of problems and to apply selected methods and tools.	...	...	...
	Ability to develop an Andon-concept for Production	Design of an Andon-system (physical implementation)	Knowledge, that visual and acoustical signals and an Andonboard are needed; knowledge of the examined workplaces; knowledge of the functionality of Andon; knowledge of the meaning of the colors
		Planning of an escalation process for the problem escalation with Andon	Knowledge of the person in charge and of the available time; knowledge of the theoretical sequence of an escalation process (point in time for information, order of notification)
	...	...	...

Fig. 6.14 Extract from the competency transformation chart of the learning module "Quality techniques of Lean Production" (Enke, Tisch, & Metternich, 2016)

- Content that existed at different points in the training without any new depth of knowledge or additional performance can be classified as unnecessary redundant content.
- Ensuring the competence-oriented definition of learning objectives by consistently linking knowledge aspects and corresponding performances.

Figure 6.14 shows an extract of the competency transformation table of the learning module "Quality techniques of Lean Production" offered in the Process Learning Factory CiP.[32] This competency transformation table is the starting point for the **technical–methodical design of the learning module**. The main intention of this learning module is the ability of reflective application of Jidoka methods and tools, which is the part of the Toyota Production System that deals with prevention and elimination of rework and defects. The extract from the competency transformation table shows the subcompetency that addresses the ability of the learners to develop an Andon concept in production systems that includes a stop of production processes, an alert, and conditions for escalation in case of an abnormal conditions. The assigned actions are addressed in the practical exploration and testing phases of the learning module[33]:

- Exploration Activities: Exploration activities are directly related to the learners' professional challenges. In this phase, the learners discover new content and problem situations in an acting manner. Exploration activities require a subsequent systematization; see Fig. 6.15.
- Testing Activities: Testing activities are as well directly related to the learners' professional challenges. In this phase, learners test already learned content and

---

[32] See Enke et al. (2016).

[33] See Abele et al. (2015), Tisch (2018).

6.1 Learning Factory Planning and Design

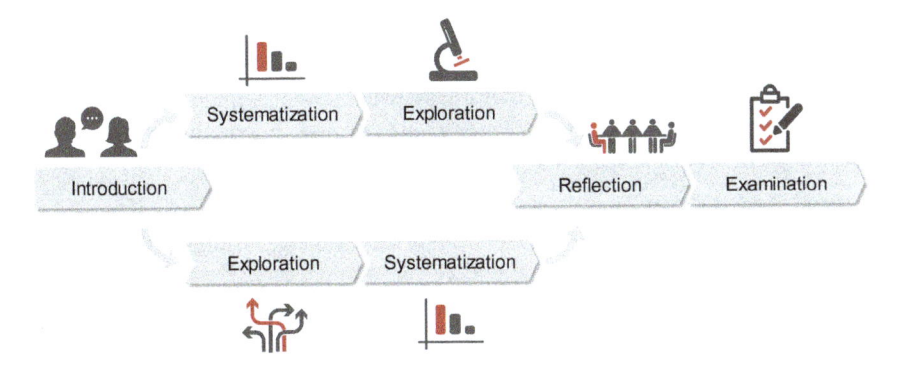

Fig. 6.15 Possible sequences of activities, according to Abele et al. (2015)

problems in an acting manner. In contrast to the exploration activity, no new areas are discovered, but already-known content is applied. This means, a problem or task is solved following given specifications, for example, a method or an approach. Testing activities require systematization beforehand. A lack of activity-oriented phases (exploration and testing activities) leads to a low contextualization and impractical learning modules.

In the quality technique use case, the competency transformation table assigns two related actions to the subcompetency that can be performed in the factory environment of the Process Learning Factory CiP:

- Designing the physical Andon system: Here, a problem scenario with a missing Andon concept in the assembly department of the CiP is used.
- Planning the escalation process for problem escalation part of the Andon concept: Here, the learners plan the escalation process for the beforehand designed physical Andon system in the assembly department.

Furthermore, Fig. 6.14 shows the required professional and conceptual knowledge base that is mostly addressed in systematization and reflection phases of the learning module.[34] Systematization and reflection activities are[35]:

- Systematization Activities: The systematization activity is closely related to the technical or scientific basis of the content. Existing knowledge systems of the learners are activated, checked, supplemented, extended, and corrected on the basis of objectified knowledge. The systematization concretizes and stabilizes the action-related knowledge gained. A lack of these activities in the learning module does not meet the requirement of professionalism and leads to actionist learning modules.

---

[34] Further information on the creation of the competency transformation table for this learning module is given in Enke et al. (2015).
[35] See Abele et al. (2015), Tisch (2018).

- Reflection Activities: Reflection is an important part of the sequences. The learner is in the focus during the reflection phases. Here, the learners themselves determine if they succeeded or not as well as which content might not have been understood. Furthermore, the reflection phases provide important hints about the learning effects of the sequence.

Figure 6.15 shows the most used sequences in learning factory learning modules: Those sequences are using either exploration-based or the testing-based strategy. After the reflection phase of both sequences an examination of the trainer regarding the effectiveness of the learning phases.

In addition to these sequencing strategies, reflection-based strategies may also be used in some cases. Reflection-based strategies start with an introduction followed directly by a discussion and reflection making use of professional experience of the learners. After the initial reflection, this strategy goes on with Systematization, Testing, 2nd Reflection or Exploration, Systematization, 2nd Reflection. Figure 6.16 shows an overview on the sequence and the application fields of the mentioned strategies.

Figure 6.17 exemplarily shows the sequence of activities used for the subcompetency "Ability to develop an Andon concept for production" of the learning module

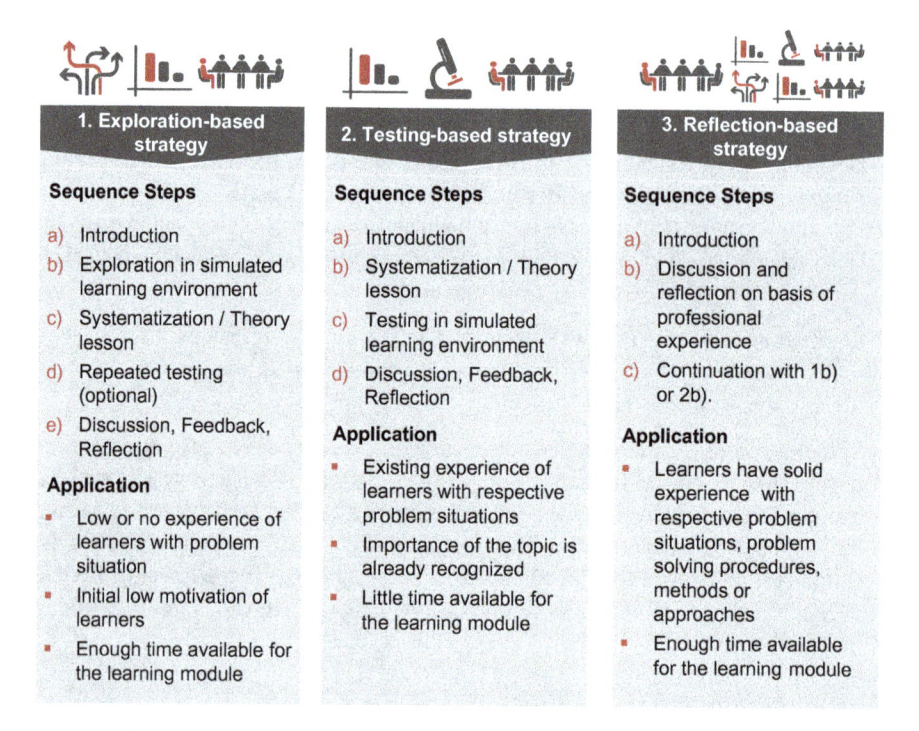

1. Exploration-based strategy	2. Testing-based strategy	3. Reflection-based strategy
Sequence Steps	**Sequence Steps**	**Sequence Steps**
a) Introduction	a) Introduction	a) Introduction
b) Exploration in simulated learning environment	b) Systematization / Theory lesson	b) Discussion and reflection on basis of professional experience
c) Systematization / Theory lesson	c) Testing in simulated learning environment	c) Continuation with 1b) or 2b).
d) Repeated testing (optional)	d) Discussion, Feedback, Reflection	
e) Discussion, Feedback, Reflection	**Application**	**Application**
Application	• Existing experience of learners with respective problem situations	• Learners have solid experience with respective problem situations, problem solving procedures, methods or approaches
• Low or no experience of learners with problem situation	• Importance of the topic is already recognized	
• Initial low motivation of learners	• Little time available for the learning module	
• Enough time available for the learning module		• Enough time available for the learning module

Fig. 6.16 Overview on sequence steps and application areas of sequencing strategies for learning factory modules (Tisch, 2018)

6.1 Learning Factory Planning and Design

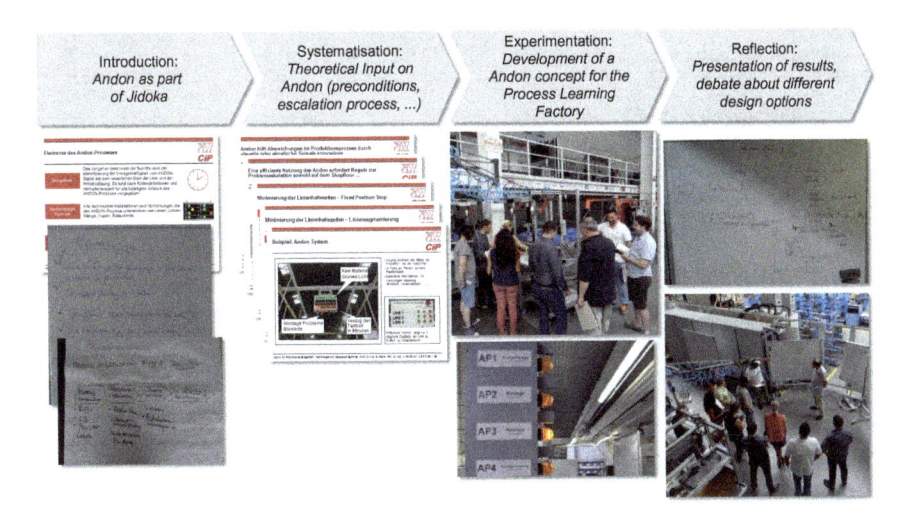

Fig. 6.17 Sequence of activities for the subcompetency "Ability to develop an Andon concept for production" (Enke et al., 2016)

"Quality techniques of Lean Production." The introduction takes the role to place the Andon concept in the overall concept of Jidoka. Based on this, the testing-based sequencing strategy is used, which is composed of theoretical input regarding the design of Andon systems, a development of the Andon concept in the Process Learning Factory, and a reflection phase that contains the presentation of group results and a discussion about results and the Andon concept in general.

Parallel to the illustrated sequencing of the learning modules, the socio-technical infrastructure, predefined at the macro level, is detailed, adapted, and expanded as required. For this purpose, in analogy to the procedure on macro level, requirements can be derived from the intended learning outcomes and transferred to concepts for extension. Likewise, a review of the addressability of competencies at the different factory levels is performed.[36]

6.1.2.3 Learning Factory Design on the Micro Level

On the micro level, the specific learning activities, which were sequenced on meso level, are created. Since the design of the learning situations takes place here, the second didactic transformation started at meso level is continued here at the micro level (Fig. 6.18).

The first step on the micro level is the **framework and target definition**. Framework conditions, requirements, and goals for the learning situations at the micro level have for the most part already been defined at the higher levels. At the micro

---
[36]See Tisch (2018).

```
┌─────────────────────────────────────────────────────────────┐
│              2nd didactic transformation (How?)              │
├──────────────────────────────────┬──────────────────────────┤
│  Framework and target definition │ Design of learning situations │
└──────────────────────────────────┴──────────────────────────┘
```

- Specification of framework, requirements and targets of learning situations
- Specific planning of the duration of learning situations

- Design of exploration and testing activities
- Design of systematization activities
- Design of reflection phases

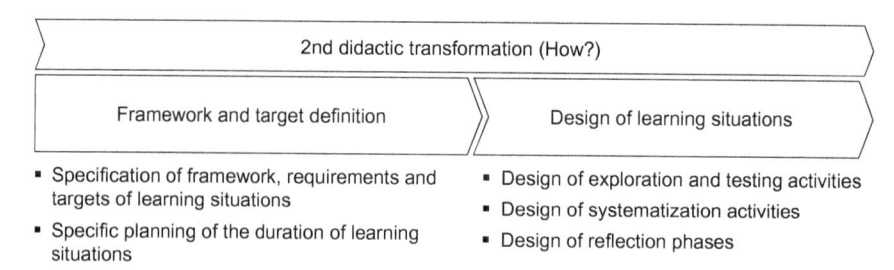

Fig. 6.18 Simplified learning factory design process on micro level according to Tisch (2018)

level, it is now necessary to distribute the defined time capacities for the individual sequences to the learning situations and to specify the individual components of the sequence. For this purpose, the following rules can be applied[37]:

- Output orientation instead of input orientation: The learning situations are adapted to the predefined learning targets. In this context, the question arises as to how learning situations have to be arranged in order to achieve the intended learning outcomes (output orientation). In contrast, the question which topics from the content-related field could also be addressed should not be the leading question (input orientation).
- Practical orientation instead of theoretical orientation: In particular, a main emphasis on systematization activities leads to input orientation. Theoretical units should therefore be planned as long as necessary, but also as compact as possible. Theory units should only contain relevant knowledge elements and avoid unwanted redundancy. Testing, development, and reflection elements should outweigh the systematization activities in terms of duration.

Consequently, at the center of the learning factory approach are the practice-related phases. Accordingly, regarding the **design of individual learning situations** of a sequence, first, exploration and testing activities are created, followed by the systematization activities and reflection elements.[38] Finally, implemented learning modules can also be reviewed and improved with a competency-based evaluation approach. The evaluation approaches are discussed in Sect. 6.3.3.

6.2 Learning Factory Built-up, Sales, and Acquisition

The content of the phases "learning factory built-up" and "Sales – Acquisition" in the learning factory life-cycle are strongly dependent on different business models that are pursued in the context of the learning factories. Three different types of

---

[37] Abele et al. (2015).

[38] A detailed description and guideline how those learning situations can be designed based on the competency transformation can be found in Abele et al. (2015) and Tisch (2018).

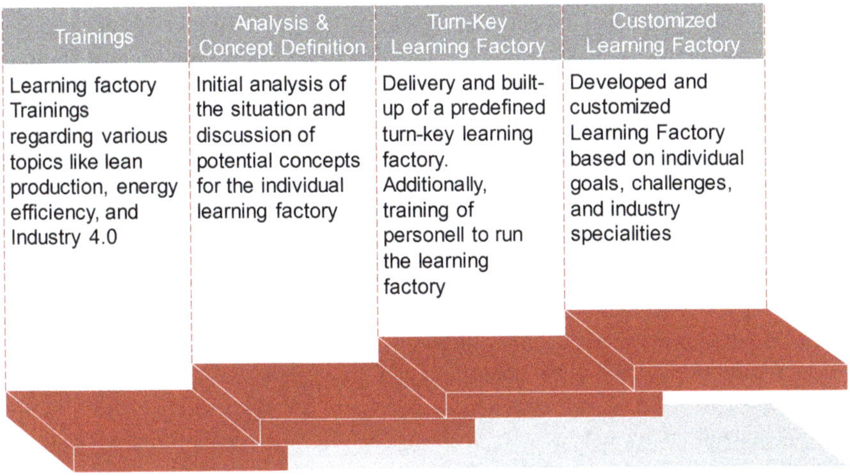

6.2 Learning Factory Built-up, Sales, and Acquisition

Trainings	Analysis & Concept Definition	Turn-Key Learning Factory	Customized Learning Factory
Learning factory Trainings regarding various topics like lean production, energy efficiency, and Industry 4.0	Initial analysis of the situation and discussion of potential concepts for the individual learning factory	Delivery and built-up of a predefined turn-key learning factory. Additionally, training of personell to run the learning factory	Developed and customized Learning Factory based on individual goals, challenges, and industry specialities

Fig. 6.19 Business models with learning factory offers for industry

business models are identified that can be seen in connection with the built-up, sales, and acquisition phases of learning factories. Those business models are[39]:

- The design and the built-up of customer-individual learning factories,
- The built-up of standardized turnkey learning factories,
- The offer of learning factory trainings for industry.

A fourth business model with "The auditioning and certification of existing learning factories" is regarded in the next section "Learning factory operation, evaluation, and improvement." In the following, the three above-mentioned business models and a downsized variation of the first business model with "Analysis and Concept definition" are described more in detail with regard to learning factory built-up, sales, and acquisition. Figure 6.19 shows an overview over the four described business models.

6.2.1 Design and Built-up of Customer-Individual Learning Factories

High qualification of personnel in different areas of production can be a decisive competitive advantage. Learning factories are a suitable and technology-adequate form of learning for this, but their intuitive design and built-up in course of pilot projects are complex and only promise limited success. Learning factories have to be designed and built-up geared to the specific requirements of individual companies. Here, the models and methods for systematic design of learning factory systems are

---
[39] According to Abele et al. (2015).

helpful.[40] These learning factories are developed on the basis of current technical didactic findings and are systematically adapted to the individual requirements of the operators.[41] The resulting learning factories are thus more effective regarding the achievement of learning objectives. Questions that have to be addressed in course of the company-specific learning factory design are described in Sect. 6.1 of this chapter. Figure 6.20 exemplarily shows some of the most important issues related to the customer-specific design of learning factories along the dimensions of learning factory systems.

Both universities and private companies[42] could appear as developers and operators of these systematic designed learning factories. Learning factories, including technical, structural, and didactical questions, can be designed according to individual requirements, and consequently, target groups can be trained according to their needs. However, developers and operators do not necessarily have to be the same organization—in recent years, learning factories have been often planned and implemented by external learning factory experts and later used by other operating organizations. Here, the scope of the developer's performance can be individually adapted to the framework conditions of the project. Projects range from simple use of design guidelines and methods over a competency-oriented revision of existing learning factories to a comprehensive development of the individual learning factories with preparation for the constructional implementation; see also Fig. 6.21.

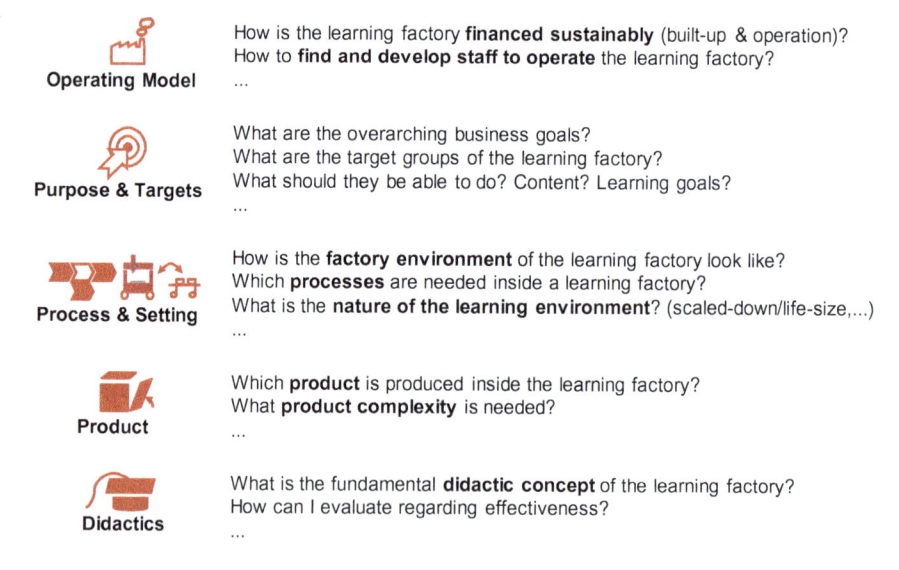

Operating Model
How is the learning factory **financed sustainably** (built-up & operation)?
How to **find and develop staff to operate** the learning factory?
...

Purpose & Targets
What are the overarching business goals?
What are the target groups of the learning factory?
What should they be able to do? Content? Learning goals?
...

Process & Setting
How is the **factory environment** of the learning factory look like?
Which **processes** are needed inside a learning factory?
What is the **nature of the learning environment**? (scaled-down/life-size,...)
...

Product
Which **product** is produced inside the learning factory?
What **product complexity** is needed?
...

Didactics
What is the fundamental **didactic concept** of the learning factory?
How can I evaluate regarding effectiveness?
...

Fig. 6.20 Exemplary questions to deal with in the planning and design of company-specific learning factories

---

[40] Those models and methods are presented in Sect. 6.1.

[41] For example, industry, university, vocational schools.

[42] Or in some cases also others like vocational schools.

6.2 Learning Factory Built-up, Sales, and Acquisition

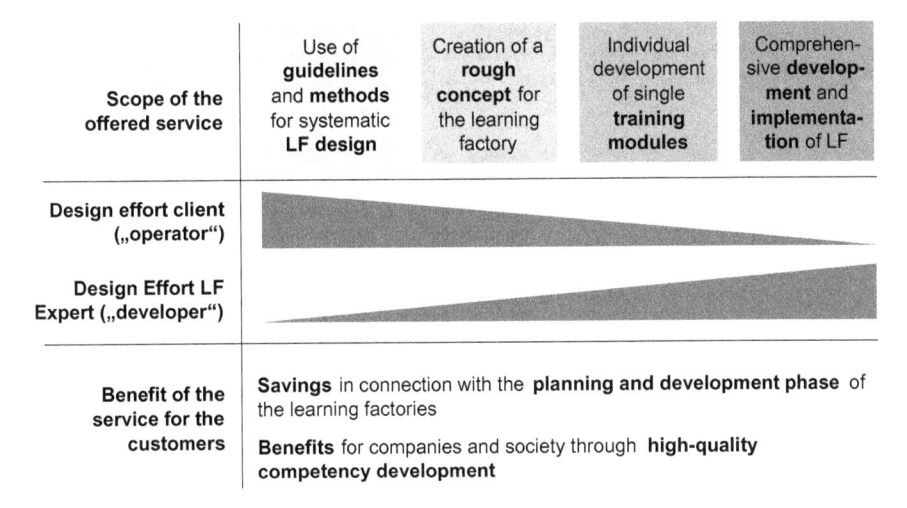

Scope of the offered service	Use of **guidelines** and **methods** for systematic **LF design**	Creation of a **rough concept** for the learning factory	Individual development of single **training modules**	Comprehensive **development** and **implementation** of LF
Design effort client („operator")	████████████████████████████████▶ (decreasing)			
Design Effort LF Expert („developer")	◀████████████████████████████████ (increasing)			
Benefit of the service for the customers	**Savings** in connection with the **planning and development phase** of the learning factories			
	Benefits for companies and society through **high-quality competency development**			

Fig. 6.21 Scope of performance related to the development and distribution of individualized learning factories (Abele et al., 2015)

Both the industry and the public sector are showing great interest in learning factories. Potential clients can accordingly belong to both sectors.

Industrial companies need technology-appropriate forms of learning in order to face the challenges of future production. First, learning factories will be of importance primarily to manufacturing industry. For example, in Germany alone, there are currently 205,028 manufacturing enterprises.[43] The share of large companies,[44] which should be of particular interest to learning factories, is 2.6%.[45] Consequently, there are at least around 5300 manufacturing companies for which an individually adapted learning factory would be interesting and relevant—in Germany alone.

Due to the great social relevance of learning factories, this form of learning is also interesting for the public sector. For universities and technical colleges, which are researching and educating close to production, a realistic form of learning could have positive effects on the learning success of the learners and the relevance of research results for industry. This would allow the students to make a greater contribution to strengthening the manufacturing sector immediately upon entering the labor market. Universities, technical colleges, and companies are involved as private or public clients in the development process. The architecture of the described value creation is shown in Fig. 6.22.

Due to the design of an optimal learning factory, which is tailored to the individual requirements, the client can avoid great efforts and a lengthy development period for planning and construction of the learning factory. This also leads to a reduction in

---
[43] See Statistisches Bundesamt (2017).
[44] More than 249 employees or more than € 50 million in annual turnover according to EU Recommendation 2003/361/EC.
[45] See Statistisches Bundesamt (2017).

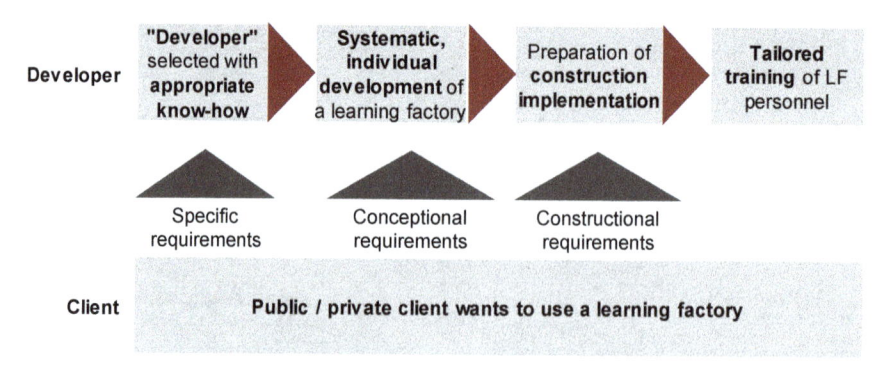

Fig. 6.22 Exemplary value-adding architecture for the design and construction of individualized learning factories (Abele et al., 2015)

development costs of the learning factory. Especially for industrial companies, an external development is attractive because an own development requires deployed highly qualified personnel on a large scale. In addition, with the systematic development the expected economic benefits of customers increase in terms of an effective and targeted competency development. Thus, the capacity of employees and thus productivity and flexibility of the industrial enterprise can be improved. In addition, with the help of the learning factories, technology, product, and process innovations can be prepared in a didactically meaningful way within a short amount of time and thus made accessible to the employees. Furthermore, public institutions also empower production workers to address production challenges through learning factories in a better way. A qualification improvement of the production personnel of today and tomorrow bears great economic potential. In the context of the business model "Design and built-up of customer-individual learning factories," income can be generated for learning factory developers for the systematic design of learning factories[46] as well as for the needs-based training of learning factory staff.[47]

6.2.2 Built-up of Standardized Turnkey Learning Factories

As already mentioned, the great benefits of customer-individual learning factories are opposed by considerable efforts for design and construction. Another business model is therefore the construction of standardized turnkey learning factories. In this case, an existing learning factory or a part thereof is replicated one-to-one at the customer, whereby the client is not or little included in the concept development phase. The correspondence between the replicated and the original learning factory must be high, for example, regarding the target group, the operator model as well as the learning

---

[46] Advisory service learning factory and training setup.

[47] Training service for the operator of the learning factory.

factory targets and purposes. Supposedly, the greater the requirement deviations between original and replicate, the more the effective competence development is reduced.

The turnkey learning factory business model

Only organizations with the necessary expertise, personnel, and the corresponding capacities can handle such projects. After agreement and ordering of the turnkey learning factory, the learning factory supplier is responsible for all orders of the needed factory equipment, assures the functionality of the learning factory concept, supports the setup, and instructs trainers and operators among others in the production processes, products, and training concepts. Advantages for the customer in this scenario are

- Lower cost compared to completely individually developed learning factories.
- No expertise and personnel regarding the establishment of learning factories in the customer organization are needed.
- The exchange between experienced providers and the operator of the learning factory guarantees a smooth technical operation as well as high-quality standards in the competency development.
- Learning Factory concept is already geared to the desired topic, and research results are already implemented as part of a continuous improvement process in the learning factory (especially for learning factories affiliated to research institutions). The customer makes use of these research results by purchasing such a learning factory.

Potential customers can be industrial companies, operators of further education institutions as well as research institutions. In particular, learning factory concepts can be offered as part of this business model, which have already been successfully implemented (possibly with small variations) elsewhere; otherwise, a fast and uncomplicated implementation of the learning factory concept cannot be ensured. As a rule, in this model, the customer will also be the operator of the learning factory; in other constellations, a corresponding high investment would be difficult to justify. On the part of the customer, appropriate employees must be organized, which (can be trained by the learning factory provider in order to) ensure the sustainable learning factory operation.

As part of the turnkey learning factory project, coordination of offers, execution of the orders, and development of necessary aids for the operation as well as selection and configuration of the learning environment take place by the supplier of the learning factory. An individual configuration of the learning factory within certain limits is possible, which is presented to the customer as a complete offer, which includes the construction, the production of the functionality, and the training of the personnel. Usually, a complete offer is made to the customer, which includes the customizing of the standardized learning factory, the construction, the production of the functionality, and the training of the personnel. The latter is particularly important because the know-how to operate the learning factory in this model is initially not available to the customer. However, as he has to gradually internalize the expertize for the learning factory operation, corresponding consulting and coaching services have to be planned in this regard.

In this business model, on the one hand of course revenues are generated by the supplier. Furthermore, on the other hand the customer and then operator of the learning factory can generate revenue offering trainings to (organization-external) industrial participants (see also business model "Offer of learning factory trainings for industrial companies"). The presented model has already been successfully implemented several times in cooperation with the Process Learning Factory CiP. The PTW (TU Darmstadt) establishment in cooperation with McKinsey & Co. learning factories in the Netherlands, South Africa, Russia, and Singapore. In those cases, mostly the assembly line including material supply for the Process Learning Factory CiP was exported and implemented on-site; for impressions of the delivered turnkey learning factories, see Fig. 6.24. Furthermore, Fig. 6.23 shows an overview on operators and target groups of the turnkey learning factory projects. Further learning factory implementations are in the planning phase. These learning factories are operated by research institutions as well as by private companies. Accordingly, the application of the turnkey learning factories is varying.

Phases of a typical turnkey learning factory project

The project duration for the construction of turnkey learning factories comprises approximately 31 weeks from accepted offer to the final acceptance by the customer. In addition, a subsequent support in the operation of about 8 weeks via telephone

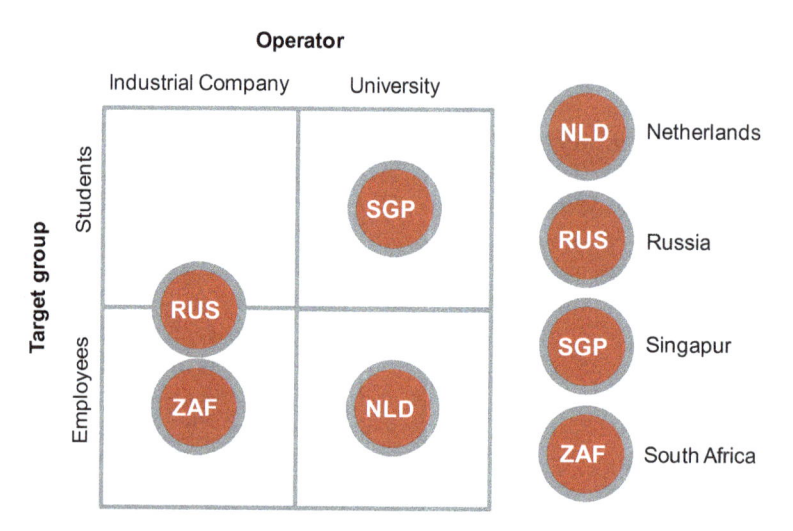

Fig. 6.23 Overview on operators and target groups of the turnkey learning factory projects mentioned (Enke, Kaiser, & Metternich, 2017)

6.2 Learning Factory Built-up, Sales, and Acquisition

NLD — *The learning factory LeaRn* is located at the university in Nyenrode, Netherlands and is used for training of office staff of a dutch bank, and a dutch insurance company as well as a consulting firm.

RUS — The learning factory in Russia is located at a university and was built for a big Russian heavy industry and manufacturing conglomerate. Students and industrial employees are trained together.

SGP — The learning factory at a university in Singapur provides trainings for students, especially with focus on systematic problem solving.

ZAF — In South Africa the learning factory is located at the training campus of one of the world's largest mining companies in Johannesburg. It is used to train worldwide operators in lean methods.

Fig. 6.24 Examples for turnkey learning factories built-up around the world after the model of the Process Learning Factory CiP

calls and mail is guaranteed.[48] Figure 6.25 shows the phases of a typical turnkey learning factory project.

Negotiation and Coordination: Expenses for materials and personnel are estimated in an offer together with the customer. Expenses depend on the specific design of the learning environment at the request of the customer. Optionally, a site inspection

---
[48] Enke, Kaiser et al. (2017) describe the construction of such a turnkey learning factory from offer to final acceptance. This section is strongly based on the descriptions available there.

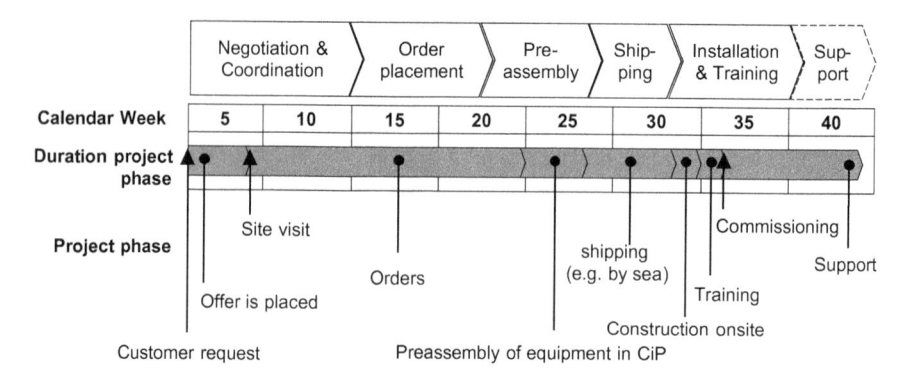

Fig. 6.25 Phases of typical turnkey learning factory projects according to Enke, Kaiser et al. (2017)

can be provided at the customer, in which the conditions on-site are checked and a detailed layout planning takes place.

Order placement: A significant part of the personnel expenses of the project is based on the scheduled selection and ordering of all necessary components. For this purpose, a part list of the original learning factory together with established providers is maintained continuously. For custom-made products (assembly devices, etc.), which result above all from prior continuous improvement of the own learning factory, internal workshops and contract manufacturers are instructed.

Preassembly: At an early stage, a team, which includes at least one expert with experience from previous construction projects, is compiled for the preassembly of the workstations and material supply equipment. The focus is at this time a functional test of all customized constructions and the assembly and testing of all functional units.

Shipping: For subsequent shipping, the learning factory equipment is partially dismantled and a detailed packaging list is created. Packages consist of functional units to simplify on-site installation and to avoid searching activities. In particular, for the shipping of the learning factory by sea requires a buffer of up to three weeks for regional customs and import regulations.

Installation and Training, Service: Two experts usually carry out the installation and training on-site. After unpacking the parts, the installation begins with the assembly of space-intensive infrastructure, such as material storages. Workplaces remain largely assembled and equipped after preassembly. If the complete infrastructure is ready, a test run follows. Here, the construction team operates the learning factory in order to make final adjustments and to be able to guarantee a smooth operation of the complete learning factory environment. Before dealing with the core training content, an introduction to technical processes, the product (pneumatic cylinder), and the basic didactical concept of the learning factory is necessary. The training of various learning factory states (predefined scenarios of the production system) is coordinated depending on number of trainees. For larger teams, subgroups can be created that rotate through the different tasks in turn. The training also addresses the transformation between different learning factory states. Here, a qualification matrix can help to keep track. Furthermore, daily performance test runs under real training conditions help to maintain the motivation of the participants and to ensure training progress. The training concludes with a test run, summary, and discussion of remaining questions. In the following, a technical briefing, the provision of spare parts, and maintenance instructions for selected personnel of the learning factory operator are provided. After the end of the project, long-term collaborations are sought in order to facilitate and consolidate the stated benefits of learning factories.

Success barriers and countermeasures for turnkey learning factory projects

Due to the complexity and duration of the project, there are numerous barriers to success on personal, organizational, communicative, and technical levels. These barriers and proven countermeasures from the extensive project experience[49] are described in detail in the following sections and shown in Fig. 6.26.

In practical application, an advance on-site inspection was helpful, especially with regard to existing layout restrictions at the construction site. In this way, planning errors and necessary adjustments in setup and operation can already be avoided in advance. Early communication with the logistics company responsible for transport

personal	organisational	personal	technical
• High travel load due to time zones • High physical strain during construction • High mental stress during training • Dependence on individual experts	• Varied product portfolio of suppliers • Lack of employee capacity • Lack of expertise regarding transport by air and sea • Loss of expertise through staff turnover	• Language barriers • Local work philosophies • Varying group size in training • Cultural diversity	• Emergency exits in the layout • Rework on the learning factory product • Transport damage & loss of equipment during transport • Missing uniform ground marking

Success barriers for projects ...

... and countermeasures

• Buffer day before start of work • Restriction of the construction days to seven hours per day • Support by the customer's staff • Alternating trainer in the process • Changing project teams for broad competence development	• Development of alternative suppliers • Early formation of an assembly team • Integration of logistics service from the beginning of the project • Knowledge Management	• Use native speakers as construction experts • Vocabulary and signage must be agreed in advance with the customer • Forming small groups for playful competition • Adaptation of the training plan	• Onsite visit to include restrictions • Disassembly and deburring of the product before delivery • Early research on retailers on site • Integrate layout markers into offer

Fig. 6.26 Success barriers and countermeasures regarding turnkey learning factory projects according to Enke, Kaiser et al. (2017)

---

[49] Presented in Enke, Kaiser et al. (2017).

helps to shorten the process of preassembly and packaging. Although existing parts lists can be used in the ordering process, the large share of purchased parts and complex custom-made equipment is risky with regard to necessary adjustments. Due to changing product portfolios of the suppliers, seasonally fluctuating delivery times, and regularly rising prices, this phase is time-consuming but crucial for economic project success. The time required from assembly over functional testing to commissioning can be reduced through the deployment of customer service personnel for packaging disposal and the construction of simple standard parts to reduce project costs.

The burden on the experts during the one-week on-site training should not be underestimated. Here, the composition of a training team of two experts and another experienced colleague for possible support can help. In particular, the project team should have sufficient language and cultural experience at the construction site. The use of an interpreter extends the training of customer personnel significantly, especially for the transfer of the conceptual approach of the learning factory. The formation of a project team involving fixed contact persons of the customer and a telephone conference every two weeks make a significant contribution to preventing the barriers to success presented.

6.2.3 Offer of Learning Factory Trainings for Industrial Companies

As already described, universities and private training providers can use learning factories as authentic learning environments for learner-active production-related training. In this business model, the training providers offer learning factory courses on different topics[50] freely accessible on the market or to individual partner companies of the learning factory; see Fig. 6.27. In the context of business and operating models, in particular the economic, the contentual, and the personnel quality must be ensured.[51]

Economic quality: Learning factory training suppliers can individually advertise trainings on the market[52] or also offer training in the form of partnership models.[53] The partnership model of the Process Learning Factory allows interested industrial companies to sign a partnership agreement. This contract grants the partners a certain number of training days, which can be freely distributed by the company to its employees. The Process Learning Factory CiP has such a partnership with approximately 20 industrial companies of different branches. The financial contribution

---

[50]For example, lean production, Industry 4.0, energy efficiency.

[51]See also Sect. 5.1.1 and Fig. 5.2: Three sustainability dimensions of learning factory operational models.

[52]See, for example, the Stuttgarter Produktionsakademie (2017).

[53]See, for example, the partnership model of the Process Learning Factory CiP (PTW, TU Darmstadt 2017).

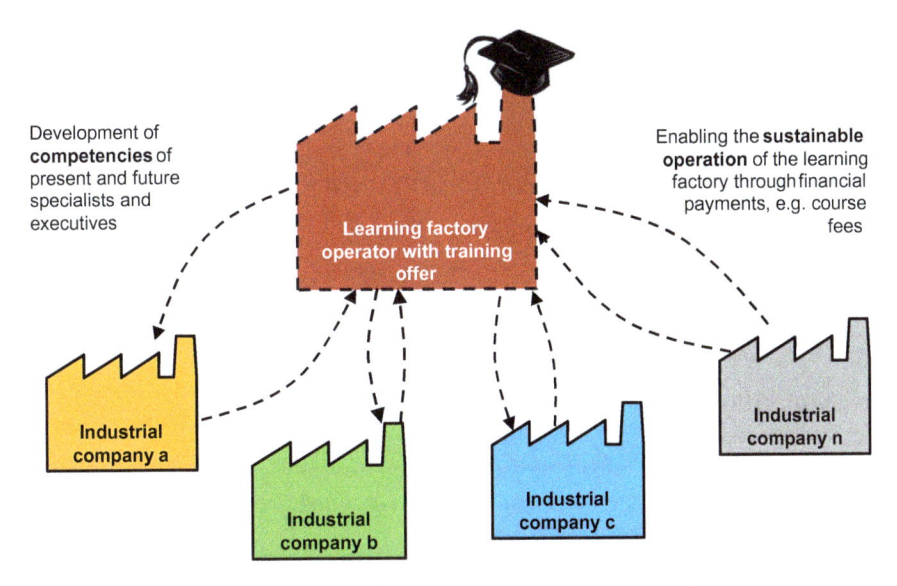

6.2 Learning Factory Built-up, Sales, and Acquisition

Fig. 6.27 Offer of learning factory trainings for industry (Abele et al., 2015)

of these companies is a substantial factor in the sustainable operation of the learning factory. For SMEs, additional compact training series are offered. Occasionally, events are also booked by German and foreign universities for students of certain degree programs.

Contentual quality: In addition, the training provider must ensure the quality of content shown in the learning factory. Academic operators have the opportunity to keep training content up to date through insights from current research and industry projects—this prevents the training contents from becoming obsolete; at present, for example, great efforts are being made to link digitalization and Industry 4.0 with existing offerings. Only constant further development and incorporation of up-to-date research results into the training courses continuously ensures relevant and innovative trainings. Of course, also training providers from the private sector[54] must ensure that their training offers and also addresses the problems and challenges of customers in the longer term in a up-to-date manner.

Personal and organizational quality: Personnel and organizational aspects play a major role in the context of the quality of learning factory concepts. In addition to the technical expertise, learning factory trainers require didactic-methodical skills for the development of the learning modules as well as for the moderation and the management of trainings and the coaching of training participants. Suitable personnel for learning factories must be well recruited and trained. Important tools for this matter can be train-the-trainer workshops, training and training development guidelines as well as a proper qualification procedure for learning factory trainings.

---

[54] such as management consultants.

6.3 Learning Factory Operation, Evaluation, and Improvement

This chapter describes approaches and concepts connected to the phase operation, evaluation, and improvement of the learning factory life cycle.

6.3.1 Training Management for Learning Factories in Operation

For the operation of learning factories, various different models exist. Of course, they vary mostly depending on the various operators[55] and target groups.[56] Regarding the training management of a learning factory, several issues regarding the operation in the fields of training support and administration, training coordination, and training delivery have to be addressed; see Fig. 6.28.

Training support & administration
- Administration of bookings
- Administration of partnership agreements
- Advertisement of learning factory trainings
- Catering during learning factory trainings
- Administration of learning factory workforce
- Supply of training materials
- ...

Training coordination
- Learning factory training program management
- Quality checks of trainings and training program
- Coordinate improvements based on feedback
- Ensure trainer qualification and coordination
- Enable continuous improvement of the learning factory
- ...

Training delivery
- Standards for the use of learning factory equipment
- Standardized training modules
- Creation of (standardized) training materials
- Coordinate online and learning factory learning
- Role of workforce in trainings
- ...

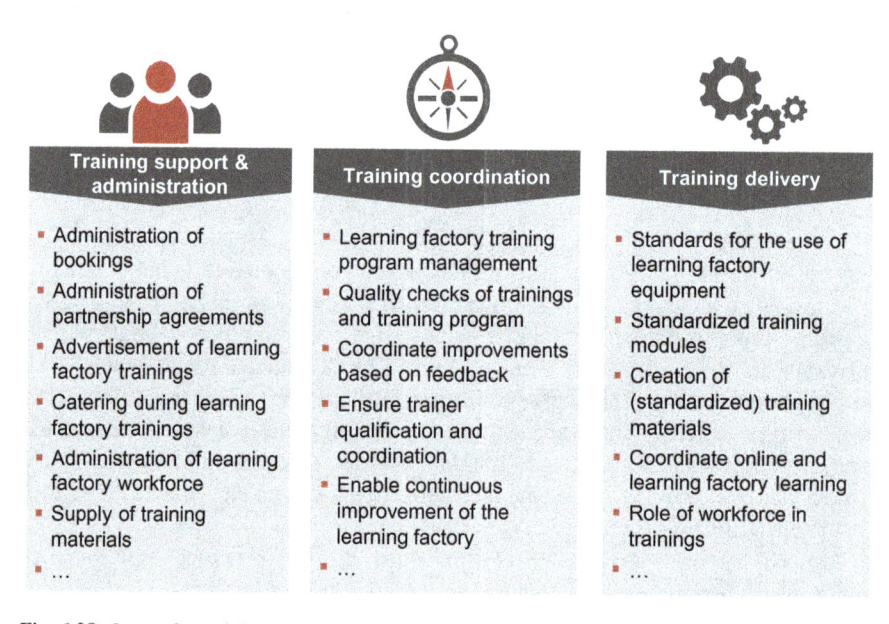

Fig. 6.28 Issues for training management in learning factories

---

[55] For example, university, consulting, industry, vocational schools.
[56] For example, students, industrial employees.

6.3 Learning Factory Operation, Evaluation, and Improvement

In the field of training support and administration, several tasks regarding the direct administration of the trainings,[57] support activities[58] as well as organizational issues[59] are fulfilled.

The training coordination includes tasks that coordinate the management of the learning factory program based on target group-specific learning targets, appropriate quality checks and evaluation as well as based on this a continuous improvement of the learning factory concept and the learning factory trainings.

The field of training delivery includes all tasks that are connected with the actual learning processes; this includes, for example, the standardization of learning factory equipment and the creation of offline and online learning materials.

The operation phase of various learning factory use cases is described in all the best practice examples shown in Chap. 11.[60]

6.3.2 Quality System for Learning Factories Based on a Maturity Model

In the usage phase of the learning factory, it is important to prevent the deterioration of efficiency after the built-up and to improve the existing learning system continuously; see Fig. 6.29.

Quality systems and maturity models for learning factories offer systematic approaches for assessing the current state, the potential for improvement in relation to the target state and for deriving improvement measures. The aim of quality definitions and a quality system for learning factories is to fully exploit the great potential of the Lernfabrik learning system in the usage phase and to prevent the learning factory processes and performance from decaying. Thus, learning factories can be further developed to higher maturity levels, the quality of the learning systems is sustainably secured, and the efficiency is increased. Such a maturity-based quality system is currently being developed in the research project RQLes.[61] Within this framework,

---

[57] e.g. booking management.

[58] e.g. catering for the trainings.

[59] e.g. advertisement for trainings and administration of partnership agreements.

[60] Further use cases regarding the operation mode of learning factories can, for example, be found in Jorgensen, Lamancusa, Zayas-Castro, and Ratner (1995), Abele, Eichhorn, and Kuhn (2007), Jäger, Mayrhofer, Kuhlang, Matyas, and Sihn (2012), Steffen, May, and Deuse (2012), Rentzos et al. (2014), Balve and Albert (2015), Bender, Kreimeier, Herzog, and Wienbruch (2015), Böhner, Weeber, Kuebler, and Steinhilper (2015), ElMaraghy and ElMaraghy (2015), Faller and Feldmüller (2015), Gebbe et al. (2015), Goerke, Schmidt, Busch, and Nyhuis (2015), Hummel, Hyra, Ranz, and Schuhmacher (2015), Kaluza et al. (2015), Kreitlein, Höft, Schwender, and Franke (2015), Krückhans et al. (2015), Lanza, Moser, Stoll, and Haefner (2015), Nöhring, Rieger, Erohin, Deuse, and Kuhlenkötter (2015), Seitz and Nyhuis (2015), and many more.

[61] Reifegradbasierte, multidimensionale Qualitätsentwicklung von komplexen Lernsystemen am Beispiel der Lernfabriken für die Produktion, literal translation: Maturity-based, multi-dimensional

Challenges of the operation phase of a learning factory

(a) Continuous improvement of existing learning systems

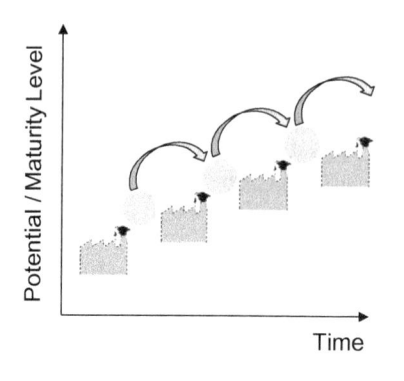

(b) Preventing the deterioration of efficiency after the learning factory built-up

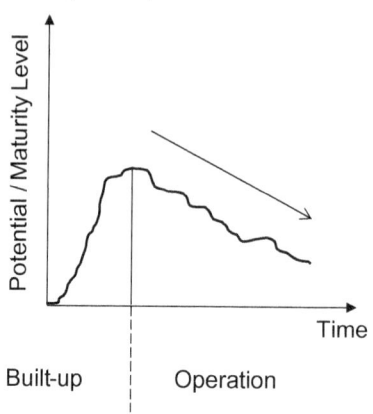

Fig. 6.29 Challenges for a quality system for learning factories in the operation phase

- quality standards for the learning factory are formulated,
- a quality system for education and training in highly complex learning environments is developed, and
- within the validation the extent to which the implemented quality system has an impact on the learning outcomes of the participants is evaluated.

Building on the acquired international learning factory understanding[62] and the competency-oriented design of learning factories,[63] the goal of the RQLes project is the creation and implementation of a maturity-based quality system for learning factories. The quality system enables quality assurance as well as further development of learning factories (Fig. 6.30). The quality system is developed especially for learning factories, but should also be applied to complex learning systems that integrate formal, non-formal, and informal learning.

The development process of the learning factory maturity model is shown in Fig. 6.31. The results and the rough process of the maturity model are described in the following with the four simplified sequential phases "analysis," "structure," "design and implementation," and "validation."[64]

---

quality development of complex learning systems using the example of learning factories for production.

[62] See Chap. 4.

[63] See Sect. 6.1.

[64] The following sections summarize briefly the results of the phases. A detailed description of the complete approach can be found in Enke, Glass et al. (2017) and Enke, Metternich, Bentz, and Klaes (2018).

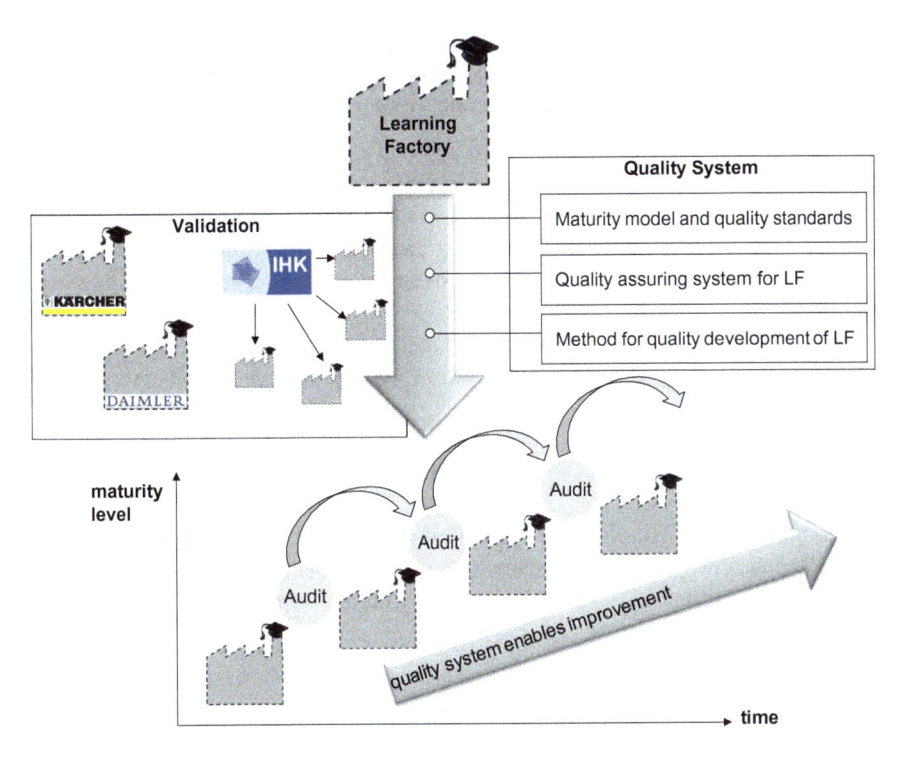

6.3 Learning Factory Operation, Evaluation, and Improvement

Fig. 6.30 Concept of the research project RQLes

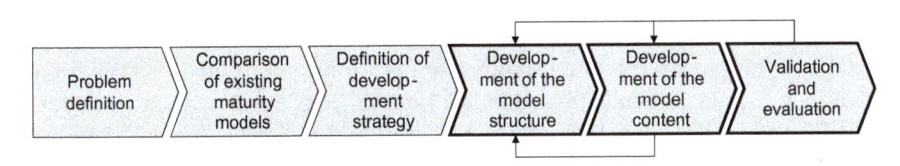

Fig. 6.31 Approach for the learning factory maturity model development (Enke et al., 2018)

In the **analysis** phase, the literature review on existing maturity models in the field is performed; this includes maturity models in learning-related as well as in production-related fields. Based on this analysis requirement for a maturity model for learning factories is derived that includes general requirements, like the comprehensibility of the model or the possibility for determination of a maturity level including deducted development paths, as well as learning factory-specific requirements, like the consideration of learning-, factory-, production-, and -content-specific issues.

The findings of this analysis phase form the basis for the maturity model development. In order to fulfill identified general and specific requirements, several existing models and approaches are combined to the **structure** of the maturity model[65]:

- CMM and CMMI: Both models have a strong process orientation. The learning factory maturity model makes use of aspects of those models, e.g. the definition of the maturity levels. In reference to the CMMI, process improvement is measured using capability levels for different areas (action fields) within the learning factory concept as well as one aggregated maturity level for overall learning factory concept.
- EFQM: The European Foundation for Quality Management model[66] is already adapted for training and education organizations, which proofs the general applicability to a maturity model for learning factories. The EFQM model offers a comprehensive examination of organizational aspects of the learning factory concept. For this reason, parts of the EFQM model are integrated in the learning factory maturity model.
- Learning Factory Morphology and Learning Factory Design Levels: In order to create a learning factory-specific maturity model, as structural elements the Learning Factory Morphology[67] and the learning factory design levels[68] are used.

Based on those models, the learning factory maturity model structure is derived. In this structure, the EFQM model elements are connected to the design dimensions and design elements of the morphology as well as to the design levels; see Fig. 6.32. This creates different action fields for the learning factory maturity model.

Those identified action fields contain various statements with which the capability level[69] of a specific learning factory action field[70] can be defined, and an overall maturity level[71] of the learning factory can be derived from this; see Fig. 6.33. The maturity level contains certain capability levels of various action fields.[72] With rising maturity level, new action fields and higher expectations on the capability levels are added. This is achieved with different target profiles,[73] e.g. in the exemplary shown assessment in order to achieve an overall maturity level of 2 the capability levels in all action fields assigned to the respective maturity level have to be at least 2 for the assessed learning factory.[74] In the given example in Fig. 6.33, the assessed learning factory achieves the overall maturity level 1 due to the low capability rating

---

[65]For further information, see Enke, Glass et al. (2017).
[66]EFQM model, see http://www.efqm.org/.
[67]See Chap. 5 and Tisch, Ranz et al. (2015).
[68]See Fig. 6.3 and Tisch, Hertle, Abele et al. (2015).
[69]See (a) in Fig. 6.33 for a general description of the capability levels 1–3.
[70]See (b) in Fig. 6.33 for exemplary assessment of action fields.
[71]See (c) in Fig. 6.33 for a general description of the overall maturity levels 1–5.
[72]See (b) in Fig. 6.33.
[73]See (d) in Fig. 6.33.
[74]See (b) and (d) in Fig. 6.33, a legend for the symbols for "actual state," "target state," and "necessary development" are given in (e) in Fig. 6.33.

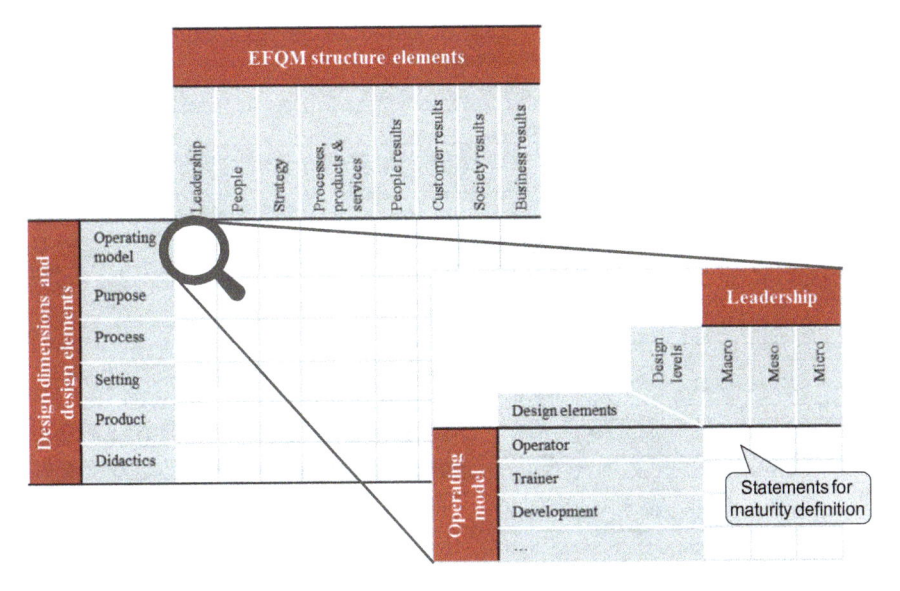

6.3 Learning Factory Operation, Evaluation, and Improvement

Fig. 6.32 Structure of the maturity model (Enke, Glass et al., 2017)

in action field 1. Necessary learning factory developments in order to reach an overall maturity level of 4 are indicated in the example. Currently, the complete maturity model comprises in total 51 action fields and 374 maturity elements.[75]

In the **design and implementation** content of the action fields and capability levels of the action fields are specified. The assessment of action fields is enabled by the assignment of maturity elements to action fields. EFQM surveys for further education as well as the Learning Factory Morphology elements and the learning factory design levels are used to identify and formulate relevant maturity elements. Furthermore, results of the learning factory stakeholder analysis[76] as well as aspects of DIN EN ISO 9001 and DIN ISO 29,990 are integrated. Figure 6.34 exemplarily shows defined maturity levels regarding the statement "Schedule for periods of reflection in learning modules." The CMMI includes five levels with rising quality: initial, managed, defined, quantitatively managed, and optimizing. In the shown example, the description of different maturity levels range from no scheduled reflections to a use of reflections as a central step for the competency development.

The **validation** of the described maturity model is executed in the coming year using a multi-perspective approach. In this approach especially four perspectives will be addressed[77]:

- Economic perspective: How are benefits and efforts of the development and application of the maturity model?

---

[75] Enke et al. (2018).
[76] See Enke et al. (2016).
[77] For further explanation, see Enke et al. (2018).

(a)

Capability level	Description of capability level(general)
1	First steps are taken to identify requirements.
2	Requirements are known.
3	Requirements are described in detail and inter dependencies between the mare considered.

(b)

Added in		Capability level
		1 2 3
Maturity level 1	Action field 1	
	Action field 2	
Maturitylevel 2	Action field 3	
	Action field 4	
Maturitylevel 3	Action field 5	
	Action field 6	
Maturity level 4	Action field 7	
Maturitylevel 5	Action field 8	

(c)

Maturity level	Description of maturity level
1	Essential basics to operate a learning factory exist.
2	The success of the learning factory can be ensured over a longer period of time.
3	Interdependencies between the action fields are considered.
4	Procedures to measure and evaluate key figures are established.
5	Based on the key figures measures for continuous improvement are defined and implemented.

(d)
- Target profile for maturity level 1
- Target profile for maturity level 2
- Target profile for maturity level 3
- Target profile for maturity level 4
- Target profile for maturity level 5

(e)
- ✗ Actual state
- ● Target state
- → Necessary developments

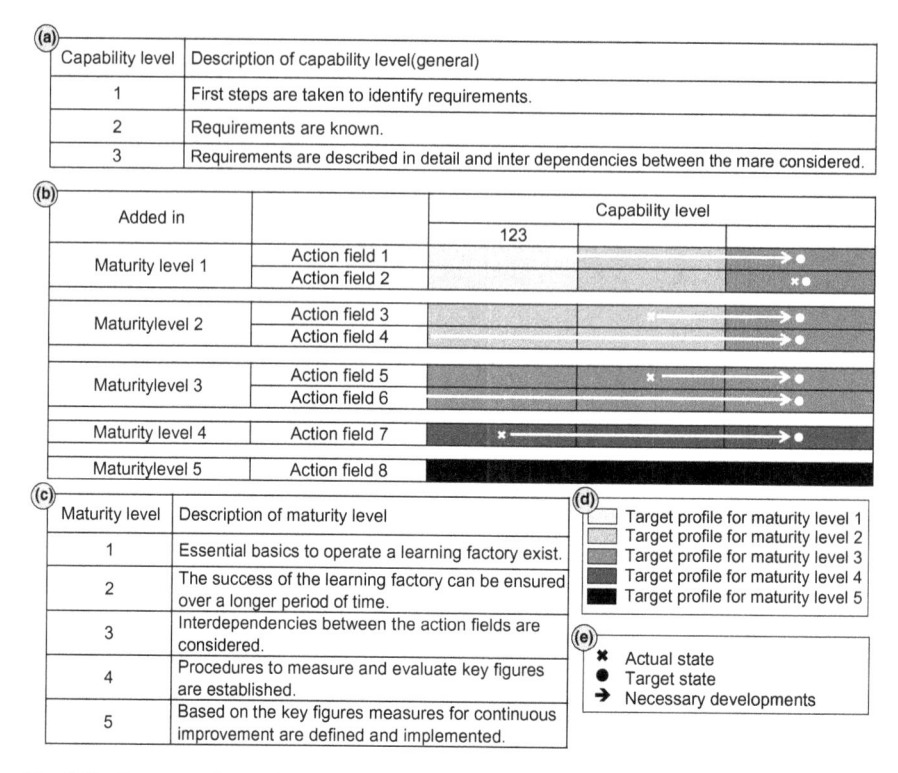

Fig. 6.33 Structure of maturity and capability level relations of the learning factory maturity model (Enke et al., 2018)

- Deployment perspective: Are users capable and willing to apply the maturity model?
- Engineering perspective: Is the maturity model accurately modeled and formally correct?
- Epistemological perspective: Are the results of the maturity model objective, reliable, and valid?

6.3.3 Evaluation of the Success of Learning Factories

Evaluation of training institutions, systems, and programs monitors and controls the quality and the success of learning activities and furthermore allows an improvement in quality.[78] Evaluation of training programs and systems is often carried out in the four fields of the widespread CIPP model[79]:

---

[78] See Solga (2011).
[79] See Stufflebeam (1972).

6.3 Learning Factory Operation, Evaluation, and Improvement

Statement for maturity definition:
"Schedule for periods of reflection in learning modules"

Initial	Managed	Defined	Quantitatively managed	Optimizing
No periods of reflection are scheduled.	It is prescribed that periods of reflection should be integrated. However this is not being checked.	Periods of reflection in learning modules are scheduled.	Periods of reflection in learning modules are scheduled and their implementation is checked regularly.	Periods of reflection are used as one of the central steps to competence development.

Fig. 6.34 Definition of capability levels for all statements to enable a classification of particular learning factories (Enke, Glass et al., 2017)

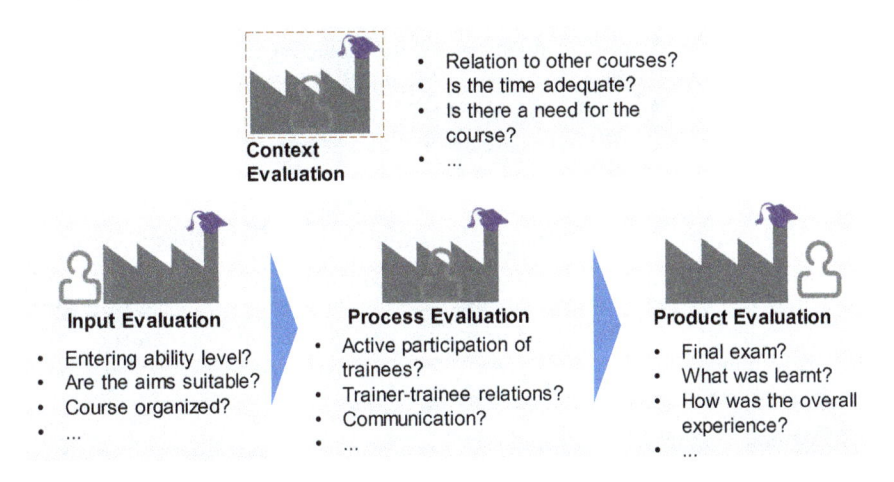

Context Evaluation
- Relation to other courses?
- Is the time adequate?
- Is there a need for the course?
- ...

Input Evaluation
- Entering ability level?
- Are the aims suitable?
- Course organized?
- ...

Process Evaluation
- Active participation of trainees?
- Trainer-trainee relations?
- Communication?
- ...

Product Evaluation
- Final exam?
- What was learnt?
- How was the overall experience?
- ...

Fig. 6.35 Evaluation in learning factories along the CIPP model according to Stufflebeam (1972)

- **Context:** Evaluation of context of learning, didactic concepts, etc.
- **Input:** Evaluation of preparation of the training activities, which is located before the actual learning.
- **Process:** Evaluation of the learning processes, which takes place during actual learning.
- **Product/Output:** Evaluation of the effects of the learning measure, which is located after the actual learning (Fig. 6.35).

This section deals with the different evaluation possibilities of products/output of learning processes in learning factories. The three quality criteria that any science-based learning success measurement forms have to meet are objectivity, validity, and reliability.[80] Further relevant criteria coming from a practice-oriented point of view are, for example, economy, opportunity equality,[81] transparency, manageability,[82] sensitivity to change, or differentiability.[83]

The possibilities to measure somehow the various effects of learning activities are almost without limits. A classification of these different instruments and methods, which is most recognized in practice,[84] is provided by the Kirkpatrick four-level model.[85] The four-level model classifies the levels of output evaluation:

- Reaction: How satisfied are the participants with the training?
- Learning: What was learned well? And what was learned not so well?
- Behavior/transfer: Can the developed competencies or skills be used (transferred) to the work environment?
- Results: What is the economic effect of the trainings?

Studies undertaken in this field were not able to confirm a (positive) effect from lower levels on the higher ones.[86] This means that, for example, learning effects cannot be estimated via participants' satisfaction. However, the model gives a intuitive and pragmatic framework for evaluation of learning results,[87]; therefore, these in this section regarded levels of evaluation are classified with the help of the presented CIPP and Kirkpatrick four-level model in Fig. 6.36. Furthermore, respective goals, evaluation questions, methods, and indicators of the different levels are given. In the next sections, the evaluation levels are described in detail.

6.3.3.1 Reaction of Participants (Satisfaction Surveys, "Happy Sheets")

In learning factory practice, training programs are most often evaluated with structured and standardized satisfaction or feedback surveys[88] which participants fill in after completing the course or program.[89] Such feedback sheets in established practice may contain in this context questions regarding:

- The overall impression of the training,
- The quality of moderation,

---

[80] See Becker (2005).

[81] See Elster, Dippl, and Zimmer (2003).

[82] See Hertle, Tisch, Kläs, and Metternich (2016).

[83] See Clasen (2010), Burlingame et al. (2006).

[84] See Smith (2001).

[85] See Kirkpatrick (1998).

[86] See Alliger, Tannenbaum, Bennett, Traver, and Shotland (1997).

[87] Nerdinger, Blickle, and Schaper (2014).

[88] Sometimes, they are also referred to with a disparaging connotation as "Happy Sheets."

[89] See Becker et al. (2010).

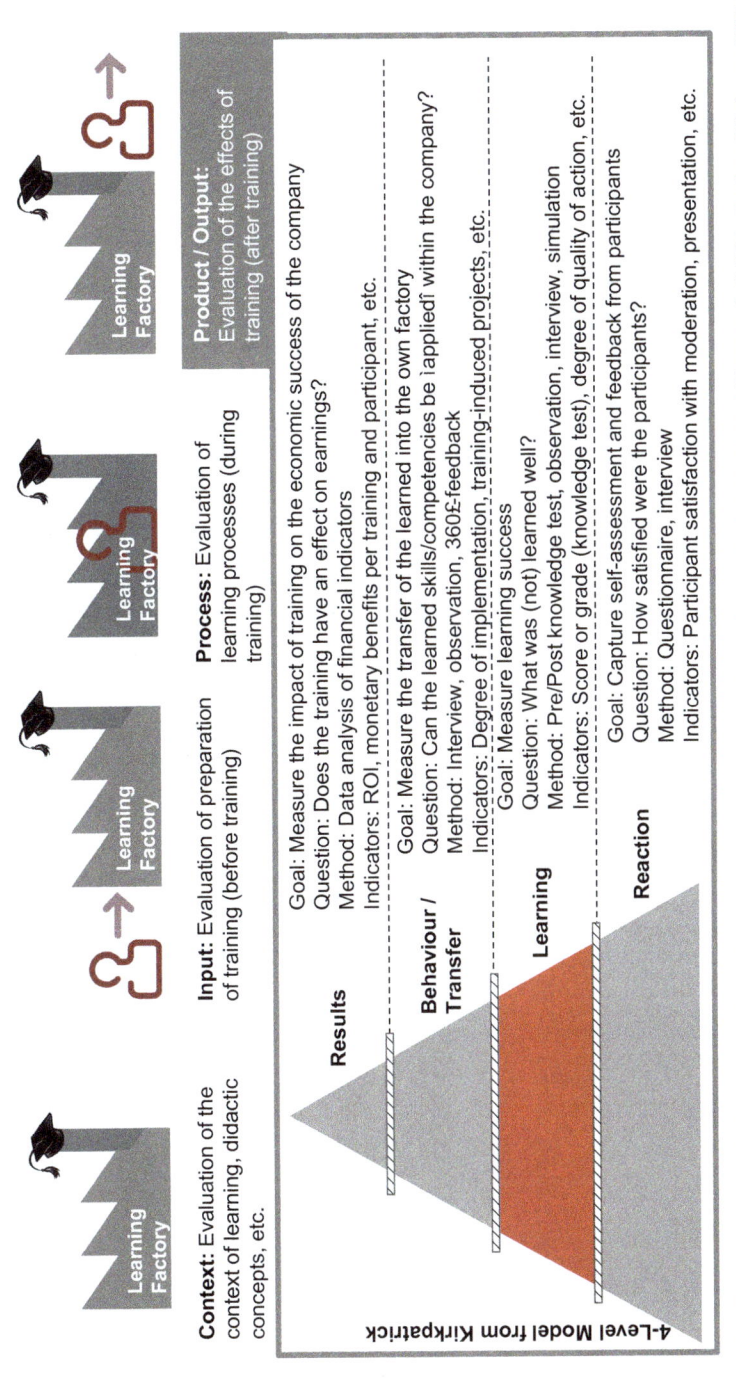

6.3 Learning Factory Operation, Evaluation, and Improvement

Fig. 6.36 Evaluation possibilities for the effects of learning factories according to Tisch et al. (2014) based on Alliger et al. (1997), Becker, Wittke, and Friske (2010), Gessler (2005), and Kirkpatrick (1998)

- The quality of presentations,
- The quality of practical exercises on the shop floor,
- Quality of theoretical content,
- The adequacy of the mix of theory and practice,
- The exchange between participants,
- The quality of specific exercises or presentations,
- Free text fields for entering anything else.

In general, the structure of these feedback sheets is as follows

(a) A superordinate view, in which all quality-related requirements for the entire learning module are answered comprehensively,[90] or
(b) A learning situation-specific evaluation in which individual lectures or exercises of the training are assessed individually by the participants (Fig. 6.37).[91]

While option (a) allows a more detailed analysis of the components of the training arrangement, the procedure in (b) supports the improvement of individual learning phases. The exact design of the feedback forms must always be coordinated with the goals associated with the evaluation, for example, improvements of trainings, quality assurance of training components. Although, as mentioned earlier, the feedback of participants immediately after the training program has no connection to the achievement of goals at higher evaluations levels, like a corresponding learning success or a successful transfer of concepts and methods to the own working environment.

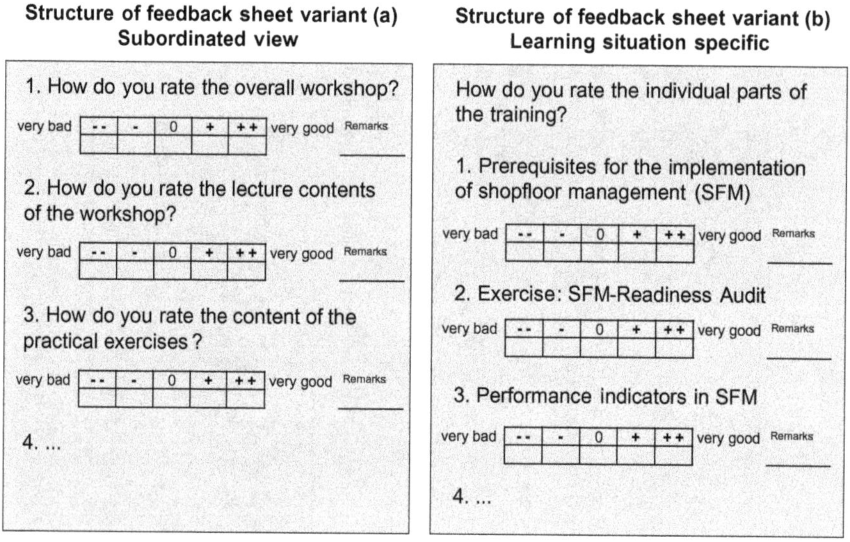

Fig. 6.37 Two variants of feedback sheets for the use in learning factory trainings

---

[90] See exemplary Fig. 6.37 on the left.
[91] See exemplary Fig. 6.37 on the right.

6.3 Learning Factory Operation, Evaluation, and Improvement

6.3.3.2 Learning of Participants

Mainly two key factors influence the learning success of learners:

1. The most influencing factor is the individual learner including his or her attitude toward the learning situation, the teacher, and the content.[92]
2. Learning situations need to be geared to requirements of predefined learning targets.[93] This comprises factors such as the arrangement of the learning environment, the behavior and the attitude of the trainer, the followed didactic principles and concepts as a whole, as well as the specific composition of the learning situation.

In the measurement of learning (the extent of) certain characteristics, traits, and behaviors associated with a learner are determined.[94] In order to implement those kinds of learning success measurements, a lot of different measurement techniques and forms of learning success evaluation can be identified. In general, no specific form of measurement in this field can be identified that is fundamentally better or worse than other forms. The forms of measurement must always be considered in the context seeing the complete arrangement. Consequently, there are only more or less suiting measuring methods in different situations and arrangements.[95] In the literature, various classifications of techniques for the measurement of learning success are identified; the most important forms are described briefly in the following sections.[96]

Classification of Learning Evaluation

In the following sections, forms of learning evaluation are classified distinguishing between

- Subjective and objective measurements,
- Open and closed measurements,
- Self-measurement and external measurement,
- Summative and formative measurements,
- Direct and indirect measurements,
- Quantitative and qualitative measurements.

Subjective and objective measurement

A simple and fast way for learning evaluation is facilitated with subjective measuring forms. On the downside, the results of subjective measuring forms have limited

---

[92] See Becker (2005).
[93] See Preussler and Baumgartner (2006).
[94] See McMillan (1997).
[95] Pietzner (2002).
[96] The descriptions are based on Clasen (2010); for further information, see also Clasen (2010).

validity since they are strongly dependent on the respective evaluator. For this reason, increasingly objective criteria form the basis for evaluations.[97] In this context, it has to be mentioned that objective and subjective ratings on the same evaluation objects show only a low-to-medium correlation of .389.[98] Other meta-studies show similar results.[99]

Example for subjective learning evaluation in learning factories:

Questions like "How do you rate the knowledge development by the learning factory training related to the topic 'value stream analyses'?" are posed in a self- or an external assessment.

Example for objective learning evaluation in learning factories:

For example, simple questions ("What are the seven types of waste?") or more complex tasks ("Please perform a value stream analysis for the displayed production system") related to learning content are evaluated on the basis of predefined objective measuring points.

Open and closed measurements

Open evaluations do not limit potential reactions or answers of the learners, while in closed evaluations reactions and answers are predefined by the evaluator and are correspondingly only selected by the learner.[100]

Example for open learning evaluation in learning factories:

Tasks like "Please describe shortly the procedure for a value stream analysis" are answered (in written or orally) by the learner.

Example for closed learning evaluation in learning factories:

Questions with predefined possible answers the learner has to select, like "What are existing symbols in the value stream analysis?
(a) Supplier,
(b) Process box,
(c) Inventory,
(d) Walking paths,
(e) Information flows."

The first two systematization sets "objective and subjective measurements" and "open and closed measurements" are used in combination in Table 6.2 to further systematize the different forms of learning success evaluation in learning factories and identify characteristics of those forms of evaluation.

Self-measurement and external measurement

---

[97] See Clasen (2010).
[98] Bommer, Johnson, Rich, Podsakoff, and Mackenzie (1995).
[99] See, for example, Mabe and West (1982) or Harris and Schaubroeck (1988).
[100] See Clasen (2010).

6.3 Learning Factory Operation, Evaluation, and Improvement

Table 6.2 Characteristics of subjective and objective measuring approaches according to Clasen (2010) with orientation among others at Bommer et al. (1995) and Harris and Schaubroeck (1988)

	Metering access			
	Subjective aggregated (closed)	Subjective global (closed)	Objective open	Objective closed
Effort				
—For construction	Low	Very low	High	Very high
—For evaluation	Low	Very low	Very high	Low
—Total	Low	Very low	High/very high	Medium/high
Processing time	Low	Very low	High	Medium
Requirements to the participant	Assessment of knowledge, skills, characteristics		High (free reproduction)	Low (recognition)
Bias tendency	Possible	Possible	–	–
Right solution due to guessing	–	–	–	Possible
At multiple measurement: substantial test effects	–	–	–	Given
Objectivity of evaluation (incl. generation of criteria)	High	High	Possibly limited	High
(Construct) Validity	High	Low	Measurement of one/many detail(s) of the construct	
	Aggregation (e.g. over a period or a field)			
Requirements to evaluator skills on the target of evaluation	Medium	Rather low	Very high	Very high

Self- and external evaluations are varieties of subjective evaluations.[101] Self-evaluation techniques, which are also referred to as internal evaluation techniques, in general consist of the learners' reflection on the bygone learning process; Fig. 6.38 shows some exemplary questions for self-evaluation with regard to different competency classes. In self- (or internal) evaluation, learners reflect on the actual learning success.[102] Self-evaluation is fraught with problems, and Harris and Schaubroeck (1988) discuss those potential problems. In contrast, external evaluations make use of persons other than the learner.

---

[101] See Clasen (2010).
[102] See, for example, Mabe and West (1982).

Exemplary competencies (competency class)	Exemplary scale for the self-evaluation
Expertise for the control of machines (technical and methodological)	0 — •————————• 10 — Non-existant Optimally available
Structured problem solving (technical and methodological)	0 — •————————• 10 — Non-existant Optimally available
Willingness to learn (Personal)	0 — •————————• 10 — Non-existant Optimally available
Capacity for teamwork (Social and communicative)	0 — •————————• 10 — Non-existant Optimally available
...	...

Fig. 6.38 Exemplary subjective self-evaluation sheet for different competency classes

Example for self-evaluation in learning factories:

Participants answer questions like "How do you rate your knowledge after the learning factory training related to the topic 'value stream analyses'?".

Example for external evaluation in learning factories:

Evaluators others than the learner answer questions like "How do you rate the knowledge development of the participants by the learning factory training related to the topic 'value stream analysis'?"

Summative and formative measurements

Summative evaluations compare (final) results with predefined goals. Summative evaluation is therefore linked to the accountability of learning factory training. In contrast to this, formative evaluation is carried out during the learning process and therefore also allows immediate changes or improvements in current learning factory trainings.[103]

Example for summative evaluation in learning factories:

For example a learning test or task or other form of evaluation at the end of a learning factory training in order to assess learning results can be used.

Example for formative evaluation in learning factories:

For example accompanying observation through externals or participant-related evaluation during the learning process in order to find improvement potentials for the (current and next) learning factory training can be used.

Direct and indirect measurements

Indirect evaluations can be used with both objective and subjective measurement criteria, while direct evaluations are in general connected to subjective evaluations. If measurements are conducted at different time points in relation to the training, for example, directly before training (pretest) and immediately after the training (posttest), and the change between the two measurements is seen as change or success

---

[103] See Dagley and Orso (1991).

6.3 Learning Factory Operation, Evaluation, and Improvement

indicator, this is called indirect evaluation. In contrast to this, direct evaluations look directly at the change or improvement, for example, using subjective and direct questions to the learning progress of participants in learning factory trainings.[104]

Example for direct evaluation in learning factories:

Often subjective, direct questions at the learning progress, like "How do you rate the knowledge development by the learning factory training related to the topic 'value stream analyses'?" are used.

Example for indirect evaluation in learning factories:

Equivalent to the example given above for the direct evaluation, an indirect question could look like this:

- (before training, pretest) "How do you rate your knowledge related to the topic 'value stream analyses'?"
- (after training, posttest) "How do you rate your knowledge level now after the learning factory training related to the topic 'value stream analyses'?"

Indirect evaluations are not limited to subjective measurement criteria. Furthermore, other evaluation designs are possible; for an overview of different evaluation designs, please see Fig. 6.39. In addition to the test designs mentioned in Fig. 6.39, an integration of follow-up tests (tests carried out after a few weeks or months after the treatment) can be used to determine the long-term success of the treatment.

Quantitative and qualitative measurements

Name	Group	Pretest	Treatment	Posttest
Post-test without control	E	-	X	O
2 Group, Post-test comparison	E	-	X	O
	C	-	-	O
1 Group, Pre-test/post-test	E	O	X	O
2Group, Pre-test/post-test	E	O	X	O
	C	O		O
Solomon 4-Group Design	E w. PT	O	X	O
	C w. PT	O		O
	E \PT		X	O
	C \PT			O

E	Experimental Group	w. PT	with pre-test
C	Control Group	X	Treatment
\PT	without pre-test	O	Test

Fig. 6.39 General possible experimental designs for learning success evaluation

---

[104] See Clasen (2010).

In general, evaluations using quantitative measures[105] are focusing on the question what is learned. In contrast, qualitative evaluations try to explore and describe learning processes in a differentiated manner.

Example for quantitative evaluation in learning factories:

Often open or closed evaluation questions answered by the participants are rated and accordingly associated with points.

Example for qualitative evaluation in learning factories:

For example, answers to open evaluation questions can be evaluated in a qualitative way. Often qualitative measures are used in the evaluation of feedback, the learning context, or the learning process.

General Approaches to Competency-Oriented Evaluation in Learning Factories

The problem with competency-oriented evaluation is that the construct "competency," which is in this case the intended focus of the evaluation, cannot be observed directly in an objective manner. Accordingly, evaluations that directly target the competency construct can only rely on subjective methods (self and external). Additionally, in the approaches to competency-based evaluation, the following two statements are of central importance:

- Competencies manifest themselves in individual actions (or also: performances) in specific problem situations. These performances are observable.[106]
- An important basis for the ability to act in unknown problem situations are knowledge elements corresponding to those competencies (in particular for the technical and methodological competency class).[107] Those knowledge elements can be evaluated directly in an objective manner.

Those two statements result in a knowledge-based and a performance-based approach for competence-oriented evaluation in learning factories; i.e. the manifestations of competencies (actions) can be observed and the requirements can be queried. Furthermore, those two approaches can be combined to get a broader view on the competency concept. Figure 6.40 shows an overview over the possibilities for competency-oriented evaluation approaches.

In order to gain access to the competency construct in this way, the (by the learning module) intended competencies are operationalized with assignable professional performances/actions as well as underlying knowledge elements (regarding professional and conceptual knowledge). The so-called competency transformation table

---

[105] Or translate qualitative measures in quantitative measures, for example, with scoring models for answers to open questions.

[106] See Chomsky (1962).

[107] See, for example, Erpenbeck and Rosenstiel (2007) and Pittich (2014).

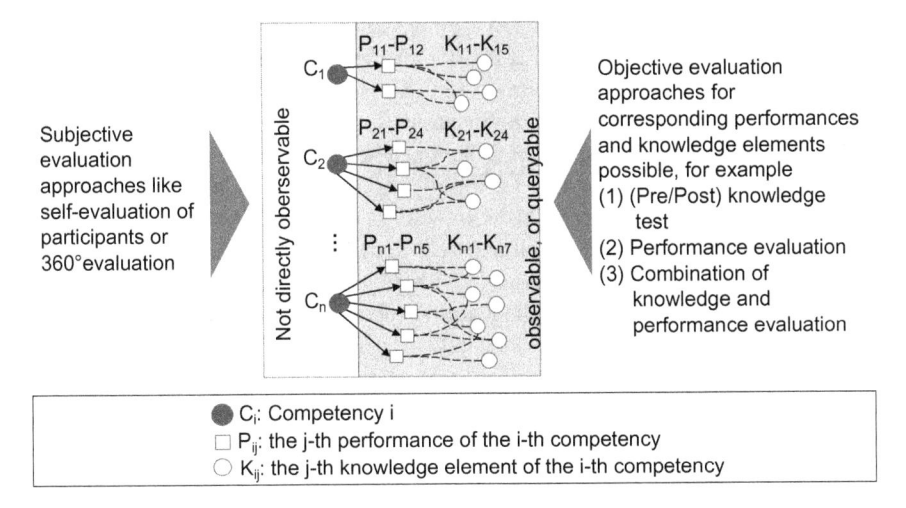

6.3 Learning Factory Operation, Evaluation, and Improvement

Fig. 6.40 Competency-oriented learning success evaluation approaches in learning factories

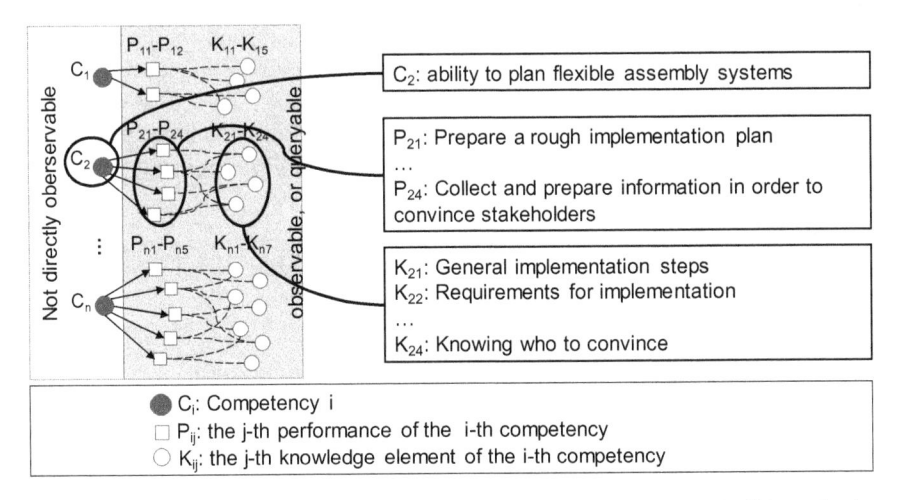

Fig. 6.41 Exemplary operationalized competencies of a learning module for "flexible production systems"

is an instrument that can be used for this operationalization of competencies.[108] Figure 6.41 demonstrates the operationalization of the competency regarding the planning of flexible assembly systems in the context of the learning module "flexible production systems."

Competency-oriented evaluation in learning factories using knowledge tests

---
[108] See Sect. 6.1 as well as Abele et al. (2015) and Tisch, Hertle, Abele et al. (2015) for further information.

The effectiveness of competency development is often evaluated with the help of knowledge tests.[109] The complex evaluation of the competency construct is in this case simplified by replacing it with a more accessible examination of transferred knowledge; see Fig. 6.40. This simplification comes with the cost of a less valid evaluation since knowledge is part of the competency concept, but only basis for self-organized professional ability to act.[110] Furthermore, in this context, it is important what kind of knowledge is queried in the knowledge test. Studies show that conceptual knowledge in particular (which contains reasoning and referential knowledge) can be a good predictor for the professional ability to act.[111] If knowledge tests are used in learning modules to determine the learning success, then either tests covering all knowledge types or tests covering especially higher forms of conceptual knowledge should be used. Knowledge can be tested in open and closed as well as in oral or written evaluations. Examples for the use of knowledge tests regarding the different knowledge levels are given in Fig. 6.42.

Competency-oriented evaluation in learning factories using simulated problem scenarios

In addition to the knowledge test described, simulated problem scenarios can be used to check the ability to act in unknown and complex situations of the learner. In this case, the evaluation is simplified in the sense that not the competency construct itself, but its manifestations (the performances) are observed. This kind of evaluation emphasizes the characteristic of technical and methodological competencies to be effective in domain-specific application situations. Here, learning factory environ-

Knowledge levels:	Professional knowledge (What? How?)
	Conceptual knowledge (Why?))
Alternatives:	Multiple choice or open questions
	Oral or written

Exemplary questions (knowledge level)	Exemplary scale
Please name the 7 types of waste. (Professional knowledge, What?)	Overproduction, Defects, Overprocessing, Movement, Inventory, Transport, Waiting
Describe the steps of a structured problem solving approach (Professional knowledge, How?)	...
Explain why the one-piece-flow promotes or prevents volume flexibility in work systems (Conceptual knowledge, Why?)	...
...	...

Fig. 6.42 Exemplary knowledge tests regarding different knowledge levels

---

[109] See Lohaus and Habermann (2011).
[110] For the distinction between competency and knowledge, see also Sect. 2.1.
[111] See Pittich (2014).

6.3 Learning Factory Operation, Evaluation, and Improvement

ments offer a great basis for this kind of learning success determination. In this case, conclusions about the competencies developed are drawn from the observed actions of trainees in new problem situations that require the intended competencies of the learning module.

The procedure for the competency-oriented evaluation using problem scenarios can be described in the three phases:

- Preparation (development of the evaluation task based on the learning targets, of the learning factory training),
- Implementation (perform evaluation task at the end of the training) and observation, and
- Follow-up/evaluation (comparison of learning targets and developed competencies, adjustment of training).[112]

Preparation of the simulated problem scenario:

Basis for the adequate preparation of the simulated problem scenario is the competency transformation table in which all actions assigned to the intended competencies can be identified. Based on this information, a problem scenario is created that as realistically as possible demands the professional ability to act of the trainees. The environment used for evaluation should differ from the environment used during the rest of the learning module. Furthermore, a high complexity of the problem scenario challenges the ability to act.

The background and predefined objectives of the simulated problem scenario are communicated to the trainees before starting with the task. Here, necessary information[113] has to be communicated without giving too much information about how to solve the problem or complete the task.[114] In order to prevent that the scenario is too simple or optimized for the application of a certain method, ill-structured problems can be used; usually, these kinds of problems predominantly occur in reality.[115]

Implementation and observation of the simulated problem scenario:

The evaluation exercise can be planned individually or in small groups of learners. On the one hand, the smaller the group, the more specific the results evaluation regarding the individual learning success. On the other hand, working and solving problems in teams resembles authentic work situations.

For the analysis of the effects of learning modules, a detailed observation following predefined observation guidelines is reasonable, in order to be able to log reliable observations and to reduce observation effort. Using these guidelines, the practice-relevant performances of participants (or participant groups) can be observed in the

---

[112] See Tisch, Hertle, Metternich et al. (2015).

[113] Such as description of initial situation of the production environment, description of the related problem situation, goals of the task.

[114] See also Tisch et al. (2014) and Tisch, Hertle, Metternich et al. (2015).

[115] Reiß (2012) proposes the so-called resilience learning approach in order to deal with this problem.

simulated, complex situation. The observation guidelines are based on the previous derived actions in the competency transformation table. Those specific, self-organized actions in new problem situations are now observed regarding comprehensive criteria using detailed observation indicators. Relevant observation criteria and indicators are described in Table 6.3.

The observation can take place in the form of a "live" accompanying observation or the observation of recorded videos of the performances.[116] The use of video recordings allows a more accurate and repeated analysis of the actions, where a video recording in some cases may inhibit the conduct of actions. In accompanying observations, observation sheets can be used in which the performances P_i of the learners can be logged regarding the observation criteria mentioned in Table 6.3. The semi-structured observation sheet supports the accompanying observer and allows the note taking on evaluation key points. In addition to the semi-structured observation sheet, further observations independent of the predefined criteria are recorded in a comment field which is provided for each defined performance.

Follow-up/evaluation of the simulated problem scenario:

The quality of performances of the associated competencies is rated based on the observed indicators. Therefore, the importance of the performances for the competencies as well as the importance of different observation criteria for the overall quality of the performance has to be weighted. This kind of evaluation includes sub-

Table 6.3 Observation criteria and indicators for the performance evaluation in the simulated, complex problem situation (Tisch, Hertle, Metternich et al., 2015)

Dimension	Description	Possible observation indicators
Self-sufficiency	Are the participants able to tackle the problem autonomously?	Participants are able to perform the task (a) autonomously, (b) hesitantly but on their own (discussion or look into script), (c) after guiding questions, (d) after a next step is pointed out, (e) they are not able to perform
Quality of approach	Do the participants follow a structured approach to solve the problem?	(a) Methodologically or analytically structured, (b) trial-and-error, (c) unstructured
Result	Are the results correct? (not applicable for all actions)	(a) Correct, (b) mainly correct, (c) mainly not correct, (d) not correct (at the end of the task, the overall solution is assessed as well)
Collaboration in team	Are the participants working together to tackle the problem?	(a) Together as a team, (b) individually

---

[116] See, e.g., Hambach, Diezemann, Tisch, and Metternich (2016).

jective components due to subjective observation as well as a subjective weighing process by the training provider. In order to reduce these subjective components, it is important that the observer and training provider have a clear understanding of learning goals, outcomes, and the performance qualities. The assessed performances finally give an indication of the competency development during the learning module.

Participants or participant groups that solve the problem independently in a structured manner would be assessed at 100% quality of performances. Defined performance indicators that are not observed lead to a devaluation of performance quality. The resulting percentage values should be regarded as relative to the expectations and goals of the learning module. Furthermore, the results depend on individual valuation and weighting of specific indicators; thus, this form of learning success evaluation is not suited for intertraining comparison. But, in repeated use of the evaluation, effects of training revisions are measurable and controllable.[117]

Competency-oriented evaluation using a combination of simulated problem scenarios and knowledge tests

Finally, the last described option for competency-oriented evaluation of learning success in learning factories integrates the in the previous sections presented knowledge- and performance-oriented evaluation approaches; see Fig. 6.43.

The disadvantage of the approaches to learning outcome evaluation described in the previous two sections is the sole focus on individual aspects of the competence concept: in the first case the underlying knowledge elements or in the second case the manifestations of competencies in professional application situations. Therefore, a more reliable evaluation combines the knowledge- and the performance-oriented

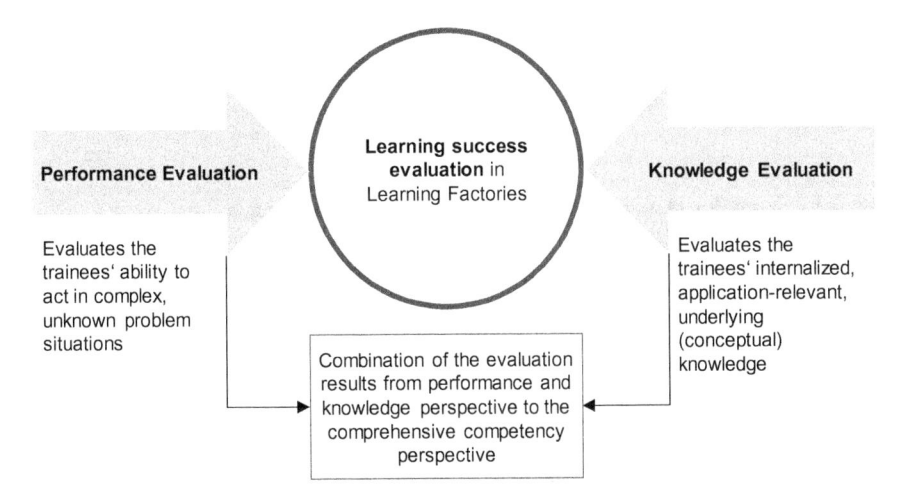

Fig. 6.43 Combination of knowledge and performance perspective to evaluate learning success in learning factories

---

[117] See also Tisch, Hertle, Metternich et al. (2015).

evaluation in one approach. More specifically this means that this combined approach tries to exclude positive evaluation results in the following two situations:

(a) that participants are only able to perform specific actions (that may be only coincidentally right, for example, because they simply remember a sequence of work steps shown in an example) without internalizing underlying knowledge that is crucial for the flexible handling of unknown situations.
(b) that participants only remember competency-relevant knowledge elements without bringing it in the professional context with the specific application of this knowledge. In this case, only inert knowledge was created that shows no effect for the ability to find solutions in unknown situations.

For this purpose, after the observation of a performance task (see previous section), the observer questions the underlying knowledge by starting a discussion regarding the reasons for the procedure shown within the previously performance task. Answers of the trainees can be recorded and analyzed afterward. Participants or participant groups that solve the problem independently, in a structured manner, and have the competency underlying knowledge would be assessed at 100% competency level. Performance indicators that are not observed or underlying conceptual knowledge that is not present in the discussion lead to a devaluation of the competency level.[118]

Alternatively, a written knowledge test can be used for this purpose. In contrast to the performance task, which is perceived as an additional exercise as a part of the learning module, the written knowledge test is perceived as a mere evaluation tool and therefore may lead to demotivation.

The competency-oriented measuring of learning success conceptually already builds a bridge to the next evaluation level "transfer"; trainees that can handle unknown problem situations and have internalized the underlying conceptual knowledge are likely to be successful with the transfer of procedures to the own factory. Although for the transfer, additional organizational obstacles must be overcome, see the following section.

6.3.3.3 Transfer (Use of Projects for Evaluation and Better Transfer)

An observable problem with learning factory trainings is that the transfer from the learning factory (despite an effective competency development) to the work environment can be hampered. The reasons for this are manifold; among others, they can be found in

- a lack of ability to transfer the learnt to the own work environment,
- a lack of motivation to apply,
- a lack of permission to implement, and
- a lack of opportunities to implement.

---

[118] An example for such a combined performance- and knowledge-based evaluation is described in Tisch, Hertle, Metternich et al. (2015), Tisch, Hertle, Abele et al. (2015), and Tisch (2018).

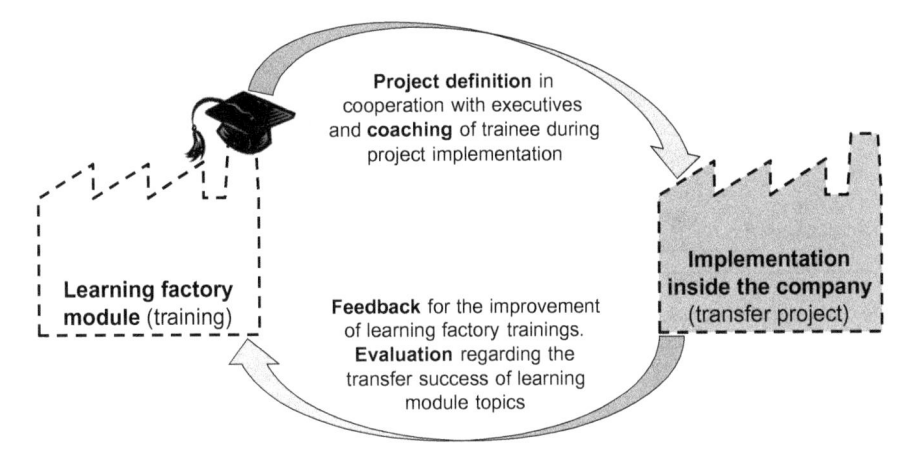

6.3 Learning Factory Operation, Evaluation, and Improvement

Fig. 6.44 Integrated learning factory training and transfer concept

In order to counteract these transfer obstacles, it is useful for learning factories used for training purposes to integrate the transfer in the training concept. The general dependencies between the training and the transfer are shown in Fig. 6.44.

In this integrated learning factory training and transfer concept, not only the selection of learning modules based on individual necessities and target group affiliation is managed, but also the implementation in the real factory is directly planned in advance. In this way, the path for the transfer of the learnt from the learning factory to the real factory is already prepared. In addition, these transfer projects allow a practice-oriented evaluation of the transfer success as well as additional certification possibilities. Here, a parallel planning of the learning module participation and the transfer projects can take place. Figure 6.45 shows a proposed process of planning, implementation, and evaluation of learning modules and transfer projects in parallel. In the following, the three phases shown in Fig. 6.45 are explained more in detail. Alternatively, transfer projects can also be planned after the attendance of learning modules; here, however, there is the disadvantage that the selection of learning modules as a first step cannot be deliberately adjusted to the identification of possible transfer projects. In this way, it can happen that the learners visit unnecessary or inappropriate learning modules.

Planning of trainings and projects

In the planning phase of the integrated training and transfer concept

1. suiting learning modules are selected, and
2. an adequate transfer project is defined for the trainee.

The selection of learning modules is of course dependent on the current and target competency profile of the trainee. In general, learning modules are selected in a way that they can close the gap between current and target competency profile. Parallel to this, project ideas that can be carried out in the near future in the sphere of influence of

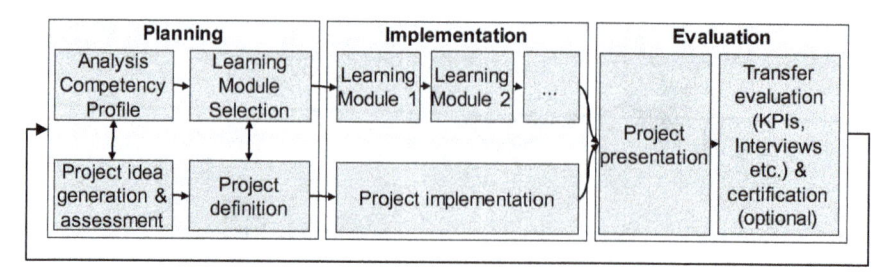

182　　6 The Life Cycle of Learning Factories for Competency Development

Fig. 6.45 Process of planning, implementation, and evaluation using learning factory modules in combination with transfer projects

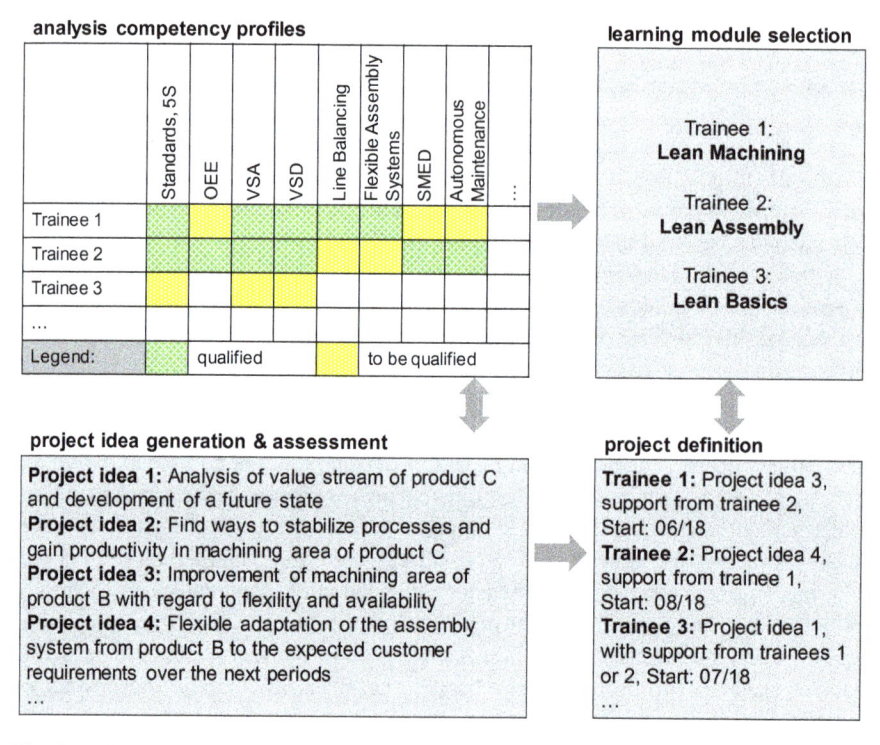

Fig. 6.46 Simplified example for the parallel planning of learning modules and transfer projects

the trainee are generated and assessed regarding the fit to the competence profiles and the importance of the projects. Of course, this does not mean that all improvement projects have to be linked to learning factory training; e.g. urgent projects may not be appropriate for this practice. Figure 6.46 shows a simplified example for the parallel planning of learning module participation and transfer projects for three trainees in the field of lean production systems.

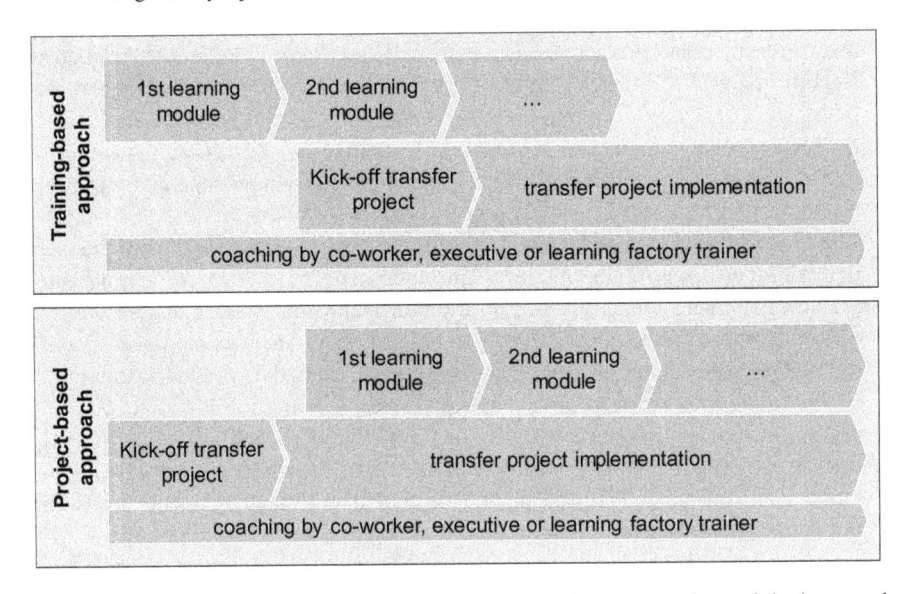

6.3 Learning Factory Operation, Evaluation, and Improvement

Fig. 6.47 Training- and project-based approach for the implementation phase of the integrated training and transfer concept

Implementation

In the implementation phase of the integrated training and transfer concept

1. selected learning modules are attended by the trainee and
2. the defined transfer project is implemented in parallel.

The implementation of the transfer project can be supported by (a) a coworker or an executive that is qualified in the defined field of action and (b) the trainer of the learning factory training that can act as a coach for the project. Additionally, it can be helpful for the success of the project that coach and trainee define a desired target state for the project that describes as precise as possible the targets and boundary conditions of the project. Those properly defined target states can be used in daily coaching routines to overcome obstacles and to improve processes and to learn.[119] Target states are again oriented on a higher-level vision for production.

Although training and transfer runs in parallel, with regard to the details of the implementation phase, it have to be decided whether the training or the transfer project forms the starting point for the integrated concept; see Fig. 6.47. Both approaches are associated with advantages and disadvantages.

- The training-based approach is to start with the (first) training module and follow with the kickoff of the transfer project later. This approach can be seen as the intuitive one for most people, because it follows the logic, which can also be recognized in instruction situations. First, the trainee learns what and how to do

---
[119]Rother (2009) describes in detail how.

and later he or she applies the learnt in real life. This order is not mandatory because the real application not only follows the learning process but can also be a part of it. Correspondingly, the weakness of the training-based approach is the missing understanding of the real problem situation. Accordingly, the trainee attends the learning module with fewer questions and knowledge about the real problem situation; see also problem pull versus theory push.[120]

- The project-based approach is to start with the project kickoff and continue after that with the (first) training module. This approach can be seen as the more integrated one, where the application in the real factory is more interwoven in the overall learning process. The advantage of the project-based approach is in analogy to the disadvantage of the training-based approach, the problem sensitization before the attendance of the first learning module. The trainee knows accordingly before starting with the first learning module which content will be particularly relevant for him or her and which questions he or she already has on individual topics. The knowledge of the problem or project to be worked on can be motivating in the context of the learning process.

Evaluation

In the evaluation phase of the integrated training and transfer concept

1. results of parallel implementation of transfer project and learning modules are presented,
2. the results of the transfer project are evaluated, and
3. the trainee receives a certificate for the successful transfer project and the attendance of the learning module.

In the presentation of the results, the project coach, executives, and involved colleagues can participate. In particular, the problem definition and the initial situation of the project, the procedure within the transfer project as well as the results of the project, which should also be able to be supported with key performance indicators (KPI) in a before-and-after-project comparison are presented. Suiting KPIs vary depending on the project orientation; for example, they may be reject rate, rework cost, lead time, development time, volume or mix flexibility, productivity metrics, or many more.

The coach and the executive can evaluate the results of the transfer project based on these key performance indicators, target states agreed in advance with the trainee as well as the described trainee's approach. Additionally, potentially positive or negative effects of other improvement measures or events (such as new machines, new employees, new products, or further improvement projects in the area) may overlap within a production area with the transfer project. This cannot be prevented completely and has to be considered in the evaluation.

As an optional last point, the implementation of the transfer projects also provides a good opportunity to reward the gained trainee's qualification with a corresponding

---

[120] See Tisch et al. (2013).

certificate. For this matter, of course certificates have to be standardized with regard to level and content in order to be meaningful for further use. On the one hand, this promotes the trainee's motivation and, on the other, within the company gives indications about the suitability of the trainee for further upcoming improvement projects.

6.3.3.4 Economy of Learning Factories

The estimation of economic risks in the implementation of learning factories in the industrial and the public sector is an important question for many leaning factory operators. The high investments and costs for the setup and the operation of a learning factory on the one hand have to be weighed against the potential benefits on the other. It must be examined in each individual case, whether the investment pays off. The following sections illustrate how to identify and weigh the costs and benefits of learning factories. A distinction is made between

- direct monetary effects that can be attributed to the investment project without analysis or estimation (level I), and
- non-direct monetary effects that can not or only with difficulty be converted into monetary benefit equivalents with a further estimate (level II).

Level 1: Direct monetary effects

Direct monetary effects occur on the cost as well as on the benefit side in the form of cash inflows and outflows. The direct costs can be divided into non-recurring investment costs and operating learning factory costs. Likewise, direct monetary benefits may be of a non-recurring or recurring nature. In order to anticipate the necessary data for the evaluation of the economic aspects as accurately as possible, checklists can be used for the cost and benefit side. Table 6.4 shows a checklist for the costs for the setup and operation of learning factories divided into non-recurring and operating costs. Table 6.5 gives a checklist for the direct monetary benefits of learning factories, also divided into non-recurring and operating monetary benefits.

If the expenses and revenues mentioned in the checklists are known, the level 1 capital budgeting can be carried out on this basis. For the capital budgeting, static and dynamic methods can be distinguished. Static methods use for the calculation only averages of incoming and outgoing payments, making it easier but also less meaningful compared to the dynamic methods that take the time aspect of cash flows into account. Due to simplicity and high significance, the application of the dynamic NPV[121] method is widespread. The NPV method calculates and compares the net present value of the investment alternatives. The net present value is the sum of all discounted cash flows, which are caused by the investment object, to the present value. The net present value can therefore be seen as the equivalent of a payment series of an investment.[122]

---

[121] Net present value.
[122] See Busse von Colbe, Lassmann, and Witte (2015).

Table 6.4 Checklist for non-recurring and operating direct costs of learning factories according to Abele et al. (2015) based on Zangemeister (1993)

	Non-recurring costs	Operating costs
Direct costs of learning factories	1. Planning of the learning system • Internal • External	1. Consumables • Raw materials and supplies • Energy costs
	2. Acquisitions • Land, buildings • Machinery and equipment • Tools and other equipment	2. Personnel costs • Direct labor cost • Overhead costing • Incidental wage costs
	3. Construction effort	3. External services (repairs, material)
	4. Staff cost • Recruitment • Training	4. Allocations • Room costs (land, building) • General works service • IT, etc.
	5. Subsequent investment in upstream and downstream sectors	5. Debt service • Imputed amortization • Imputed interest

Table 6.5 Checklist for direct monetary benefits of learning factories, non-recurring and operating benefits (Abele et al., 2015)

	Non-recurring benefits	Operating benefits
Direct benefits of learning factories	1. Grants/subsidies • Public • Private	1. Trainings • Internal • External
	2. Tax relief • Spending on…	2. Product sales • Contract • On the market
	3. Advertising revenue • Internal • External	3. Use as test bed • Internal • External
	4. Investments • Sale of learning factory share	4. Advertising revenue • Internal • External

6.3 Learning Factory Operation, Evaluation, and Improvement

$$\text{NPV} = \sum_{t=0}^{T} \frac{(\text{ci}_t - \text{co}_t)}{(1+i)^t} = R_0 + \sum_{t=1}^{T} \frac{(R_t)}{(1+i)^t}$$

NPV	Net present value
t	Time of the cash flow
T	Last time of relevant payments
ci_t	Cash inflow in the year t
co_t	Cash outflow in the year t
R_t	Resulting cash flow in the year t (Cash inflow − Cash outflow)
i	Discount rate

In the following, an exemplary calculation of the Learning Factory X at the Fantasy University is presented that calculates in the first year a difference between estimated non-recurring incoming and outgoing cash flows R_0 of a hypothetical learning factory of two million euros (investments in machinery, etc., less one-off income). The detailed estimated initial costs and benefits of the Learning Factory X as basis for the calculation are shown in Table 6.6.

Beyond the initial payments, an annual positive cash flow of 500,000 euros is calculated based on the checklists in Tables 6.4 and 6.5. The detailed estimated operating costs and benefits of the Learning Factory X are listed in Table 6.7.

Having the estimated resulting cash flows R_t of a specific learning factory or learning factory alternatives, the NPV can be calculated by specifying the observation period (T) and the interest rate i with which the investment is compared. If the NPV of an investment is bigger than zero, it is absolutely advantageous compared to the comparative investment with the interest rate i. Furthermore, the investment object whose NPV is bigger than the NPV of all other alternatives is relatively advantageous.[123] Table 6.8 gives calculated NPV for the Learning Factory X regarding the number of regarded periods in years and various interest rates i.

In this example, for the Learning Factory X, independent from the interest rate i in the range from 1 to 10% the payment pays for itself (amortization) after 4 years (highlighted in bold in Table 6.8). Consequently, if the period of observation is 4 years or longer, investing in the Learning Factory X can be considered to be absolutely beneficial. This is achieved in the present example, inter alia with a comparatively high utilization of the learning factory capacity for industrial training; in order to achieve the one million euros in training mode, the training factory must be used for industrial training in 100–150 days per year. Mathematically, the amortization can be calculated using the formula for the NPV and find the period t for which the NPV is positive (or zero) for the first time.

---

[123] See Götze (2014).

Table 6.6 Detailed overview of expected non-recurring costs and direct monetary benefits of the learning factory X

Non-recurring costs calculated/estimated cost in Million €		Non-recurring benefits calculated/estimated benefits in Million €	
1. Planning of the learning system	0.35	1. Grants/subsidies	1
2. Acquisitions	1.5	2. Tax relief	–
3. Construction effort	1.5	3. Advertising revenue	0.15
4. Staff cost (Recruitment, Training)	0.05	4. Investments	0.35
5. Subsequent investment in upstream and downstream sectors	–		
Sum of non-recurring costs (in Million €)	3.4	Sum of non-recurring direct monetary benefits (in Million €)	1.4
Balance of non-recurring costs and direct monetary benefits (R_0)	−2 Million Euro		

Table 6.7 Detailed overview of expected operating costs and direct monetary benefits of the learning factory X

Operating cost calculated/estimated cost in Million €		Operating benefits calculated/estimated benefits in Million €	
1. Consumables	0.1	1. Trainings	1
2. Personnel costs	0.3	2. Product sales	–
3. External services	0.05	3. Use as test bed	0.2
4. Allocations	0.1	4. Advertising revenue	–
5. Debt service	–		
Sum of all operating cost	0.55	Sum of all operating direct monetary benefits	1.2
Balance of operating costs and direct monetary benefits (R_t)	650,000 Euro		

6.3 Learning Factory Operation, Evaluation, and Improvement

Table 6.8 Calculated NPV for the learning factory X depending on interest rate i and the number of regarded periods

Period t	R_t in Million €	NPV for $i =$ 1%	NPV for $i =$ 2%	NPV for $i =$ 5%	NPV for $i =$ 10%
0	−2	−2	−2	−2	−2
1	0.65	−1.36	−1.36	−1.38	−1.41
2	0.65	−0.72	−0.74	−0.79	−0.87
3	0.65	−0.09	−0.13	−0.23	−0.38
4	0.65	**0.54**	**0.48**	**0.30**	**0.06**
5	0.65	1.15	0.98	0.81	0.46
6	0.65	1.77	1.52	1.30	0.83
7	0.65	2.37	2.05	1.76	1.16
8	0.65	2.97	2.56	2.20	1.47
9	0.65	3.57	3.06	2.62	1.74
10	0.65	4.16	3.54	3.02	2

$\min\{t\}$, for which

$$\text{NPV} = \sum_{t=0}^{T} \frac{(ci_t - co_t)}{(1+i)^t} = R_0 + \sum_{t=1}^{T} \frac{(R_t)}{(1+i)^t} \geq 0$$

Of course, in addition to the discount rate, the time in which the investment in a learning factory pays off changes with

- deviating initial net investments R_0 and
- different annual cash flows R_t.

Therefore, Table 6.9 gives an overview on the amortization periods in years depending on R_0 and R_t for an discount rate of 5%. Table 6.9 is visualized regarding the following rules

- Learning factory investments with a calculated amortization period of 5 years or shorter are marked in green (fast amortization, low risk investment).
- Learning factory investments with a calculated amortization period from 6 to including 10 years are marked in yellow (medium amortization, medium risk investment).
- Learning factory investments with a calculated amortization period of 11 years or longer are marked in red (slow amortization, high risk investment).

In the following, besides the direct monetary effects of learning factories regarded at level 1 of the economic financial analysis, level 2 of the economic analysis looks at indirect and non-monetary effects of learning factories.

Level 2: Indirect and non-monetary effects of learning factories

The economic success of learning factories should not be valued solely in directly monetary terms. In addition, efficacy and impact of different learning outcomes

Table 6.9 Amortization periods in years for learning factory investments depending on R_0 and R_t, based on a discount rate of 5%

Yearly cash flow R_t in Euro \ Initial net investment R_0 in Euro	250,000	500,000	1,000,000	1,500,000	2,000,000	2,500,000	3,000,000	4,000,000	5,000,000	7,500,000	10,000,000
50,000	6	15	>20	>20	>20	>20	>20	>20	>20	>20	>20
100,000	3	6	15	>20	>20	>20	>20	>20	>20	>20	>20
150,000	2	4	9	15	>20	>20	>20	>20	>20	>20	>20
200,000	2	3	6	10	15	>20	>20	>20	>20	>20	>20
300,000	1	2	4	6	9	12	15	>20	>20	>20	>20
400,000	1	2	3	5	6	8	10	15	>20	>20	>20
500,000	1	2	3	4	5	6	8	11	15	>20	>20
600,000	1	1	2	3	4	5	6	9	12	>20	>20
700,000	1	1	2	3	4	5	5	7	10	16	>20
800,000	1	1	2	3	3	4	5	6	8	13	>20
900,000	1	1	2	2	3	4	4	6	7	12	17
1,000,000	1	1	2	2	3	3	4	5	6	10	15
1,500,000	1	1	1	2	2	2	3	3	4	6	9
2,000,000	1	1	1	1	2	2	2	3	3	5	6

must be considered in the analysis. In general, it is possible to differentiate between monetary, indirect monetary, and non-monetary effects on the cost and benefit side; see also Fig. 6.48. Based on this classification, extended, multi-dimensional, and decision-oriented procedures can be applied. The aim of this procedure may be to quantify as much of the effects as possible and then monetize them. Table 6.10 defines direct monetary, indirect monetary, and non-monetary effects, gives examples from the manufacturing field, and names assessment tools that can be used for analysis.

Diverse effects of learning factories can be detected in general. Those effects of learning factories have an impact on an individual level, university or industrial level, and societal level—furthermore, those different levels are mutually dependent. Figure 6.49 shows an insight into complex cause-and-effect relationships on the different levels related to the use of learning factories in industry and academia. In order to consider these effects within the framework of a profitability analysis, it must be described in detail and specifically which indirect monetary or non-monetary effects are to be analyzed. Each of the arrows depicted (supposed effects) could be analyzed in detail.

6.4 Remodeling Learning Factory Concepts

Fig. 6.48 Monetary, indirect monetary, and non-monetary effects of learning factories on the cost and benefit side

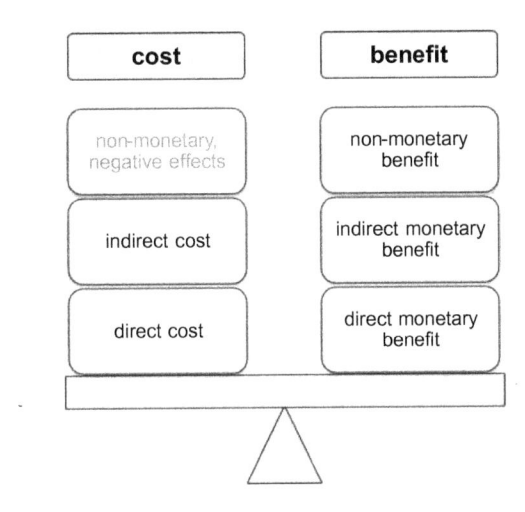

Table 6.10 Distinction between directly monetary, indirectly monetary, and non-monetary effects based on Heinrich and Lehner (2005) and Hanssen (2010)

	Definition	Examples	Assessment tools
Direct monetary effects	Direct monetary effects can be assessed directly in monetary terms	Effective savings, sales increase, increase in production costs, etc.	Capital budgeting, traditional economic analysis
Indirect monetary effects	Indirect monetary effects cannot be assessed directly in monetary terms. In order to quantify the monetary effects, other factors, for example, temporal quantities, must be consulted	Increased employee productivity, higher customer satisfaction, lower absenteeism of employees, etc.	Cost accounting; an analysis of the indirect monetary effects is usually done by capturing quantities
Non-monetary effects	For non-monetary effects, there are no valuation standards available for quantification	Image gains, increased employee morale, increased versatility, increased expertise	Utility analysis, work performance analysis, qualitative analysis

6.4 Remodeling Learning Factory Concepts

The last phase of the learning factory life cycle contains the remodeling or in some cases also the recycling of learning factory parts. This section focuses only on the remodeling of existing learning factory concepts. The need for the remodeling of learning factories can have various sources:

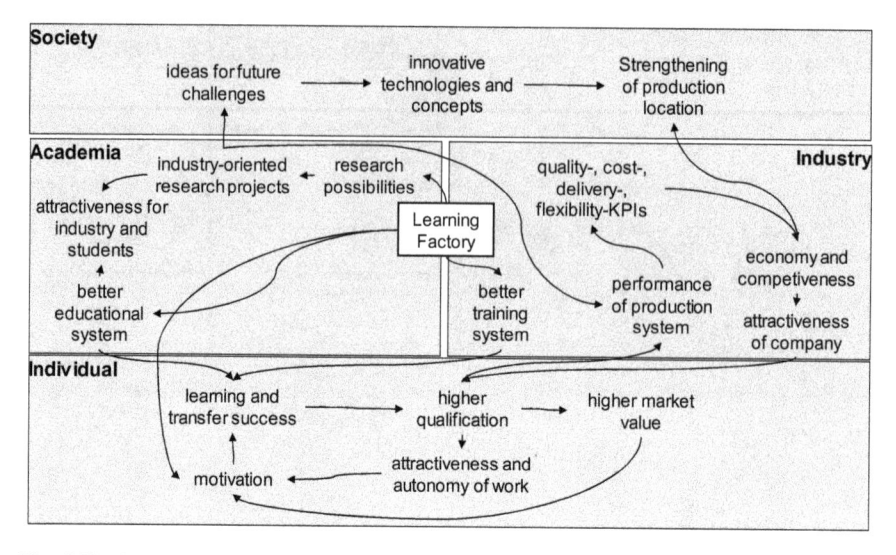

Fig. 6.49 Cause-and-effect diagram regarding indirect and non-monetary effects of learning factories on individual, academic, industrial, and societal levels

- New trends or technologies: Learning factories have to react on new trends or innovative technologies that are coming up. Most recently, this can be recognized in the case of Industrie 4.0 or the Internet of things. With this trends coming up, various learning factories were remodeled in order to be able to open up possibilities for the demonstration, experimentation, and testing of new ideas regarding this trend.[124]
- As a result of evaluation: Remodeling of learning factories and learning factory modules can be a result of the evaluation of learning processes, the learning outcome, or the learning concept.[125]
- Extension of the contentual or target group scope of the learning factory: When new topics or target groups are added to the learning factory program, an extension of the learning factory program as well as most likely an extension of the learning factory environment is needed.

In order to remodel learning factory concepts including the learning environment as well as the learning factory program consisting of various learning factory modules, learning factory design approaches[126] can be used. Often those design

---

[124] See Sect. 8.2 or e.g. Faller and Feldmüller (2015), Wank et al. (2016), Prinz et al. (2016), and Erol, Jäger, Hold, Ott, and Sihn (2016).

[125] Different types of those evaluation approaches are presented in this book with the evaluation of reaction, learning, transfer, and economy (see Sect. 6.3.3) or the evaluation of the complete learning factory concept based on a quality system and a maturity model for learning factories (see Sect. 6.3.2).

[126] Presented in Sect. 6.1.

6.4 Remodeling Learning Factory Concepts

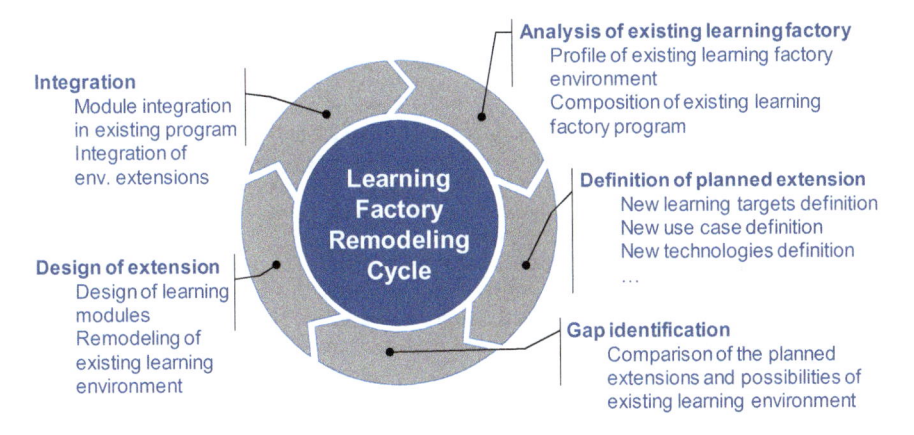

Fig. 6.50 Learning factory remodeling cycle

approaches already contain remodeling aspects for existing learning factories.[127] Figure 6.50 shows[128] a learning factory remodeling cycle that consists of

- the analysis of existing learning factories,
- the definition of the planned extensions, based on this
- a gap identification,
- the design of the extension, and finally,
- the integration of the designed extension into the existing learning factory concept.

6.5 Wrap-Up of This Chapter

This chapter systematizes models and methods in the field of learning factories regarding the planning, the design, the evaluation, improvement, and remodeling of learning factories. All methods and models are structured along the learning factory life cycle starting from the customer requirements and the company goals until the remodeling of existing learning factory concepts.

References

Abel, M., Czajkowski, S., Faatz, L., Metternich, J., & Tenberg, R. (2013). Kompetenzorientiertes Curriculum für Lernfabriken. *Werkstattstechnik Online: Wt, 103*(3), 240–245.

---

[127] See, for example, Tisch, Hertle, Abele et al. (2015) or Plorin et al. (2015).

[128] Derived from design approaches.

Abele, E., Chryssolouris, G., Sihn, W., Metternich, J., ElMaraghy, H. A., Seliger, G., et al. (2017). Learning factories for future oriented research and education in manufacturing. *CIRP Annals—Manufacturing Technology, 66*(2), 803–826.

Abele, E., Eichhorn, N., & Kuhn, S. (2007). Increase of productivity based on capability building in a learning factory. In *Computer Integrated Manufacturing and High Speed Machining: 11th International Conference on Production Engineering*, Zagreb (pp. 37–41).

Abele, E., Metternich, J., Tenberg, R., Tisch, M., Abel, M., Hertle, C., et al. (2015). *Innovative Lernmodule und -fabriken: Validierung und Weiterentwicklung einer neuartigen Wissensplattform für die Produktionsexzellenz von morgen*. Darmstadt: tuprints.

Abele, E., Tenberg, R., Wennemer, J., & Cachay, J. (2010). Kompetenzentwicklung in Lernfabriken für die Produktion. *Zeitschrift für wirtschaftlichen Fabrikbetrieb (ZWF), 105*(10), 909–913.

Alliger, G. M., Tamnenbaum, S. I., Bennett, W., Jr., Traver, H., & Shotland, A. (1997). A meta-analysis of the relations among training criteria. *Personnel Psychology, 50*(2), 341–358.

Anderson, L. W., Krathwohl, D. R., & Airasian, P. W. (Eds.). (2001). *A taxonomy for learning, teaching, and assessing: A revision of bloom's taxonomy of educational objectives* (Complete ed.). New York: Longman.

Balve, P., & Albert, M. (2015). Project-based learning in production engineering at the Heilbronn learning factory. In *5th CIRP-Sponsored Conference on Learning Factories. Procedia CIRP, 32*, 104–108. https://doi.org/10.1016/j.procir.2015.02.215.

Becker, F. G., Wittke, I., & Friske, V. (2010). *"Happy Sheets": Empirische Befragung von Bildungsträgern*. Retrieved from http://nbn-resolving.de/urn/resolver.pl?urn=urn:nbn:de:0070-bipr-48208.

Becker, M. (2005). *Systematische Personalentwicklung: Planung, Steuerung und Kontrolle im Funktionszyklus*. Stuttgart: Schäffer-Poeschel.

Bender, B., Kreimeier, D., Herzog, M., & Wienbruch, T. (2015). Learning factory 2.0—Integrated view of product development and production. In *5th CIRP-sponsored Conference on Learning Factories. Procedia CIRP, 32*, 98–103. https://doi.org/10.1016/j.procir.2015.02.226.

Bloom, B. S., Engelhart, M. D., Furst, E. J., Hill, W. H., & Krathwohl, D. R. (1956). *Taxonomy of educational objectives: The classification of educational goals* (Book I: Cognitive Domain). New York: McKay.

Böhner, J., Weeber, M., Kuebler, F., & Steinhilper, R. (2015). Developing a learning factory to increase resource efficiency in composite manufacturing processes. In *5th CIRP-Sponsored Conference on Learning Factories. Procedia CIRP, 32*, 64–69. https://doi.org/10.1016/j.procir.2015.05.003.

Bommer, W. H., Johnson, L. J., Rich, G. A., Podsakoff, P. M., & Mackenzie, S. B. (1995). On the interchangeability of objective and subjective measures of employee performance: A meta-analysis. *Personnel Psychology, 48*(1), 587–605.

Burlingame, G. M., Seaman, S., Johnson, J. E., Whipple, J., Richardson, E., Rees, F., et al. (2006). Sensitivity to change of the Brief Psychiatric Rating Scale- Extended (BPRS- E): An item and subscale analysis. *Psychological Services, 3*(2), 77–87.

Busse von Colbe, W., Lassmann, G., & Witte, F. (2015). *Investitionstheorie und Investitionsrechnung* (4., vollständig überarbeitete Auflage). Berlin, Heidelberg: Springer.

Cachay, J., Wennemer, J., Abele, E., & Tenberg, R. (2012). Study on action-oriented learning with a learning factory approach. *Procedia—Social and Behavioral Sciences, 55*, 1144–1153.

Chomsky, N. (1962). Explanatory models in linguistics. In E. Nagel, P. Suppes, & A. Tarski (Eds.), *Logic, methodology and philosophy of science* (pp. 528–550). Stanford: Stanford University Press.

Clasen, H. (2010). *Die Messung von Lernerfolg: Eine grundsätzliche Aufgabe der Evaluation von Lehr- bzw. Trainingsinterventionen* (Dissertation). Dresden.

Dagley, D. L., & Orso, J. K. (1991). Integrating summative, formative modes of evaluation. *NASSP Bulletin, 75*(536), 72–82. https://doi.org/10.1177/019263659107553610.

Dinkelmann, M. (2016). *Methode zur Unterstützung der Mitarbeiterpartizipation im Change Management der variantenreichen Serienproduktion durch Lernfabriken* (Dissertation). Stuttgart.

References

Doch, S., Merkler, S., Straube, F., & Roy, D. (2015). Aufbau und Umsetzung einer Lernfabrik. Produktionsnahe Lean-Weiterbildung in der Prozess- und Pharmaindustrie [lit. trans.: Construction and implementation of a learning factory. Lean advanced training in the process and pharmaceutical industry.]. *Industrie Management* (03), 26–30.

ElMaraghy, H. A., & ElMaraghy, W. (2015). Learning integrated product and manufacturing systems. In *5th CIRP-Sponsored Conference on Learning Factories. Procedia CIRP, 32*, 19–24.

Elster, F., Dippl, Z., & Zimmer, G. (2003). *Wer bestimmt den Lernerfolg? Leistungsbeurteilung in projektorientierten Lernarrangements.* Bielefeld: Bertelsmann.

Enke, J., Glass, R., & Metternich, J. (2017). Introducing a maturity model for learning factories. In *7th CIRP-Sponsored Conference on Learning Factories. Procedia Manufacturing, 9*, 1–8.

Enke, J., Kaiser, J., & Metternich, J. (2017b). Die Lernfabrik als Export-Erfolg. *Zeitschrift für wirtschaftlichen Fabrikbetrieb (ZWF), 112*(10), 644–647.

Enke, J., Kraft, K., & Metternich, J. (2015). Competency-oriented design of learning modules. In *5th CIRP-Sponsored Conference on Learning Factories. Procedia CIRP, 32*, 7–12. https://doi.org/10.1016/j.procir.2015.02.211.

Enke, J., Metternich, J., Bentz, D., & Klaes, P.-J. (2018). Systematic learning factory improvement based on maturity level assessment. In *8th CIRP Conference on Learning Factories—Advanced Engineering Education & Training for Manufacturing Innovation* (Vol. 23, pp. 51–56).

Enke, J., Tisch, M., & Metternich, J. (2016). Learning factory requirements analysis—Requirements of learning factory stakeholders on learning factories. In *5th CIRP Global Web Conference (CIRPe 2016). Procedia CIRP, 55*, 224–229.

Erol, S., Jäger, A., Hold, P., Ott, K., & Sihn, W. (2016). Tangible industry 4.0: A scenario-based approach to learning for the future of production. In: *6th CIRP-Sponsored Conference on Learning Factories. Procedia CIRP, 54*, 13–18.

Erpenbeck, J., & von Rosenstiel, L. (2007). *Handbuch Kompetenzmessung: Erkennen, verstehen und bewerten von Kompetenzen in der betrieblichen, pädagogischen und psychologischen Praxis* (2., überarb. und erw. Aufl.). Stuttgart: Schäffer-Poeschel.

Faller, C., & Feldmüller, D. (2015). Industry 4.0 learning factory for regional SMEs. In *5th CIRP-Sponsored Conference on Learning Factories. Procedia CIRP, 32*, 88–91. https://doi.org/10.1016/j.procir.2015.02.117.

Gebbe, C., Hilmer, S., Götz, G., Lutter-Günther, M., Chen, Q., Unterberger, E., et al. (2015). Concept of the Green Factory Bavaria in Augsburg. In *5th CIRP-Sponsored Conference on Learning Factories. Procedia CIRP, 32*, 53–57. https://doi.org/10.1016/j.procir.2015.02.214.

Gessler, M. (2005). Gestaltungsorientierte Evaluation und der Return on Investment von Weiterbildungsprogrammen. *bwp@* (9), 1–25.

Goerke, M., Schmidt, M., Busch, J., & Nyhuis, P. (2015). Holistic approach of lean thinking in learning factories. In *5th CIRP-Sponsored Conference on Learning Factories. Procedia CIRP, 32*, 138–143. https://doi.org/10.1016/j.procir.2015.02.221.

Götze, U. (2014). *Investitionsrechnung: Modelle und Analysen zur Beurteilung von Investitionsvorhaben* (7. Aufl.). *Springer-Lehrbuch*. Berlin: Springer Gabler.

Hambach, J., Diezemann, C., Tisch, M., & Metternich, J. (2016). Assessment of students' lean competencies with the help of behavior video analysis—Are good students better problem solvers? In *5th CIRP Global Web Conference (CIRPe 2016). Procedia CIRP, 55*, 230–235.

Hanssen, S.-C. (2010). *Bestimmung und Bewertung der Wirkungen von Informationssystemen* (Dissertation).

Harris, M. M., & Schaubroeck, J. (1988). A metaanalysis of self-supervisor, self-peer, and peer-supervisor ratings. *Personnel Psychology, 41*(1), 43–62.

Hart, J. (2015). *Modern workplace learning: A resource book for L&D.* Corsham, Wiltshire, England: Centre for Learning & Performance Technologies.

Heinrich, L. J., & Lehner, F. (2005). *Informationsmanagement: Planung, Überwachung und Steuerung der Informationsinfrastruktur* (8., vollst. überarb. und erg. Aufl.). *Wirtschaftsinformatik.* München, Wien: Oldenbourg.

Hertle, C., Tisch, M., Kläs, H., & Metternich, J. (2016). Recording shop floor management competencies—A guideline for a systematic competency gap analysis. In *49th CIRP Conference on Manufacturing Systems. Procedia CIRP, 57*, 625–630.

Hummel, V., Hyra, K., Ranz, F., & Schuhmacher, J. (2015). Competence development for the holistic design of collaborative work systems in the logistics learning factory. In *5th CIRP-Sponsored Conference on Learning Factories. Procedia CIRP, 32*, 76–81. https://doi.org/10.1016/j.procir.2015.02.111.

Jäger, A., Mayrhofer, W., Kuhlang, P., Matyas, K., & Sihn, W. (2012). The "Learning Factory": An immersive learning environment for comprehensive and lasting education in industrial engineering. In *16th World Multi-Conference on Systemics, Cybernetics and Informatics, 16*(2), 237–242.

Jäger, A., Mayrhofer, W., Kuhlang, P., Matyas, K., & Sihn, W. (2013). Total immersion: Hands and heads-on training in a learning factory for comprehensive industrial engineering education. *International Journal of Engineering Education, 29*(1), 23–32.

Jorgensen, J. E., Lamancusa, J. S., Zayas-Castro, J. L., & Ratner, J. (1995). The learning factory: Curriculum integration of design and manufacturing. In *4th World Conference on Engineering Education* (pp. 1–7).

Kaluza, A., Juraschek, M., Neef, B., Pittschellis, R., Posselt, G., Thiede, S., et al. (2015). Designing learning environments for energy efficiency through model scale production processes. In *5th CIRP-Sponsored Conference on Learning Factories. Procedia CIRP, 32*, 41–46. https://doi.org/10.1016/j.procir.2015.02.114.

Kirkpatrick, D. L. (1998). *Evaluating training programs: The four levels* (2nd ed.). San Francisco, California: Berrett-Koehler Publishers.

Kreitlein, S., Höft, A., Schwender, S., & Franke, J. (2015). Green factories Bavaria: A network of distributed learning factories for energy efficient production. In *5th CIRP-Sponsored Conference on Learning Factories. Procedia CIRP, 32*, 58–63. https://doi.org/10.1016/j.procir.2015.02.219.

Krückhans, B., Wienbruch, T., Freith, S., Oberc, H., Kreimeier, D., & Kuhlenkötter, B. (2015). Learning factories and their enhancements—A comprehensive training concept to increase resource efficiency. In *5th CIRP-sponsored Conference on Learning Factories. Procedia CIRP, 32*, 47–52. https://doi.org/10.1016/j.procir.2015.02.224.

Lanza, G., Moser, E., Stoll, J., & Haefner, B. (2015). Learning factory on global production. In *5th CIRP-sponsored Conference on Learning Factories. Procedia CIRP, 32*, 120–125. https://doi.org/10.1016/j.procir.2015.02.081.

Lohaus, D., & Habermann, W. (2011). *Weiterbildung im Mittelstand: Personalentwicklung und Bildungscontrolling in kleinen und mittleren Unternehmen*. München: Oldenbourg.

Mabe, P. A., & West, S. G. (1982). Validity of self-evaluation of ability: A review and meta-analysis. *Journal of Applied Psychology, 67*(3), 280. https://doi.org/10.1037/0021-9010.67.3.280.

Matt, D. T., Rauch, E., & Dallasega, P. (2014). Mini-factory—A learning factory concept for students and small and medium sized enterprises. In *47th CIRP Conference on Manufacturing Systems. Procedia CIRP, 17*, 178–183.

McMillan, J. H. (1997). *Classroom assessment: Principles and practice for effective instruction*. Boston: Allyn and Bacon.

Metternich, J., Abele, E., & Tisch, M. (2013). Current activities and future challenges of the process learning factory CiP. In G. Reinhart, P. Schnellbach, C. Hilgert, & S. L. Frank (Eds.), *3rd Conference on Learning Factories*, Munich, May 7th, 2013 (pp. 94–107). Augsburg.

Nerdinger, F. W., Blickle, G., & Schaper, N. (2014). *Arbeits- und Organisationspsychologie: Mit 51 Tabellen* (3., vollst. überarb. Aufl. 2014). *Springer-Lehrbuch*. Berlin: Springer.

Nöhring, F., Rieger, M., Erohin, O., Deuse, J., & Kuhlenkötter, B. (2015). An interdisciplinary and hands-on learning approach for industrial assembly systems. In *5th CIRP-Sponsored Conference on Learning Factories. Procedia CIRP, 32*, 109–114. https://doi.org/10.1016/j.procir.2015.02.112.

Pedrazzoli, P., Rovere, D., Constantinescu, C., Bathelt, J., Pappas, M., Dépincé, P., et al. (2007). High value adding VR tools for networked customer-driven. In *4th International Conference on Digital Enterprise Technology (DET2007)* (pp. 19–21).

References

Pietzner, V. (2002). *Lernkontrolle im "Vernetzten Studium Chemie": Entwicklung und Evaluation eines Konzepts am Beispiel des Kapitels "Addition von Halogenen an Doppelbindungen"* (Dissertation). Braunschweig.

Pittich, D. (2014). *Diagnostik fachlich-methodischer Kompetenzen. Wissenschaft* (Vol. 37). Stuttgart: Fraunhofer IRB Verlag.

Pittschellis, R. (2015). Multimedia support for learning factories. In *5th CIRP-sponsored Conference on Learning Factories. Procedia CIRP, 32*, 36–40. https://doi.org/10.1016/j.procir.2015.06.001.

Plorin, D. (2016). *Gestaltung und Evaluation eines Referenzmodells zur Realisierung von Lernfabriken im Objektbereich der Fabrikplanung und des Fabrikbetriebes* (Dissertation). Chemnitz. *Wissenschaftliche Schriftenreihe des Instituts für Betriebswissenschaften und Fabriksysteme: Heft 120*. Chemnitz: Technische Universität Institut für Betriebswissenschaften und Fabriksysteme.

Plorin, D., Jentsch, D., Hopf, H., & Müller, E. (2015). Advanced learning factory (ALF)—Method, implementation and evaluation. In *5th CIRP-Sponsored Conference on Learning Factories. Procedia CIRP, 32*, 13–18. https://doi.org/10.1016/j.procir.2015.02.115.

Preussler, A., & Baumgartner, P. (2006). Qualitätssicherung in mediengestützten Lernprozessen - sind theoretische Konstrukte messbar? In A. Sindler, C. Bremer, U. Dittler, P. Hennecke, C. Sengstag, & J. Wedekind (Eds.), *Qualitätssicherung im E-Learning* (pp. 73–85). Münster: Waxmann.

Prinz, C., Morlock, F., Freith, S., Kreggenfeld, N., Kreimeier, D., & Kuhlenkötter, B. (2016). Learning factory modules for smart factories in industrie 4.0. In *6th CIRP-Sponsored Conference on Leanring Factories. Procedia CIRP, 54*, 113–118. https://doi.org/10.1016/j.procir.2016.05.105.

PTW, TU Darmstadt. (2017). *Prozesslernfabrik CiP: Der Weg zur operativen Exzellenz*. Retrieved from http://www.prozesslernfabrik.de/.

Reiner, D. (2009). *Methode der kompetenzorientierten Transformation zum nachhaltig schlanken Produktionssystem* (Dissertation). Darmstadt. *Schriftenreihe des PTW: "Innovation Fertigungstechnik"*. Aachen: Shaker.

Reiß, M. (2012). *Change management: A balanced and blended approach*. Norderstedt: Books on Demand.

Rentzos, L., Doukas, M., Mavrikios, D., Mourtzis, D., & Chryssolouris, G. (2014). Integrating manufacturing education with industrial practice using teaching factory paradigm: A construction equipment application. In *47th CIRP Conference on Manufacturing Systems. Procedia CIRP, 17*, 189–194.

Riffelmacher, P. (2013). *Konzeption einer Lernfabrik für die variantenreiche Montage* (Dissertation). Stuttgart. *Stuttgarter Beiträge zur Produktionsforschung* (Vol. 15). Stuttgart: Fraunhofer Verlag.

Rother, M. (2009). *Toyota Kata: Managing people for improvement, adaptiveness and superior results*. New York: McGraw-Hill Education.

Schermutzki, M. (2007). *Lernergebnisse—Begriffe, Zusammenhänge, Umsetzung und Erfolgsermittlung: Lernergebnisse und Kompetenzvermittlung als elementare Orientierungen des Bologna-Prozesses*. Aachen: Fachhochschule Aachen.

Seitz, K.-F., & Nyhuis, P. (2015). Cyber-physical production systems combined with logistic models—A learning factory concept for an improved production planning and control. In *5th CIRP-sponsored Conference on Learning Factories. Procedia CIRP, 32*, 92–97. https://doi.org/10.1016/j.procir.2015.02.220.

Seufert, S., & Euler, D. (2005). *Learning Design: Gestaltung eLearning-gestützter Lernumgebungen in Hochschulen und Unternehmen. SCIL-Arbeitsbericht* (Vol. 5). St. Gallen: SCIL.

Smith, A. (2001). *Return on investment in training: Research readings*. Leabrook, S. Aust.: NCVER.

Solga, M. (2011). Evaluation der Personalentwicklung. In *Praxishandbuch Personalentwicklung: Instrumente, Konzepte, Beispiele* (pp. 369–399). Wiesbaden: Gabler.

Statistisches Bundesamt. (2017). *Statistisches Jahrbuch 2017: Gewerbe und Dienstleistungen im Überblick*. Retrieved from https://www.destatis.de/DE/Publikationen/StatistischesJahrbuch/ProdGewerbeDienstleistungen.pdf?__blob=publicationFile.

Steffen, M., Frye, S., & Deuse, J. (2013a). The only source of knowledge is experience: Didaktische Konzeption und methodische Gestaltung von Lehr-Lern-Prozessen in Lernfabriken zur Aus- und

Weiterbildung im Industrial Engineering. *TeachING LearnING. EU. Innovationen für die Zukunft der Lehre in den Ingenieurwissenschaften*, 117–129.

Steffen, M., Frye, S., & Deuse, J. (2013b). Vielfalt Lernfabrik - Morphologie zu Betreibern, Zielgruppen und Ausstattungen von Lernfabriken im Industrial Engineering. *Werkstattstechnik Online: Wt, 103*(3), 233–239.

Steffen, M., May, D., & Deuse, J. (2012). The industrial engineering laboratory: Problem based learning in industrial engineering education at TU Dortmund University. In *Global Engineering Education Conference (EDUCON), IEEE—Collaborative Learning & New Pedagogic Approaches in Engineering Education*, Marrakesch, Marokko, 17–20.04.2012 (pp. 1–10).

Stufflebeam, D. L. (1972). *Educational evaluation decision making* (3rd ed.). Bloomington: PDK National Study Committee on Evaluation.

Stuttgarter Produktionsakademie. (2017). *Produktion strategisch gestalten: vom Produkt bis zur Wertschöpfungsverteilung*. Retrieved from http://www.stuttgarter-produktionsakademie.de/.

Tisch, M. (2018). *Modellbasierte Methodik zur kompetenzorientierten Gestaltung von Lernfabriken für die schlanke Produktion* (Dissertation). Darmstadt. Aachen: Shaker.

Tisch, M., Hertle, C., Abele, E., Metternich, J., & Tenberg, R. (2015a). Learning factory design: A competency-oriented approach integrating three design levels. *International Journal of Computer Integrated Manufacturing, 29*(12), 1355–1375. https://doi.org/10.1080/0951192X.2015.1033017.

Tisch, M., Hertle, C., Cachay, J., Abele, E., Metternich, J., & Tenberg, R. (2013). A systematic approach on developing action-oriented, competency-based learning factories. In *46th CIRP Conference on Manufacturing Systems. Procedia CIRP, 7*, 580–585.

Tisch, M., Hertle, C., Metternich, J., & Abele, E. (2014). Lernerfolgsmessung in Lernfabriken: Kompetenzorientierte Weiterentwicklung praxisnaher Schulungen. *Industrie Management, 30*(3), 39–42.

Tisch, M., Hertle, C., Metternich, J., & Abele, E. (2015b). Goal-oriented improvement of learning factory trainings. *The Learning Factory, An Annual Edition from the Network of Innovative Learning Factories, 1*(1), 7–12.

Tisch, M., & Metternich, J. (2017). Potentials and limits of learning factories in research, innovation transfer, education, and training. In *7th CIRP-Sponsored Conference on Learning Factories. Procedia Manufacturing* (in Press).

Tisch, M., Ranz, F., Abele, E., Metternich, J., & Hummel, V. (2015). Learning factory morphology: Study on form and structure of an innovative learning approach in the manufacturing domain. In *TOJET*, July 2015 (Special Issue 2 for *International Conference on New Horizons in Education*, 2015) (pp. 356–363).

Tvenge, N., Martinsen, K., & Kolla, S. S. V. K. (2016). Combining learning factories and ICT-based situated learning. In *6th CIRP-Sponsored Conference on Leanring Factories. Procedia CIRP, 54*, 101–106.

VDI. (1993). *VDI 2221/Methodik zum Entwickeln und Konstruieren technischer Systeme und Produkte. VDI-Richtlinien*. Berlin: Beuth.

Wagner, U., AlGeddawy, T., ElMaraghy, H. A., & Müller, E. (2014). Product family design for changeable learning factories. In *47th CIRP Conference on Manufacturing Systems. Procedia CIRP, 17*, 195–200. https://doi.org/10.1016/j.procir.2014.01.119.

Wagner, U., AlGeddawy, T., ElMaraghy, H. A., & Müller, E. (2015). Developing products for changeable learning factories. *CIRP Journal of Manufacturing Science and Technology, 9*, 146–158.

Wank, A., Adolph, S., Anokhin, O., Arndt, A., Anderl, R., & Metternich, J. (2016). Using a learning factory approach to transfer industrie 4.0 approaches to small- and medium-sized enterprises. In *6th CIRP-Sponsored Conference on Leanring Factories. Procedia CIRP, 54*, 89–94. https://doi.org/10.1016/j.procir.2016.05.068.

Zangemeister, C. (1993). *Erweiterte Wirtschaftlichkeits-Analyse (EWA): Grundlagen und Leitfaden für ein "3-Stufen-Verfahren" zur Arbeitssystembewertung. Schriftenreihe der Bundesanstalt für Arbeitsschutz Forschung* (Vol. 676). Dortmund: Wirtschaftsverl. NW Verl. für Neue Wiss.

Chapter 7
Overview on Existing Learning Factory Application Scenarios

As shown in earlier chapters, learning factory implementations spread in recent years around the globe, particularly in Europe. Those learning factories vary in size, purpose, scope, and sophistication targeting an enhanced learning experience for trainees from industry and academia in areas of manufacturing engineering knowledge. The following three chapters try to map the variety of existing learning factories, for example, regarding the general concept, the used equipment, or the industries addressed. Therefore, learning factory concepts are presented and structured regarding the most common application scenarios of learning factories,[1] the content which is addressed in learning factories,[2] and further variations of the general learning factory concept.[3] Additionally to presented classifications and descriptions, more than thirty learning factory best practice examples are presented.[4] Many learning factory operators very active in the field of learning factories describe their best practice learning factory implementation as guest authors. Those essential contributions to the book demonstrate the relevance learning factories in academia and industry.

Table 7.1 gives an overview of established learning factories around the globe, among those are of course also the ones presented in the guest contributions; the last column contains a cross-reference to the section in the book in which the respective learning factory is presented in detail. Furthermore, the list gives some important information about those learning factories like the location of the learning factory, the key topics or the operator of the learning factory. Additionally, Fig. 7.1 gives an overview of the structure of the following chapters laying out the variety of existing learning factories described in this chapter.

This chapter takes closer look at the most common application scenarios of existing learning factories. The primary purposes of learning factories and accordingly the most common application scenarios of learning factories are education, training,

---

[1] See this chapter.
[2] See Chap. 8.
[3] See Chap. 9.
[4] See Chap. 11.

© Springer Nature Switzerland AG 2019
E. Abele et al., *Learning Factories*, https://doi.org/10.1007/978-3-319-92261-4_7

Table 7.1 Overview on existing learning factories around the globe (extract)

Name of learning factory	Location	Operated by	Key topics	Research	Education	Training	Transfer	See also
AAU Smart Production Laboratory	Aalborg, Denmark	DMP, Aalborg University	Industrie 4.0, smart production		x	x		Madsen and Møller (2017)
Alberta Learning Factory	Edmonton, Canada	Mechanical Engineering, University of Alberta	Industry 4.0, lean management	x	x			Ahmad, Masse, Jituri, Doucette, and Mertiny (2018)
AM model factory	Nuremberg, Germany	FAPS/KTmfK, FAU Erlangen-Nürnberg, McKinsey	Additive manufacturing	x		x	x	Yoo et al. (2016)
Anglo American Training Center	Johannesburg, South Africa	Anglo American	Business improvement, lean production			x		Makumbe, Hattingh, Plint, and Esterhuizen (2018)
AutoFab	Darmstadt, Germany	University of Applied Sciences Darmstadt	Automation, Industrie 4.0	x	x	x	x	Best Practice Example 1

(continued)

7 Overview on Existing Learning Factory Application Scenarios 201

Table 7.1 (continued)

Name of learning factory	Location	Operated by	Key topics	Research	Education	Training	Transfer	See also
Bernard M. Gordon Learning Factory	State College, Pennsylvania, USA	Pennsylvania State University	Engineering design		x			See, for example, Lamancusa and Simpson (2004) and Penn State University (2017)
BERTHA	Bremen, Germany	Bremen Institute for Mechanical Engineering, University Bremen	Manual assembly	x	x	x		Schreiber, Funke, and Tracht (2016)
CETPM Akademie	Ansbach, Germany	HS Ansbach	Lean production			x		https://www.cetpm.de
CubeFactory	no fixed location	iwf, TU Berlin	Sustainability			x		Muschard and Seliger (2015)
Demonstration Factory	Aachen, Germany	wzl, RWTH Aachen	Industrie 4.0, prototypes and industrialization	x		x	x	Best Practice Example 2

(continued)

Table 7.1 (continued)

Name of learning factory	Location	Operated by	Key topics	Research	Education	Training	Transfer	See also
Die Lernfabrik	Braunschweig, Germany	iwf, TU Braunschweig	Sustainable production, cyber-physical production, urban production	x	x	x	x	Best Practice Example 3
ElDrive-Center	Nuremberg, Germany	FAPS, FAU Erlangen-Nürnberg	Production technology	x	x	x	x	Best Practice Example 4
E^3-Forschungsfabrik	Chemnitz, Germany	Fraunhofer IWU	Resource efficiency, Industrie 4.0	x			x	Putz (2013)
EDF (Experimentier- und Digitalfabrik)	Chemnitz, Germany	FPL, TU Chemnitz	Factory planning and operation	x	x	x		https://www.tu-chemnitz.de/mb/FabrPlan/edf.php
Electronics Production	Nuremberg, Germany	FAPS, FAU Erlangen-Nürnberg	Electronics production	x	x	x	x	Best Practice Example 15
ESB logistics learning factory	Reutlingen, Germany	ESB Business School, Reutlingen University	Intralogistics, Industrie 4.0, process and product planning	x	x	x	x	Best Practice Example 5

(continued)

7 Overview on Existing Learning Factory Application Scenarios

Table 7.1 (continued)

Name of learning factory	Location	Operated by	Key topics	Research	Education	Training	Transfer	See also
ETA-Factory	Darmstadt, Germany	PTW, TU Darmstadt	Energy efficiency, energy flexibility	x	x	x	x	Best Practice Example 6
Festo Learning Factory Scharnhausen	Scharnhausen, Germany	Festo AG	Workplace-oriented trainings, Industry 4.0, lean production	x		x		Best Practice Example 8
FMS Training Center	Tampere, Finland	Laboratory of Mechanical Engineering and Industrial Systems, Tampere University of Technology	Automation, digital twin		x			Toivonen, Lanz, Nylund, and Nieminen (2018)

(continued)

Table 7.1 (continued)

Name of learning factory	Location	Operated by	Key topics	Research	Education	Training	Transfer	See also
Green Factory Bavaria	Amberg; Ansbach; Augsburg; Bayreuth; Coburg; Erlangen; Fürth; Hof; Ingolstadt; Munich; Nuremberg; Schweinfurt, Germany	12 universities and Fraunhofer institutes in Bavaria	Resource and energy efficiency	x		x	x	FAPS (2018)
iFactory, iDesign, iPlan	Windsor, Canada	Intelligent Manufacturing Systems (IMS) Center, University of Windsor	Integrated products-systems learning, Industry 4.0	x	x	x	x	Best Practice Example 9
IFA-Learning Factory	Hanover, Germany	IFA, LU Hannover	Lean production, factory planning, PPC	x	x	x		Best Practice Example 10

(continued)

7 Overview on Existing Learning Factory Application Scenarios 205

Table 7.1 (continued)

Name of learning factory	Location	Operated by	Key topics	Research	Education	Training	Transfer	See also
Industrial Engineering Laboratory	Dortmund, Germany	IPS, TU Dortmund	Design of work systems, industrial engineering		x			See, for example, Steffen, May, and Deuse (2012)
Integrated Learning Factory	Seattle, Washington, USA	University of Washington	Engineering design		x			University of Washington (2018)
Integrated Learning Factory	Bochum, Germany	LPE/LPS, RU Bochum	Product development and production	x	x			Best Practice Example 11
Laboratory for flexible industrial automation	Paderborn, Germany	Heinz Nixdorf Institute, University of Paderborn	CPPS, holistic product creation	x	x			Gräßler, Pöhler, and Pottebaum (2016) and Gräßler, Taplick, and Yang (2016)
Lean Academy	Winnenden, Germany and many locations worldwide	Kärcher	Lean production, industrie 4.0			x		See, for example, Kärcher (2018) and Thomar (July, 8th, 2015)

(continued)

Table 7.1 (continued)

Name of learning factory	Location	Operated by	Key topics	Research	Education	Training	Transfer	See also
Lean Laboratory	Gjøvik, Norway	NTNU	Lean production		x	x		See, for example, Tvenge, Martinsen, and Kolla (2016)
Lean Manufacturing Laboratory	Luxembourg, Luxembourg	University of Luxembourg	Lean manufacturing	x	x			Oberhausen and Plapper (2015)
LEAN-Factory	Berlin, Germany	Pharmaceutical Company, TU Berlin	Lean management, sustainability		x	x		Best Practice Example 12
LeanLaboratory	Graz, Austria	Institute of Innovation and Industrial Management, Graz University of Technology	Lean, Industrie 4.0, industrial engineering, logistics	x	x	x		Karre, Hammer, Kleindienst, and Ramsauer (2017)
Learning & Innovation Factory	Vienna, Austria	imw, TU Wien	Integrated product and process planning, optimization		x			Best Practice Example 13

(continued)

7 Overview on Existing Learning Factory Application Scenarios

Table 7.1 (continued)

Name of learning factory	Location	Operated by	Key topics	Research	Education	Training	Transfer	See also
Learning Factory aIE	Stuttgart, Germany	IFF, University Stuttgart	Lean production, quality management	x	x	x	x	Best Practice Example 14
Learning Factory at the Campus Velbert/Heiligenhaus	Heiligenhaus, Germany	University of Applied Sciences Bochum	Automation, Industrie 4.0, energy efficiency		x	x		Faller and Feldmüller (2015)
Learning Factory for Global Production	Karlsruhe, Germany	wbk, KIT Karlsruhe	Global production networks, lean production, assembly planning, industry 4.0	x	x	x	x	Best Practice Example 17
Learning Factory for Innovation, Manufacturing and Cooperation	Heilbronn, Germany	Faculty of Industrial and Process Engineering, Heilbronn University of Applied Sciences	Engineering methods along product creation, cooperation	x	x		x	Best Practice Example 16

(continued)

Table 7.1 (continued)

Name of learning factory	Location	Operated by	Key topics	Research	Education	Training	Transfer	See also
Learning Factory Split	Split, Croatia	FESB, University Split	Lean manufacturing		x			See, for example, Veza, Gjeldum, and Mladineo (2015) and Veza et al. (2017)
Lernfabrik 4.0	16 locations in Baden-Württemberg	Vocational Schools in Baden-Württemberg	Industrie 4.0		x			See Sect. 7.1.9
Lernfabrik für Schlanke Produktion (LSP)	Munich, Germany	iwb, TU München	Lean philosophy, lean assembly		x	x	x	Best Practice Example 18
Lernfabrik für vernetzte Produktion	Augsburg, Germany	Fraunhofer IGCV	Digitization, paperless production			x		Best Practice Example 19
Live Training Center	Bruchsal, Germany	SEW Eurodrive	Lean production			x		Markus Reichert (2011)
LMS Factory	Patras, Greece	LMS, University Patras	Engineering		x	x	x	Best Practice Example 20

(continued)

7 Overview on Existing Learning Factory Application Scenarios

Table 7.1 (continued)

Name of learning factory	Location	Operated by	Key topics	Research	Education	Training	Transfer	See also
LPS Learning Factory	Bochum, Germany	LPS, RU Bochum	Resource efficiency, lean production, Industrie 4.0, workers' participation, labor 4.0	x	x	x	x	Best Practice Example 21
MAN Learning Factory	Berlin, Germany	MAN Diesel & Turbo SE	Assembly and maintenance of compressors		x	x		Best Practice Example 22
Micromanu	Stellenbosch, South Africa	IBi, Stellenbosch University	Rapid prototyping, manufacturing, quality management		x	x		
Model Factories	Approx. 30 locations worldwide	McKinsey	Lean production, design, Industrie 4.0, energy productivity, service operations etc.			x		See, for example, McKinsey & Company (2017) and Hammer (2014)
Model Factory	Mayagüez, Puerto Rico	University of Puerto Rico	Manufacturing systems		x			UPRM (2018)

(continued)

Table 7.1 (continued)

Name of learning factory	Location	Operated by	Key topics	Research	Education	Training	Transfer	See also
Move academy	Herzogenaurach, Germany	Schaeffler	Lean production			x		See, for example, Beauvais (2013) and Helleno et al. (2013)
MPS Lernplattform	Sindelfingen, Germany	Daimler AG	Lean		x	x	x	Best Practice Example 23
MTA SZTAKI Learning Factory Győr	Győr, Hungary	MTA SZTAKI	CPPS, Human-robot collaboration, production planning, and scheduling	x	x	x	x	Best Practice Example 24
Pilot Factory Industrie 4.0	Vienna, Austria	imw, TU Wien	Factory virtualization, adaptive manufacturing, cyber-physical assembly and logistics	x		x	x	Best Practice Example 25
Process Learning Factory CiP	Darmstadt, Germany	PTW, TU Darmstadt	Lean production, Industrie 4.0	x	x	x	x	Best Practice Example 26

(continued)

7 Overview on Existing Learning Factory Application Scenarios 211

Table 7.1 (continued)

Name of learning factory	Location	Operated by	Key topics	Research	Education	Training	Transfer	See also
Product Development Process Learning Factory	Piracicaba, Brazil	Laboratory for Computer Integrated Design and Manufacturing, Methodist University of Piracicaba	Product development		x			Schützer, Rodrigues, Bertazzi, Durão, and Zancul (2017)
Railway Operation Research Center	Darmstadt, Germany	Chair of Railway Engineering, TU Darmstadt	Railway operation	x	x	x		Streitzig and Oetting (2016)
SEPT Learning Factory	Hamilton, Canada	Booth School of Engineering Practice and Technology, McMaster University	Industry 4.0	x	x	x		Elbestawi, Centea, Singh, and Wanyama (2018)
Smart Factory	Kaiserslautern, Germany	Technologie-Initiative SmartFactory KL e.V.	Automation, Industrie 4.0, Digitization	x			x	Best Practice Example 28

(continued)

Table 7.1 (continued)

Name of learning factory	Location	Operated by	Key topics	Research	Education	Training	Transfer	See also
Smart Factory MTA SZTAKI	Budapest, Hungary	MTA SZTAKI	CPPS, production planning, scheduling, and execution, mechatronics, automation	x	x		x	Best Practice Example 27
Smart Mini factory	Bolzano, Italy	IEA, FU Bozen-Bolzano	Smart manufacturing systems, automation	x	x	x	x	Best Practice Example 29
Teaching Factory	Patras, Greece and varying industry partner locations	LMS, University Patras	Industrial problems, real-life engineering practices, R&D outputs	x	x	x	x	Best Practice Example 30
Textile Learning Factorie 4.0	Aachen, Germany	ITA Aachen	Industrie 4.0 and the textile industry	x		x	x	Küsters, Praß, and Gloy (2017)

(continued)

7 Overview on Existing Learning Factory Application Scenarios 213

Table 7.1 (continued)

Name of learning factory	Location	Operated by	Key topics	Research	Education	Training	Transfer	See also
The PuLL® Learning Factory	Landshut, Germany	Competence Center for Production and Logistics, University of Applied Sciences Landshut	Lean production/lean logistics	x	x			See, for example, Blöchl and Schneider (2016) and Blöchl, Michalicki, and Schneider (2017)
Value Stream Academy	Several locations	Knorr Bremse	Lean production			x		u-Quadrat (2018)
VPS Center of the Production Academy	Munich, Germany	BMW Group	Lean production	x		x	x	Best Practice Example 31
World Class Manufacturing Academy	Michigan, USA	Chrysler	World class manufacturing			x		UAW-Chrysler National Training Center (2016)
XPRES Laboratory	Stockholm, Sweden	KTH Stockholm	Production research	x	x			Sivard and Lundholm (2013)

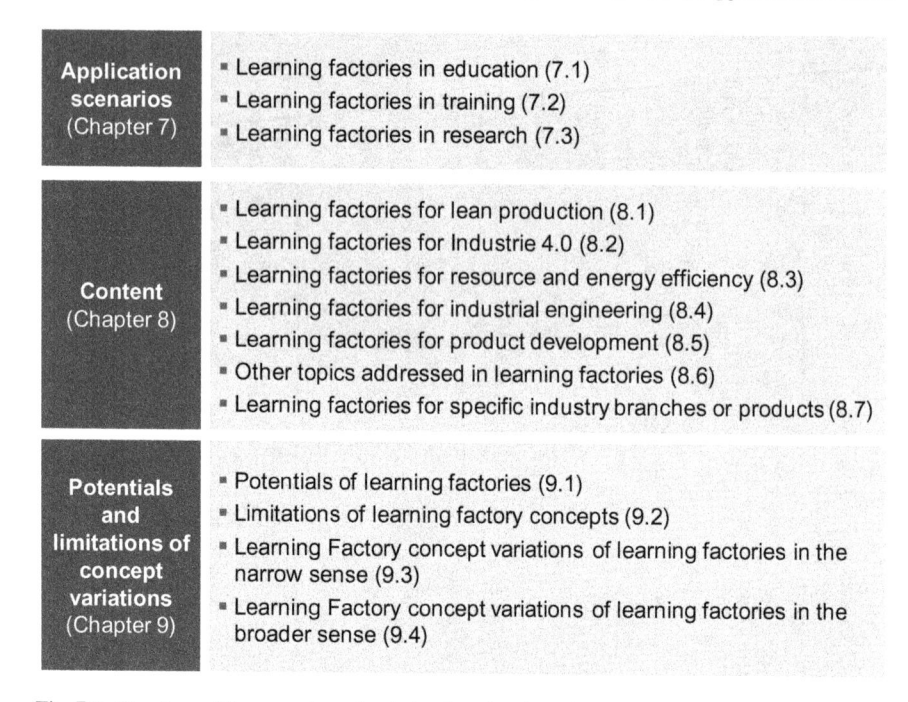

Application scenarios (Chapter 7)	• Learning factories in education (7.1) • Learning factories in training (7.2) • Learning factories in research (7.3)
Content (Chapter 8)	• Learning factories for lean production (8.1) • Learning factories for Industrie 4.0 (8.2) • Learning factories for resource and energy efficiency (8.3) • Learning factories for industrial engineering (8.4) • Learning factories for product development (8.5) • Other topics addressed in learning factories (8.6) • Learning factories for specific industry branches or products (8.7)
Potentials and limitations of concept variations (Chapter 9)	• Potentials of learning factories (9.1) • Limitations of learning factory concepts (9.2) • Learning Factory concept variations of learning factories in the narrow sense (9.3) • Learning Factory concept variations of learning factories in the broader sense (9.4)

Fig. 7.1 Structure of the overview of existing learning factories in this chapter

and research. In the following sections, the variety of learning factory concepts with those application scenarios are further described, explained, and structured. Furthermore, in each section best practice examples from academia and industry are given for the different types of learning factories. Section 7.1 looks at learning factories in education, Sect. 7.2 focuses learning factories in training or further education practices, and Sect. 7.3 finally addresses the learning factory concept in research. Figure 7.2 gives an overview of the detailed structure of this chapter.

7.1 Learning Factories in Education

Beginning from the early learning factory implementation at Penn State University and other locations,[5] education of students often plays a key role in the learning factory concept. In general, two different main models for the use of learning factories in engineering education programs can be identified:

- Open student projects,
- Steered and closed courses.

---

[5] See Jorgensen, Lamancusa, Zayas-Castro, and Ratner (1995).

7.1 Learning Factories in Education

Overview on existing learning factory application scenarios (Chapter 7)
- Education (7.1)
  - Open, steered, and closed student projects
  - Active learning in learning factories
  - Action-oriented learning in learning factories
  - Experiential learning and learning factories
  - Game-based learning in learning factories and gamification
  - Problem-based learning in learning factories
  - Project-based learning in learning factories
  - Research-based learning in learning factories
- Training / further education (7.2)
  - Scenario-based learning
  - Change management through learning factories
  - Information assimilation and experiential learning in training
- Research (7.3)
  - Learning factories as research objects
  - Learning factories as research enablers (focus)

Fig. 7.2 Detailed structure of the overview on application scenarios of existing learning factories

Student projects in general take at least several weeks or even months. In those projects, students or student groups are working on finding or designing technical or organizational solutions for predefined requirements or problem situations. As a rule, the students' solution finding process is open to new creative solutions and is not restricted to a certain path in advance. Student projects are in general somehow connected to current research questions. In some concepts, the student projects are industry partnered.[6] In those projects, students are most of the time loosely coached by supervisors. The project work is in general not directly dependent on specific courses. The concepts of the use of learning factories for educational purposes are visualized for stand-alone student projects and industry-partnered student projects; see Fig. 7.3.

In contrast, steered and closed courses usually take several hours or a maximum of a few days. Courses can take place in regular sessions over a certain period of time, for example, as part of a lecture. In those courses, students or student groups are working on predefined problem scenarios that are suiting a connected theoretical fundament. In general, the students' solution finding process is steered or the solution space is somehow restricted. Characteristic of those courses is that theory and practical exploration and/or testing is somehow coordinated and planned in advance. The alternation between theory and practice can be carried out in short or in long cycles, see also Fig. 7.4, for a visualization of the concepts. The concepts of this use of learning factories for educational purposes are visualized for short- and long-cycled steered courses integrating learning factories; see Fig. 7.4.

---

[6]See, e.g., Jorgensen et al. (1995).

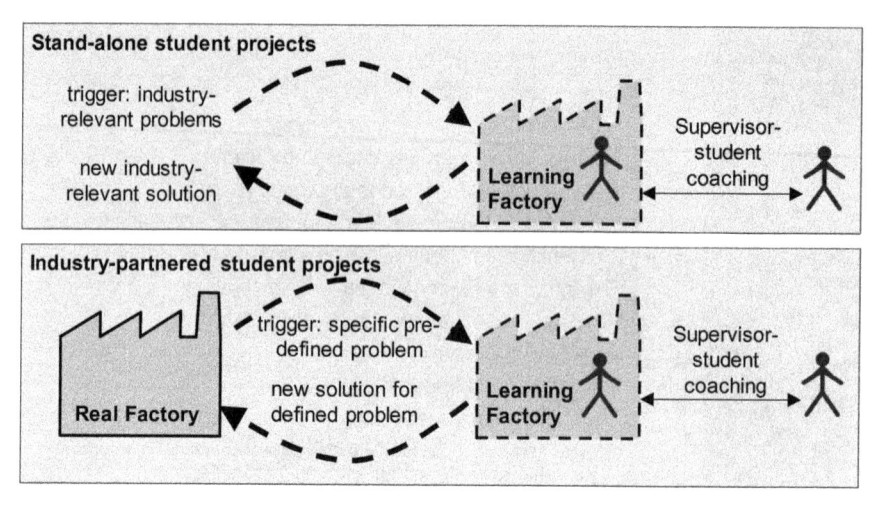

Fig. 7.3 The use of learning factories in education in connection with stand-alone and industry-partnered projects

Using these main models for the integration of learning factories in education, additionally several fundamental theoretical concepts can be identified that explain why and how learning factories can be embedded and used in course of the learning process. Basic characteristic of the concepts is an active role of the learner in the learning process. In the following, first the basic concept of active learning is presented briefly. Furthermore, specific orientations inside the active learning field in learning factory literature are identified and described.

7.1.1 Active Learning in Learning Factories

Active learning tries to avoid the purely passive information transfer and instead concentrates on the active involvement of learners, who analyze situations and intervene by testing their own ideas as well as evaluate results. This type of learning is not focused on the mere information reproduction but the comprehensive understanding of problem situations and concepts.[7] Active learning encourages students to do things and think about what they do.[8] Traditionally, active learning can be stimulated using learning methods like writing, debates, cooperative learning, learning games, role playing, and many more. In engineering education, learning factories are a very complex form of learning that enables active learning with manifold possibilities. Active learning can be seen as the basic concept, which is further subdivided by many other

---

[7]See Crawley, Malmqvist, Ostlund, and Brodeur (2007).
[8]See Bonwell and Eison (1991).

7.1 Learning Factories in Education

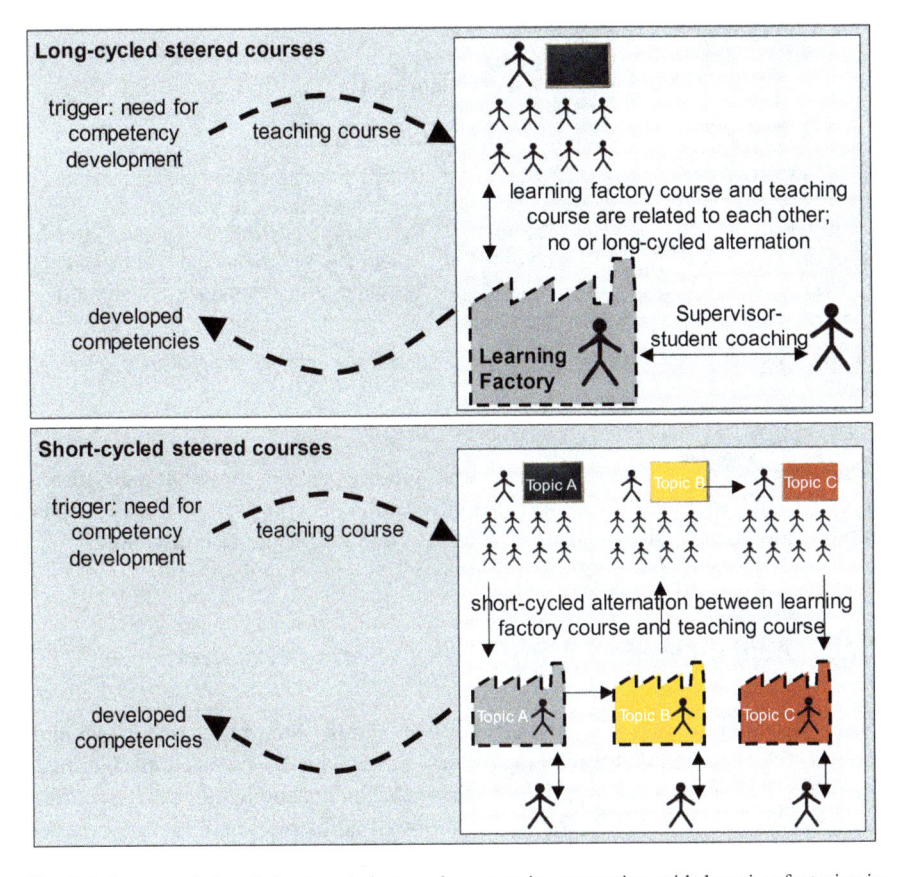

Fig. 7.4 Long-cycled and short-cycled steered courses in connection with learning factories in education

concepts, depending on the specific orientation. Examples of such concepts with specific orientation are

- Research-based learning, which intends to integrate education and research activities for mutual benefits,
- Problem-based learning, which bases all learning processes on a comprehensive problem scenario, or
- Experiential learning, which creates knowledge on the basis of own experiences.

Figure 7.5 gives an overview of the most important active learning sub-concepts in the field of learning factories for engineering education. The structure of the concepts presented could as well be modified to a three-level structure; for example, problem-based learning and project-based learning could be interpreted as sub-concept of experiential learning. Since the drawn structure depends on the perspective and is getting only more complex with further subordination, all the concepts are shown in only one level as sub-concept of active learning. In the majority of cases, learning

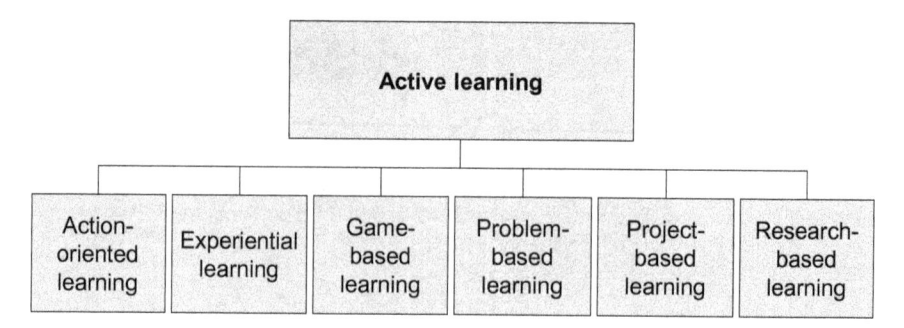

Fig. 7.5 Most important active learning concepts in the field of engineering education in learning factories

factory approaches do not follow a specific sub-concept but represent a mixture of several forms. In the following sections, those sub-concepts are described in alphabetical order. Connections and similarities of the concepts are also discussed.

7.1.2 Action-Oriented Learning in Learning Factories

Action-oriented learning is a subtype of active learning focusing on the active integration of the learner into the learning process through his/her own actions. It aims to improve conceptual knowledge, which allows the understanding of cause-and-effect relationships as a prerequisite for the resolution of problems.[9] In the action-oriented learning process, learners have to deal with complex problems independently while teachers keep themselves in the background and rather play the role of a moderator.[10] The action-oriented learning approach is not limited to simple actions but includes planning activities and reflection in order to enhance experiences gained in the learning process.[11] Consequently, action-oriented learning is closely related to problem-based learning and experiential learning, whereby action-oriented learning focuses clearly on the actions and related observations of the learner during the problem-solving procedure. A suitable learning environment in this case is indicated by a close proximity to the working context; besides the use of learning factories this can be achieved by further simulations, role-plays or the use of virtual reality.[12] This type of learning should be seen in the context of activity theory, where conscious learning emerges from activities, rather than activities emerging from learning, as widely believed.[13]

---

[9] See Abele et al. (2017).
[10] See Cachay, Wennemer, Abele, and Tenberg (2012).
[11] See Jank and Meyer (2002).
[12] See Lindemann (2002), Cachay et al. (2012).
[13] See Jonassen and Rohrer-Murphy (1999).

7.1.3 Experiential Learning and Learning Factories

One of the most acknowledged concepts for learning processes is the experiential learning.[14] A characteristic of experiential learning in engineering education is that students adopt simulated roles of the engineering profession. Also in learning factory-specific literature, the experiential learning model is often used to describe how learning processes within the learning factory approach are arranged.[15] As a fundamental understanding of the experiential learning concept, learning is defined as "the process whereby knowledge is created through the transformation of experience."[16] Consequently, the experiential learning concept does not make use of a traditional teacher role but relies on meaningful learners' experiences. In the experiential learning process, knowledge and competencies of the learner are created through an inherent transformation starting with

(a) The **concrete experience**,[17] which is, for example, gained through actions in a specific field followed by phases of
(b) **Reflective observation**,[18] in which the made experiences are reviewed and reflected,
(c) **Abstract conceptualization**,[19] which contains cognitive processes like analysis, interpretation, cause-and-effect relationships, and even first ideas for improvement, and
(d) **New active experimentation**,[20] i.e. a transfer of the recently developed concepts to a new situation based on the previous phases of the cycle, which leads again to new concrete experiences (see step a).[21]

With the realistic learning environment and the possibility to create an active learning experience, the learning factory is well-suited to implement this kind of experiential learning processes for manufacturing education. Figure 7.6 sketches the learning process as well as a possible implementation of this learning process inside learning factories. Other forms of learning in engineering education that allow competency development through experiential learning processes are simulations, learning in projects, and case studies, which are also well-suited to be combined in one course with further active learning approaches.[22]

---

[14] Experiential learning according to Kolb (1984).
[15] See, for example, Steffen, Frye, and Deuse (2013), Müller, Menn, and Seliger (2017).
[16] Kolb (1984).
[17] See Fig. 7.6a.
[18] See Fig. 7.6b.
[19] See Fig. 7.6c.
[20] See Fig. 7.6d.
[21] See Kolb (1984).
[22] See Crawley et al. (2007).

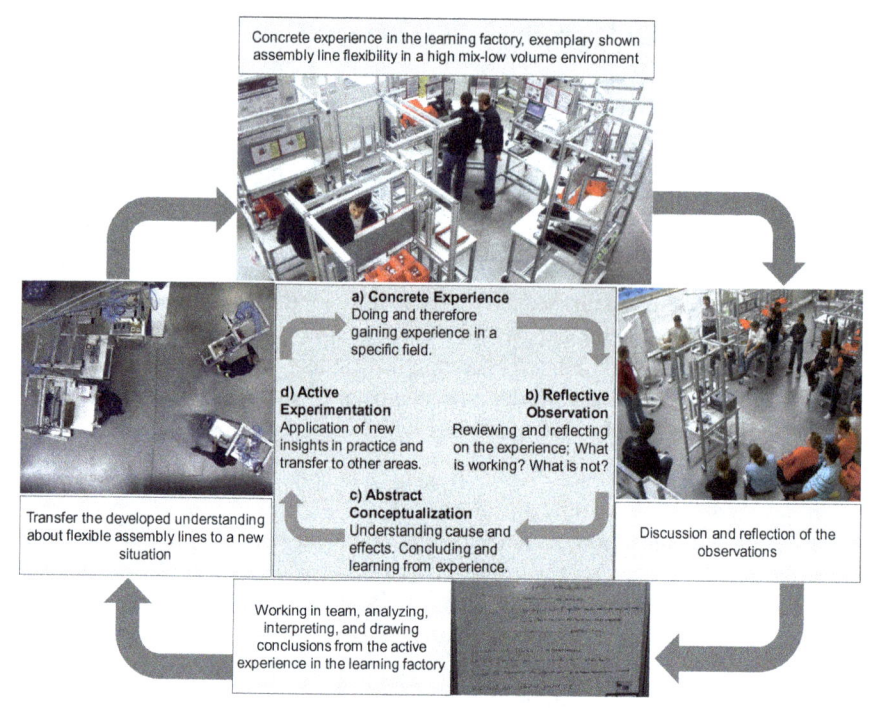

Fig. 7.6 The experiential learning cycle in learning factories

7.1.4 Game-Based Learning in Learning Factories and Gamification

Forms of learning that integrate game elements into the serious purpose of learning can be differentiated in

- Serious games and
- Gamification approaches.[23]

Game-based learning, "[...] the product of **serious games**[24] [...]" (Felicia, 2014) makes use of structured playing situations with the aim of enabling learning toward explicitly, clearly and carefully defined learning objectives/educational purposes.[25] Game-based learning can help the learner to understand certain topics and concepts or to help learning and improving certain capabilities by actively incorporating the

---

[23] See Deterding, Dixon, Khaled, and Nacke (2011).

[24] Or very similar used terms like educational games see also De Freitas (2007).

[25] See Abt (1987), Felicia (2014).

7.1 Learning Factories in Education

Dynamics
Constraints: limitations and forced trade-offs
Emotions: competition, curiosity, frustrations, happiness
Narrative: Continuous, persistent storyline
Progression: Player development in the learning situation
Relationships: social interactions, status, altruism
...

Mechanics
Challenges that require the player's effort
Chance: random elements in the learning situation
Competition: Environments with winners and losers
Cooperation: Teamwork among players
Feedback on progress in learning situation
Rewards on actions in learning situation
...

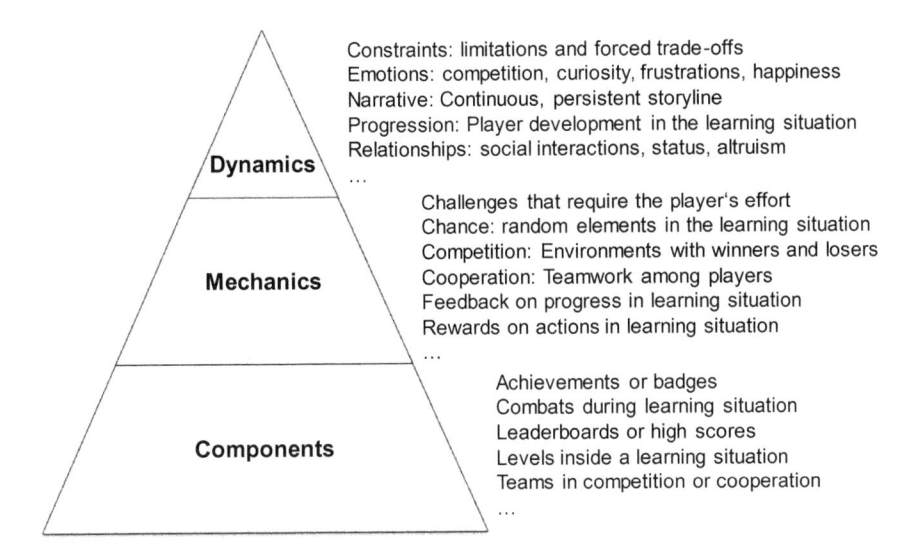

Components
Achievements or badges
Combats during learning situation
Leaderboards or high scores
Levels inside a learning situation
Teams in competition or cooperation
...

Fig. 7.7. Framework and example of game elements used for gamification purposes according to Werbach and Hunter (2012)

learner's own actions.[26] Consequently, the pleasure of playing, which is the primary purpose of classic games, is only of secondary importance in serious games.[27]

In contrast to the above-described game-based learning which can be interpreted as the use of complete (serious) games in a non-entertainment context, the term **gamification** describes approaches that take specific game design elements and put them into a non-game context.[28] Gamification can be used to make non-game situations more motivating, enjoyable, and engaging.[29] An often cited framework for those gamification elements is given with a pyramid containing different types of elements on the levels "components," "mechanics," and "dynamics."[30] Figure 7.7 gives an overview of the framework containing some exemplary game elements.

In the learning factory context, both game-based learning and gamification can be implemented. An example of a game-based learning approach in learning factories is given with the creation of a game in which participants have to work together in a team to reach certain targets of the game, e.g. produce enough products in a predefined quality; see the example of the paper plane game in one of the next paragraphs. Another frequently used example of gamification in learning factories is the formation of several teams within the group of participants, all of whom have the same task and work on it in competition with each other. In the Process Learning

---

[26] See Gredler (2004).
[27] See Djaouti, Alvarez, Jessel, and Rampnoux (2011).
[28] See Deterding et al. (2011).
[29] Deterding et al. (2011).
[30] See Werbach and Hunter (2012).

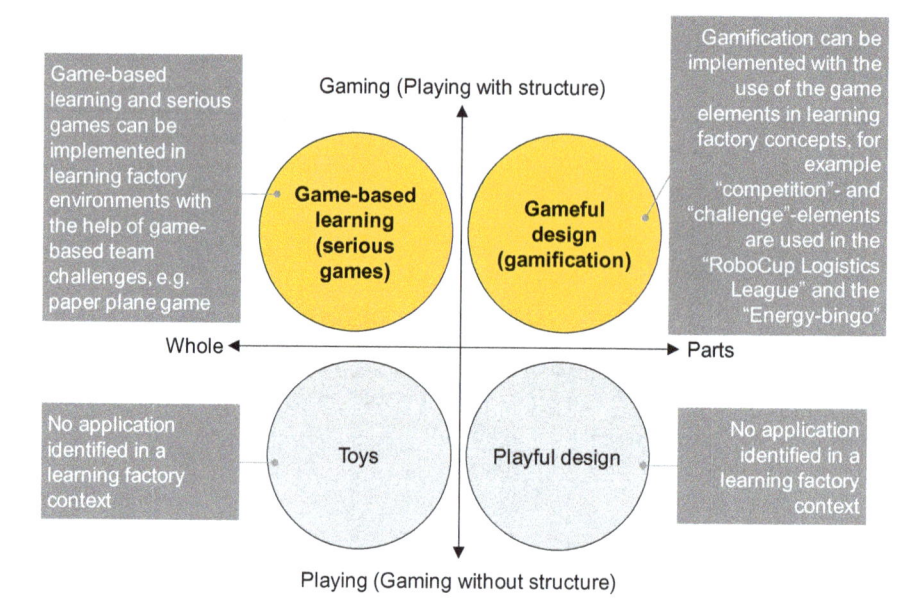

222   7 Overview on Existing Learning Factory Application Scenarios

Fig. 7.8 Game-based learning and gamification in learning factories. Classification according to Deterding et al. (2011)

Factory CiP, this is used, among other things, in connection with the SMED[31] method, in which the participants compete in the change-over activity of an abstract machine model within a simulation game using the SMED method. Further examples of the use of gamification in a learning factory context are given with

- The so-called RoboCup Logistics League,[32] or with
- the so-called energy-bingo.[33]

Game-based learning and gamification are structured in Fig. 7.8. The three-mentioned approaches of the "energy-bingo," the "RoboCup Logistics League," and the paper plane game are presented in the next paragraphs.

In many learning factories, simple simulation games are implemented. The paper plane game is one example, which is used in the Process Learning Factory CiP to sensitize the learners for the topics material flow and pull systems in production. This paper plane game is an example of game-based learning. At the beginning, goals, rules of the game, and the rough course of the three simulation rounds are coordinated with the participants and two competing teams are formed. A simulation round always consists the steps of

- Preparation for the new round,

---

[31] "Single Minute Exchange of Dies" is a method to systematically reduce change-over times.
[32] See Pittschellis (2015).
[33] See Böhner et al. (2015).

7.1 Learning Factories in Education

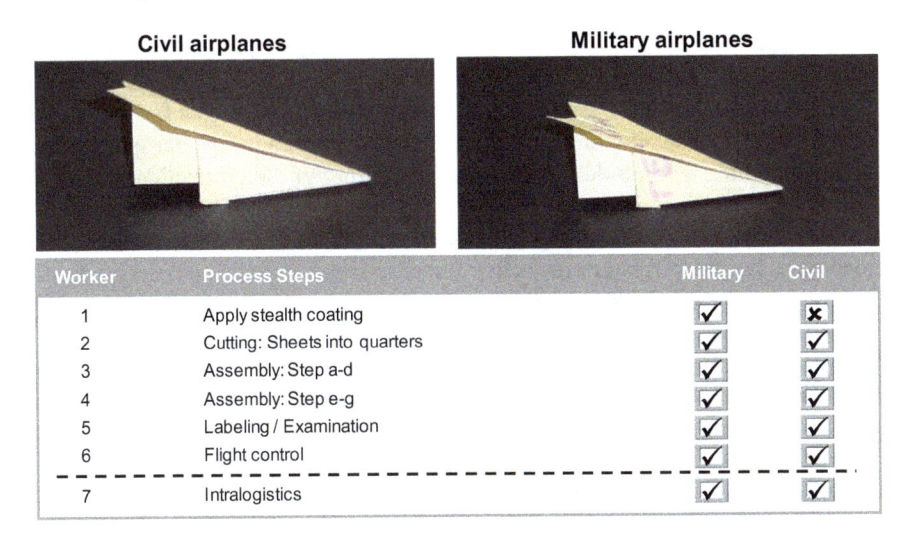

Worker	Process Steps	Military	Civil
1	Apply stealth coating	✓	✗
2	Cutting: Sheets into quarters	✓	✓
3	Assembly: Step a-d	✓	✓
4	Assembly: Step e-g	✓	✓
5	Labeling / Examination	✓	✓
6	Flight control	✓	✓
7	Intralogistics	✓	✓

Fig. 7.9 Civil and military paper airplanes and respective process steps of the paper airplane game

- Execution of the actual simulation,
- The subsequent analysis based on performance measurement,
- And the improvement of the manufacturing system.

After the work instructions for the production of paper planes have been explained to both teams, the first ten-minute production round starts in the initial production system. Based on a performance measurement regarding quality, cost, and delivery, the production round is analyzed and first improvement possibilities are discussed inside the two teams. Throughout the three simulation rounds, learners will playfully gain access to topics like waste reduction, the benefits of flow, and take time within a production system, and the use of pull systems. Figure 7.9 shows the civil and military airplanes and respective process steps during the simulation game.[34]

Furthermore, there is a great variety of simulation games that are used for teaching lean manufacturing.[35] Those simple simulation games of course can be and are implemented separated from the learning factory concept. Such simulation games are also carried out in learning factory environments—although in the phase of sensitization it may help that used production system are not too complex.[36]

For the "RoboCup Logistics League," a learning factory setting consisting of several Festo MPS® stations[37] that represent certain steps in a multistage production process is used. Furthermore, mobile robots transport workpieces between the sta-

---

[34] An example of a similar simulation game is, for example, described in Stier (2003).
[35] See Badurdeen, Marksberry, Hall, and Gregory (2010).
[36] An example of the use of a simulation game in a learning factory environment can be found in Blöchl and Schneider (2016).
[37] Modular Production System, scaled down production machines, see also Best Practice Example 7.

tions. The learning factory approach for the "RoboCup Logistics League" is used in classical project learning (see also the Sect. 1.6): University students participating in the course have to develop and implement transportation routes for the robots meeting the prior unknown factory layout. And in a second step, transport routes have to be optimized in order to maximize the output of the production system in a given time. Furthermore, the optimized robot routes must be able to react to randomly occurring express orders with higher priority.[38] As already mentioned, to this point the RoboCup looks like many other student projects executed in learning factories. But additionally, the RoboCup is a competition between student groups coming from all around the world. Also, the locations for the championship in the last years were picked worldwide: Leipzig, Germany (2016), Hefei, China (2015), João Pessoa, Brasil (2014), Eindhoven, Netherlands (2013), and so on. This gamification element of global competition between the different universities motivates to send the best students, who are learning in an intense and motivating way (Fig. 7.10).[39]

Another gamification approach that uses the game element "challenge" in combination with a learning factory setting is the "energy-bingo."[40] The learning method used in this approach is based on the PDCA[41] cycle. For the learning method, the phases of the cycle are altered to Experience–Estimate–Measure–Transfer; see also Fig. 7.11. Those phases describe the sequence of the implemented learning course, while especially the "estimate" phase makes use of game elements. Here, the participants make educated guesses about certain parameters regarding the energy consumption of a machine.

Fig. 7.10 RoboCup Logistics League 2016 in Leipzig Germany, screenshot taken from RoboCup (2016)

---

[38] See Pittschellis (2015).
[39] See Pittschellis (2015).
[40] Introduced by Böhner, Weeber, Kuebler, and Steinhilper (2015).
[41] Plan-Do-Check-Act.

7.1 Learning Factories in Education

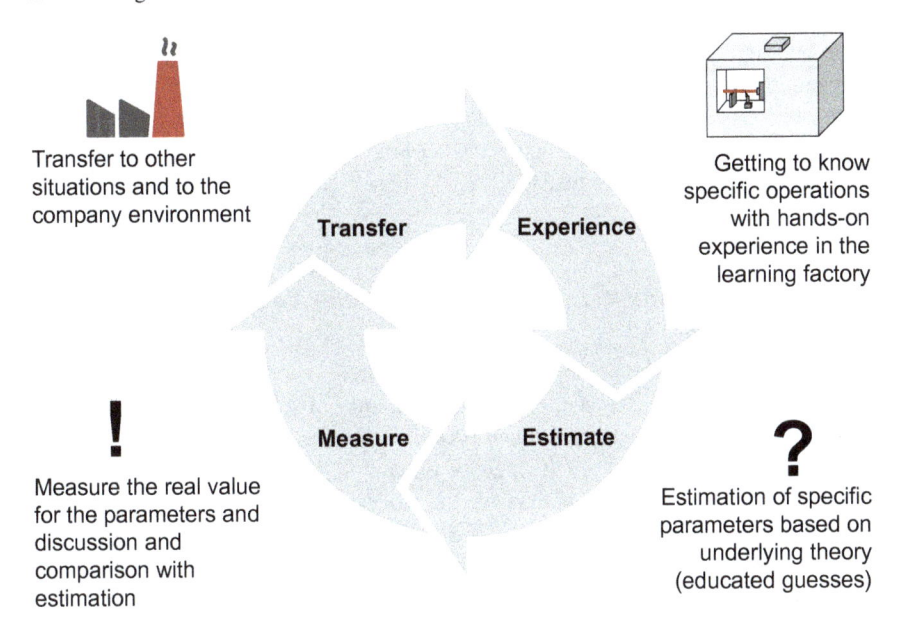

Fig. 7.11 Steps of a learning method using gamification elements, according to Böhner et al. (2015)

In the following, a short example of the visualized game-based learning process for energy consumption is described in which learners investigate the energy consumption of a cutting machine in order to identify the most relevant energy consumers[42]:

- Experience: The learning process starts with a demonstration of a cutting process typically used in composite manufacturing by a supervisor. Additionally, the functions of different machine components relevant for energy consumption (including a vacuum pump, a conveyor motor, a cabinet fan, servomotors moving the cutter knife, and a machine control unit) are explained.
- Estimation: Based on experiences made in the previous learning phase, participants evaluate and estimate the expected energy consumption for the previously explained machine components. This game element is also referred to as "energy-bingo." For this purpose, the participants are asked to indicate there educated guesses of the energy consumption for each component on a prepared energy consumption chart using adhesive tags.
- Measure: In the measuring phase, suitable equipment to measure the electric power consumption of the above-mentioned components is installed. Doing this, additional knowledge about measuring systems and concepts is imparted. Furthermore, the participants get feedback on their educated guesses regarding the power consumption of various machine components. A comparison of estimations and measured data leads to discussion and reflection. A surprise for most learners is

---

[42]Böhner et al. (2015).

the fact that in this example the vacuum pump consumes 95% of the total energy consumption.
- Transfer: Goal of the learning module is that participants are able and motivated to transfer what they have learned in the learning factory to other machines in their own sphere of action in industry.

7.1.5 Problem-Based Learning in Learning Factories

The problem-based learning approach has originated from the field of medical education in response to insufficient clinical performance of students which was caused by an emphasis on the fragmented memorization of biomedical knowledge in medical education to this point.[43] In the medical education field, the learning approach is defined as "[...] learning that results from the process of working toward understanding or resolution of a problem. [...]."[44] The learning process in this approach starts with encountering the problem,[45] i.e. the problem should act as a stimulus for active learning and is therefore given to the students before they receive information about the theoretical background or the principles connected to the problem. Especially in the 1990s, the problem-based learning approach was widely adopted outside of medical education,[46] e.g. also in engineering education.[47] A frequently used procedure for the learning process is described by the so-called Maastricht Seven Jump method. The method consists of the steps[48]:

1. Clarify unfamiliar terms or concepts in the problem description.
2. Define the problem; that is, list the phenomena to be explained.
3. Analyze the problem; "brainstorm"; try to produce as many different explanations for the phenomena as you can. Use prior knowledge and common sense.
4. Criticize the explanations proposed and try to produce a coherent description of the processes that, according to what you think, underlie the phenomena.
5. Formulate learning issues for self-directed learning.
6. Fill the gaps in your knowledge through self-study.
7. Share your findings with your group and try to integrate the knowledge acquired into a comprehensive explanation for the phenomena. Check whether you know enough now.

Additionally to this sequence in the learning process, seven characteristics of effective problem-based learning can be summarized[49]:

---

[43] See Hung, Jonassen, and Liu (2008).
[44] Barrows and Tamblyn (1980).
[45] See Barrows and Tamblyn (1980).
[46] See Hung et al. (2008).
[47] See Cawley (1989).
[48] See, for example, Evensen and Hmelo-Silver (2000).
[49] See Wood (2003).

7.1 Learning Factories in Education

1. Learning objectives defined by the learners after analyzing the problem situation should be consistent with intended objectives of the course.
2. The presented problem should be appropriate (to the level of understanding and stage of curriculum).
3. Scenarios (problem situations) should be of interest for students and/or relevant for future practice.
4. Fundamental theory is presented in the context of problem scenarios to motivate to knowledge integration.
5. Problem scenarios stimulate the discussion among the learners and deliver motivation to seek explanations.
6. Problem scenarios are (sufficiently) open in order to leave room for discussion along the learning process.
7. Problem scenarios promote active participation by learners in finding relevant information from various resources.

Learning factories provide an excellent starting point for problem-based learning in engineering education. The simulated, authentic learning environments can be used to define and prescribe problem situations, without them being only formulated linguistically in an uncontextualized way. Likewise, learning factories offer the opportunity to test and, if necessary, revise developed explanations that address previously analyzed problems.

The use of problem-based learning in learning factories is also described.[50] Although at some point the mentioned implementations represent slight variations of the original problem-based learning approach, since, for example, the learning process does not start with the problem occurrence as a stimulus for active learning but a self-directed literature research to a predefined topic.[51] Or, for example, in the TU Wien Learning and Innovation Factory no pure problem-based approach, but a combination of the problem-based learning with the very similar action-oriented[52] and experiential learning[53] approaches as well as with an traditional lecture in the initial preparation phase of the students are used.[54] In the Best Practice Example 13, the TU Wien Learning and Innovation Factory is described by Prof. Dr.-Ing. Wilfried Sihn, Fabian Ranz, and Philipp Hold.

7.1.6 Project-Based Learning in Learning Factories

Project-based learning is a student-centered form of learning in which real-world or authentic projects are worked on in student groups in order to create motivation

---

[50] See, for example, Jäger, Mayrhofer, Kuhlang, Matyas, and Sihn (2012, 2013), and Tietze, Czumanski, Braasch, and Lödding (2013).
[51] See the problem-based learning approach described in Tietze et al. (2013).
[52] See Sect. 7.1.2.
[53] See Sect. 7.1.3.
[54] See Jäger et al. (2012, 2013).

to deal with the learning content while solving problems, answering questions, and completing the projects.[55] Project-based learning is not a new form of learning; early implementations of project-based learning at the Massachusetts Institute of Technology are developed already in mid-nineteenth century.[56]

In the context of teaching activities in the field of production, learning factories can be a good starting point for project-based learning.[57] In the Process Learning Factory CiP, small groups of students can also carry out projects in the authentic production environment in so-called Advanced Design Projects. An overview of one of these Advanced Design Projects regarding the optimization of a lean machining line regarding quality and autonomation aspects is given in Fig. 7.12.

7.1.7 Research-Based Learning in Learning Factories

Learning in the context of education in learning factories can or should be organized in mutual dependence to research. The roots of this idea are going back to the German Philosopher Wilhelm von Humboldt.[58] For Humboldt, universities are primarily research institutions passing on knowledge and discussing it critically. Humboldt's vision for university education aims at a unity of teaching and research—learning in the context of the university should consequently be organized in a research mode, solving not yet fully solved problems enabling an open culture for discussion between teachers and learners as well as between the various disciplines.[59] It becomes clear that both traditional teacher-centered teaching and isolated research routines are not suitable for this kind of university education.[60] The concept of research-based learning is founded on this vision.[61]

In the literature, four different concepts linking research and education (research-based learning being one of them) are described.[62] The classification is further used to define and specify the concept of research-based learning[63]:

- Research-led learning: Learning subjects are picked for education based on current research findings. Students are not actively involved in the research process but take the role of a passive audience.
- Research-oriented learning: The understanding of the research process and corresponding techniques and instruments are in the focus of this concept. Students

---

[55] See Bender (2012), Wurdinger (2016).

[56] See Knoll (1997).

[57] Examples of the use of project-based learning with the help of learning factories are for example described in Balve and Albert (2015) and Jorgensen et al. (1995).

[58] See Blume et al. (2015).

[59] See Blume et al. (2015), Euler (2005), von Humboldt (1957).

[60] See von Humboldt (1957).

[61] See Blume et al. (2015), Euler (2005), von Humboldt (1957).

[62] See Healey (2005).

[63] Healey (2005).

7.1 Learning Factories in Education

Team of students:

Challenge:
Improvement of machining process in cellular manufacturing

Quality
- 100% source inspection of critical specifications
- Failure proof design

Autonomation
- Integration of simple intelligence into the machine (Jidoka)
- Simple and cost efficient Autonomation to reduce manual work

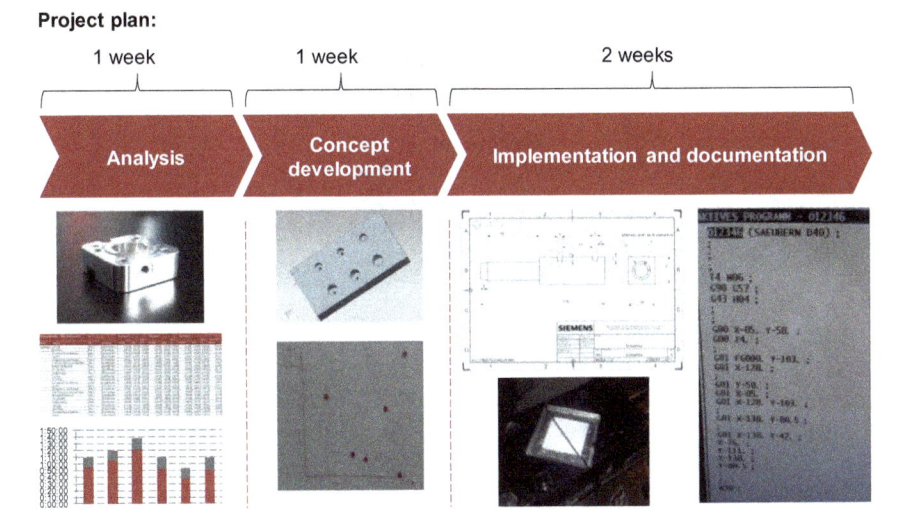

Project plan:

1 week — Analysis
1 week — Concept development
2 weeks — Implementation and documentation

Fig. 7.12 Advanced design project regarding the optimization of a lean machining line in the process learning factory CiP

are not actively involved in the research process, but take the role of a passive audience.
- Research-tutored learning: Active discussion of research content between learners and teachers. Although the students are active participants in the discussion, they are not immediately involved in the research process.
- Research-based learning: Students learn through actively designing, experiencing, and reflecting the process of research projects, actively generating results, and new findings. Therefore, the students are actively and directly involved in the research process.

Figure 7.13 visualizes the classification of the different concepts related to the questions whether the students play the role of active participants or passive audience and whether the research content or the research process and problems are in the

230   7 Overview on Existing Learning Factory Application Scenarios

Student-focused
(students as participants)

Research-tutored

Curriculum emphasizes
learning focused on
students writing and
discussing papers or
essays

Research-based

Curriculum emphasizes
students undertaking
inquiry-based learning

Emphasis on research content

Emphasis on research processes and problems

Research-led

Curriculum is structured
around teaching subject
content

Research-oriented

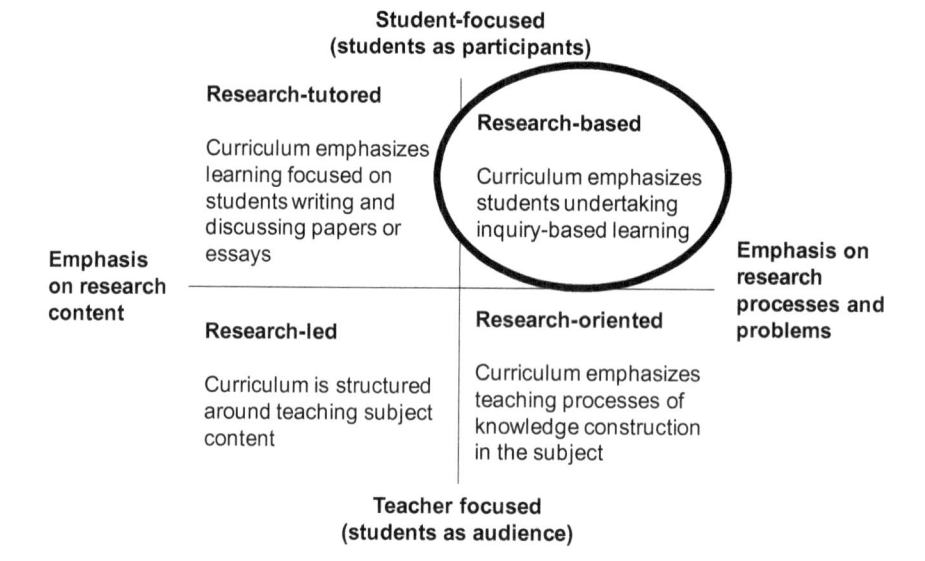

Curriculum emphasizes
teaching processes of
knowledge construction
in the subject

Teacher focused
(students as audience)

Fig. 7.13 Classification of forms of teaching linked with research—the research–teaching nexus (Healey, 2005)

focus. It is mentioned that good education programs should integrate the different concepts and find a good balance between them depending on the goals of educational activities.[64] Although, for university education it is recommended to spend most of the time with student-focused activities, which means that concepts from the top half of Fig. 7.13 should be used primarily.[65]

Learning factories are very well-suited to implement research-based learning concepts regarding factories and production processes, since learning factories allow students access to industry-equivalent processes. Without this possibility, data collection or experiments can only be carried out under laboratory conditions that are less realistic.[66] Figure 7.14 shows the research process used as an orientation for the research-based learning in learning factories.[67]

7.1.8 Best Practice Examples for Education

Improving education and enabling a practice-oriented manufacturing education is a fundamental motivation for building learning factories. Correspondingly, many

---

[64] Healey (2005).

[65] See Healey (2005), Blume et al. (2015).

[66] As two examples of research-based learning in learning factories, Blume et al. (2015) and Schreiber et al. (2016) can be named.

[67] Research process according to Blume et al. (2015).

7.1 Learning Factories in Education

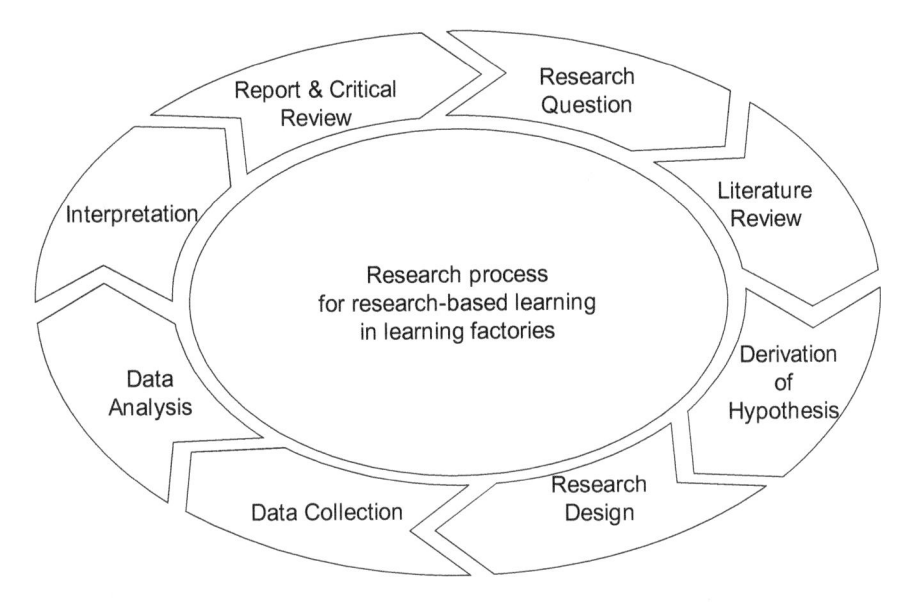

Fig. 7.14 Research process for research-based learning in learning factories according to Blume et al. (2015). Adapted from Creswell (2008)

examples can be given for the use of learning factories in manufacturing education. At this point, reference can be made to the corresponding best practice examples in Chap. 11:

- Best Practice Example 1: AutoFab at the Faculty of Electrical Engineering of the University of Applied Sciences Darmstadt, Germany,
- Best Practice Example 3: Die Lernfabrik at IWF, TU Braunschweig, Germany,
- Best Practice Example 4: E|Drive-Center at FAPS, Friedrich-Alexander University Erlangen-Nürnberg,
- Best Practice Example 5: ESB Logistics Learning Factory at ESB Business School at Reutlingen University, Germany,
- Best Practice Example 6: ETA-Factory at PTW, TU Darmstadt,
- Best Practice Example 7: Festo Didactic Learning Factories,
- Best Practice Example 9: iFactory at the Intelligent Manufacturing Systems (IMS) Center, University of Windsor, Canada,
- Best Practice Example 10: IFA Learning Factory at IFA, Leibniz Universität Hannover,
- Best Practice Example 11: Integrated Learning Factory at LPE & LPS, Ruhr-University Bochum, Germany,
- Best Practice Example 12: LEAN-Factory for a pharmaceutical company in Berlin, Germany,
- Best Practice Example 13: Learning and Innovation Factory at IMW, IFT, and IKT, TU Wien, Austria,
- Best Practice Example 14: Learning Factory aIE at IFF, University of Stuttgart,

- Best Practice Example 15: Learning Factory for electronics production at FAPS, Friedrich-Alexander University Erlangen-Nürnberg, Germany,
- Best Practice Example 16: Learning Factory for Innovation, Manufacturing, and Cooperation at the Faculty of Industrial and Process Engineering of Heilbronn University of Applied Sciences,
- Best Practice Example 17: Learning Factory for Global Production at wbk, KIT Karlsruhe,
- Best Practice Example 18: "Lernfabrik für schlanke Produktion," Learning Factory for Lean Production at iwb, TU München, Germany,
- Best Practice Example 20: LMS Factory at the Laboratory for Manufacturing Systems & Automation (LMS), University Patras, Greece,
- Best Practice Example 21: LPS Learning Factory at LPS, Ruhr-Universität Bochum,
- Best Practice Example 22: MAN learning factory at MAN Diesel & Turbo SE in Berlin, Germany,
- Best Practice Example 23: MPS Lernplattform at Daimler AG in Sindelfingen, Germany,
- Best Practice Example 24: MTA SZTAKI Learning Factory at the Research Laboratory on Engineering and Management Intelligence, MTA SZTAKI in Györ,
- Best Practice Example 26: Process Learning Factory CiP,
- Best Practice Example 27: Smart Factory at the Research Laboratory on Engineering & Management, MTA SZTAKI,
- Best Practice Example 29: Smart Mini Factory at IEA, Free University of Bolzano,
- Best Practice Example 30: Teaching Factory: An emerging paradigm for manufacturing education.

7.1.9 Example: Learning Factories for Industrie 4.0 Vocational Education in Baden-Württemberg

Beyond university education, learning factories are also deployed in vocational education. Learning factories are used extensively for this context in Germany in the federal state of Baden-Württemberg. With a total of 6.8 million euros, the Ministry of Economics, Labor, and Housing of Baden-Württemberg is supporting the establishment of 16 learning factories 4.0 at vocational schools. The aim of these learning factories is to prepare professionals and junior staff for digitalization. The "Lernfabrik 4.0" is a laboratory in which processes and equipment are similar to industrial automation solutions. In the learning factories, the basics of practice-oriented Industrie 4.0 processes can be taught. Mechanical engineering and electrical engineering are linked by professional production control systems.[68] The 16 learning factories spread over the federal state are shown on the map in Fig. 7.15.

---

[68] See Ministerium für Wirtschaft, Arbeit und Wohnungsbau Baden-Württemberg (2017).

7.1 Learning Factories in Education

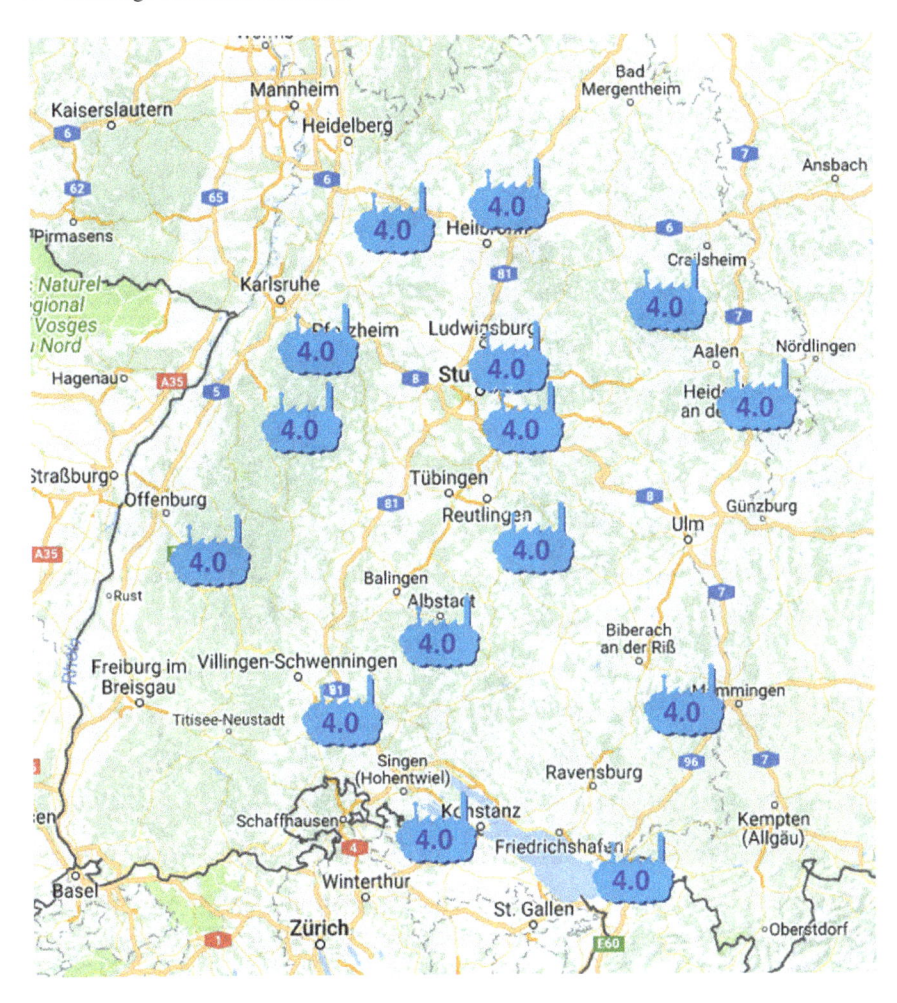

Fig. 7.15 The 16 learning factories in Baden-Württemberg (Germany) on the map (Ministerium für Wirtschaft, Arbeit und Wohnungsbau Baden-Württemberg, 2017)

One of the vocational schools that was equipped with a learning factorie 4.0 in 2016 was the Philipp-Matthäus-Hahn-Schule in Balingen.[69] In this learning factory in Balingen, engineering students as well as vocational students are trained to be ready for digitalized production environments. The learning factory contains Festo Didactic Learning Factory modules that contain latest industrial technology in the fields of CNC milling and turning processes, Industrie 4.0, hydraulics, and mechatronics (Fig. 7.16).[70]

---

[69] See Festo Didactic (2017).
[70] See Festo Didactic (2017).

Fig. 7.16 Impression of learning factorie 4.0 in Balingen as example of one of the 16 learning factories 4.0 in Baden-Württemberg, pictures taken from Festo Didactic (2017)

7.2 Learning Factories in Training

Learning factories in training can play different roles or pursue different goals. In general, three different application types of learning factories in training can be identified:

- In many cases, when learning factories are used for training purposes competency development can be identified as the main goal.[71] Here, similar concepts as for the purpose education can be applied, with the difference that in general the time frame available for training activities is strongly limited.

---

[71] See Tisch, Hertle, Abele, Metternich, and Tenberg (2015b).

7.2 Learning Factories in Training

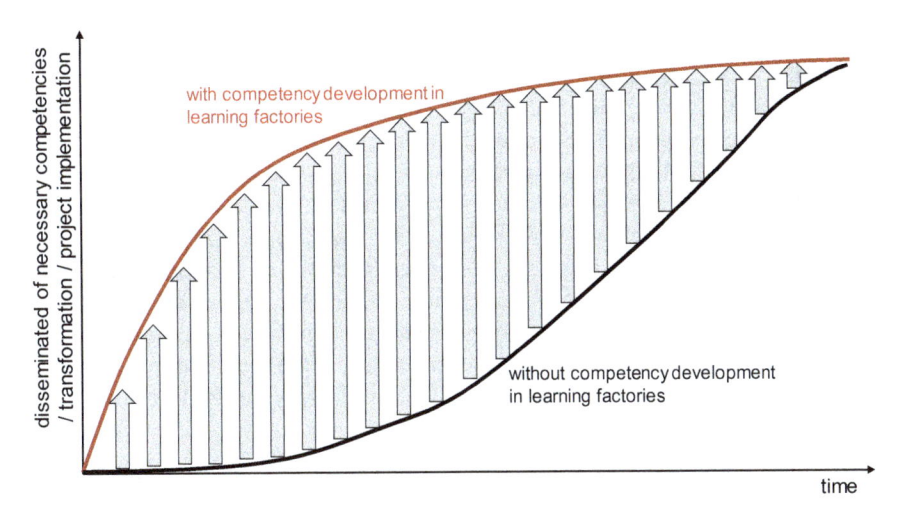

Fig. 7.17 Learning factories in training to speed up transformation and project implementation

- Besides that learning factories in training can be used as a part of change management approaches.[72] In addition to the technical topics, the motivation and the overcoming of internal barriers of the participants play a key role.
- In recent years, and mostly in context with digitalization and Industrie 4.0, learning factories in training show frequent application in the field of demonstration of new technologies or other innovations, see, for example, a scenario-based approach[73] and an approach based on elaborated use cases for Industrie 4.0.[74]

7.2.1 Competency Development in Course of Trainings in Learning Factories

The use of learning factories for competency development in trainings is very similar to the already described learning factory implementations for education. The aim of using the learning factories for training is in many cases the faster dissemination of knowledge and abilities to act within the company in order to enable the most successful implementation of improvement or transformation projects (Fig. 7.17).

For work-oriented learning in course of further education and training in learning factories in general, mainly two opposing learning process sequences can be identi-

---

[72] See Reiner (2009), Wagner, Heinen, Regber, and Nyhuis (2010), Dinkelmann, Siegert, and Bauernhansl (2014) and Dinkelmann (2016).
[73] See Erol, Jäger, Hold, Ott, and Sihn (2016).
[74] See Wank et al. (2016).

fied: the so-called information assimilation approach[75] and the experiential learning approach.[76]

Characteristic of the information assimilation is that learning content is theoretically derived, explained, and after that applied and tested. For a learning factory course, this could mean that certain methods or experiences of others are first approached theoretically and afterward applied, tested, and observed in a learning factory environment. Opposed to this, the experiential learning approach starts with the experience through own actions and experiences which are seen as the basis to understand learning content. For a learning factory course, this means that first experiences are made in the learning factory environment based on own actions and observations of the learner followed by understanding of the underlying principles. In connection with learning factory trainings, the two concepts are also labeled: "theory push" and "problem pull."[77] The benefit of the learning factory in this context is that it enables the active part in both types of learning processes for production-related topics.

On the one hand, without the learning factory concept the theory-based learning process persists in theory without the possibility of a transition to practical implementation and reflection—as a consequence, the learning loop where new information can be based on experiences made cannot be closed, see also Fig. 7.18. On the other hand, for the experiential learning regarding production-related issues without the learning factory concept the basis for planned experiences in a formal but close-to-reality learning setup is inexistent. This means that experiential learning processes cannot be initiated or must use unauthentic simulations of production processes, hampering the later transfer to the real production. Figure 7.18 depicts the integration of the learning factory concept for the support and the improvement of both opposed types of learning processes in production-related education and training courses.

In literature learning processes as well as benefits and trade-offs influencing which learning processes to use are widely discussed.[78] Table 7.2 summarizes respective advantages and disadvantages of both learning process types mentioned in the literature.

Consequently, being aware of the different pros and cons of the two opposed learning process sequences, it is becoming clear that both sequences can be beneficial in context of learning factory trainings and courses depending on targets and the framework regarding among others time, target group, and skills of the trainers.

In summary, the experiential learning sequence is particularly useful if the learners have had little or no experience with the corresponding problems and the motivation to deal with the contents is correspondingly low. A necessary prerequisite for the application of experiential learning is sufficient available time for the learning module. The information assimilation sequence, on the other hand, has an advantage

---

[75] According to Coleman (1982).
[76] According to Kolb (1984).
[77] See Tisch et al. (2013).
[78] See, e.g., Wurdinger (2005), Keeton, Sheckley, Griggs, and Council for Adult and Experiential Learning (2002), Coleman (1982).

7.2 Learning Factories in Training

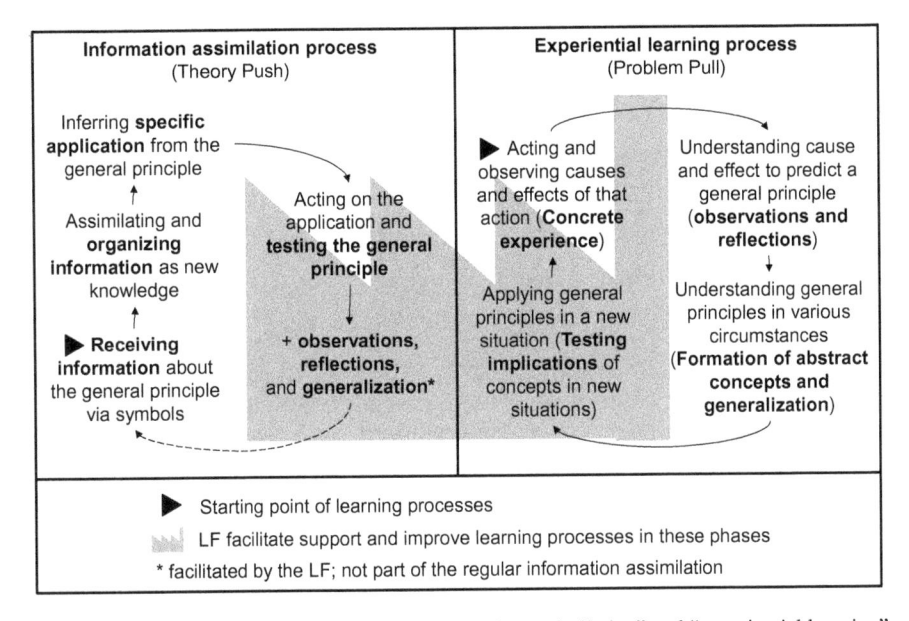

Fig. 7.18 The role of learning factories in "information assimilation" and "experiential learning" (Tisch & Metternich, 2017)

even with stricter time limitations. It is advantageous for the application of this learning sequence if the learners have already gained experience with the corresponding problems in their own company and the importance of the topic has correspondingly already been recognized.[79]

7.2.2 Best Practice Examples for Training

Learning factories are widely used for competency development of employees of industry. Examples for competency development in training can be found in this book in several chapters; see:

- Best Practice Example 1: AutoFab at the Faculty of Electrical Engineering of the University of Applied Sciences Darmstadt, Germany,
- Best Practice Example 2: Demonstration Factory at WZL, RWTH Aachen, Germany,
- Best Practice Example 3: Die Lernfabrik at IWF, TU Braunschweig, Germany,
- Best Practice Example 4: E|Drive-Center at FAPS, Friedrich-Alexander University Erlangen-Nürnberg,

---

[79] See also Tisch (2018).

Table 7.2 Advantages and disadvantages of the two opposed learning process sequences based on Coleman (1982), Keeton, Sheckley, Griggs, and Council for Adult and Experiential Learning (2002), Kolb (1984), Tisch (2018)

	Information assimilation	Experiential learning
Advantages	Great information density in a short amount of time is possible	Context-specific perception of knowledge is enabled automatically
	Learning processes can be planned better in advance compared to experiential learning processes	Potential to develop leadership capabilities inside a learning group
	Direct and systematic classification of information without uncertainty for the learner	Application-oriented learning motivates and creates a link between learning and real life, facilitates the transfer, and generates motivation
	(Unexperienced) learners are brought to an appropriate level of knowledge	Knowledge transfer tends to be more sustainable than in information assimilation learning processes
Disadvantages	Knowledge is presented mostly based on speech, if concepts and principles behind words are not understood, problems are the result	The knowledge transfer is less efficient and the preparation of courses are more complex
	Transition to the application phase often does not take place—can be prevented with the use of the application phases in learning factories	Systematization of experiences and observations often doesn't take place
	Often there is a time gap before the application phase—can be prevented with the use of learning factories for application	Might not be helpful for inexperienced learners; negative experience may lead to demotivation
	Straightforward learning processes may encourage experienced learners too little	Learning outcomes are less predictable; learning process tends to be less focused

- Best Practice Example 5: ESB Logistics Learning Factory at ESB Business School at Reutlingen University, Germany,
- Best Practice Example 6: ETA-Factory at PTW, TU Darmstadt,
- Best Practice Example 7: Festo Didactic Learning Factories,
- Best Practice Example 8: Festo Learning Factory in Scharnhausen, Germany,
- Best Practice Example 9: iFactory at the Intelligent Manufacturing Systems (IMS) Center, University of Windsor, Canada,

7.2 Learning Factories in Training   239

- Best Practice Example 10: IFA-Learning Factory at IFA, Leibniz-Universität Hannover,
- Best Practice Example 12: LEAN-Factory for a pharmaceutical company in Berlin, Germany,
- Best Practice Example 14: Learning Factory aIE at IFF, University of Stuttgart,
- Best Practice Example 15: Learning Factory for electronics production at FAPS, Friedrich-Alexander University Erlangen-Nürnberg, Germany,
- Best Practice Example 17: Learning Factory for Global Production at wbk, KIT Karlsruhe,
- Best Practice Example 18: "Lernfabrik für schlanke Produktion," Learning Factory for Lean Production at iwb, TU München, Germany,
- Best Practice Example 19: "Lernfabrik für vernetzte Produktion" at Fraunhofer IGCV, Augsburg, Germany,
- Best Practice Example 20: LMS Factory at the Laboratory for Manufacturing Systems & Automation (LMS), University Patras, Greece,
- Best Practice Example 21: LPS Learning Factory at LPS, Ruhr-Universität Bochum,
- Best Practice Example 22: MAN learning factory at MAN Diesel & Turbo SE in Berlin, Germany,
- Best Practice Example 23: MPS Lernplattform at Daimler AG in Sindelfingen, Germany,
- Best Practice Example 24: MTA SZTAKI Learning Factory at the Research Laboratory on Engineering and Management Intelligence, MTA SZTAKI in Györ,
- Best Practice Example 25: Pilot Factory Industrie 4.0 at IMW, IFT, and IKT, TU Wien, Austria,
- Best Practice Example 26: Process Learning Factory CiP,
- Best Practice Example 29: Smart Mini Factory at IEA, Free University of Bolzano,
- Best Practice Example 30: Teaching Factory: An emerging paradigm for manufacturing education,
- Best Practice Example 31: VPS Center of the Production Academy at BMW in Munich, Germany.

7.2.3 Success Factors of Learning Factories for Education and Training

Conducive learning processes are often interpreted as feedback loops,[80] consisting of a learner modifying the physical world (work environment) which is changing in a certain way. The exact way in which the environment alters gives learner information about the system he or she is acting in. The acting person integrates this information into an updated understanding of the respective system, which influences the future

---

[80] See Sterman (1994), Forrester (1961), Argyris and Schön (1996).

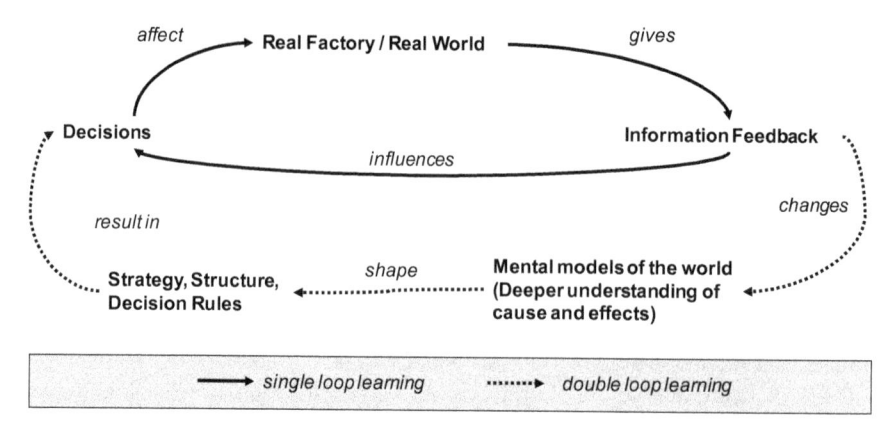

Fig. 7.19 Learning as a feedback process according to Sterman (1994)

attempts to shape the environment according to particular goals and intentions.[81] Behavior and cognition are interrelated,[82] while action is steered by comprehension, the comprehension itself is corrected by the observations made based on actions.[83] The outlined feedback loop is visualized in Fig. 7.19. In this context, in the organizational learning literature the direct link between "information feedback" and "decision" is referred to as "single-loop learning," the indirect link overcorrected "mental models of the world" and connected "decision rules" as a fundament for decisions is called "double-loop learning."[84] In the context of competency development, it can be stated that only the double loop learning path creates the dispositions that are referred to as competencies in the cognitive domain. Double-loop learning on the one hand improves the capability of the learner in general to deal with (other) complex and unknown situations that are related to the changed deeper understanding of cause and effects. While on the other hand in single-loop learning, the information feedback is directly linked to this single situation/decision and reasonably cannot be transferred to other situations; although the learner may try to do so this (usually without success).

Single- and double-loop learning based on feedback happens when persons act in real factory environments. In the various steps of the feedback process, certain problems arise in this context, for example, they are:

- False decisions in real factory environments may be connected with high risks regarding economic and safety issues.
- Factories are complex systems and consequently cannot be grasped easily and in a short amount of time.

---

[81] See Sterman (1994).
[82] See Neisser (1976).
[83] See Brown and Duguid (1991), Weick (1979), Crossan, Lane, and White (1999).
[84] Argyris and Schön (1996).

7.2 Learning Factories in Training

- Feedback on actions may be delayed, superimposed by other developments or cannot be perceived properly.
- Learning processes are hardly controllable in real work environments; therefore learning is not following certain learning goals (informal learning).

In the manufacturing domain, a further possibility to build up this feedback loop for learning addressing the problems of informal learning processes mentioned above can be seen in the use of learning factories. In this case, the learning factory environment can be interpreted as a "virtual world"[85] that provides high-quality, reality-like feedback and experience in industry-relevant issues, problems or transformation processes linked to the manufacturing domain. In this virtual world learning factory, the focus is shifted from the primary need to produce goods to learning and gaining experience in the respective fields. Since the practical experience phases within the virtual world of the learning factory are enriched with systematization phases, it becomes easier for the learner to classify the experiences made in the overall context (change mental models of the world) and, accordingly, to support the double loop learning. Accordingly, through effective competence development, learning factories can help to ensure that the feedback received not only influences individual decisions repeatedly, but also permanently changes the underlying understanding and approaches of the learners. The learning factory extended feedback cycle is depicted in Fig. 7.20.

Furthermore, in different fields of science[86] various success factors of effective learning processes and the modeling of those processes are identified and discussed, for example, in literature regarding active learning,[87] situated learning,[88] problem-based learning[89], or constructivist learning in general.[90] In those approaches and models, success factors are identified that should be followed in order to facilitate effective competency development. In the following, an overview of the summarized and structured main success factors of methodical modeling of learning processes including indications in which way and to what extent the learning factory concept enforces these success factors is given[91]:

- Contextualization, situated context[92] in learning factories: Partial model of real factory included in the learning factory concept provides a rich and authentic learning context, in which situated learning approaches can be embedded.
- Activation of learner[93] in learning factories: Generation and application of knowledge in the learning factory in learner active phases is a key element of the learning factory concept.

---

[85] See also Sterman (1994).
[86] Such as psychology, didactics, and learning design.
[87] See Johnson, Johnson, and Smith (1991).
[88] See Lave and Wenger (1991).
[89] See Boud and Feletti (1999).
[90] See Jonassen (1999).
[91] According to Tisch and Metternich (2017).
[92] See Jonassen (1999), Lave and Wenger (1991).
[93] See Bonwell and Eison (1991), Johnson et al. (1991).

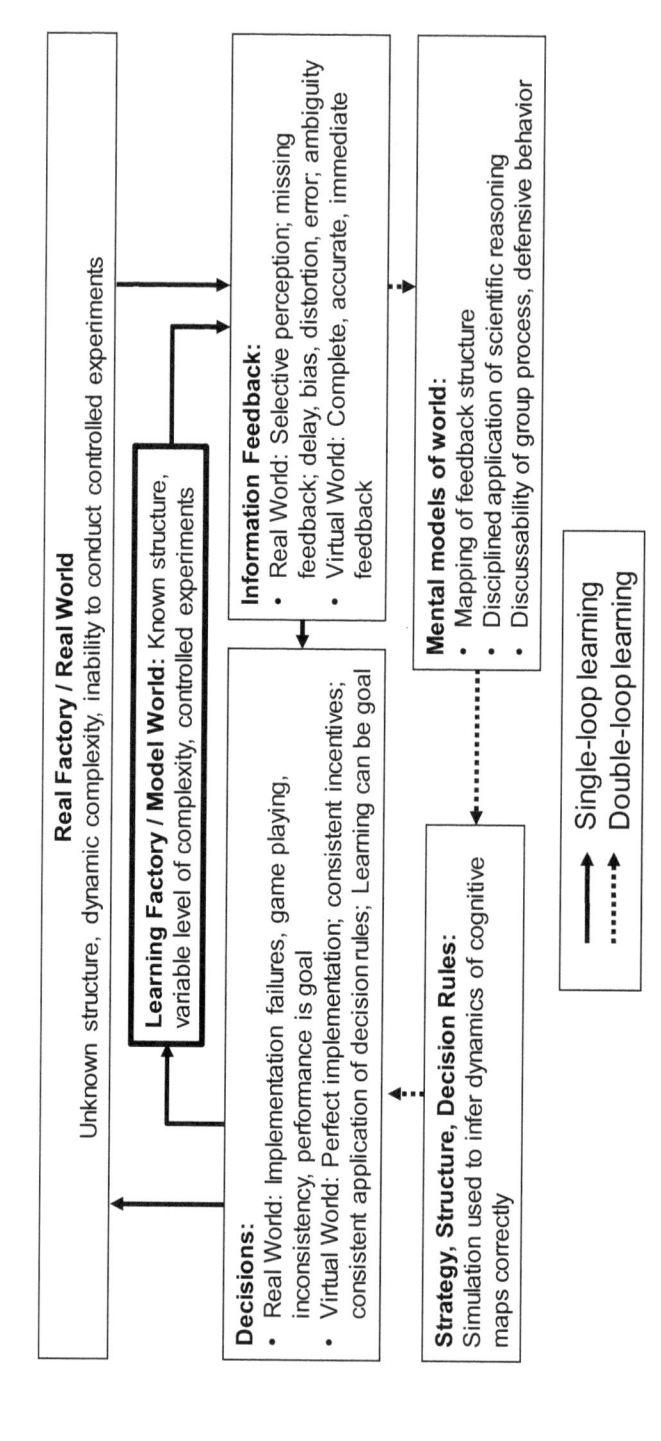

Fig. 7.20 Extended feedback loop using the learning factory as virtual world, shown in Abele et al. (2017) inspired by Sterman (1994)

7.2 Learning Factories in Training

- Problem solving[94] in learning factories: Solving of real problem situations in the learning factory environment is in many cases the basis for learner active learning phases.
- Motivation[95] in learning factories: Motivation to learn is generated by the real character and the possibility to act hands-on immediately in the learning factory. Furthermore, experienced success in the learning factory boosts motivation to apply learning content in real-life situations.
- Collectivization[96] in learning factories: Self-organized learning in groups is a suitable model in exploration and testing phases as well as in systematization and reflection phases in learning factories.
- Integration of thinking and doing[97] in learning factories: The alternation of hand-on phases in the simulated factory environment and systematization phases part of the learning factory concept supports a controllable alternation and integration between thinking and doing.
- Self-regulation[98] and self-direction[99] in learning factories: External and self-controlled learning processes are enabled in learning factories—the type of learning processes used depends on the prerequisites of the learner and the goals of the courses.

Looking at those success factors of effective learning processes, it is becoming clear that the learning factory concept possesses the potential for high-quality competency development. Of course in order to make use of this potential, it is necessary to design, built-up, use, and improve learning factories in an appropriate way. In order to be able to achieve this, consequently, Chap. 6 deals with topics connected to this along the learning factory life cycle from leaning factory planning to the modification or recycling of learning factories. Among others, the learning factory design is addressed, starting from an early conceptualizing phase, learning environment design, learning module generation, and an iterative improvement of the design based on suiting evaluation methods.

7.2.4 Learning Factory Trainings as a Part of Change Management Approaches

Furthermore, learning factories can be used in order to facilitate change processes in organizations.[100] A survey with executives from Austria, Switzerland, and Germany

---

[94] See Boud and Feletti (1999).
[95] See Deci, Vallerand, Pelletier, and Ryan (1991).
[96] See Greeno, Collins, and Resnick (1996).
[97] See Aebli (1994).
[98] See Schunk (1990).
[99] See Garrison (1997).
[100] See Dinkelmann et al. (2014).

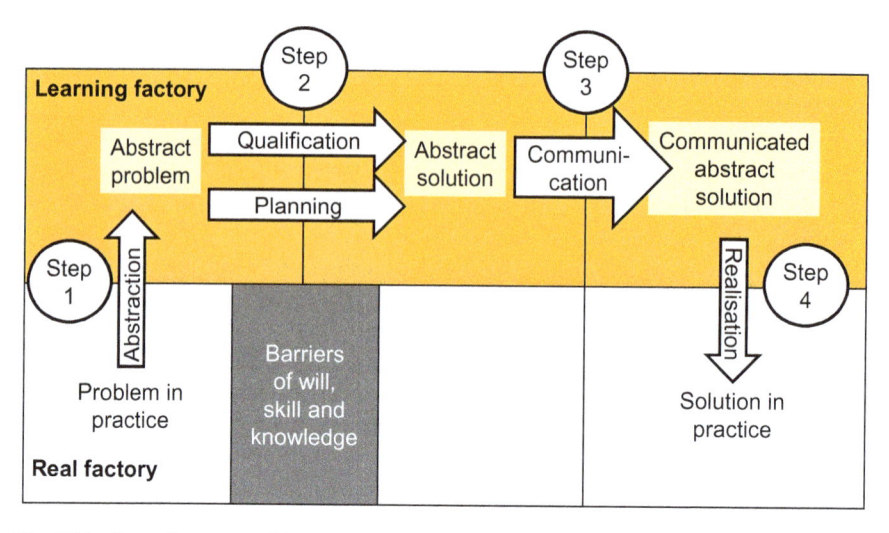

Fig. 7.21 General process of integrating the learning factory concept in a change management approach according to Dinkelmann et al. (2014) and Dinkelmann (2016)

on the main failing causes for change projects found as primary cause the resistance of employees against change.[101] Sources for this resistance can be either found in barriers related to will or barriers related to skill and knowledge.[102] Learning factory concepts can be used to surmount these barriers.[103] Figure 7.21 shows the general process for the learning factory supported change management approach: First, the problem is abstracted in the learning factory, followed by the qualification of personnel in the learning factory as well as planning and solution finding of the employees in the learning factory. Finally, the solution found by the employees is transferred to the own factory of the employees.[104]

7.2.5 Technology and Innovation Transfer in Course of Learning Factory Trainings

A specific form of learning factory training has become extremely important, particularly in connection with the emergence of Industry 4.0. In this form, the learning environment is used to demonstrate innovative technologies and procedures. The focus of the use of the learning factory is accordingly not on the learners' own actions,

---

[101] See Hernstein/Hernstein International Management Institute (2003).

[102] See Reiß (2012), Dinkelmann et al. (2014).

[103] See Dinkelmann et al. (2014) and Dinkelmann (2016).

[104] A detailed procedure for using the learning factory concept as part of a change management approach is described in Dinkelmann (2016).

7.2 Learning Factories in Training   245

but rather on the tangible perception of the technology in a production environment. The demonstration is used as the predominant form of learning. As a rule, the aim of this form of learning is not to form high-quality competences, but to inform and stimulate. Often those kinds of demonstrations are combined with hands-on learning factory trainings. Examples for this kind of use of learning factories can be found in the literature.[105]

Furthermore, several learning factories shown in the best practice examples are used for innovation transfer:

- Best Practice Example 1: AutoFab at the Faculty of Electrical Engineering of the University of Applied Sciences Darmstadt, Germany,
- Best Practice Example 2: Demonstration Factory at WZL, RWTH Aachen,
- Best Practice Example 3: Die Lernfabrik at IWF, TU Braunschweig, Germany,
- Best Practice Example 4: E|Drive-Center at FAPS, Friedrich-Alexander University Erlangen-Nürnberg,
- Best Practice Example 5: ESB Logistics Learning Factory at ESB Business School at Reutlingen University, Germany,
- Best Practice Example 6: ETA-Factory at PTW, TU Darmstadt,
- Best Practice Example 9: iFactory at the Intelligent Manufacturing Systems (IMS) Center, University of Windsor, Canada,
- Best Practice Example 12: LEAN-Factory for a pharmaceutical company in Berlin, Germany,
- Best Practice Example 14: Learning Factory aIE at IFF, University of Stuttgart,
- Best Practice Example 15: Learning Factory for electronics production at FAPS, Friedrich-Alexander University Erlangen-Nürnberg, Germany,
- Best Practice Example 16: Learning Factory for Innovation, Manufacturing, and Cooperation at the Faculty of Industrial and Process Engineering of Heilbronn University of Applied Sciences,
- Best Practice Example 17: Learning Factory for Global Production at wbk, KIT Karlsruhe,
- Best Practice Example 18: "Lernfabrik für schlanke Produktion," Learning Factory for Lean Production at iwb, TU München, Germany,
- Best Practice Example 20: LMS Factory at the Laboratory for Manufacturing Systems & Automation (LMS), University Patras, Greece,
- Best Practice Example 21: LPS Learning Factory at LPS, Ruhr-Universität Bochum
- Best Practice Example 22: MAN learning factory at MAN Diesel & Turbo SE in Berlin, Germany,
- Best Practice Example 23: MPS Lernplattform at Daimler AG in Sindelfingen, Germany,
- Best Practice Example 24: MTA SZTAKI Learning Factory at the Research Laboratory on Engineering and Management Intelligence, MTA SZTAKI in Györ,
- Best Practice Example 25: Pilot Factory Industrie 4.0 at IMW, IFT, and IKT, TU Wien,

---

[105] See, for example, Wank et al. (2016) or Erol et al. (2016).

- Best Practice Example 26: Process Learning Factory CiP,
- Best Practice Example 27: Smart Factory at the Research Laboratory on Engineering & Management, MTA SZTAKI,
- Best Practice Example 28: Smart Factory at DFKI, TU Kaiserslautern, Germany,
- Best Practice Example 29: Smart Mini Factory at IEA, Free University of Bolzano,
- Best Practice Example 30: Teaching Factory: An emerging paradigm for manufacturing education,
- Best Practice Example 31: VPS Center of the Production Academy at BMW in Munich,

7.3 Learning Factories in Research

Learning factories are linked in two ways to research. As shown in the learning factory morphology[106], the connection between learning factories and research is established as

(a) Learning factories as research objects,
(b) Learning factories as research enablers,

(a) Learning factories as research objects.

In recent years, research on the learning factory system was dealing with questions like,

- How learning factories can be described and defined holistically,[107]
- How learning factories can be classified according to their changeability,[108]
- How learning factories, learning modules, and learning situations in learning factories can be designed,[109]
- How products for changeable learning factories can be designed,[110]
- How miniaturized factories can be used as learning environments in learning factories,[111]
- How non-visible topics such as energy efficiency can be addressed in learning factories,[112]
- How learning success can be measured in learning factories,[113]

---

[106] See Chap. 4 or Tisch, Ranz, Abele, Metternich, and Hummel (2015c).
[107] See Tisch et al. (2015d), Abele (2016).
[108] See Wagner, AlGeddawy, ElMaraghy, and Müller (2012).
[109] See Tisch, Hertle, Abele, Metternich, and Tenberg (2015b).
[110] See Wagner, AlGeddawy, ElMaraghy, and Müller (2015).
[111] See Kaluza et al. (2015).
[112] See Abele, Bauerdick, Strobel, and Panten (2016).
[113] See Tisch, Hertle, Metternich, and Abele (2014, 2015).

7.3 Learning Factories in Research 247

- How learning factories can be improved holistically based on a maturity model and a quality system,[114]
- What are the essential requirements on learning factories for the different stakeholders,[115] and many more.

Regarding the learning factory as a research object, research is only at the beginning of a wide field. The topic learning factory is now mostly structured, described, and first research regarding the learning factory system is published.[116] But there are many more topics that can and should be addressed (e. g., scalability of learning factories), deepened (e. g., effectiveness of learning factories), or newly discovered (e. g., Industrie 4.0 in learning factories in recent years) in order to further evolve and expand the possibilities and the potential of learning factories. For further research topics on the system learning factory; see also Chap. 6.

(b) **Learning factories as research enabler.**

Production engineering as applied science strives for the discovery of new, practically applicable knowledge. In applied sciences, research questions typically emerge from industrial practice.[117] The research process of applied sciences[118] describes process of knowledge creation from the identification of practice-relevant problems to the consultation of practice with newly created knowledge. Figure 7.22 illustrates this research process as a pendulum movement between theory and practice, i.e., between desk research and analytics on the one hand and empiricism on the other. One problem of this research process in the production-related research is that the direct link between research and daily industrial practice is associated with the risk of losing the factory's basic stability. In addition, the direct transfer of research results into industrial production is accompanied by high complexity and correspondingly high costs. Especially for small- and medium-sized companies that typically neither have a research infrastructure nor research experts from various fields, this is problematic.[119] Therefore, in the field of production engineering, learning factories can support the research process with facilitated risk-free integration of practical experience at lower cost and complexity.

Learning factories in research are well-suited to be used as laboratory environments for manufacturing systems. A description of a concept for the integration of learning factories into research as research enablers[120] is shown in Fig. 7.23. The process is starting from existing knowledge going on with

- Problem identification,
- Abstraction of the problem with real data,

---

[114] See Enke, Glass, and Metternich (2017).
[115] See Enke, Tisch, and Metternich (2016).
[116] For an overview, see also Abele et al. (2017).
[117] See Ulrich, Dyllick, and Probst (1984).
[118] According to Ulrich et al. (1984).
[119] See Abele et al. (2017).
[120] See Seifermann, Metternich, and Abele (2014).

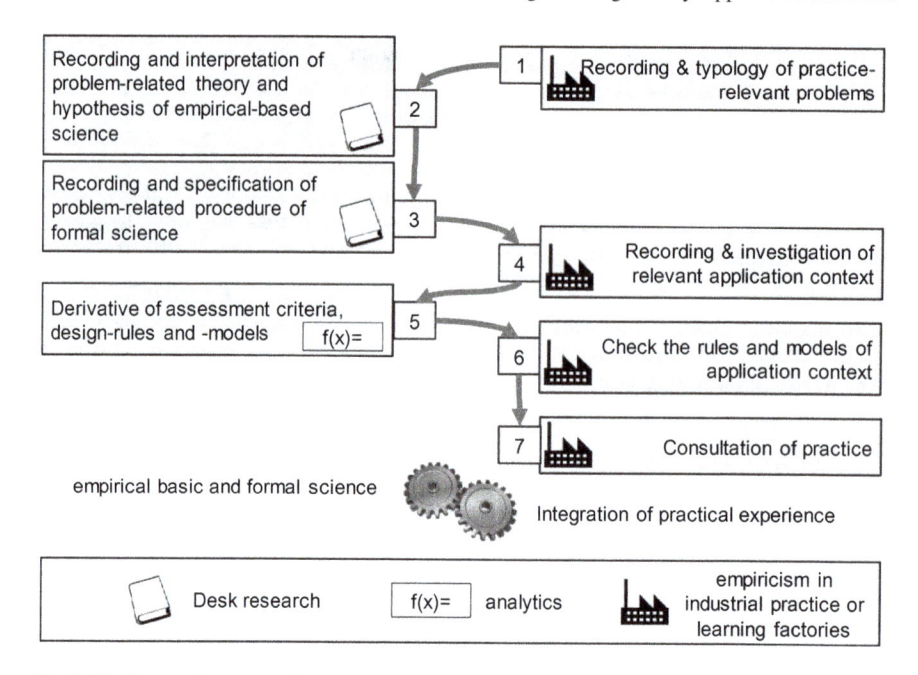

Fig. 7.22 Research process of applied sciences and the integration of practical experience through learning factories (Abele et al., 2017) according to Schuh and Warschat (2013) on the basis of Ulrich et al. (1984)

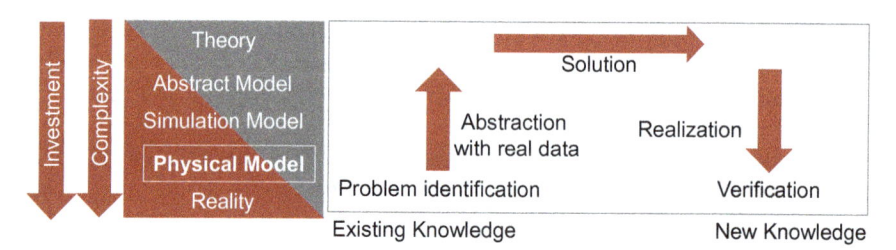

Fig. 7.23 Learning factories as research enablers according to Seifermann et al. (2014)

- Theoretically based solution finding,
- Solution realization into practice, and verification of new knowledge.[121]

Based on this concept, the learning factory as research enabler helps to identify research problems in the quasi-realistic environment, to test solutions with the physical factory model at reduced complexity and costs in comparison to solution testing in reality.[122] In the following, this concept of integration of the learning factory in research as research enabler is illustrated exemplarily for the topic "cellular manu-

---

[121] See Seifermann et al. (2014).

[122] See Abele et al. (2017).

7.3 Learning Factories in Research

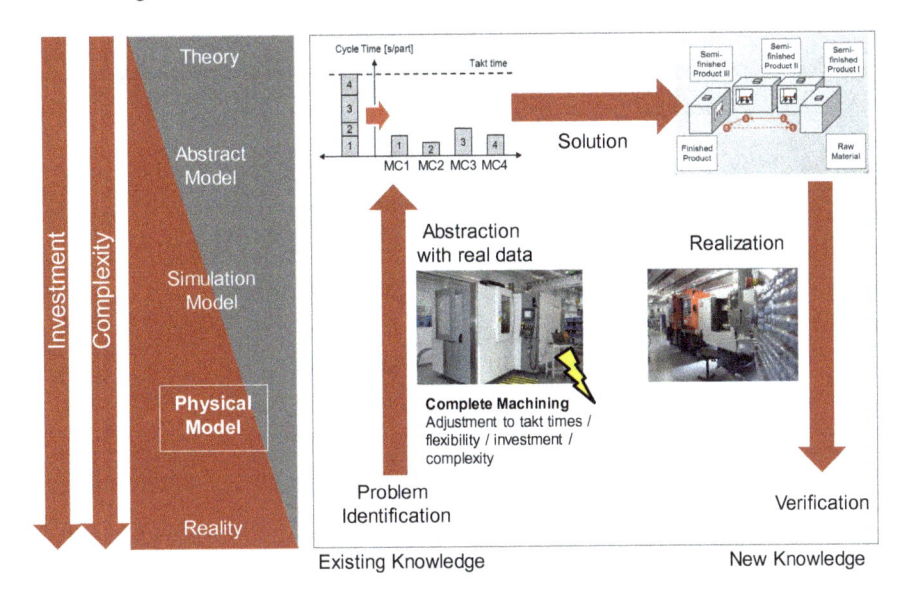

Fig. 7.24 Learning factories as a research enabler with the example of the research on the cellular manufacturing approach at process learning factory CiP at PTW, TU Darmstadt, according to Seifermann et al. (2014)

facturing" at the PTW and the Process Learning Factory CiP at TU Darmstadt. For a summary of this example; see also Fig. 7.24 .

Step 1: "Problem identification"[123]:

Originally, only classic complete machining was used in the machining area of the Process Learning Factory CiP, which can also produce complex components in just a few setups. The complete machining approach, which is based on maximum capital productivity, is the most economical solution for high volume low mix environments when the individual machining system is considered. However, due to limited flexibility this approach presents for certain applications not an optimal solution, let alone a lean solution.[124] This disadvantage of the complete machining approach was also identified in the production scenario of the Process Learning Factory CiP: The required components produced on the milling machine would require 23 shifts per week. Since this could not be achieved even in three-shift operation seven days a week, a second large and expensive machine would have to be purchased while retaining the complete machining approach.

Step 2: "Abstraction of the problem with real data"[125]:

---

[123] Described in detail, for example, in Abele, Bechtloff, and Krause (2011) and Abele, Bechtloff, and Seifermann (2012).

[124] See Abele et al. (2011), Abele et al. (2012).

[125] Described in detail, for example, in Abele et al. (2012) and Seifermann, Böllhoff, Adolph, Abele, and Metternich (2017).

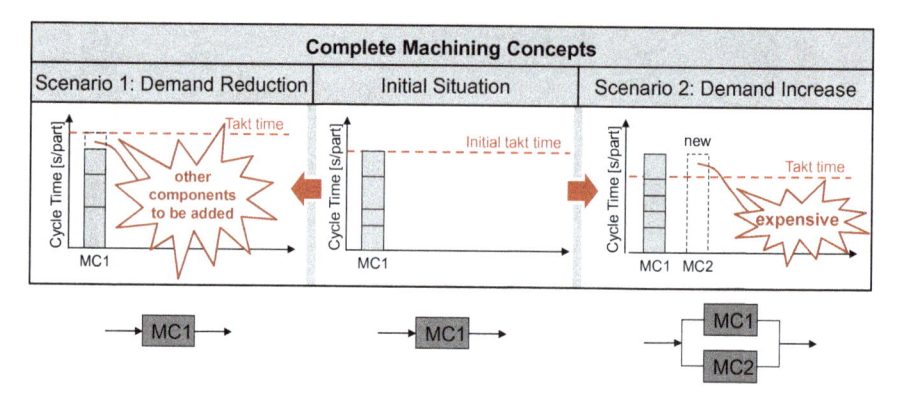

Fig. 7.25 Reaction of complete machining systems on a volume increase or decrease with slight adaptions according to Seifermann et al. (2017)

In machining areas, the complete machining processes of a component in a complex machine tool do not correspond to the ideal of the one-piece flow, since cycle times cannot be aligned with customer take times. When using the complete machining approach, an increase in volume is merely possible with a parallel installation of an additional complex and expensive machine in the system; see also Fig. 7.25. This inflexible reaction with respect to volume is associated with high investment and no approximation of the intended one piece flow. An appropriate reaction from a lean point of view would be to distribute necessary work operations on several stations (or machining center (MC) in this case)—just like in assembly.

Step 3: "Theoretically based solution finding"[126]:

The solution approach to the present problem situation is inspired by the ideas of lean assembly systems design. In contrast to the complete machining of a component in a complex machine tool, the principle of so-called cellular manufacturing is based on the separability of machining operations. The cycle times to be realized are always viewed in comparison with the customer take and are distributed as evenly as possible to several machine tools with comparatively little complexity. The machines can be arranged in a compact U-, V-, L-, and O-shaped layout for reduced transport and better communication possibilities.[127] Furthermore, manual and machine operations can be separated. Work organization and operation are therefore similar to a lean assembly cell.[128] The cellular manufacturing concepts for a flexible reaction on a change in demand as well as the layout and the balancing of a cell are visualized in Fig. 7.26.

Step 4: "Solution realization into practice and verification of new knowledge":

---

[126] Described in detail, for example, in Abele et al. (2011, 2012), and Seifermann et al. (2017).
[127] See Shingo (1989).
[128] See Abele et al. (2011, 2012), Seifermann et al. (2017).

7.3 Learning Factories in Research

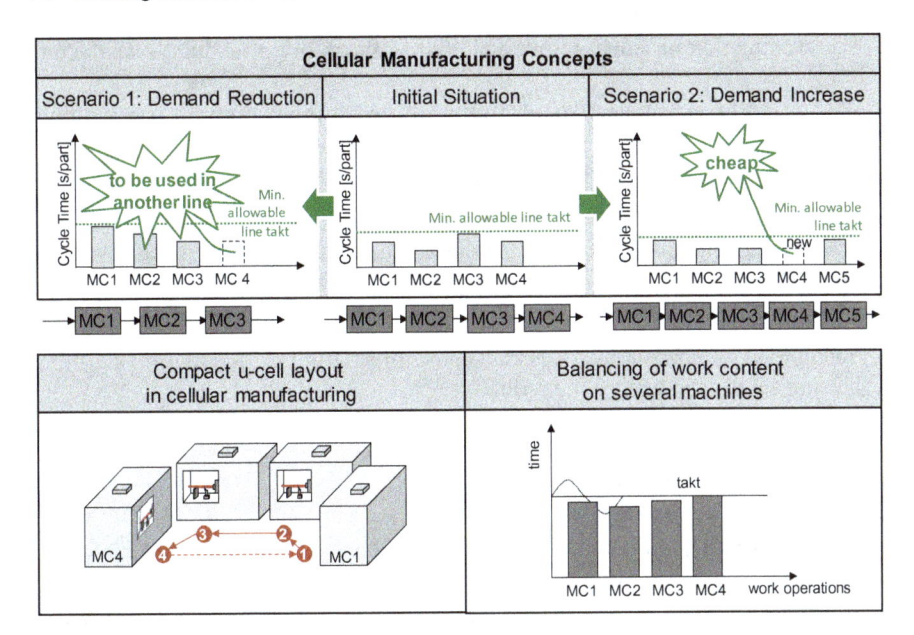

Fig. 7.26 Theoretical solution of cellular manufacturing for a lean machining with slight adaptions according to Seifermann et al. (2017) and Abele et al. (2012)

Until this point, the idea of cellular manufacturing is an ideal concept based on the vision of a lean production system. So far, it is not clear if this concept only works on paper or if it can be successfully transferred to the real factory. Here, the research enabler "learning factory" provides helpful support for research by installing the so far theoretical concept and testing it extensively. In the present example, the value stream of the pneumatic cylinder in the Process Learning Factory CiP, six small turning, and milling machines were installed in a production cell parallel to the done-in-one machines. Both machining areas (the cellular as well as the complete machining) are used for the manufacturing of the bottom of the cylinder as well as the piston rod. Having this environment, the general concept of cellular manufacturing has been tested and verified.

Today, on the basis of this created cellular manufacturing environment, new research processes[129] are triggered in various research fields around the cellular manufacturing approaches. Research in this field in the Process Learning Factory CiP deals with

- The evaluation of general economic application fields of cellular manufacturing,[130]

---

[129] Consisting of problem identification, abstraction, solution finding, realization, and verification.
[130] See Bechtloff (2014), Metternich, Abele, Bechtloff, and Seifermann (2015), Metternich, Bechtloff, and Seifermann (2013).

- The examination of gains in mix and volume flexibility of cellular manufacturing,[131]
- Quality issues in cellular manufacturing related to human errors and multiple clampings,[132]
- Low-cost automation solutions as an enabler for cellular manufacturing[133].

In addition, independently of the presented research process, learning factories are (more) frequently used as validation environment for developed research results. From the recent past, numerous examples can be identified, while only a few of those coming from different research areas are mentioned here at this point:

- Continuous improvement processes and systems,[134]
- The impact of goal-setting in production,[135]
- Online control of assembly processes,[136]
- Video analysis of production-relevant competencies,[137]
- Technologies for smart factories,[138]
- Value stream mapping 4.0,[139]
- Defect prevention with optical object identification,[140]
- Lean stress sensitization.[141]

Furthermore, additional possibilities of learning factories for research are opened up by the generation of so-called embedded experiments. In this concept, learning factories offer an ideal framework for research by having the possibility to combine advantages of field experiments and experiments in the laboratory.[142] Figure 7.27 shows this by comparing characteristics of field and laboratory experiments and linking them to the research possibilities in learning factories. This way, learning factories may combine authentic social realities, a high influence as well as a high internal and external validity of the embedded experiments.

In addition, learning factory environments are increasingly being used to enable companies to learn about the latest technologies and innovative methods. Learning factories have a high potential for the demonstration and transfer of innovations. In addition to an application-oriented technology and innovation platform, which offer

---

[131] See Metternich, Böllhoff, Seifermann, and Beck (2013), Seifermann et al. (2017).

[132] See Böllhoff, Metternich, Frick, and Kruczek (2016), Böllhoff, Seifermann, Metternich, and Heß (2015).

[133] See Seifermann, Böllhoff, Metternich, and Bellaghnach (2014), Seifermann (2018).

[134] See Cachay and Abele (2012).

[135] See Asmus, Karl, Mohnen, and Reinhart (2015).

[136] See Tracht, Funke, and Schottmayer (2015).

[137] See Hambach, Tenberg, and Metternich (2015), Hambach, Diezemann, Tisch, and Metternich (2016).

[138] See Kemény et al. (2016).

[139] See Meudt, Metternich, and Abele (2017).

[140] See Wiech, Böllhoff, and Metternich (2017).

[141] See Dombrowski, Wullbrandt, and Reimer (2017).

[142] See Schuh, Gartzen, Rodenhauser, and Marks (2015).

7.3 Learning Factories in Research

	field experiment	laboratory experiment
Environment	social reality	arranged
Influence	low	high
Internal validity	low	high
External validity	high	low

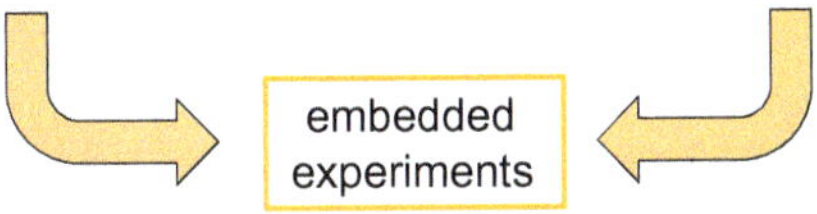

embedded experiments

Fig. 7.27 Combining the advantages of field and laboratory experiments for research in learning factories according to Schuh et al. (2015)

learning factories for research and development until market maturity of production processes, production technologies, and products. Furthermore, learning factories support the subsequent transmission of those innovations. Examples for research, demonstration, and transfer in learning factories can be found especially in the following best practice examples:

- The Smart Factory in Kaiserslautern is described by Dr. Haike Frank, Dennis Kolberg, Prof. Detlef Zühlke as a best practice example of research regarding Industrie 4.0, digitalization, and factory automation; see Best Practice Example 28.
- The Pilot Factory Industrie 4.0 as a research, testing, and demonstration facility at TU Wien is described by Prof. Dr.-Ing. Wilfried Sihn, Fabian Ranz, and Philipp Hold; see Best Practice Example 25.

Beyond those examples learning factories are widely used in research. Examples of research in learning factories can be found in various sections:

- Best Practice Example 1: AutoFab at the Faculty of Electrical Engineering of the University of Applied Sciences Darmstadt, Germany,
- Best Practice Example 2: Demonstration Factory at WZL, RWTH Aachen,
- Best Practice Example 3: Die Lernfabrik at IWF, TU Braunschweig, Germany,
- Best Practice Example 4: E|Drive-Center at FAPS, Friedrich-Alexander University Erlangen-Nürnberg,
- Best Practice Example 5: ESB Logistics Learning Factory at ESB Business School at Reutlingen University, Germany,
- Best Practice Example 6: ETA-Factory at PTW, TU Darmstadt,
- Best Practice Example 7: Festo Didactic Learning Factories,
- Best Practice Example 8: Festo Learning Factory in Scharnhausen,
- Best Practice Example 9: iFactory at the Intelligent Manufacturing Systems (IMS) Center, University of Windsor, Canada,
- Best Practice Example 10: IFA-Learning Factory at IFA, Leibniz-Universität Hannover,

- Best Practice Example 11: Integrated Learning Factory at LPE & LPS, Ruhr-University Bochum, Germany,
- Best Practice Example 14: Learning Factory aIE at IFF, University of Stuttgart,
- Best Practice Example 15: Learning Factory for electronics production at FAPS, Friedrich-Alexander-University Erlangen-Nürnberg, Germany,
- Best Practice Example 16: Learning Factory for Innovation, Manufacturing, and Cooperation at the Faculty of Industrial and Process Engineering of Heilbronn University of Applied Sciences,
- Best Practice Example 17: Learning Factory for Global Production at wbk, KIT Karlsruhe,
- Best Practice Example 21: LPS Learning Factory at LPS, Ruhr-Universität Bochum,
- Best Practice Example 24: MTA SZTAKI Learning Factory at the Research Laboratory on Engineering and Management Intelligence, MTA SZTAKI in Györ,
- Best Practice Example 25: Pilot Factory Industrie 4.0 at IMW, IFT, and IKT, TU Wien,
- Best Practice Example 26: Process Learning Factory CiP,
- Best Practice Example 27: Smart Factory at the Research Laboratory on Engineering & Management, MTA SZTAKI,
- Best Practice Example 28: Smart Factory at DFKI, TU Kaiserslautern, Germany,
- Best Practice Example 29: Smart Mini Factory at IEA, Free University of Bolzano,
- Best Practice Example 30: Teaching Factory: An emerging paradigm for manufacturing education,
- Best Practice Example 31: VPS Center of the Production Academy at BMW in Munich.

7.4 Wrap-up of This Chapter

This chapter provides an overview of the ways in which learning factories are used in teaching, training, and research. In this context, the role of the learning factory for the most diverse concepts in the context of the application areas is clarified. For the different application fields and concepts, various examples from existing learning factories are given.

References

Abele, E. (2016). Learning factory. *CIRP Encyclopedia of Production Engineering.*
Abele, E., Bauerdick, C., Strobel, N., & Panten, N. (2016). ETA learning factory: A holistic concept for teaching energy efficiency in production. In *6th CIRP-Sponsored Conference on Leanring Factories. Procedia CIRP, 54*, 83–88.
Abele, E., Bechtloff, S., & Krause, F. (2011). Flexible Serienfertigung im Kundentakt. *Werkstatt Und Betrieb-Munchen, 144*(6), 24–27.

References

Abele, E., Bechtloff, S., & Seifermann, S. (2012). Sequenzfertigung für flexible und schlanke Zerspanung. *Productivity Management, 17*(1), 45–48.

Abele, E., Chryssolouris, G., Sihn, W., Metternich, J., ElMaraghy, H. A., Seliger, G., et al. (2017). Learning factories for future oriented research and education in manufacturing. *CIRP Annals—Manufacturing Technology, 66*(2), 803–826.

Abt, C. C. (1987). *Serious games*. USA: University Press of America.

Aebli, H. (1994). *Denken: das Ordnen des Tuns* (2. Aufl). Stuttgart: Klett-Cotta.

Ahmad, R., Masse, C., Jituri, S., Doucette, J., & Mertiny, P. (2018). Alberta Learning factory for training reconfigurable assembly process value stream mapping. *Procedia Manufacturing, 23*, 237–242. https://doi.org/10.1016/j.promfg.2018.04.023.

Argyris, C., & Schön, D. A. (1996). *Organizational learning* (Reprinted with corr). *Organization development series*. Reading, Mass: Addison-Wesley Pub. Co.

Asmus, S., Karl, F., Mohnen, A., & Reinhart, G. (2015). The impact of goal-setting on worker performance—empirical evidence from a real-effort production experiment. In *12th Global Conference on Sustainable Manufacturing*. Procedia CIRP, 26, 127–132.

Badurdeen, F., Marksberry, P., Hall, A., & Gregory, B. (2010). Teaching lean manufacturing with simulations and games: A survey and future directions. *Simulation & Gaming, 41*(4), 465–486.

Balve, P., & Albert, M. (2015). Project-based learning in production engineering at the Heilbronn learning factory. In *5th CIRP-Sponsored Conference on Learning Factories*. Procedia CIRP, 32, 104–108. https://doi.org/10.1016/j.procir.2015.02.215.

Barrows, H. S., & Tamblyn, R. M. B. S. N. (1980). *Problem-based learning: An approach to medical education*. USA: Springer Publishing Company.

Beauvais, W. (2013). Qualification as an effective tool to support the implementation of lean. In G. Reinhart, P. Schnellbach, C. Hilgert, & S. L. Frank (Eds.), *3rd Conference on Learning Factories*, Munich, 7th May 2013 (pp. 108–117). Augsburg.

Bechtloff, S. (2014). *Identifikation wirtschaftlicher Einsatzgebiete der Sequenzfertigung in der Bohr- und Fräsbearbeitung von Kleinserien*. Dissertation, Darmstadt. *Schriftenreihe des PTW: "Innovation Fertigungstechnik"*. Aachen: Shaker.

Bender, W. N. (2012). *Project-based learning: differentiating instruction for the 21st century*. USA: Sage Publications.

Blöchl, S. J., & Schneider, M. (2016). Simulation game for intelligent production logistics—The PuLL® learning factory. *Procedia CIRP, 54*, 130–135. https://doi.org/10.1016/j.procir.2016.04.100.

Blöchl, S. J., Michalicki, M., & Schneider, M. (2017). Simulation game for lean leadership—Shopfloor management combined with accounting for lean. *Procedia Manufacturing, 9*, 97–105. https://doi.org/10.1016/j.promfg.2017.04.031.

Blume, S., Madanchi, N., Böhme, S., Posselt, G., Thiede, S., & Herrmann, C. (2015). Die Lernfabrik—Research-based learning for sustainable production engineering. In *5th CIRP-Sponsored Conference on Learning Factories*. Procedia CIRP, 32, 126–131. https://doi.org/10.1016/j.procir.2015.02.113.

Böhner, J., Weeber, M., Kuebler, F., & Steinhilper, R. (2015). Developing a learning factory to increase resource efficiency in composite manufacturing processes. In *5th CIRP-Sponsored Conference on Learning Factories*. Procedia CIRP, 32, 64–69. https://doi.org/10.1016/j.procir.2015.05.003.

Böllhoff, J., Metternich, J., Frick, N., & Kruczek, M. (2016). Evaluation of the human error probability in cellular manufacturing. In *5th CIRP Global Web Conference (CIRPe 2016)*. Procedia CIRP, 55, 218–223.

Böllhoff, J., Seifermann, S., Metternich, J., & Heß, T. (2015). Qualität in der Sequenzfertigung: Bewertung und Diskussion der Prozessfähigkeit einer schlanken Zerspanungszelle. *Werkstattstechnik online: wt, 105*(1/2), 78–83.

Bonwell, C. C., & Eison, J. A. (1991). *Active learning: Creating excitement in the classroom. ASHE-ERIC higher education report: 1, 1991*. Washington, D.C.: School of Education and Human Development, George Washington University.

Boud, D., & Feletti, G. (Eds.). (1999). *The challenge of problem-based learning* (2nd ed., reprinted.). London: Kogan Page.
Brown, J. S., & Duguid, P. (1991). Organizational learning and communities-of-practice: Toward a unified view of working. Learning and innovation. *Organization Science, 2*(1), 40–57.
Cachay, J., & Abele, E. (2012). Developing competencies for continuous improvement processes on the shop floor through learning factories—Conceptual design and empirical validation. In *45th CIRP Conference on Manufacturing Systems. Procedia CIRP, 3*(3), 638–643.
Cachay, J., Wennemer, J., Abele, E., & Tenberg, R. (2012). Study on action-oriented learning with a learning factory approach. *Procedia—Social and Behavioral Sciences, 55,* 1144–1153.
Cawley, P. (1989). The introduction of a problem-based option into a conventional engineering degree course. *Studies in Higher Education, 14*(1), 83–95. https://doi.org/10.1080/0307507891 2331377632.
Coleman, J. S. (1982). Experiential learning and information assimilation: Toward an appropriate mix. In D. Conrad & D. Hedin (Eds.), *Child & youth services: Volume 4, numbers 3/4. Youth participation and experimental education* (pp. 13–20). New York: The Haworth Press.
Crawley, E., Malmqvist, J., Ostlund, S., & Brodeur, D. (2007). *Rethinking engineering education: The CDIO approach* (1st ed.). New York: Springer.
Creswell, J. W. (2008). *Educational research: Planning, conducting, and evaluating quantitative and qualitative research* (3rd ed.). Upper Saddle River, N.J.: Pearson/Merrill Prentice Hall.
Crossan, M. M., Lane, H. W., & White, R. E. (1999). An organizational learning framework: From intuition to institution. *The Academy of Management Review, 24*(3), 522–537.
De Freitas, S. (2007). *Learning in immersive worlds: A review of game based learning.* Prepared for the JISC e-Learning Programme. Retrieved from http://researchrepository.murdoch.edu.au/id/eprint/35774/1/gamingreport_v3.pdf.
Deci, E. L., Vallerand, R. J. M., Pelletier, L. G., & Ryan, R. M. (1991). Motivation and education: The self-determination perspective. *Educational Psychologist, 26*(3 & 4), 325–346.
Deterding, S., Dixon, D., Khaled, R., & Nacke, L. (2011). From game design elements to gamefulness: Defining "gamification". *MindTrek', 11,* 9–15, 28–30 September 2011. Tampere, Finland.
Dinkelmann, M. (2016). *Methode zur Unterstützung der Mitarbeiterpartizipation im Change Management der variantenreichen Serienproduktion durch Lernfabriken.* Dissertation, Stuttgart.
Dinkelmann, M., Siegert, J., & Bauernhansl, T. (2014). Change management through learning factories. In M. F. Zäh (Ed.), *Enabling manufacturing competitiveness and economic sustainability* (pp. 395–399). Springer. https://doi.org/10.1007/978-3-319-02054-9_67.
Djaouti, D., Alvarez, J., Jessel, J. P., & Rampnoux, O. (2011). Origins of serious games. In M. Ma, A. Oikonomou, & L. Jain (Eds.), *Serious games and edutainment applications* (pp. 25–43). London: Springer.
Dombrowski, U., Wullbrandt, J., & Reimer, A. (2017). Lean stress sensitization in learning factories. *Procedia Manufacturing, 9,* 339–346. https://doi.org/10.1016/j.promfg.2017.04.016.
Elbestawi, M., Centea, D., Singh, I., & Wanyama, T. (2018). SEPT learning factory for industry 4.0 education and applied research. *Procedia Manufacturing, 23,* 249–254. https://doi.org/10.1016/j.promfg.2018.04.025.
Enke, J., Glass, R., & Metternich, J. (2017). Introducing a maturity model for learning factories. In *7th CIRP-Sponsored Conference on Learning Factories. Procedia Manufacturing, 9,* 1–8.
Enke, J., Tisch, M., & Metternich, J. (2016). Learning factory requirements analysis—Requirements of learning factory stakeholders on learning factories. In *5th CIRP Global Web Conference (CIRPe 2016). Procedia CIRP, 55,* 224–229.
Erol, S., Jäger, A., Hold, P., Ott, K., & Sihn, W. (2016). Tangible industry 4.0: A scenario-based approach to learning for the future of production. In *6th CIRP-Sponsored Conference on Leanring Factories. Procedia CIRP, 54,* 13–18.
Euler, D. (2005). Forschendes Lernen. In S. Spoun & W. Wunderlich (Eds.), *Studienziel Persönlichkeit: Beiträge zum Bildungsauftrag der Universität heute* (pp. 253–272). Frankfurt, New York: Campus-Verlag.

Evensen, D. H., & Hmelo-Silver, C. E. (2000). *Problem-based learning: A research perspective on learning interactions*. UK: Taylor & Francis.

Faller, C., & Feldmüller, D. (2015). Industry 4.0 learning factory for regional SMEs. In *5th CIRP-Sponsored Conference on Learning Factories. Procedia CIRP, 32*, 88–91. https://doi.org/10.1016/j.procir.2015.02.117.

FAPS. (2018). *Green factory bavaria*. Retrieved from http://www.greenfactorybavaria.de/.

Felicia, P. (2014). *Game-based learning: Challenges and opportunities*. UK: Cambridge Scholars Publisher.

Festo Didactic. (2017). *Learning factory 4.0, Philipp-Matthäus-Hahn-Schule*. Balingen: Reference Project Global Project Solutions.

Forrester, J. W. (1961). *Industrial dynamics*. Cambridge: MIT Press.

Garrison, D. R. (1997). Self-directed learning: Toward a comprehensive model. *Adult Education Quarterly, 48*(1), 18–33. https://doi.org/10.1177/074171369704800103.

Gräßler, I., Pöhler, A., & Pottebaum, J. (2016). Creation of a learning factory for cyber physical production systems. In *6th CIRP-Sponsored Conference on Leanring Factories. Procedia CIRP, 54*, 107–112. https://doi.org/10.1016/j.procir.2016.05.063.

Gräßler, I., Taplick, P., & Yang, X. (2016b). Educational learning factory of a holistic product creation process. *Procedia CIRP, 54*, 141–146. https://doi.org/10.1016/j.procir.2016.05.103.

Gredler, M. E. (2004). Games and simulations and their relationships to learning. In D. Jonassen (Ed.), *Handbook of research on educational communications and technology* (2nd ed., pp. 571–581). Mahwah, New Jersey: Lawrence Erlbaum Associates Publishers.

Greeno, J., Collins, A., & Resnick, L. (1996). Cognition and learning. In D. C. Berliner & R. C. Calfee (Eds.), *Handbook of educational psychology* (pp. 15–46). New York, London: Macmillan; Prentice Hall.

Hambach, J., Diezemann, C., Tisch, M., & Metternich, J. (2016). Assessment of students' lean competencies with the help of behavior video analysis—Are good students better problem solvers? In *5th CIRP Global Web Conference (CIRPe 2016). Procedia CIRP, 55*, 230–235.

Hambach, J., Tenberg, R., & Metternich, J. (2015). Guideline-based video analysis of competencies for a target-oriented continuous improvement process. In *5th CIRP-Sponsored Conference on Learning Factories. Procedia CIRP, 32*, 25–30. https://doi.org/10.1016/j.procir.2015.02.212.

Hammer, M. (2014, August). Making operational transformations successful with experiential learning. In *CIRP collaborative working group—Learning factories for future oriented research and education in manufacturing*. France: CIRP General Assembly, Nantes.

Healey, M. (2005). Linking research and teaching: Exploring disciplinary spaces and the role of inquiry-based learning. In R. Barnett (Ed.), *Reshaping the university: New relationships between research, scholarship and teaching* (pp. 67–78). UK: McGraw-Hill/Open University Press.

Helleno, A. L., Simon, A. T., Papa, M. C. O., Ceglio, W. E., Rossa Neto, A. S., & Mourad, R. B. A. (2013). Integration university-industry: Laboratory model for learning lean manufacturing concepts in the academic and industrial environments. *International Journal of Engineering Education, 29*(6), 1387–1399.

Hernstein/Hernstein International Management Institute. (2003). *Management report: Befragung von Führungskräften in Österreich, Schweiz und Deutschland*. Wien: Hernstein.

Hung, W., Jonassen, D. H., & Liu, R. (2008). Problem-based learning. In M. Spector, D. Merrill, J. van Merrienböer, & M. Driscoll (Eds.), *Handbook of research on educational communications and technology* (pp. 485–506). Erlbaum.

Jäger, A., Mayrhofer, W., Kuhlang, P., Matyas, K., & Sihn, W. (2012). The "learning factory": An immersive learning environment for comprehensive and lasting education in industrial engineering. In *16th World Multi-Conference on Systemics. Cybernetics and Informatics, 16*(2), 237–242.

Jäger, A., Mayrhofer, W., Kuhlang, P., Matyas, K., & Sihn, W. (2013). Total immersion: Hands and heads-on training in a learning factory for comprehensive industrial engineering education. *International Journal of Engineering Education, 29*(1), 23–32.

Jank, W., & Meyer, H. (2002). *Didaktische Modelle* (5., völlig überarb. Aufl.). Berlin: Cornelsen-Scriptor.

Johnson, D. W., Johnson, R. T., & Smith, K. A. (1991). *Active learning: Cooperation in the college classroom*. Edina, MN: Interaction Book Co.

Jonassen, D. (1999). Designing constructivist learning environments. In C. M. Reigeluth (Ed.), *Instructional theories and models: A new paradigm of instructional theory* (pp. 215–239). Mahwah, New Jersey: Lawrence Erlbaum Associates Publishers.

Jonassen, D. H., & Rohrer-Murphy, L. (1999). Activity theory as a framework for designing constructivist learning environments. *Educational Technology Research and Development, 47*(1), 61–79. https://doi.org/10.1007/BF02299477.

Jorgensen, J. E., Lamancusa, J. S., Zayas-Castro, J. L., & Ratner, J. (1995). The learning factory: Curriculum integration of design and manufacturing. In *4th World Conference on Engineering Education* (pp. 1–7).

Kaluza, A., Juraschek, M., Neef, B., Pittschellis, R., Posselt, G., Thiede, S., & Herrmann, C. (2015). Designing learning environments for energy efficiency through model scale production processes. In *5th CIRP-Sponsored Conference on Learning Factories*. *Procedia CIRP, 32*, 41–46. https://doi.org/10.1016/j.procir.2015.02.114.

Kärcher. (2018). *Kärcher lean consulting: Lean erleben und verstehen*. Retrieved from https://www.kaercher.com/de/services/professional/service-angebote/kaercher-lean-consulting.html.

Karre, H., Hammer, M., Kleindienst, M., & Ramsauer, C. (2017). Transition towards an industry 4.0 state of the LeanLab at Graz University of Technology. *Procedia Manufacturing, 9*, 206–213. https://doi.org/10.1016/j.promfg.2017.04.006.

Keeton, M. T., Sheckley, B. G., Griggs, J. K., & Council for Adult and Experiential Learning. (2002). *Effectiveness and efficiency in higher education for adults: A guide for fostering learning*. Dubuque, Iowa: Kendall/Hunt Pub.

Kemény, Z., Nacsa, J., Erdős, G., Glawar, R., Sihn, W., Monostori, L., & Ilie-Zudor, E. (2016). Complementary research and education opportunities—A comparison of learning factory facilities and methodologies at TU Wien and MTA SZTAKI. In *6th CIRP-Sponsored Conference on Leanring Factories*. *Procedia CIRP, 54*, 47–52. https://doi.org/10.1016/j.procir.2016.05.064.

Knoll, M. (1997). The project method: Its vocational education origin and international development. *Journal of Industrial Teacher Education, 34*(3).

Kolb, D. A. (1984). *Experiential learning: Experience as the source of learning and development*. Englewood Cliffs, N.J.: Prentice Hall.

Küsters, D., Praß, N., & Gloy, Y.-S. (2017). Textile learning factory 4.0—Preparing Germany's textile industry for the digital future. *Procedia Manufacturing, 9*, 214–221. https://doi.org/10.1016/j.promfg.2017.04.035.

Lamancusa, J. S., & Simpson, T. (2004). The learning factory: 10 years of impact at Penn State. *Procedings International Conference on Engineering Education*, 1–8.

Lave, J., & Wenger, E. (1991). *Situated learning: Legitimate peripheral participation. Learning in doing*. Cambridge: Cambridge University Press.

Lindemann, H.-J. (2002). *The principle of action-oriented learning*. Retrieved from http://www.halinco.de/html/docde/HOL-prinzip02002.pdf.

Madsen, O., & Møller, C. (2017). The AAU smart production laboratory for teaching and research in emerging digital manufacturing technologies. *Procedia Manufacturing, 9*, 106–112. https://doi.org/10.1016/j.promfg.2017.04.036.

Makumbe, S., Hattingh, T., Plint, N., & Esterhuizen, D. (2018). Effectiveness of using learning factories to impart lean principles in mining employees. *Procedia Manufacturing, 23*, 69–74. https://doi.org/10.1016/j.promfg.2018.03.163.

McKinsey & Company. (2017). *Model factories and offices: Building operations excellence*. Retrieved from https://capability-center.mckinsey.com/files/mccn/2017-03/emea_model_factories_brochure_1.pdf.

Metternich, J., Abele, E., Bechtloff, S., & Seifermann, S. (2015). Static total cost comparison model to identify economic fields of application of cellular manufacturing for milling and drilling processes versus done-in-one-concepts. *CIRP Annals—Manufacturing Technology, 64*(1), 471–474.

References

Metternich, J., Bechtloff, S., & Seifermann, S. (2013). Efficiency and economic evaluation of cellular manufacturing to enable lean machining. In *46th CIRP Conference on Manufacturing Systems. Procedia CIRP, 7,* 592–597.

Metternich, J., Böllhoff, J., Seifermann, S., & Beck, S. (2013b). Volume and mix flexibility evaluation of lean production systems. In *2nd CIRP Global Web Conference (CIRPe 2013). Procedia CIRP, 9,* 79–84.

Meudt, T., Metternich, J., & Abele, E. (2017). Value stream mapping 4.0: Holistic examination of value stream and information logistics in production. *CIRP Annals—Manufacturing Technology, 66*(1), 413–416. https://doi.org/10.1016/j.cirp.2017.04.005.

Ministerium für Wirtschaft, Arbeit und Wohnungsbau Baden-Württemberg. (2017). *Lernfabriken 4.0 in Baden-Württemberg: Digitalisierung BW*. Retrieved from https://wm.baden-wuerttemberg.de/de/innovation/schluesseltechnologien/industrie-40/lernfabrik-40/.

Müller, B. C., Menn, J. P., & Seliger, G. (2017). Procedure for experiential learning to conduct material flow simulation projects, enabled by learning factories. *Procedia Manufacturing, 9,* 283–290. https://doi.org/10.1016/j.promfg.2017.04.047.

Muschard, B., & Seliger, G. (2015). Realization of a learning environment to promote sustainable value creation in areas with insufficient infrastructure. In *5th CIRP-Sponsored Conference on Learning Factories. Procedia CIRP, 32,* 70–75. https://doi.org/10.1016/j.procir.2015.04.095.

Neisser, U. (1976). *Cognition and reality: Principles and implications of cognitive psychology*. San Francisco: W.H. Freeman.

Oberhausen, C., & Plapper, P. (2015). Value stream management in the "lean manufacturing laboratory". In *5th CIRP-sponsored Conference on Learning Factories. Procedia CIRP, 32,* 144–149. https://doi.org/10.1016/j.procir.2015.02.087.

Reichert, Markus. (2011). Qualification of employees in the development department with SEW live training center. In E. Abele, J. Cachay, A. Heb, & S. Scheibner (Eds.), *1st Conference on learning factories, Darmstadt* (pp. 100–117). Darmstadt: Institute of Production Management, Technology and Machine Tools (PTW).

Penn State University. (2017). *Bernard M. Gordon learning factory: We bring the real world into the classroom*. Retrieved from http://www.lf.psu.edu/.

Pittschellis, R. (2015). Multimedia support for learning factories. In *5th CIRP-Sponsored Conference on Learning Factories. Procedia CIRP, 32,* 36–40. https://doi.org/10.1016/j.procir.2015.06.001.

Putz, M. (2013). The concept of the new research factory at Fraunhofer IWU—To objectify energy and resource efficiency R&D in the E3-factory. In G. Reinhart, P. Schnellbach, C. Hilgert, & S. L. Frank (Eds.), *3rd Conference on Learning Factories*, Munich, 7th May 2013 (pp. 62–77). Augsburg.

Reiner, D. (2009). *Methode der kompetenzorientierten Transformation zum nachhaltig schlanken Produktionssystem*. Dissertation, Darmstadt. *Schriftenreihe des PTW: "Innovation Fertigungstechnik"*. Aachen: Shaker.

Reiß, M. (2012). *Change management: A balanced and blended approach*. Norderstedt: Books on Demand.

RoboCup. (2016). *RoboCup logistics league 2016 final—Carologistics vs. solidus*. Retrieved from https://www.youtube.com/watch?time_continue=300&v=od1oEeHl8k8.

Schreiber, S., Funke, L., & Tracht, K. (2016). BERTHA—A flexible learning factory for manual assembly. *Procedia CIRP, 54,* 119–123. https://doi.org/10.1016/j.procir.2016.03.163.

Schuh, G., Gartzen, T., Rodenhauser, T., & Marks, A. (2015, July). Promoting work-based learning through industry 4.0. RU Bochum. In *5th Conference on Learning Factories*, Bochum, Germany.

Schuh, G., & Warschat, J. (2013). *Potenziale einer Forschungsdisziplin Wirtschaftsingenieurwesen. acatech DISKUSSION*. Munich: Herbert Utz Verlag.

Schunk, D. H. (1990). Goal setting and self-efficacy during self-regulated learning. *Educational Psychologist, 25,* 71–86.

Schützer, K., Rodrigues, L. F., Bertazzi, J. A., Durão, L. F. C. S., & Zancul, E. (2017). Learning environment to support the product development process. *Procedia Manufacturing, 9,* 347–353. https://doi.org/10.1016/j.promfg.2017.04.018.

Seifermann, S. (2018). *Methode zur angepassten Erhöhung des Automatisierungsgrades hybrider, schlanker Fertigungszellen.* Dissertation. *Schriftenreihe des PTW: "Innovation Fertigungstechnik".* Aachen: Shaker.

Seifermann, S., Böllhoff, J., Adolph, S., Abele, E., & Metternich, J. (2017). Flexible design of lean production systems in response to fluctuations due to logistics and traffic. In E. Abele, M. Boltze, & H.-C. Pfohl (Eds.), *Dynamic and seamless integration of production, logistics and traffic: Fundamentals of interdisciplinary decision support* (pp. 51–82). Cham, S.L.: Springer International Publishing. https://doi.org/10.1007/978-3-319-41097-5_4.

Seifermann, S., Böllhoff, J., Metternich, J., & Bellaghnach, A. (2014). Evaluation of work measurement concepts for a cellular manufacturing reference line to enable low cost automation for lean machining. In *47th CIRP Conference on Manufacturing Systems. Procedia CIRP, 17,* 588–593.

Seifermann, S., Metternich, J., & Abele, E. (2014, January). *Learning factories—Benefits for research and exemplary results.* CIRP, CIRP January Meeting, STC-O Technical Presentation, Paris, France.

Shingo, S. (1989). *A study of the Toyota production system from an industrial engineering viewprint* (Rev. ed.). Cambridge (Mas) [etc]: Productivity.

Sivard, G., & Lundholm, T. (2013). XPRES—A digital learning factory for adaptive and sustainable manufacturing of future products. In G. Reinhart, P. Schnellbach, C. Hilgert, & S. L. Frank (Eds.), *3rd Conference on Learning Factories,* Munich, 7th May, 2013 (pp. 132–154). Augsburg.

Steffen, M., Frye, S., & Deuse, J. (2013). The only source of knowledge is experience: Didaktische Konzeption und methodische Gestaltung von Lehr-Lern-Prozessen in Lernfabriken zur Aus- und Weiterbildung im Industrial Engineering. *TeachING LearnING.EU. Innovationen für die Zukunft der Lehre in den Ingenieurwissenschaften,* pp. 117–129.

Steffen, M., May, D., & Deuse, J. (2012). The industrial engineering laboratory: Problem based learning in industrial engineering education at TU Dortmund University. In *Global Engineering Education Conference (EDUCON), IEEE,—Collaborative Learning & New Pedagogic Approaches in Engineering Education,* Marrakesch, Marokko, 17–20 April 2012, pp. 1–10.

Sterman, J. D. (1994). *Learning in and about complex systems. Working paper/Alfred P. Sloan School of Management: WP# 3660-94-MSA.* Cambridge: Alfred P. Sloan School of Management, Massachusetts Institute of Technology.

Stier, K. W. (2003). Teaching lean manufacturing concepts through project-based learning and simulation. *Journal of Industrial Technology, 19*(4), 1–6.

Streitzig, C., & Oetting, A. (2016). Railway operation research centre—A learning factory for the railway sector. *Procedia CIRP, 54,* 25–30. https://doi.org/10.1016/j.procir.2016.05.071.

Thomar, W. (2015, July 8). Kaerchers global lean academy approach: Incentive talk (industry). In *5th Conference on Learning Factories,* Bochum, Germany.

Tietze, F., Czumanski, T., Braasch, M., & Lödding, H. (2013). Problembasiertes Lernen in Lernfabriken. *Werkstattstechnik online: wt, 103*(3), 246–251.

Tisch, M. (2018). *Modellbasierte Methodik zur kompetenzorientierten Gestaltung von Lernfabriken für die schlanke Produktion.* Dissertation, Darmstadt. Aachen: Shaker.

Tisch, M., Hertle, C., Abele, E., Metternich, J., & Tenberg, R. (2015a). Learning factory design: A competency-oriented approach integrating three design levels. *International Journal of Computer Integrated Manufacturing, 29*(12), 1355–1375. https://doi.org/10.1080/0951192X.2015.1033017.

Tisch, M., Hertle, C., Cachay, J., Abele, E., Metternich, J., & Tenberg, R. (2013). A systematic approach on developing action-oriented, competency-based learning factories. In *46th CIRP Conference on Manufacturing Systems. Procedia CIRP, 7,* 580–585.

Tisch, M., Hertle, C., Metternich, J., & Abele, E. (2014). Lernerfolgsmessung in Lernfabriken: Kompetenzorientierte Weiterentwicklung praxisnaher Schulungen. *Industrie Management, 30*(3), 39–42.

References

Tisch, M., Hertle, C., Metternich, J., & Abele, E. (2015a). Goal-oriented improvement of learning factory trainings. *The Learning Factory, An Annual Edition From the Network of Innovative Learning Factories, 1*(1), 7–12.

Tisch, M., & Metternich, J. (2017). Potentials and limits of learning factories in research, innovation transfer, education, and training. In *7th CIRP-Sponsored Conference on Learning Factories. Procedia Manufacturing.* (In Press).

Tisch, M., Ranz, F., Abele, E., Metternich, J., & Hummel, V. (2015c). Learning factory morphology: Study on form and structure of an innovative learning approach in the manufacturing domain. In *TOJET, July 2015 (Special Issue 2 for International Conference on New Horizons in Education 2015)*, pp. 356–363.

Toivonen, V., Lanz, M., Nylund, H., & Nieminen, H. (2018). The FMS training center—A versatile learning environment for engineering education. *Procedia Manufacturing, 23,* 135–140. https://doi.org/10.1016/j.promfg.2018.04.006.

Tracht, K., Funke, L., & Schottmayer, M. (2015). Online-control of assembly processes in paced production lines. *CIRP Annals—Manufacturing Technology., 64,* 395–398.

Tvenge, N., Martinsen, K., & Kolla, S. S. V. K. (2016). Combining learning factories and ICT-based situated learning. In *6th CIRP-Sponsored Conference on Leanring Factories. Procedia CIRP, 54,* 101–106.

UAW-Chrysler National Training Center. (2016). *World class manufacturing academy.* Retrieved from http://www.uaw-chrysler.com/world-class-mfg-academy/.

Ulrich, H., Dyllick, T., & Probst, G. (1984). Management. *Schriftenreihe Unternehmung und Unternehmungsführung: Bd. 13.* Bern: P. Haupt.

University of Washington. (2018). *Integrated learning factory.* Retrieved from https://www.washington.edu/change/proposals/factory.html.

UPRM. (2018). *Model factory.* Retrieved from http://uprm.edu/p/model_factory/about.

U-Quadrat. (2018). *Knorr-Bremse und U^2 sind partner.* Retrieved from http://www.u-quadrat.de/knorr-bremse-und-u%C2%B2-sind-partner/.

Veza, I., Gjeldum, N., & Mladineo, M. (2015). Lean learning factory at FESB—University of Split. In *5th CIRP-Sponsored Conference on Learning Factories. Procedia CIRP, 32,* 132–137. https://doi.org/10.1016/j.procir.2015.02.223.

Veza, I., Gjeldum, N., Mladineo, M., Celar, S., Peko, I., Cotic, M., et al. (2017). Development of assembly systems in lean learning factory at the University of Split. *Procedia Manufacturing, 9,* 49–56. https://doi.org/10.1016/j.promfg.2017.04.038.

Wagner, U., AlGeddawy, T., ElMaraghy, H. A., & Müller, E. (2012). The state-of-the-art and prospects of learning factories. In *45th CIRP Conference on Manufacturing Systems. Procedia CIRP, 3,* 109–114.

Wagner, U., AlGeddawy, T., ElMaraghy, H. A., & Müller, E. (2015). Developing products for changeable learning factories. *CIRP Journal of Manufacturing Science and Technology, 9,* 146–158.

Wagner, C., Heinen, T., Regber, H., & Nyhuis, P. (2010). Fit for change—Der Mensch als Wandlungsbefähiger. *Zeitschrift für wirtschaftlichen Fabrikbetrieb (ZWF), 100*(9), 722–727.

Wank, A., Adolph, S., Anokhin, O., Arndt, A., Anderl, R., & Metternich, J. (2016). Using a learning factory approach to transfer industrie 4.0 approaches to small- and medium-sized enterprises. In *6th CIRP-Sponsored Conference on Leanring Factories. Procedia CIRP, 54,* 89–94. https://doi.org/10.1016/j.procir.2016.05.068.

Weick, K. E. (1979). The social psychology of organizing. In *Topics in social psychology* (2nd ed.). New York: McGraw-Hill.

Werbach, K., & Hunter, D. (2012). *For the win: How game thinking can revolutionize your business.* Philadelphia, PA: Wharton Digital Press.

Wiech, M., Böllhoff, J., & Metternich, J. (2017). Development of an optical object detection solution for defect prevention in a learning factory. *Procedia Manufacturing, 9,* 190–197. https://doi.org/10.1016/j.promfg.2017.04.037.

Wood, D. F. (2003). Problem based learning. *BMJ: British Medical Journal, 326*(7384), 328–330.

Wurdinger, S. D. (2005). *Using experiential learning in the classroom: Practical ideas for all educators*. Lanham, Md: Scarecrow Education.

Wurdinger, S. D. (2016). *The power of project-based learning: Helping students develop important life skills*. USA: Rowman & Littlefield Publishers.

von Humboldt, W. (1957). Über die innere und äußere Organisation der höheren wissenschaftlichen Anstalten in Berlin. In H. Weinstock (Ed.) (pp. 229–241). Frankfurt: Fischer Bücherei.

Yoo, I. S., Braun, T., Kaestle, C., Spahr, M., Franke, J., Kestel, P., et al. (2016). Model factory for additive manufacturing of mechatronic products: Interconnecting world-class technology partnerships with leading AM players. *Procedia CIRP, 54,* 210–214. https://doi.org/10.1016/j.procir.2016.03.113.

Chapter 8
Overview on the Content of Existing Learning Factories

This chapter takes a closer look at learning content of existing learning factories. In this context, an overview on existing learning factory concepts from the content point of view is given. Figure 8.1 shows the structure of this chapter.

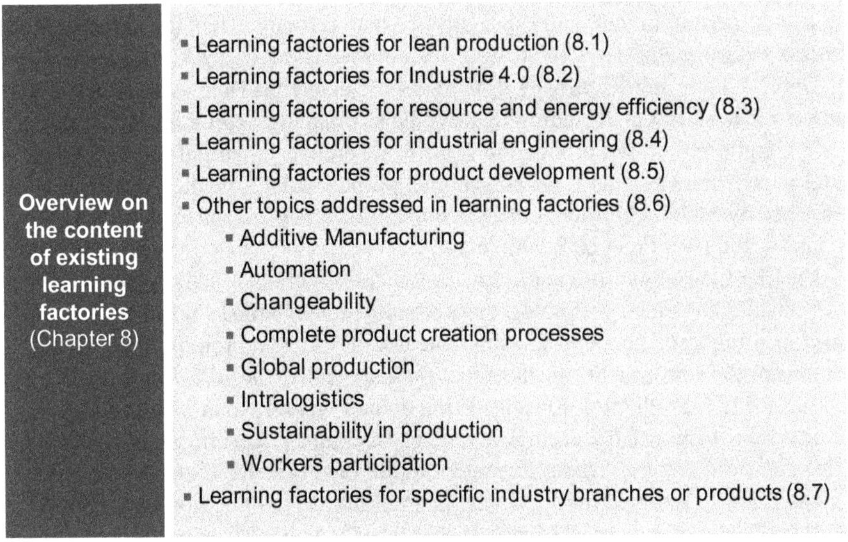

Overview on the content of existing learning factories (Chapter 8)
Learning factories for lean production (8.1)
Learning factories for Industrie 4.0 (8.2)
Learning factories for resource and energy efficiency (8.3)
Learning factories for industrial engineering (8.4)
Learning factories for product development (8.5)
Other topics addressed in learning factories (8.6)
Additive Manufacturing
Automation
Changeability
Complete product creation processes
Global production
Intralogistics
Sustainability in production
Workers participation
Learning factories for specific industry branches or products (8.7)

Fig. 8.1 Detailed structure of the overview on content of existing learning factories

E. Abele et al., *Learning Factories*, https://doi.org/10.1007/978-3-319-92261-4_8

8.1 Learning Factories for Lean Production

Learning factories for lean production or production process improvement empower the learner to apply or further develop lean methods and principles, like line balancing, problem solving, value stream analysis and design, just-in-time, or job optimization. Literature reviews showed that the vast majority of existing learning factories are addressing the topic lean management or lean production,[1] while approaches coming from industry[2] and academia[3] can be identified. The following sections exemplarily describe some of those approaches.

Lean Learning Factories in Academia

One of the early implementations of learning factories for lean production was initiated by PTW, TU Darmstadt, with the Process Learning Factory CiP (Center for industrial Productivity) in 2007 (Fig. 8.2). The learning factory includes nine machine tools and two assembly lines used for training and education in industrial engineering and in particularly lean manufacturing methodologies and Industrie 4.0. The close-to-reality factory environment includes a holistic, multistage value stream of a pneumatic cylinder. Beyond that, the learning factory is as a test bed for the identification of research gaps and the implementation of research results. The history, the learning environment, and especially the lean training offers for industry in the Process Learning Factory CiP are described in the Best Practice Example 26.

The IFA Learning Factory at IFA, Leibniz Universität Hannover, offers a wide range of lean trainings including factory layout planning, lean thinking, workplace design, ergonomics, and production control.[4] Moreover, a virtual representation of factory environment is used for educational purposes. The IFA Learning Factory is described by Melissa Quirico, Lars Nielsen, Niklas Rochow, and Prof. Dr.-Ing. Peter Nyhuis in the Best Practice Example 10.

The LPS Learning Factory at Ruhr University of Bochum is active in the interfaces between human beings, technology, and organizations. In the LPS Learning Factory, developed theoretic concepts are implemented in order to promote the technology demonstration and transfer to industry. As a specialty, the LPS Learning Factory focuses the lean production concepts in the light of workers' participation.[5] The LPS Learning Factory at LPS, Ruhr Universität Bochum, is described by Christopher Prinz and Prof. Dr.-Ing. Dieter Kreimeier in the Best Practice Example 21.

The LSP (Lernfabrik für die Schlanke Produktion) operated by the Institute for Machine Tools and Industrial Management (iwb, TU Munich) focuses also on streamlined production management and lean management. The LSP is presented described

---

[1] See Micheu and Kleindienst (2014), Plorin (2016).
[2] See, for example, Block, Bertagnolli, and Herrmann (2011), Hammer (2014).
[3] See, for example, Abele, Eichhorn, and Kuhn (2007), Goerke, Schmidt, Busch, and Nyhuis (2015), Tietze, Czumanski, Braasch, and Lödding (2013).
[4] See Wagner, Heinen, Regber, and Nyhuis (2010), Goerke et al. (2015), Seitz and Nyhuis (2015).
[5] See Prinz, Morlock, Wagner, Kreimeier, and Wannöffel (2014), Wagner, Prinz, Wannöffel, and Kreimeier (2015).

8.1 Learning Factories for Lean Production

Fig. 8.2 Process Learning Factory CiP at PTW, TU Darmstadt

later in the book by Harald Bauer, Felix Brandl, and Prof. Dr.-Ing. Gunther Reinhart as a best practice example for a mobile use of the learning factory concept (see Best Practice Example 18).

The Lean Lab at NTNU in Gjøvik, Norway, is a learning factory primarily for teaching flow production, line balancing, and workplace design. The learning factory is built up and operated in close cooperation with a nearby industrial park that is home to various companies.[6]

Throughout the whole book, even more best practice examples from academia addressing the lean topic can be identified:

- Best Practice Example 5: ESB Logistics Learning Factory at ESB Business School at Reutlingen University,
- Best Practice Example 13: Learning and Innovation Factory at IMW, IFT, and IKT, TU Wien,
- Best Practice Example 14: Learning Factory aIE at IFF, University of Stuttgart,
- Best Practice Example 17: Learning Factory for Global Production at wbk, KIT Karlsruhe.

Best Practice Example 29: Smart Mini-Factory at IEA, Free University of Bolzano.

Although industrial employees are frequently trained in academic learning factories, some industrial companies have set up by themselves or in cooperation with universities or consulting firms learning factories that focus especially on the most relevant topics and technologies for their business. In those cases, in general, company-specific learning environments and products are used.

---

[6]See Tvenge, Martinsen, and Kolla (2016).

Lean Learning Factories in Industry

The Value-Oriented Production System (VPS) Center, a learning factory at BMW opened in Munich, Germany, in 2012. The concept integrates interactive learning stations that visualize the VPS/LEAN principles.[7] The learning factory concept is the pilot for a global rollout to all BMW production sites. The VPS Center is also described by Carolin Lorber and Tobias Stäudel in the Best Practice Example 31.

Similar to this rollout plan from BMW, the company Kärcher uses a similar strategy for the global qualification of employees: The first Kärcher Learning Factory was piloted in 2013 based on the concept of the Process Learning Factory CiP in Winnenden, Germany. The Kärcher Learning Factory concept has been exported according to this model but individually adapted to site-typical products to all Kärcher sites worldwide.[8]

The MPS Lernplattform at Daimler AG in Sindelfingen is a learning center or executives, production planners, plant engineers, and improvement managers on 3000 m^2 around the lean topic f. From 2011 up to now, over 10,000 Daimler employees have been trained in various topics over the entire production process. Michael Schwarz presents the MPS Lernplattform in the Best Practice Example 23.

The LEAN-Factory in Berlin is a learning factory for process optimization geared to the needs of the pharmaceutical industry integrating social topics like health, safety in factory environments or the worldwide participation in value creation. A big pharmaceutical company, Fraunhofer IPK, and TU Berlin jointly planned and operate the learning factory. The LEAN-Factory is presented by Prof. Dr.-Ing. Holger Kohl, Prof. Dr.-Ing. Roland Jochem, Felix Sieckmann, and Christoffer Rybski in the Best Practice Example 12.

The Chrysler World Class Manufacturing Academy[9] in Michigan, USA, includes experiential learning related to required manufacturing competencies in two life-size physical learning factories. Furthermore, online learning courses that can be accessed remotely by the employees are offered as a support.

The MOVE academy disseminates the lean philosophy of the industrial, automotive, and aerospace components supplier Schaeffler. The learning factory includes real drilling, deburring, assembly, logistics, and quality assurance processes.[10]

The McKinsey Learning Factory network consists of globally distributed learning factories that are addressing various topics. Initiation and operation of those learning factories are achieved through cooperation between McKinsey and various academic and industrial partners.[11]

In Scharnhausen, Festo is operating a learning factory in the center of the plant that among others deals with lean topics. The Festo Learning Factory in Scharnhausen is presented by Holger Regber in the Best Practice Example 8.

---

[7] See Herrmann and Stäudel (2014).
[8] See Thomar (July, 8th, 2015).
[9] UAW-Chrysler National Training Center (2016).
[10] See Beauvais (2013); Helleno et al. (2013).
[11] See Hammer (2014); McKinsey & Company (2017).

8.2 Learning Factories for Industrie 4.0

Industry today faces massive problems with the transfer and implementation of Industrie 4.0.[12] There is a broad consensus in the literature that industry needs highly educated professionals and executives to design and adapt the complex interdisciplinary dependencies regarding all facets of cyber-physical production systems.[13] For this task, learning factories are identified as particularly well suited,[14] as they allow a unique connection of thinking and doing in the context of production. This way, the learner is actively involved in the design of systems at the intersection of manufacturing, information and communication technologies. Consequently, in recent years, numerous learning factory approaches aiming at facilitation and transfer of Industrie 4.0 solutions can be recognized.[15]

Mostly, those learning factory concepts are addressing the needs of small- and medium-sized companies (SMEs). For example in Germany, the initiative "Mittelstand 4.0—Digitale Produktions- und Arbeitsprozesse" (literal translation: "Mittelstand 4.0—Digital Production and Work Processes") supports SMEs and crafts in the digitization, networking, and introduction of Industrie 4.0 applications. German companies are assisted by 23 Mittelstand 4.0 Competence Centers throughout Germany as of 2018; see Fig. 8.3. With the latest know-how and learning factories used as demonstration and testing facilities, this nationwide comprehensive network for SMEs offers hands-on digitization support. In addition, Mittelstand 4.0 agencies are working on overarching digitization topics such as cloud computing, communications, trade and processes and are spreading them with the help of multipliers.[16]

The focus of learning factories in this area today is more on research and technology transfer compared to education and training.

Based on the "Learning and Innovation Factory for Integrative Production Education" (LIF),[17] the Industrie 4.0 Pilot Factory (I40PF) is built up as both a research platform and an environment for teaching and training for a human-centered cyber-physical production systems, virtualized production systems, and adaptive manufacturing at the Vienna University of Technology. The I40PF gives insights into Industrie 4.0 equipment and technologies, such as intelligent assembly technologies, collaborative robotic systems, and assistance systems with augmented reality technologies.[18] In Best Practice Example 25, the I40PF is presented by Prof. Wilfried Sihn, Fabian Ranz, and Philipp Hold.

---

[12] Monostori et al. (2016).

[13] See, e.g. Acatech (2016a, b), Wank et al. (2016), Thiede, Juraschek, and Herrmann (2016), Prinz et al. (2016).

[14] See Prinz et al. (2016), Wank et al. (2016), Thiede et al. (2016), Gräßler, Pöhler, and Pottebaum (2016), Seitz and Nyhuis (2015).

[15] Faller and Feldmüller (2015), Wank et al. (2016), Prinz et al. (2016), Schuh, Gartzen, Rodenhauser, and Marks (2015), Baena, Guarin, Mora, Sauza, and Retat (2017).

[16] See BMWi (2018).

[17] See Best Practice Example 13.

[18] See Erol, Jäger, Hold, Ott, and Sihn (2016).

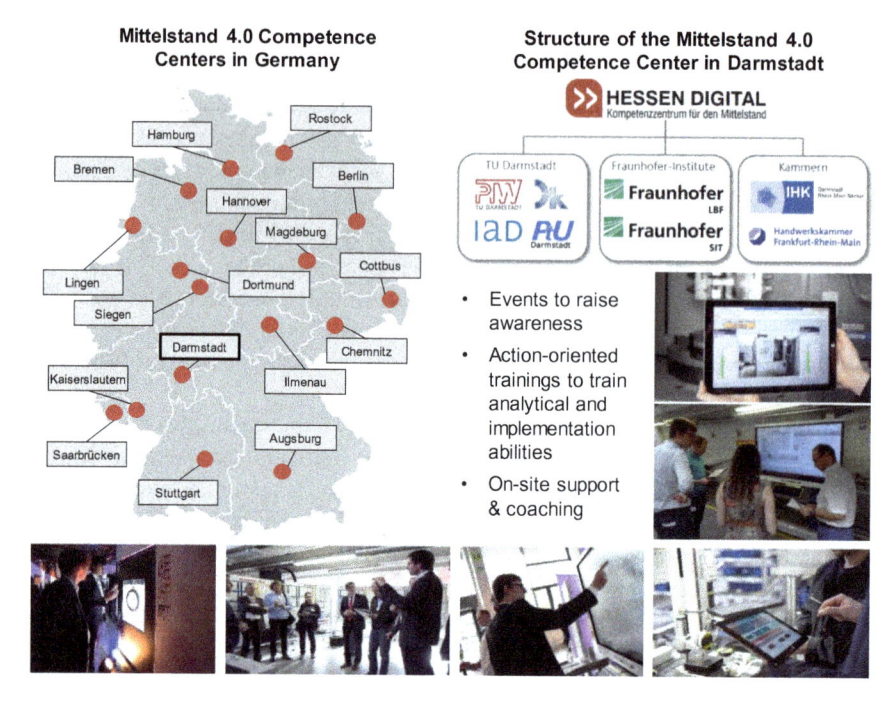

Fig. 8.3 Overview of the regional Mittelstand 4.0 Competence Centers (left) and the structure and impressions of the Hessen Digital Competence Center in Darmstadt

The Demonstration Factory Aachen (DFA) contains a small-scale production with a high vertical range of manufacture on 1600 m^2 in which products that are sold on the market are manufactured with the e.GO Mobile AG as lead customer. Processes include the sheet metal forming, the joining of automotive body structures as well as a manual assembly. Processes not only are similar to industrial production but also have the same requirements in terms of complexity and quality.[19] Furthermore, the Demonstration Factory is specialized on prototype construction and industrialization. Additionally to the real production, the Demonstration Factory serves as experiment-based production research and training center. In addition to research and trainings, the DFA supports the development of methods and tools for production-ready design and cost-efficient production. A unique characteristic of the DFA is that marketable products are manufactured. The Demonstration Factory DFA is presented by Prof. Günther Schuh, Jan-Philipp Prote, Bastian Fränken, and Julian Ays in the Best Practice Example 2.

The Mini-Factory was installed in 2012 at the University of Bolzano, Italy, in order to augment practice orientation in the education of engineers first in the field of lean production and in recent years also in the field of Industrie 4.0. A pneumatic

---

[19]Schuh et al. (2015).

cylinder and a camp stove oven serve as realistic products. Prof. Dominik Matt and Dr.-Ing. Erwin Rauch describes the Mini-Factory in the Best Practice Example 29.

The Process Learning Factory CiP was since the inauguration in 2007 used for training, education, and research in the field of lean manufacturing. In recent years, lean methods are further developed using the new opportunities of digitalization and Industrie 4.0. The Process Learning Factory CiP works on bringing together the lean and the Industrie 4.0 world.[20] The Process Learning Factory CiP is described in the Best Practice Example 26.

Institute for Computer Science and Control, Hungarian Academy of Sciences (MTA SZTAKI), is operating two learning factories, one in Budapest and one in Győr. The Smart Factory in Budapest is a scaled-down model of a factory on 30 m^2 that focuses on production planning and scheduling in CPPS as well as mechatronics and automation technology. The learning environment contains a warehouse, four automated workstations, a collaborative robot cell, a loading/unloading station, a closed-path conveyor system, and floor space for two mobile robots. The Smart Factory in Budapest is presented in the Best Practice Example 27 by Dr. Zsolt Kemény, Richárd Beregi, Dr. Gábor Erdős, and Dr. János Nacsa. The MTA SZTAKI Learning Factory is primarily used in education for human–robot collaboration as well as production planning and scheduling in CPPS. The life-size factory environment contains flexibly configurable robotic workstations, AGV for intralogistics, indoor positioning devices, a 3D printer, and reconfigurable human–machine interfaces. The MTA SZTAKI Learning Factory in Győr is presented in the Best Practice Example 24 by Dr. Zsolt Kemény, Richárd Beregi, Dr. Gábor Erdős, and Dr. János Nacsa.

Building up a learning factory, of course it has to be decided what kind of equipment to use. Several learning factory concepts predominantly in the field of Industrie 4.0 are based on Festo Didactic Learning Factory equipment.[21] The Festo Didactic Learning Factory equipment that is developed since 1989 was originally used in the education and training for factory automation and automation technology. Most famous learning factories offered by Festo Didactic, which are today predominantly used in the field of Industrie 4.0, are called MPS, iCIM, and CP Factory. Dr. Reinhard Pittschellis and Dr. Dirk Pensky present the various Festo Didactic Learning Factories in the Best Practice Example 7.

In Scharnhausen, Festo is operating a learning factory by themselves in the center of the Scharnhausen plant that among others deals with factory automation and Industrie 4.0 issues. Holger Regber in the Best Practice Example 8 presents the Festo Learning Factory in Scharnhausen.

Throughout the whole book, even more best practice examples from academia addressing the topic Industrie 4.0 can be identified:

- Best Practice Example 1: AutoFab at the Faculty of Electrical Engineering of the University of Applied Sciences Darmstadt, Germany,
- Best Practice Example 3: Die Lernfabrik at IWF, TU Braunschweig,

---

[20] See also Metternich, Müller, Meudt, and Schaede (2017).
[21] See, e.g. Madsen and Møller (2017), Mayer, Rabel, and Sorko (2017).

- Best Practice Example 5: ESB Logistics Learning Factory at ESB Business School at Reutlingen University,
- Best Practice Example 6: ETA-Factory at PTW, TU Darmstadt,
- Best Practice Example 8: Festo Learning Factory in Scharnhausen,
- Best Practice Example 17: Learning Factory for Global Production at wbk, KIT Karlsruhe,
- Best Practice Example 19: "Lernfabrik für vernetzte Produktion" at Fraunhofer IGCV, Augsburg,
- Best Practice Example 21: LPS Learning Factory at LPS, Ruhr Universität Bochum,
- Best Practice Example 28: Smart Factory at DFKI, TU Kaiserslautern,
- See also Sect. 8.6.2 on learning factories for automation.

8.3 Learning Factories for Resource and Energy Efficiency

Another relevant topic in learning factories is energy and resource efficiency[22] while also learning factory approaches in industry can be identified (Fig. 8.4 shows the Festo learning factory in Scharnhausen).[23] Learning factories for energy efficiency are used in education, training, and research,[24] whereby the use of energy-efficient learning factories for the teaching and in particular training purposes is complicated by the lack of visibility and tangibility of energy flows.[25] Among other things, the learning factories for resource and energy efficiency deal with development and testing of strategies and measures, e.g. KPI monitoring and metering techniques, for the optimization of the relation between production output and energy or resource consumption. The following sections exemplarily describe some of the learning factories for energy and resource efficiency.

The ETA-Factory (Fig. 8.5) opened in 2016 at PTW, TU Darmstadt. The ETA-Factory is a greenfield factory that is planned in an interdisciplinary approach with the goal to reduce energy consumption as well as CO_2 emissions of industrial production processes. For this purpose, machines and building shell interact with each other. The ETA-Factory is used mainly for research and demonstration, but also for education and training in the fields of energy efficiency as well as energy flexibility. The ETA-Factory is described in the Best Practice Example 6.

The Green Factory Bavaria is a network of 12 learning factories at different locations in Bavaria, Germany, that contribute to the ability of industry to improve their resource consumption (Fig. 8.6). This involves both the transfer of know-how from applied research to industry and the organization of experience exchange among

---

[22] Abele, Bauerdick, Strobel, and Panten (2016), Plorin, Poller, and Müller (2013), Gebbe et al. (2015), Kreitlein, Höft, Schwender, and Franke (2015).
[23] McKinsey & Company (2017).
[24] See, e.g. Plorin et al. (2013).
[25] Abele et al. (2016).

8.3 Learning Factories for Resource and Energy Efficiency

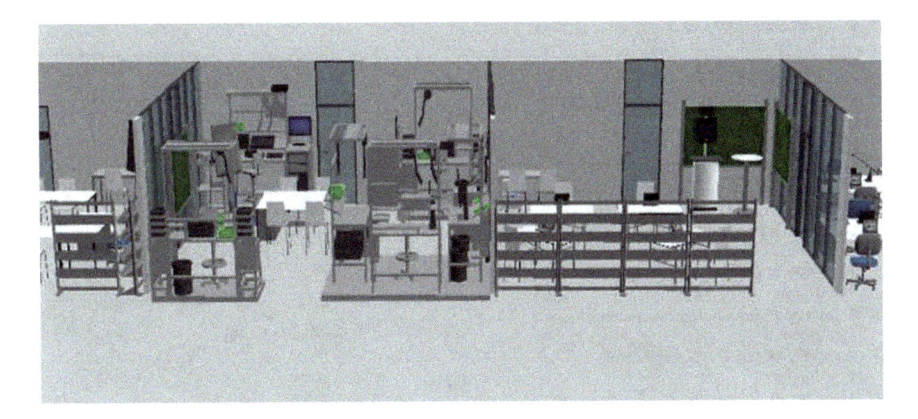

Fig. 8.4 Part of the Festo Learning Factory, Scharnhausen

Fig. 8.5 Learning factory ETA at TU Darmstadt (PTW, TU Darmstadt, 2016)

companies. The learning factories for energy-efficient production serve as demonstration, learning, and research platforms.[26]

The LPS Learning Factory at LPS, Ruhr Universität Bochum, addresses not only the topics process optimization as well as management and organization but also the resource efficiency.[27] Christopher Prinz and Prof. Dr.-Ing present the LPS Learning Factory in the Best Practice Example 21.

The "E^3-Factory" at Fraunhofer IWU in Chemnitz maps a realistic power train and car body production and focuses on energy research. The learning factory is used for teaching and prototyping as well as for technology transfer and upscaling.[28]

Further learning factories presented in this book also address the topics resource and energy efficiency:

- Best Practice Example 1: AutoFab at the Faculty of Electrical Engineering of the University of Applied Sciences Darmstadt, Germany,

---

[26] See FAPS (2018).

[27] See Kreimeier et al. (2014).

[28] Stoldt, Franz, Schlegel, and Putz (2015), Putz (2013).

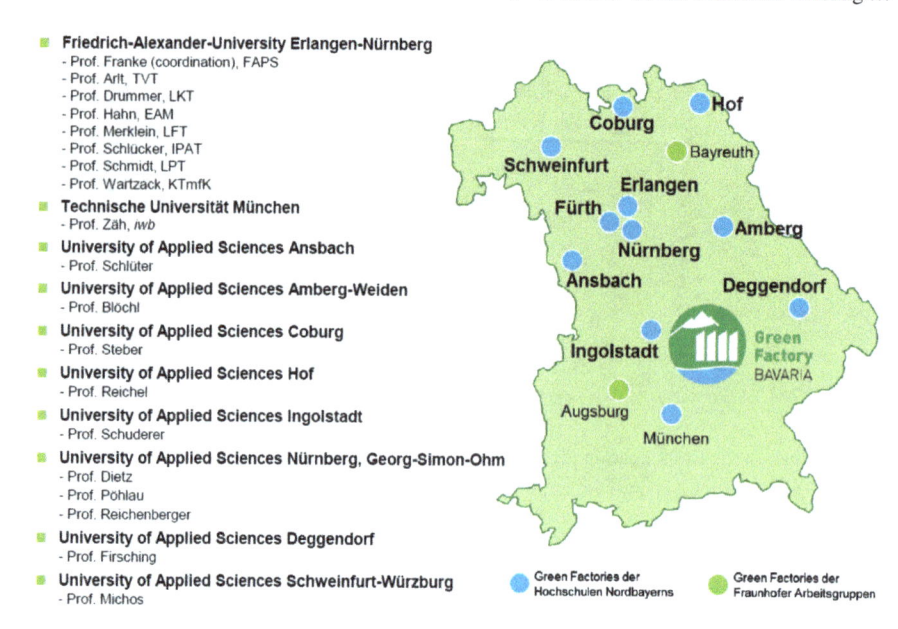

- Friedrich-Alexander-University Erlangen-Nürnberg
  - Prof. Franke (coordination), FAPS
  - Prof. Arlt, TVT
  - Prof. Drummer, LKT
  - Prof. Hahn, EAM
  - Prof. Merklein, LFT
  - Prof. Schlücker, IPAT
  - Prof. Schmidt, LPT
  - Prof. Wartzack, KTmfK
- Technische Universität München
  - Prof. Zäh, iwb
- University of Applied Sciences Ansbach
  - Prof. Schlüter
- University of Applied Sciences Amberg-Weiden
  - Prof. Blöchl
- University of Applied Sciences Coburg
  - Prof. Steber
- University of Applied Sciences Hof
  - Prof. Reichel
- University of Applied Sciences Ingolstadt
  - Prof. Schuderer
- University of Applied Sciences Nürnberg, Georg-Simon-Ohm
  - Prof. Dietz
  - Prof. Pöhlau
  - Prof. Reichenberger
- University of Applied Sciences Deggendorf
  - Prof. Firsching
- University of Applied Sciences Schweinfurt-Würzburg
  - Prof. Michos

Fig. 8.6 Twelve Green Factory Bavaria locations for resource and energy-efficient production (FAPS, 2018)

- Best Practice Example 8: Festo Learning Factory in Scharnhausen,
- Best Practice Example 19: "Lernfabrik für vernetzte Produktion" at Fraunhofer IGCV, Augsburg.

8.4 Learning Factories for Industrial Engineering

In the field of learning factories for the broad field of industrial engineering, several approaches can be identified.[29] Typical for those learning factories for industrial engineering is that they are predominantly used for educational purposes in order to give students insights into industrial engineering problems[30] and sometimes at the same time also into holistic, complex product creation processes.[31] In the following, exemplarily some learning factories of this type are presented briefly.

The Bernard M. Gordon Learning Factory is a learning factory for industrial engineering education in general at Penn State University. The learning factory is used in capstone design courses, research projects as well as in other courses and projects.

---

[29] Among others, they are described in Steffen, May, and Deuse (2012), Jäger, Mayrhofer, Kuhlang, Matyas, and Sihn (2012), Jäger, Mayrhofer, Kuhlang, Matyas, and Sihn (2013), Jorgensen, Lamancusa, Zayas-Castro, and Ratner (1995).

[30] See, e.g. Jäger et al. (2013) and Gräßler, Taplick, and Yang (2016).

[31] See also Sect. 8.6.4 regarding "Complete Product Creation Processes."

8.4 Learning Factories for Industrial Engineering

Modern design, prototyping, and manufacturing equipment are part of the learning factory. The learning factory concept is based on a university–industry partnership in which students are working on design projects on industry-relevant problems. Until today, the learning factory was home to more than 1800 design projects, sponsored by more than 500 companies, involving nearly 9000 engineering students.[32]

The Industrial Engineering Laboratory at TU Dortmund is integrated into teaching courses of the Institute for Production Systems. In those courses, students are planning a complete assembly line including the specific workplace design. Depending on the focus of the teaching course, learners are dealing, for example, with ergonomic workplace designs or time studies for the assembly process. The product assembled in these courses is a gearbox that consists among others of housings, gearwheels, gearshift levers, drive, and drive shafts.[33]

Also at Stellenbosch University, the industrial engineering department initiated in 2015 with the help of several NIL[34] partners a learning factory concept for undergraduates' education in the field of industrial engineering. Here, for example, a learning module for Methods-Time Measurement (MTM) is created in the learning factory in Stellenbosch.[35]

Furthermore, in Chap. 11 several additional examples addressing also the industrial engineering topic can be found:

- Best Practice Example 5: ESB Logistics Learning Factory at ESB Business School at Reutlingen University,
- Best Practice Example 9: iFactory at the Intelligent Manufacturing Systems (IMS) Center, University of Windsor, Canada,
- Best Practice Example 10: IFA Learning Factory at IFA, Leibniz Universität Hannover,
- Best Practice Example 13: Learning and Innovation Factory at IMW, IFT, and IKT, TU Wien,
- Best Practice Example 14: Learning Factory aIE at IFF, University of Stuttgart,
- Best Practice Example 16: Learning Factory for Innovation, Manufacturing and Cooperation at the Faculty of Industrial and Process Engineering of Heilbronn University of Applied Sciences,
- Best Practice Example 17: Learning Factory for Global Production at wbk, KIT Karlsruhe,
- Best Practice Example 21: LPS Learning Factory at LPS, Ruhr Universität Bochum,
- Best Practice Example 30: Teaching Factory: An emerging paradigm for manufacturing education.

---

[32] See PennState University (2017).

[33] See Steffen et al. (2012).

[34] Netzwerk Innovativer Lernfabriken, literal translation: Network of Innovative Learning Factories, also see: https://www.esb-business-school.de/forschung-alt/forschungsprojekte/nil-netzwerk-innovativer-lernfabriken/.

[35] See Morlock, Kreggenfeld, Louw, Kreimeier, and Kuhlenkötter (2017).

8.5 Learning Factories for Product Development

Existing learning factories have so far mainly addressed the actual production. Learning factories for product development or at the interface between product development and production are still not widespread today. In this context, two obstacles or difficulties of the learning factories for product development are to be stated as reasons:

- First, the learning factory concept is strongly dependent on the integration of the learners' own actions in the area of application. In the area of product development, the problems can arise that either the actions to be performed take too long or the reaction of the environment to these actions takes too long. In the first case, the problem is that product development actions within compact training sessions are difficult to carry out, and in the second case, the learner is not aware of the effects of his product development actions (e.g. on production). Here, concept is needed that enable fast-forward-like simulations.
- Second, the learning factory concept is strongly based on making the actions visible and tangible in authentic environments. In the area of product development, this is more difficult because many processes are software-based or take place in the minds of product developers. Ways must be found to make these product development processes, which are from the outside hardly visible and tangible, experienceable.

One of the rare examples for the integration of the topics product development and production is given with the Integrated Learning Factory in Bochum which is a collaboration between LPE (Chair for Product Development) and LPS (Chair for Production Systems).[36] The physical setting for the Integrated Learning Factory is the existing LPS Learning Factory at RU Bochum that also addresses mere production-relevant topics like lean production, resource efficiency, or Industrie 4.0.[37] Up to now, the Integrated Learning Factory has been used to teach students about topics and problem areas at the interface between product development and production (Fig. 8.7). Some of the students take on the role of product developers, the other part the role of production employees. The Integrated Learning Factory is presented by Prof. Beate Bender in the Best Practice Example 11.

Another learning factory approach in the field of product development is built up with the Learning Factory for Product Development Process at Methodist University of Piracicaba in Brazil.[38] The learning environment created here simulates a complete product development process for crosshead axes and is used for undergraduate and postgraduate education. The infrastructure of the learning factory includes

---

[36] See Bender, Kreimeier, Herzog, and Wienbruch (2015).

[37] See also Best Practice Example 21 for further information on the LPS Learning Factory at RU Bochum.

[38] See Schützer, Rodrigues, Bertazzi, Durão, and Zancul (2017).

8.5 Learning Factories for Product Development

Fig. 8.7 Impression of the teaching course in the Integrated Learning Factory at RU Bochum

CAD/CAM/CAE systems that are integrated into a Product Lifecycle Management (PLM) environment.[39]

In industry, individual applications of the learning factory concept can also be identified in the area of product development. The Daimler Trucks Process Learning Factory in Mannheim offers a learning module Process Learning Factory for developers. This module is of great importance for Daimler Trucks because the influence of product development and production planning on subsequent manufacturing costs is immense. The module aims to convey the philosophy of lean production also in the areas of development and production planning.[40]

Furthermore, also parts of the product development are addressed

- in the iFactory presented in the Best Practice Example 9,
- in the Learning and Innovation Factory presented in the Best Practice Example 13 as well as
- in the LMS Factory presented in the Best Practice Example 20.

8.6 Other Topics Addressed in Learning Factories

In addition to the five main topics presented herein learning factories, further specific topic blocks can be identified which are addressed and worked on by individual learning factories.[41] The topics are presented in the following in alphabetical order:

---

[39] A use case description of the Learning Factory for Product Development Process can be found in Schützer et al. (2017).

[40] See Abele et al. (2015).

[41] These topics are identical to the specific foci of competency identified in Chap. 2, Table 7.

- Additive manufacturing,
- Automation,
- Changeability,
- Complete product creation processes,
- Global production,
- Intralogistics,
- Sustainability in production, and
- Workers participation.

8.6.1 Learning Factories for Additive Manufacturing

In the coming years, one of the biggest changes of applied manufacturing technologies could supposedly come from the advances in the field of additive manufacturing. For this reason, already today, individual learning factories are planned, which are to research the technology, the organization, and the production processes around the topic of additive production and to transfer it into practice through education and training.[42] Although, until today, the authors are not aware of any learning factory in operation focusing specifically on additive manufacturing processes.

8.6.2 Learning Factories for Automation

Already in the 1980s, Festo Didactic used learning factory concepts for (vocational) education and training of students for automation basics, use of sensors, industrial networking, and PLC technology.[43] In addition to the use of the learning factories for automation in the vocational school sector, approaches in universities can be identified. In these academic learning factories, not only the fundamentals of automation, but rather automation-related research areas are addressed. Examples include scalable automation,[44] low-cost automation,[45] and smart automation[46] as research fields. Exemplarily two suiting learning factory examples in the field of automation are described briefly in the following.

The AutoFab at Hochschule Darmstadt is a learning factory for Industrie 4.0 and factory automation with focus on education for automation technologies. The AutoFab learning factory is presented by Prof. Dr.-Ing. Stephan Simons and Prof. Dr. Stephan Neser in Best Practice Example 1.

---

[42] See Yoo et al. (2016).
[43] See Pittschellis (2015).
[44] See Buergin, Minguillon, Wehrle, Haefner, and Lanza (2017).
[45] See Seifermann, Böllhoff, Metternich, and Bellaghnach (2014).
[46] See Zuehlke (2008).

8.6 Other Topics Addressed in Learning Factories

At the Technical University of Kaiserslautern, the SmartFactory^{KL} aims to find the way to the intelligent factory of tomorrow that is flexible, networked, self-organizing, and user-oriented. Besides the Industrie 4.0 production plant, in the demo center of the SmartFactory^{KL} experiences and tests with the help of demonstrators are enabled. Among others, key topics are scalable automation, cyber-physical systems as well as augmented reality. Dr. Haike Frank, Dennis Kolberg, and Prof. Detlef Zühlke present the SmartFactory^{KL} in the Best Practice Example 28.

8.6.3 Changeability

Another important topic in the field of manufacturing addressed by learning factories in recent years is the changeability of production systems. In this field, approaches can be identified that focus

- on training of engineers,[47]
- on education of students,[48] and also
- on changeability-related research.[49]

For example, the iFactory at the Intelligent Manufacturing Systems (IMS) Center, Windsor, Canada, is a changeable learning factory containing a modular assembly system with among others robotic and manual assembly stations, computer vision inspection station, and Automated Storage and Retrieval System (ASRS). The iFactory is presented by Prof. Hoda ElMaraghy in the Best Practice Example 9.

The Experimental and Digital Factory (EDF) is a learning factory at the Department of Factory Planning and Factory Management at the TU Chemnitz that consists of a network of laboratories for innovation, ergonomics, CAD, biometry, and usability. The EDF is a research and teaching facility dealing with changeability issues regarding products, processes, and resources. Equipment manufacturers use the EDF also as test environment.[50]

Further learning factory concepts described in this book dealing with this topic are:

- Best Practice Example 14: Learning Factory aIE at IFF, University of Stuttgart,
- Best Practice Example 29: Smart Mini-Factory at IEA, Free University of Bolzano.

---

[47] See, e.g. Wagner et al. (2010), Bauernhansl, Dinkelmann, and Siegert (2012).

[48] See, e.g. ElMaraghy, Moussa, ElMaraghy, and Abbas (2017).

[49] See, e.g. Wagner, AlGeddawy, ElMaraghy, and Müller (2012).

[50] See Abele et al. (2017).

Fig. 8.8 Impressions from the learning module at the Learning and Innovation Factory, TU Wien

8.6.4 Complete Product Creation Processes

Learning factories are ideal environments for addressing different topics along the entire product life cycle. In some learning factories, the focus is precisely on this comprehensive view on the complex and interdisciplinary product emergence processes. Two examples, from the TU Wien and the HS Heilbronn, are briefly described below.

The Learning and Innovation Factory (LIF) for Integrative Production Education at TU Wien is an interactive, hands-on education, and research center for methods, process, and production innovations in the areas of product development, production, and logistics. The LIF is operated by three institutes from TU Wien: the IMW, the IFT, and the MIVP. In a yearly organized learning module, the goal is that undergraduate students experience all the activities along the product emergence process and therefore understand the product life cycle and related processes.[51] In the learning module a slotcar is developed and produced by the students (Fig. 8.8). The Learning and Innovation Factory is described by Prof. Wilfried Sihn, Fabian Ranz, and Philipp Hold in the Best Practice Example 13.

In addition, the "Learning Factory for Innovation, Manufacturing and Cooperation" at the Faculty of Industrial and Process Engineering at Heilbronn University aims at a comprehensive view of their students on the product creation process. The learning factory module at the end of the study program entails product development, industrial engineering as well as management methods. In this learning module, each year a new product is created by participating students; furthermore, along with this, necessary small series production processes (manufacturing, quality assurance, logistics) are implemented by the students. The "Learning Factory for

---

[51] See Sihn, Gerhard, and Bleicher (2012), Jäger et al. (2012).

8.6 Other Topics Addressed in Learning Factories 279

Innovation, Manufacturing and Cooperation" is presented by Prof. Patrick Balve in the Best Practice Example 16.

8.6.5 Global Production Networks

The design and operation of global production networks is a very rare, complex, and difficult-to-implement issue in learning factories. One learning factory approach addressing topics in this field can be found at wbk at KIT Karlsruhe.[52] The learning factory for global production shows effects of the site location on the implementation of production systems. Participants adapt the different production system components as well as the fundamental principles of production systems to the respective production location. Prof. Gisela Lanza, Constantin Hofmann, Dr.-Ing. Benjamin Haefner, and Dr.-Ing. Nicole Stricker describes this learning factory concept for global production in the Best Practice Example 17.

8.6.6 Intralogistics and Logistics

Only a few learning factories that are focusing their activities on the topics logistics and intralogistics can be identified. In the following, exemplarily, a few of those learning factories for intralogistics and logistics are presented briefly.

One learning factory addressing logistics concepts is created at ESB Reutlingen.[53] The ESB Logistics Learning Factory represents core aspects of modern production environments integrating logistics issues. The goal of the ESB Logistics learning factory is to provide close-to-reality teaching, research, and training in the logistics area. Research in this learning factory is, for example, conducted in the field of intelligent bin systems[54] or a collaborative tugger train system. Prof. Vera Hummel, Fabian Ranz, and Jan Schuhmacher in the Best Practice Example 5 present the ESB Logistics Learning Factory.

In the Green Factories Bavaria in Bayreuth and Erlangen, the impact of intralogistics issues on the resource efficiency of factories is addressed in education and research. In Bayreuth, learners get familiar with the transportation systems' electrical energy consumption and related measures for better energy efficiency. In Erlangen, the design and implementation of various transport systems fitting respective constraints and application areas are addressed.[55]

A learning factory at the PuLL® Competence Centre at the University of Applied Sciences in Landshut is used for practice-oriented education and research inter alia

---

[52] See Lanza, Moser, Stoll, and Haefner (2015), Lanza, Minges, Stoll, Moser, and Haefner (2016).
[53] See Hummel, Hyra, Ranz, and Schuhmacher (2015).
[54] See Schuhmacher, Baumung, and Hummel (2017).
[55] See Scholz et al. (2016).

in the fields of Lean Logistics. The learning factory is recently updated with the new possibilities of Industrie 4.0 to address the new field of intelligent production logistics.[56]

Other learning factory concepts not focusing on logistics and intralogistics topics but addressing different facets of this field are described in the Best Practice Example section:

- Best Practice Example 1: **AutoFab at the Faculty 1 of Electrical Engineering of the University of Applied 2 Sciences Darmstadt,**
- Best Practice Example 2: Demonstration Factory at WZL, RWTH Aachen,
- Best Practice Example 4: E|Drive-Center at FAPS, Friedrich-Alexander University Erlangen-Nürnberg,
- Best Practice Example 6: ETA-Factory at PTW, TU Darmstadt,
- Best Practice Example 7: Festo Didactic Learning Factories,
- Best Practice Example 8: Festo Learning Factory in Scharnhausen,
- Best Practice Example 10: IFA Learning Factory at IFA, Leibniz Universität Hannover,
- Best Practice Example 12: LEAN-Factory for a pharmaceutical company in Berlin,
- Best Practice Example 14: Learning Factory aIE at IFF, University of Stuttgart,
- Best Practice Example 16: Learning Factory for Innovation, Manufacturing and Cooperation at the Faculty of Industrial and Process Engineering of Heilbronn University of Applied Sciences,
- Best Practice Example 17: Learning Factory for Global Production at wbk, KIT Karlsruhe,
- Best Practice Example 18: "Lernfabrik für schlanke Produktion," Learning Factory for Lean Production at iwb, TU München,
- Best Practice Example 19: "Lernfabrik für vernetzte Produktion" at Fraunhofer IGCV, Augsburg,
- Best Practice Example 21: LPS Learning Factory at LPS, Ruhr Universität Bochum,
- Best Practice Example 23: MPS Lernplattform at Daimler AG in Sindelfingen,
- Best Practice Example 24: MTA SZTAKI Learning Factory at the Research Laboratory on Engineering and Management Intelligence, MTA SZTAKI in Győr,
- Best Practice Example 25: Pilot Factory Industrie 4.0 at IMW, IFT, and IKT, TU Wien,
- Best Practice Example 26: Process Learning Factory CiP,
- Best Practice Example 27: Smart Factory at the Research Laboratory on Engineering and Management, MTA SZTAKI,
- Best Practice Example 29: Smart Mini-Factory at IEA, Free University of Bolzano,
- Best Practice Example 30: Teaching Factory: An emerging paradigm for manufacturing education,
- Best Practice Example 31: VPS Center of the Production Academy at BMW in Munich.

---

[56]See Blöchl and Schneider (2016).

8.6.7 Sustainability

Sustainability is a subject of great interest in the field of manufacturing research in recent years.[57] Consequently, the topic is also addressed in several learning factories. In the broad field of sustainability in manufacturing, learning factories have to address various topics that are connected to the comparatively less illuminated social dimension in order to create the right behaviors and mindsets.[58] In the following, some of those learning factory approaches are described.

"Die Lernfabrik" at TU Braunschweig that contains an education laboratory, a research laboratory, and an experience laboratory addresses several topics in the field of sustainability in manufacturing.[59] "Die Lernfabrik" and its different laboratories are presented in the Best Practice Example 3 by Dr. Gerrit Posselt.

In addition, a joint learning factory of TU Berlin and a pharmaceutical company also addresses economical, ecological, and social sustainability in combination with Lean Production Systems. This LEAN-Factory is presented by Prof. Holger Kohl, Prof. Roland Jochem, Felix Sieckmann, and Christoffer Rybski in the Best Practice Example 12.

Especially in the field of sustainable manufacturing, the learning factory approach is enriched with the use of learning-conducive artifacts. These so-called learnstruments[60] enable on the fly, automatic learning. Learnstruments attempt to improve the efficiency of learning processes and to create awareness for sustainability in social, economic, and environmental regards.[61] These so-called learnstruments are, for example, described by Jan Philipp Menn and Carsten Ulbrich in the Best Practice Example 22 for the MAN Learning Factory 1635 at MAN Diesel & Turbo SE in Berlin.

8.6.8 Workers Participation

The LPS Learning Factory of the Chair of Production Systems (LPS) at the Ruhr Universität Bochum addresses in addition to the classic learning factory topics of lean production, energy efficiency, and Industrie 4.0, the LPS Learning Factory also issues of company co-determination and the communication and organizational processes between shop floor level and middle management. The created interdisciplinary course in this field is called "Management and Organization of Labor." The LPS

---

[57] See Sutherland et al. (2016).

[58] For the potential and role of manufacturing regarding the social dimension, see also Sutherland et al. (2016).

[59] See, for example, Herrmann (2013), Blume et al. (2015).

[60] See Gausemeier, Seidel, Riedelsheimer, and Seliger (2015).

[61] The learnstruments concept is implemented in several use cases; see Müller et al. (2016), Muschard and Seliger (2015), Heyer, Nishino, Muschard, and Seliger (2014), Menn and Seliger (2016), Müller, Reise, Duc, and Seliger (2016).

Learning Factory is presented by Dr. Christopher Prinz and Prof. Dieter Kreimeier in the Best Practice Example 21.

8.7 Learning Factories for Specific Industry Branches or Products

Additionally to the focus on the topics mentioned above, some learning factories are not specifically geared to certain methodologies but on certain industry branches or products. In the following, some examples are given for that kind of learning factories.

The LEAN-Factory, a learning factory addressing Lean Management topics geared to the needs of the pharmaceutical industry, is presented by Prof. Holger Kohl, Prof. Roland Jochem, Felix Sieckmann, and Christoffer Rybski in the Best Practice Example 12.

The Textile Learning Factory 4.0 at the Institut für Textiltechnik der RWTH Aachen University is a capability building and testing center for digital solutions for the textile industry in Germany.[62]

Further Best Practice Examples addressing specific industry branches' needs can be found in this book:

- A learning factory concept for E-Drives described by Prof. Jörg Franke and Alexander Kühl in the Best Practice Example 4,
- A learning factory concept for electronics production described by Thomas Reitberger and Prof. Jörg Franke in the Best Practice Example 15,
- A learning factory concept for the assembly and maintenance of compressors described by Jan Philipp Menn and Carsten Ulbrich in the Best Practice Example 22,
- Learning factory concepts for the automotive industry are described in the Best Practice Example 23 ("MPS Lernplattform at Daimler AG in Sindelfingen, Germany" by Michael Schwarz) as well as in the Best Practice Example 31 ("VPS Center of the Production Academy at BMW in Munich, Germany" by Carolin Lorber and Tobias Stäudel).

8.8 Wrap-up of This Chapter

This chapter provides an overview of the different topics and contents addressed in learning factories. For the different content field, various examples from existing learning factories are given.

---

[62] A detailed description of the Textile Learning Factory 4.0 including first learning modules can be found in Küsters, Praß, and Gloy (2017).

References

Abele, E., Bauerdick, C., Strobel, N., & Panten, N. (2016). ETA learning factory: A holistic concept for teaching energy efficiency in production. In *6th CIRP-Sponsored Conference on Leanring Factories. Procedia CIRP, 54,* 83–88.

Abele, E., Chryssolouris, G., Sihn, W., Metternich, J., ElMaraghy, H. A., Seliger, G., et al. (2017). Learning factories for future oriented research and education in manufacturing. *CIRP Annals—Manufacturing Technology, 66*(2), 803–826.

Abele, E., Eichhorn, N., & Kuhn, S. (2007). Increase of productivity based on capability building in a learning factory. In *Computer Integrated Manufacturing and High Speed Machining: 11th International Conference on Production Engineering*, Zagreb, 37–41.

Abele, E., Metternich, J., Tenberg, R., Tisch, M., Abel, M., Hertle, C., Eißler, S., Enke, J., & Faatz, L. (2015). *Innovative Lernmodule und -fabriken: Validierung und Weiterentwicklung einer neuartigen Wissensplattform für die Produktionsexzellenz von morgen.* Darmstadt: tuprints.

Acatech. (2016a). *Industrie 4.0: International Benchmark, Options for the Future and Recommendations for Manufacturing Research.* Paderborn, Aachen: HNI, Paderborn University; WZL, RWTH Aachen University.

Acatech. (2016b). *Kompetenzen für Industrie 4.0. Qualifizierungsbedarfe und Lösungsansätze.* acatech POSITION. Munich: Herbert Utz Verlag.

Baena, F., Guarin, A., Mora, J., Sauza, J., & Retat, S. (2017). Learning factory: The path to industry 4.0. *Procedia Manufacturing, 9,* 73–80. https://doi.org/10.1016/j.promfg.2017.04.022.

Bauernhansl, T., Dinkelmann, M., & Siegert, J. (2012). Lernfabrik advanced Industrial Engineering Teil 1: Lernkonzepte und Struktur. *Werkstattstechnik online: wt, 102*(3), 80–83.

Beauvais, W. (2013). Qualification as an effective tool to support the implementation of lean. In G. Reinhart, P. Schnellbach, C. Hilgert, & S. L. Frank (Eds.), *3rd Conference on Learning Factories*, Munich. May 7th, 2013 (pp. 108–117). Augsburg.

Bender, B., Kreimeier, D., Herzog, M., & Wienbruch, T. (2015). Learning factory 2.0—Integrated view of product development and production. In *5th CIRP-Sponsored Conference on Learning Factories. Procedia CIRP, 32,* 98–103. https://doi.org/10.1016/j.procir.2015.02.226.

Blöchl, S. J., & Schneider, M. (2016). Simulation game for intelligent production logistics—The PuLL® learning factory. *Procedia CIRP, 54,* 130–135. https://doi.org/10.1016/j.procir.2016.04.100.

Block, M., Bertagnolli, F., & Herrmann, K. (2011). Lernplattform—Eine neue Dimension des Lernens von schlanken Abläufen/Learning Plattform—A new dimension to learn about lean processes. *Productivity Management, 4,* 52–55.

Blume, S., Madanchi, N., Böhme, S., Posselt, G., Thiede, S., & Herrmann, C. (2015). Die Lernfabrik—Research-based Learning for Sustainable Production Engineering. In *5th CIRP-Sponsored Conference on Learning Factories. Procedia CIRP, 32,* 126–131. https://doi.org/10.1016/j.procir.2015.02.113.

BMWi. (2018). *Mittelstand 4.0—Digitale Produktions- und Arbeitsprozesse.* Retrieved from http://www.mittelstand-digital.de/DE/Foerderinitiativen/mittelstand-4-0.html.

Buergin, J., Minguillon, F. E., Wehrle, F., Haefner, B., & Lanza, G. (2017). Demonstration of a concept for scalable automation of assembly systems in a learning factory. *Procedia Manufacturing, 9,* 33–40. https://doi.org/10.1016/j.promfg.2017.04.026.

ElMaraghy, H. A., Moussa, M., ElMaraghy, W., & Abbas, M. (2017). Integrated product/system design and planning for new product family in a changeable learning factory. In *7th CIRP-Sponsored Conference on Learning Factories. Procedia Manufacturing, 9,* 65–72. https://doi.org/10.1016/j.promfg.2017.04.008.

Erol, S., Jäger, A., Hold, P., Ott, K., & Sihn, W. (2016). Tangible industry 4.0: A scenario-based approach to learning for the future of production. In *6th CIRP-Sponsored Conference on Leanring Factories. Procedia CIRP, 54,* 13–18.

Faller, C., & Feldmüller, D. (2015). Industry 4.0 learning factory for regional SMEs. In *5th CIRP-Sponsored Conference on Learning Factories. Procedia CIRP, 32*, 88–91. https://doi.org/10.1016/j.procir.2015.02.117.

FAPS. (2018). *Green Factory Bavaria*. Retrieved from http://www.greenfactorybavaria.de/.

Gausemeier, P., Seidel, J., Riedelsheimer, T., & Seliger, G. (2015). Pathways for sustainable technology development—The case of bicycle mobility in Berlin. In *12th Global Conference on Sustainable Manufacturing. Procedia CIRP, 26*, 202–207. https://doi.org/10.1016/j.procir.2014.07.164.

Gebbe, C., Hilmer, S., Götz, G., Lutter-Günther, M., Chen, Q., Unterberger, E., Glasschröder, J., Schmidt, V., Riss, F., Kamps, T., Tammer, C., Seidel, C., Braunreuther, S., & Reinhart, G. (2015). Concept of the green factory Bavaria in Augsburg. In *5th CIRP-Sponsored Conference on Learning Factories. Procedia CIRP, 32*, 53–57. https://doi.org/10.1016/j.procir.2015.02.214.

Goerke, M., Schmidt, M., Busch, J., & Nyhuis, P. (2015). Holistic approach of lean thinking in learning factories. In *5th CIRP-Sponsored Conference on Learning Factories. Procedia CIRP, 32*, 138–143. https://doi.org/10.1016/j.procir.2015.02.221.

Gräßler, I., Pöhler, A., & Pottebaum, J. (2016). Creation of a learning factory for cyber physical production systems. In *6th CIRP-Sponsored Conference on Leanring Factories. Procedia CIRP, 54*, 107–112. https://doi.org/10.1016/j.procir.2016.05.063.

Gräßler, I., Taplick, P., & Yang, X. (2016b). Educational learning factory of a holistic product creation process. *Procedia CIRP, 54*, 141–146. https://doi.org/10.1016/j.procir.2016.05.103.

Hammer, M. (2014, August). Making operational transformations successful with experiential learning. In *CIRP Collaborative Working Group—Learning Factories for Future Oriented Research and Education in Manufacturing*, CIRP General Assembly, Nantes, France.

Helleno, A. L., Simon, A. T., Papa, M. C. O., Ceglio, W. E., Rossa Neto, A. S., & Mourad, R. B. A. (2013). Integration university-industry: Laboratory model for learning lean manufacturing concepts in the academic and industrial environments. *International Journal of Engineering Education, 29*(6), 1387–1399.

Herrmann, C. (2013). Die Lernfabrik—Research and education for sustainability in manufacturing. In G. Reinhart, P. Schnellbach, C. Hilgert, & S. L. Frank (Eds.), *3rd Conference on Learning Factories*. Munich. May 7th, 2013 (pp. 48–61). Augsburg.

Herrmann, S., & Stäudel, T. (2014). Learn and experience VPS in the BMW learning factory. In *4th Conference on Learning Factories*, Stockholm, Sweden, 1–18.

Heyer, S., Nishino, N., Muschard, B., & Seliger, G. (2014). Enabling of local value creation via openness for emergent synthesis. *International Journal of Precision Engineering and Manufacturing, 15*(7), 1489–1493. https://doi.org/10.1007/s12541-014-0496-5.

Hummel, V., Hyra, K., Ranz, F., & Schuhmacher, J. (2015). Competence development for the holistic design of collaborative work systems in the logistics learning factory. In *5th CIRP-Sponsored Conference on Learning Factories. Procedia CIRP, 32*, 76–81. https://doi.org/10.1016/j.procir.2015.02.111.

Jäger, A., Mayrhofer, W., Kuhlang, P., Matyas, K., & Sihn, W. (2012). The "learning factory": An immersive learning environment for comprehensive and lasting education in industrial engineering. *16th World Multi-Conference on Systemics, Cybernetics and Informatics, 16*(2), 237–242.

Jäger, A., Mayrhofer, W., Kuhlang, P., Matyas, K., & Sihn, W. (2013). Total immersion: Hands and heads-on training in a learning factory for comprehensive industrial engineering education. *International Journal of Engineering Education, 29*(1), 23–32.

Jorgensen, J. E., Lamancusa, J. S., Zayas-Castro, J. L., & Ratner, J. (1995). The learning factory: Curriculum integration of design and manufacturing. In *4th World Conference on Engineering Education*, 1–7.

Kreimeier, D., Morlock, F., Prinz, C., Krückhans, B., Bakir, D. C., & Meier, H. (2014). Holistic learning factories—A concept to train lean management, resource efficiency as well as management and organization improvement skills. In *47th CIRP Conference on Manufacturing Systems. Procedia CIRP, 17*, 184–188.

References

Kreitlein, S., Höft, A., Schwender, S., & Franke, J. (2015). Green factories bavaria: A network of distributed learning factories for energy efficient production. In *5th CIRP-Sponsored Conference on Learning Factories. Procedia CIRP, 32*, 58–63. https://doi.org/10.1016/j.procir.2015.02.219.

Küsters, D., Praß, N., & Gloy, Y.-S. (2017). Textile learning factory 4.0—Preparing Germany's textile industry for the digital future. *Procedia Manufacturing, 9*, 214–221. https://doi.org/10.10 16/j.promfg.2017.04.035.

Lanza, G., Minges, S., Stoll, J., Moser, E., & Haefner, B. (2016). Integrated and modular didactic and methodological concept for a learning factory. In *6th CIRP-Sponsored Conference on Leanring Factories. Procedia CIRP, 54*, 136–140. https://doi.org/10.1016/j.procir.2016.06.107.

Lanza, G., Moser, E., Stoll, J., & Haefner, B. (2015). Learning factory on global production. In *5th CIRP-Sponsored Conference on Learning Factories. Procedia CIRP, 32*, 120–125. https://doi.or g/10.1016/j.procir.2015.02.081.

Madsen, O., & Møller, C. (2017). The AAU smart production laboratory for teaching and research in emerging digital manufacturing technologies. *Procedia Manufacturing, 9*, 106–112. https://d oi.org/10.1016/j.promfg.2017.04.036.

Mayer, B., Rabel, B., & Sorko, S. R. (2017). Modular smart production lab. *Procedia Manufacturing, 9*, 361–368. https://doi.org/10.1016/j.promfg.2017.04.025.

McKinsey & Company. (2017). *Model factories and offices: Building operations excellence*. Retrieved from https://capability-center.mckinsey.com/files/mccn/2017-03/emea_model_factor ies_brochure_1.pdf.

Menn, J. P., & Seliger, G. (2016). Increasing knowledge and skills for assembly processes through interactive 3D-PDFs. In *The 23rd CIRP Conference on Life Cycle Engineering. Procedia CIRP, 48*, 454–459. https://doi.org/10.1016/j.procir.2016.02.093.

Metternich, J., Müller, M., Meudt, T., & Schaede, C. (2017). Lean 4.0—zwischen Widerspruch und Vision. *Zeitschrift für wirtschaftlichen Fabrikbetrieb (ZWF), 112*(5), 346–348. https://doi.org/1 0.3139/104.111717.

Micheu, H.-J., & Kleindienst, M. (2014). Lernfabrik zur praxisorientierten Wissensvermittlung: Moderne Ausbildung im Bereich Maschinenbau und Wirtschaftswissenschaften. *Zeitschrift für wirtschaftlichen Fabrikbetrieb (ZWF), 109*(6), 403–407.

Monostori, L., Kádár, B., Bauernhansl, T., Kondoh, S., Kumara, S., Reinhart, G., et al. (2016). Cyber-physical systems in manufacturing. *CIRP Annals—Manufacturing Technology, 65*(2), 621–641. https://doi.org/10.1016/j.cirp.2016.06.005.

Morlock, F., Kreggenfeld, N., Louw, L., Kreimeier, D., & Kuhlenkötter, B. (2017). Teaching methods-time measurement (MTM) for workplace design in learning factories. *Procedia Manufacturing, 9*, 369–375. https://doi.org/10.1016/j.promfg.2017.04.033.

Müller, B. C., Nguyen, T. D., Dang, Q.-V., Duc, B. M., Seliger, G., Krüger, J., & Kohl, H. (2016). Motion tracking applied in assembly for worker training in different locations. In *The 23rd CIRP Conference on Life Cycle Engineering. Procedia CIRP, 48*, 460–465. https://doi.org/10.1016/j.p rocir.2016.04.117.

Müller, B. C., Reise, C., Duc, B. M., & Seliger, G. (2016). Simulation-games for learning conducive workplaces: A case study for manual assembly. In *13th Global Conference on Sustainable Manufacturing. Procedia CIRP, 40*, 353–358. https://doi.org/10.1016/j.procir.2016.01.063.

Muschard, B., & Seliger, G. (2015). Realization of a learning environment to promote sustainable value creation in areas with insufficient infrastructure. In *5th CIRP-Sponsored Conference on Learning Factories, Procedia CIRP, 32*, 70–75. https://doi.org/10.1016/j.procir.2015.04.095.

PennState University. (2017). *Bernard M. gordon learning factory: We bring the real world into the classroom*. Retrieved from http://www.lf.psu.edu/.

Pittschellis, R. (2015). Multimedia support for learning factories. In *5th CIRP-Sponsored Conference on Learning Factories. Procedia CIRP, 32*, 36–40. https://doi.org/10.1016/j.procir.2015.0 6.001.

Plorin, D. (2016). Gestaltung und Evaluation eines Referenzmodells zur Realisierung von Lernfabriken im Objektbereich der Fabrikplanung und des Fabrikbetriebes. *Dissertation, Chemnitz*.

Wissenschaftliche Schriftenreihe des Instituts für Betriebswissenschaften und Fabriksysteme: Heft 120. Chemnitz: Techn. Univ. Inst. für Betriebswiss. und Fabriksysteme.

Plorin, D., Poller, R., & Müller, E. (2013). advanced Learning Factory (aLF): Integratives Lernfabrikkonzept zur praxisnahen Kompetenzentwicklung am Beispiel der Energieeffizienz. *Werkstattstechnik online: wt, 103*(3), 226–232.

Prinz, C., Morlock, F., Freith, S., Kreggenfeld, N., Kreimeier, D., & Kuhlenkötter, B. (2016). Learning factory modules for smart factories in industrie 4.0. In *6th CIRP-Sponsored Conference on Leanring Factories. Procedia CIRP, 54*, 113–118. https://doi.org/10.1016/j.procir.2016.05.105.

Prinz, C., Morlock, F., Wagner, P., Kreimeier, D., & Wannöffel, M. (2014). Lernfabrik zur Vermittlung berufsfeldrelevanter Handlungskompetenzen: Fragen der Gestaltung und des Managements von Arbeit theoretisch kennenlernen und in einer Lernfabrik realitätsnah erproben. *Industrie Management, 30*(3), 39–42.

PTW, TU Darmstadt. (2016). *Welcome to ETA-factory: The energy efficient model factory of the future*. Retrieved from http://www.eta-fabrik.tu-darmstadt.de/eta/index.en.jsp.

Putz, M. (2013). The concept of the new research factory at fraunhofer IWU—to objectify energy and resource efficiency R&D in the E3-factory. In G. Reinhart, P. Schnellbach, C. Hilgert, & S. L. Frank (Eds.), *3rd Conference on Learning Factories*, Munich. May 7th, 2013 (pp. 62–77). Augsburg.

Scholz, M., Kreitlein, S., Lehmann, C., Böhner, J., Franke, J., & Steinhilper, R. (2016). Integrating intralogistics into resource efficiency oriented learning factories. *Procedia CIRP, 54*, 239–244. https://doi.org/10.1016/j.procir.2016.05.067.

Schuh, G., Gartzen, T., Rodenhauser, T., & Marks, A. (2015). Promoting work-based learning through industry 4.0. In *5th CIRP-Sponsored Conference on Learning Factories. Procedia CIRP, 32*, 82–87. https://doi.org/10.1016/j.procir.2015.02.213.

Schuhmacher, J., Baumung, W., & Hummel, V. (2017). An intelligent bin system for decentrally controlled intralogistic systems in context of industrie 4.0. *Procedia Manufacturing, 9*, 135–142. https://doi.org/10.1016/j.promfg.2017.04.005.

Schützer, K., Rodrigues, L. F., Bertazzi, J. A., Durão, L. F. C. S., & Zancul, E. (2017). Learning environment to support the product development process. *Procedia Manufacturing, 9*, 347–353. https://doi.org/10.1016/j.promfg.2017.04.018.

Seifermann, S., Böllhoff, J., Metternich, J., & Bellaghnach, A. (2014). Evaluation of work measurement concepts for a cellular manufacturing reference line to enable low cost automation for lean machining. In *47th CIRP Conference on Manufacturing Systems. Procedia CIRP, 17*, 588–593.

Seitz, K.-F., & Nyhuis, P. (2015). Cyber-physical production systems combined with logistic models—A learning factory concept for an improved production planning and control. In *5th CIRP-Sponsored Conference on Learning Factories. Procedia CIRP, 32*, 92–97. https://doi.org/10.1016/j.procir.2015.02.220.

Sihn, W., Gerhard, D., & Bleicher, F. (2012). Vision and implementation of the learning and innovation factory of the vienna university of technology. In W. Sihn & A. Jäger (Eds.), *2nd Conference on Learning Factories—Competitive Production in Europe Through Education and Training* (pp. 160–177).

Steffen, M., May, D., & Deuse, J. (2012). The industrial engineering laboratory: problem based learning in industrial engineering education at TU dortmund university. In *Global Engineering Education Conference (EDUCON), IEEE, —Collaborative Learning and New Pedagogic Approaches in Engineering Education*, Marrakesch, Marokko, April 17–20, 2012 (1–10).

Stoldt, J., Franz, E., Schlegel, A., & Putz, M. (2015). Resource networks: Decentralised factory operation utilising renewable energy sources. In *12th Global Conference on Sustainable Manufacturing. Procedia CIRP, 26*, 486–491.

Sutherland, J. W., Richter, J. S., Hutchins, M. J., Dornfeld, D., Dzombak, R., Mangold, J., et al. (2016). The role of manufacturing in affecting the social dimension of sustainability. *CIRP Annals—Manufacturing Technology, 65*(2), 689–712. https://doi.org/10.1016/j.cirp.2016.05.003.

References

Thiede, S., Juraschek, M., & Herrmann, C. (2016). Implementing cyber-physical production systems in learning factories. In *6th CIRP-Sponsored Conference on Leanring Factories. Procedia CIRP, 54*, 7–12. https://doi.org/10.1016/j.procir.2016.04.098.

Thomar, W. (2015, July 8). Kaerchers global lean academy approach: incentive talk (industry). In *5th Conference on Learning Factories*, Bochum, Germany.

Tietze, F., Czumanski, T., Braasch, M., & Lödding, H. (2013). Problembasiertes Lernen in Lernfabriken. *Werkstattstechnik online: wt, 103*(3), 246–251.

Tvenge, N., Martinsen, K., & Kolla, S. S. V. K. (2016). Combining learning factories and ICT-based situated learning. In *6th CIRP-Sponsored Conference on Leanring Factories. Procedia CIRP, 54*, 101–106.

UAW-Chrysler National Training Center. (2016). *World class manufacturing academy*. Retrieved from http://www.uaw-chrysler.com/world-class-mfg-academy/.

Wagner, C., Heinen, T., Regber, H., & Nyhuis, P. (2010). Fit for Change—Der Mensch als Wandlungsbefähiger. *Zeitschrift für wirtschaftlichen Fabrikbetrieb (ZWF), 100*(9), 722–727.

Wagner, P., Prinz, C., Wannöffel, M., & Kreimeier, D. (2015). Learning factory for management, organization and workers' participation. In: *5th CIRP-Sponsored Conference on Learning Factories. Procedia CIRP, 32*, 115–119. https://doi.org/10.1016/j.procir.2015.02.118.

Wagner, U., AlGeddawy, T., ElMaraghy, H. A., & Müller, E. (2012). The state-of-the-art and prospects of learning factories. In *45th CIRP Conference on Manufacturing Systems. Procedia CIRP, 3*, 109–114.

Wank, A., Adolph, S., Anokhin, O., Arndt, A., Anderl, R., & Metternich, J. (2016). Using a learning factory approach to transfer industrie 4.0 approaches to small- and medium-sized enterprises. In *6th CIRP-Sponsored Conference on Leanring Factories. Procedia CIRP, 54*, 89–94. https://doi.org/10.1016/j.procir.2016.05.068.

Yoo, I. S., Braun, T., Kaestle, C., Spahr, M., Franke, J., Kestel, P., et al. (2016). Model factory for additive manufacturing of mechatronic products: Interconnecting World-class technology partnerships with leading AM players. *Procedia CIRP, 54*, 210–214. https://doi.org/10.1016/j.procir.2016.03.113.

Zuehlke, D. (2008). Smartfactory–from vision to reality in factory technologies. In *17th International Federation of Automatic Control (IFAC) World Congress*, 82–89.

Chapter 9
Overview on Potentials and Limitations of Existing Learning Factory Concept Variations

In practice and in literature, numerous variations of learning factory concepts can be identified. Here it becomes unclear which advantages and disadvantages individual variations offer. This chapter therefore provides an overview of the characteristics, advantages and disadvantages, potentials, and limitations of distinguishable concepts (Fig. 9.1).

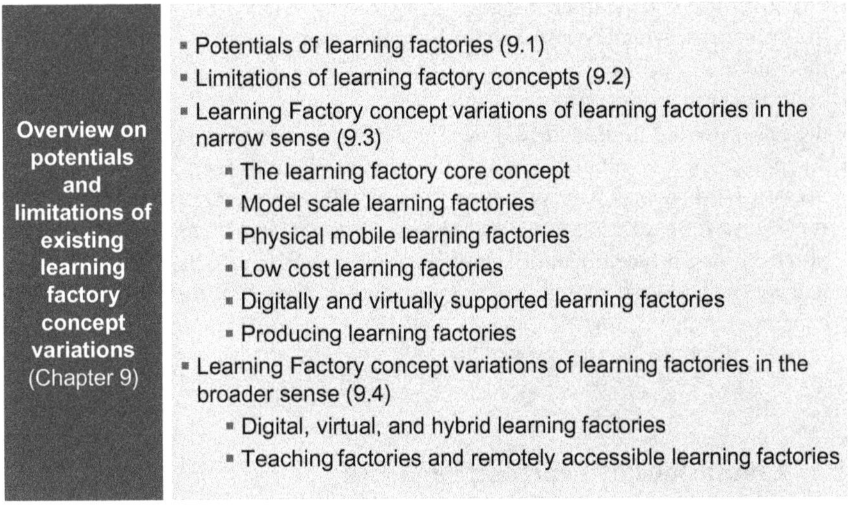

Overview on potentials and limitations of existing learning factory concept variations (Chapter 9)
- Potentials of learning factories (9.1)
- Limitations of learning factory concepts (9.2)
- Learning Factory concept variations of learning factories in the narrow sense (9.3)
  - The learning factory core concept
  - Model scale learning factories
  - Physical mobile learning factories
  - Low cost learning factories
  - Digitally and virtually supported learning factories
  - Producing learning factories
- Learning Factory concept variations of learning factories in the broader sense (9.4)
  - Digital, virtual, and hybrid learning factories
  - Teaching factories and remotely accessible learning factories

Fig. 9.1 Structure of the overview over concept variations of existing learning factories

9.1 Potentials of Learning Factories

The main objectives for the use of learning factories are

- the effective development and motivation of learners,[1]
- the facilitation of practical innovations,[2] and
- the transfer of competences and innovations to industry.[3]

The potentials of learning factories with regard to the achievement of those goals are described in the following sections:

- Sections 7.1 and 7.2: The general success factors of learning factories for education and training are described. Section 7.1 gives an overview over the various effective learning concepts that can be promoted using learning factories; examples are problem-based learning, project-based learning, or research-based learning. Furthermore, Sect. 7.2.1 especially focuses on the potentials regarding different learning processes; here the use of the two most recognized information assimilation process and the experiential learning process in learning factories are discussed. Additionally, Sect. 7.2.3 deals with success factors learning factories for education and training in general:
- feedback cycles for the learner,
- the contextualization of the learning environment,
- an activation of the learner,
- the problem-solving processes in learning factories,
- motivational aspects,
- collectivization of learning processes,
- the integration of thinking and doing during learning as well as
- an appropriate self-regulation and self-direction of the learners.
- Section 7.3: In Sect. 7.3 the potential of learning factories for research is described. It is shown how learning factories can be used as research enables in the research process along problem identification, abstraction with real data, solution finding, realization, and verification. Furthermore, the advantages of the use of embedded experiments in learning factories as well as the use of learning factories for the transfer of innovative methods and technologies to industry are shown.

9.2 Limitations of Learning Factories

On the way to achieving these goals, some obstacles of building up learning factories or limitations of the learning factory concept itself can be identified[4]:

---

[1] In education and training in learning factories.
[2] In research in learning factories.
[3] In innovation transfer in learning factories.
[4] Tisch and Metternich (2017).

9.2 Limitations of Learning Factories

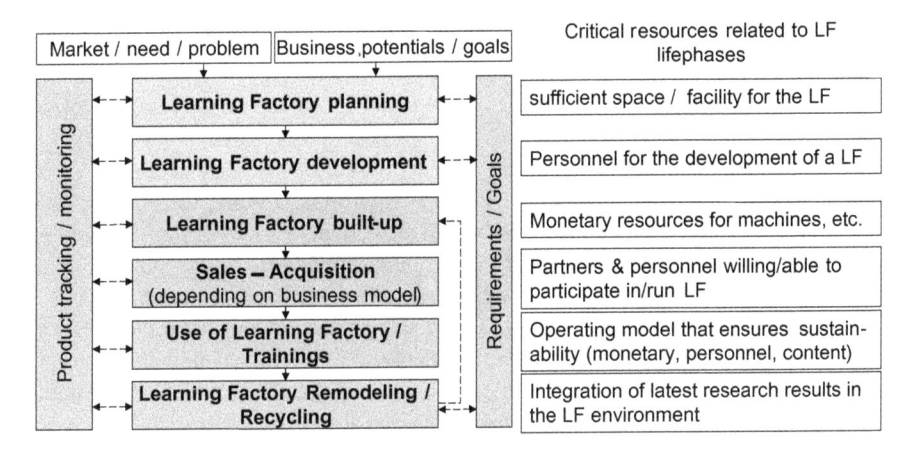

Fig. 9.2 Required resources along the learning factory lifecycle (Tisch & Metternich, 2017), life cycle similar to general product life cycle according to VDI (1993)

- The resources needed for learning factories along the learning factory life cycle
- The mapping ability in learning factories related to content and object, space, and cost, and time
- The scalability of learning factory approaches
- The mobility of learning factory approaches
- The effectiveness of learning factories.

The resources needed for learning factories along the learning factory life cycle

A massive obstacle that prevents more learning factories from being built around the globe is the fact that the planning and development as well as the construction and operation of learning factories require immense effort. This effort is not limited to the financial resources that are needed. Equally important are the suitable personnel for the learning factory initiation and operation, the necessary content that can be addressed in the learning factory as well as access to the factory equipment, enough space, and a suitable factory hall. Especially in the early phases of the learning factory life cycle, any lack of resources can bring the entire learning factory project to a standstill. Figure 9.2 lists the critical resources along the learning factory life cycle.

The mapping ability in learning factories related to content and object, space and cost, and time

An important limit of today's learning factory concepts is the fact that learning factories can only look at limited sections of industrial production. The boundaries of these sections can be related to specific industrial sectors, addressed topics, single production processes, company departments, or even target groups. A single learning factory can therefore only map a small section of a complex industrial reality since each facility has to emphasis specific topics or environments; this is also referred to as **content- and object-related mapping ability limitation**.[5]

In theory, challenges and problems of all factory levels can be considered within learning factory concepts; from the process and station level to the factory network level. However, since the factory levels whose problems need to be examined more closely have to be part of the tangible learning environment of the learning factory, **space- and cost-related mapping limitations** arise; upper factory levels in particular are difficult to address within learning factory concepts. Individual approaches can be observed in which the problems of global production systems are addressed.[6]

A further limitation of the learning factory concept arises in the temporal relation of the feedback of the learning environment to the actions performed by the learners. If it takes too long from the actions to get a natural feedback from the environment, the learning factory concept cannot easily be used for topics of this kind. Here, fast-forward mechanisms have to be integrated for bridging the gap between actions and feedback. The shorter the duration of the learning modules, the faster these limits are reached.[7] In these cases, where these **time-related mapping ability limitations** are reached, simulations must be used as a tool to shorten the feedback cycles. Otherwise, the feedback cycles and thus the experiential component of the learning factory concept cannot be fully utilized.

Figure 9.3 classifies different methods and topics from the lean area in relation to addressed factory level and duration of the feedback cycle. Topics in the lower left corner, i.e. topics addressing the lower factory levels with immediate feedback cycles, are typical topics of today's learning factories. Examples for topics with fast feedback cycles, and therefore well suited to be addressed in learning factories, are basic lean production and management topics such as 5S,[8] SMED,[9] line balancing,[10] and Poka Yoke[11] solutions. Although in the case of Poka Yoke, another limitation of learning factories is observable: In the practical phases, learners design and propose solutions for the improvement of the factory environment, e.g. on the topic of Poka Yoka, which could also be implemented directly in the learning factory. In many cases, this ad hoc implementation is connected with **solution-related mapping limitations**. This solution-related mapping ability is dependent on the flexibility and the changeability of the learning factory environment concerning several dimensions.[12]

The scalability of learning factory approaches

---

[5] See Tisch and Metternich (2017), Abele et al. (2015).

[6] See Lanza, Moser, Stoll, and Haefner (2015) and Sect. 8.6.5.

[7] As examples for topics or learning content with long feedback loops product development, supplier development or planning of maintenance activities can be named.

[8] Method to design workplaces and their environment safely, cleanly and transparently.

[9] Single Minute Exchange of Dies, a lean method to reduce changeover times.

[10] Method for a balanced work content over several stations or workers, for example, in assembly lines.

[11] Method to design mistake proof products and processes.

[12] Among others, the changeability dimensions process, product, organization, and layout are mentioned by Wiendahl et al. (2007).

9.2 Limitations of Learning Factories

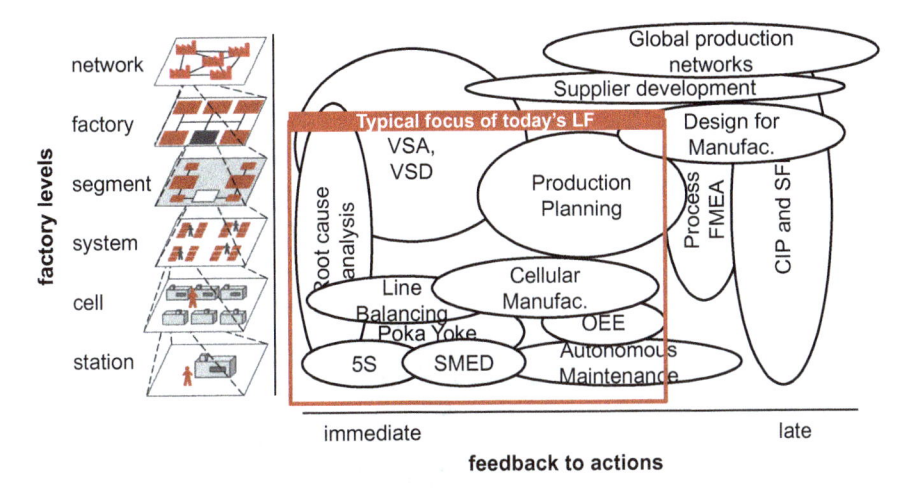

Fig. 9.3 Exemplary limits regarding space- and time-related mapping ability (Abele et al., 2017b; Tisch, 2018)

A massive limitation of today's learning factory concepts is the scalability of the learning events that can be carried out. This problematic scalability becomes particularly clear in comparison to other forms of learning: For example, several hundred students can take part in one lecture without major problems; only one teacher is required. As a rule, a maximum of fifteen to twenty students can participate in a single learning factory course, while at least two trainers are regularly assigned. But even if several courses are started in parallel to allow many students to learn in the learning factory, the learning factory facility itself is another limiting factor for capacity: At the same time, often only one course can run in learning factories, sometimes two if planned carefully.

The mobility of learning factory approaches

Another limiting factor arises from the fact that learning factories with physical learning environments are set up and operated at a defined location. Consequently, the learning factory training can only take place in this one place; the offers are therefore only accessible in a certain region; learning factories in the narrower sense are generally immobile as a whole. In order to achieve mobility of the learning factory approach, there are basically three approaches:

1. A space-saving and transport-friendly design of the entire learning factory equipment that enables a transport to different locations without bigger efforts. In those cases, often only certain selected processes (e.g. assembly processes) can be mapped inside the learning factory.[13]

---

[13] A learning factory using this approach to gain mobility is described in the Best Practice Example 18 "'Lernfabrik für schlanke Produktion,' Learning Factory for Lean Production at iwb, TU München, Germany."

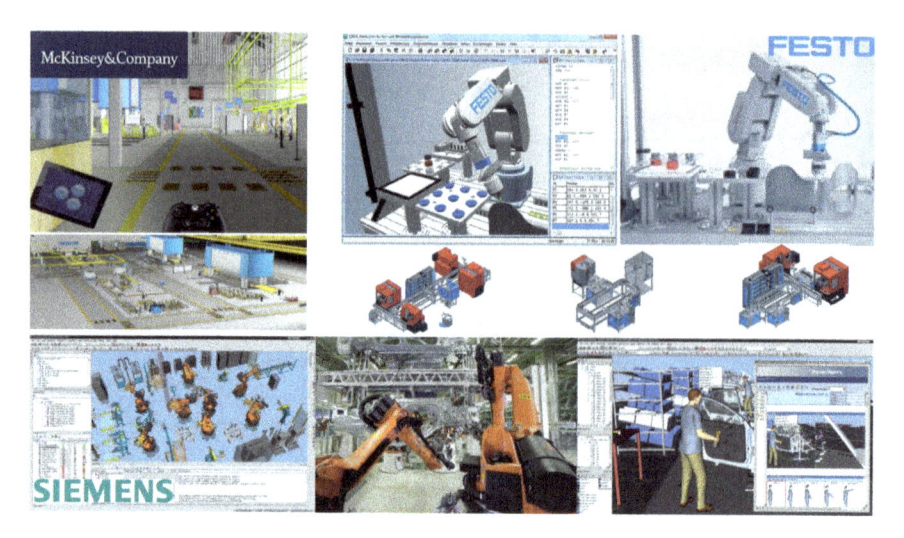

Fig. 9.4 Virtual factories and learning factories from McKinsey & Company, Festo, and Siemens (Abele et al., 2017b)

2. The use of virtual learning factories, which accordingly do not include immobile, physical factory equipment.[14] In those approaches, the challenge is to preserve the hands-on characteristic of physical learning factory concepts (see Fig. 9.4).
3. On-site learning in the factory environment is replaced by remote learning using new ICT equipment. In these cases, image and sound of the factory environment are transferred to another learning location.[15] In those approaches, the challenge is to create an immersive learning environment and allow the students to get access to the factory environment.

The effectiveness of learning factory approaches

Learning factories often aim to enable high-quality competence development. In many cases, there is no meaningful integration of these objectives in the design of the learning factory and the learning modules or a review of the achievement of the objectives. The creation of effective learning factories calls for the consideration of competency-oriented learning targets in the design phase[16] as well as a target-oriented evaluation phase.[17]

---

[14] Digital and virtual learning factory approaches are discussed in the Sect. 9.4.1.

[15] The Teaching Factory concept uses this approach to obtain mobility. Two approaches are presented in the Best Practice Example 20 "LMS Factory at the Laboratory for Manufacturing Systems and Automation (LMS), University Patras, Greece" and the Best Practice Example 30 "Teaching Factory: An emerging paradigm for manufacturing education. LMS, University of Patras, Greece."

[16] For approaches regarding learning factory planning and design, see Sect. 6.1.

[17] For the success evaluation in learning factories, see Sect. 6.3.3.

9.3 Learning Factory Concept Variations of Learning Factories in the Narrow Sense—Advantages and Disadvantages

What different variations of the learning factory are there, which try to overcome one or the other current limitation (see previous section) of the learning factory concept? On the one hand, are these new learning factory concepts such as virtual learning factories superior to the classic learning factory concept? On the other hand, are the advantages in one area of the variation linked with disadvantages in other areas? This section gives an overview of these questions and the most important learning factory concept variations of learning factories in the narrow sense, namely

- the learning factory core concept,
- model scale learning factories,
- physical mobile learning factories,
- low-cost learning factories,
- digitally and virtually supported learning factories, and
- producing learning factories.

9.3.1 The Learning Factory Core Concept

The core concept of the learning factory has already been described in Sect. 4.3. This core concept consists of a realistic, physical life-size factory environment in which a physical product is created that learners can experience directly on site. On the one hand, learning factories that follow this core concept provide best prerequisites for an effective competency development in the broad field of production technology and organization, namely through

- hands-on experience and own actions of the learners and connected feedback,
- a high contextualization of the learning environment,
- activation of the learners in practical learning tasks,
- use of realistic problem situations,
- motivational benefits of the immersive learning environment,
- possibilities for collectivization of on-site learning processes,
- possibilities for the integration of thinking and doing, and
- self-regulation and self-direction of the learners can be used appropriately.

On the other hand, these kinds of learning factory approaches come especially with some of the described learning factory obstacles and limitations, namely

- a large resource requirement for construction and operation,
- the mapping of only certain production processes and topics that represents a small part of complex manufacturing systems,

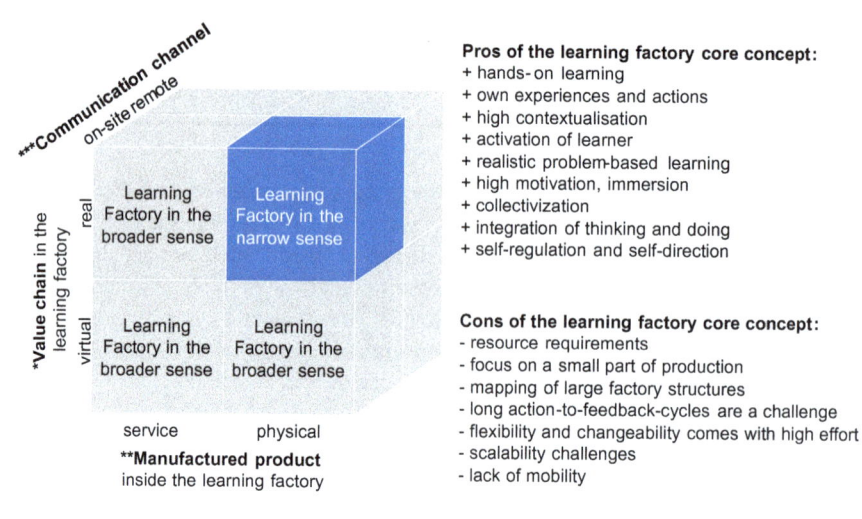

Fig. 9.5 Advantages and disadvantages of the learning factory core concept (learning factories in the narrow sense)

- the challenge in mapping large factory structures, e.g. entire factories or factory networks,
- the challenges with learning content that does not allow direct feedback from the learning environment to the learners' actions,
- the challenge with ad hoc representation of participants' ideas and solutions in the learning factory (changeability and flexibility of the factory environment),
- the difficulties with scalability, e.g. in the use of learning factories for educational purposes in lectures with many participants, and
- the lack of mobility of the learning factory facility and equipment (Fig. 9.5).

9.3.2 Model Scale Learning Factories

In contrast to the learning factories in the previous section (regarding the learning factory core concept), scaled-down or model scale learning factories do not use original factory equipment but smaller equivalents, which should differ as little as possible from the original factory equipment except for the smaller dimensions. Since the model scale equipment is an abstraction of industrial equipment, this can be a challenge, for example, when it comes to making load profiles of industrial machines and model scaled machines as comparable as possible.[18] The advantage of these model scale learning factories is that even less available space, e.g. in a classroom, is sufficient to map a factory in a scaled-down fashion. Furthermore, also

---
[18]For a design approach for model scale learning factory environments for energy efficiency, see also Kaluza et al. (2015).

9.3 Learning Factory Concept Variations of Learning Factories ... 297

Fig. 9.6 Advantages and disadvantages of model scale learning environments compared to the learning factory core concept

less financial resources are needed to build up a learning factory at model scale. Additionally, the design of the learning environment is specifically geared to the learning purpose, which means that equipment can be created in a way that they are easy accessible to the learners; of course this advantage comes with a loss of authenticity.[19] Advantages and disadvantages of the use of scaled-down learning factory environments compared to the learning factory core concept in Sect. 9.3.1 are summarized in Fig. 9.6.

In many cases, the equipment for these scaled-down learning factories comes from the manufacturer Festo Didactic[20]; Festo Didactic learning factories are mainly used in universities[21] and vocational schools,[22] but also, for example, in Festo's own company.[23]

The Learning Factory for advanced Industrial Engineering (aIE) at the Institute of Industrial Manufacturing and Management (IFF), Universität Stuttgart, Germany, contains scaled-down Festo Didactic factory equipment that addresses the link between the digital production planning and the implementation of physical models

---

[19]For a discussion of pros and cons of model scale learning factories, see also Kaluza et al. (2015).

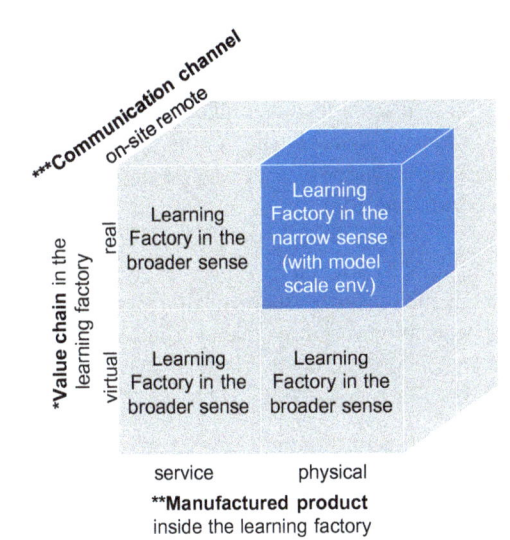

[20]See also Best Practice Example 7 "Festo Didactic Learning Factories."

[21]See for example Best Practice Example 9 "iFactory at the Intelligent Manufacturing Systems (IMS) Center, University of Windsor, Canada" and Best Practice Example 14 "Learning Factory aIE at IFF, University of Stuttgart, Germany."

[22]See Sect. 7.1.9 "Example: Learning factories for Industrie 4.0 vocational education in Baden-Württemberg."

[23]See also Best Practice Example 8 "Festo Learning Factory in Scharnhausen, Germany."

in the laboratory.[24] The transformable production platform of the aIE learning factory consists of mobile plug and play modules for assembly, inspection, coating, storage, and transportation. The reconfiguration into different layouts of the learning factory is possible. Festo Didactic designed and implemented the aIE learning factory. Prof. Thomas Bauernhansl, Erwin Gross, and Prof. Onorific. Jörg Siegert describe the learning factory aIE in the Best Practice Example 14.

The iFactory is a similar learning factory concept with Festo Didactic equipment at the Intelligent Manufacturing Systems (IMS) Center, Windsor, Canada.[25] The learning factory was set up as the first of its kind in North America; it is a modular and changeable assembly system that comprises robotic and manual assembly stations, computer vision inspection station, Automated Storage and Retrieval System (ASRS) and several material handling modules. The iFactory is extended by

- the iDesign module, which is an design innovation studio,
- the iPlan module, which contains process and production planning tools,
- a 3D printing facility, and
- a Coordinate Measuring Machine (CMM) facility.

Prof. Hoda ElMaraghy also describes the iFactory at the Intelligent Manufacturing Systems Center in Windsor, Canada, in the Best Practice Example 9.

The Festo Scharnhausen Learning Factory is placed in the middle of the Festo Technology Factory in Scharnhausen and is surrounded by production departments. The Scharnhausen Learning Factory entails four rooms (mechanical processing, valve and valve terminal assembly, automation, CPPS and process improvements, administration of the learning factory). The learning factory at the plant that uses Festo Didactic equipment is used to train employees exclusively; mainly new employees come to the learning factory but also advanced qualification takes place. Holger Regber describes the Festo Scharnhausen Learning Factory in the Best Practice Example 8.

Furthermore, in the Best Practice Example Section, further learning factories using model scale learning environments can be identified:

- Best Practice Example 1: AutoFab at the Faculty of Electrical Engineering of the University of Applied Sciences Darmstadt, Germany,
- Best Practice Example 3: Die Lernfabrik at IWF, TU Braunschweig, Germany,
- Best Practice Example 7: Festo Didactic Learning Factories,
- Best Practice Example 12: LEAN-Factory for a pharmaceutical company in Berlin, Germany,
- Best Practice Example 23: MPS Lernplattform at Daimler AG in Sindelfingen, Germany,
- Best Practice Example 27: Smart Factory at the Research Laboratory on Engineering and Management, MTA SZTAKI in Budapest, Hungary,
- Best Practice Example 28: Smart Factory at DFKI, TU Kaiserslautern, Germany.

---

[24] Hummel and Westkämper (2007).
[25] ElMaraghy and ElMaraghy (2015).

9.3 Learning Factory Concept Variations of Learning Factories … 299

Pros of physical mobile learning factories:
+ location independent use of learning factory equipment (e.g. for in-house trainings)
+ in general less and not permanent space needed

Cons of physical mobile learning factories:
- restrictions due to mobility
- lower contextualisation and authenticity of environment (small part of production is mapped)
- in general scope of learning environment is limited

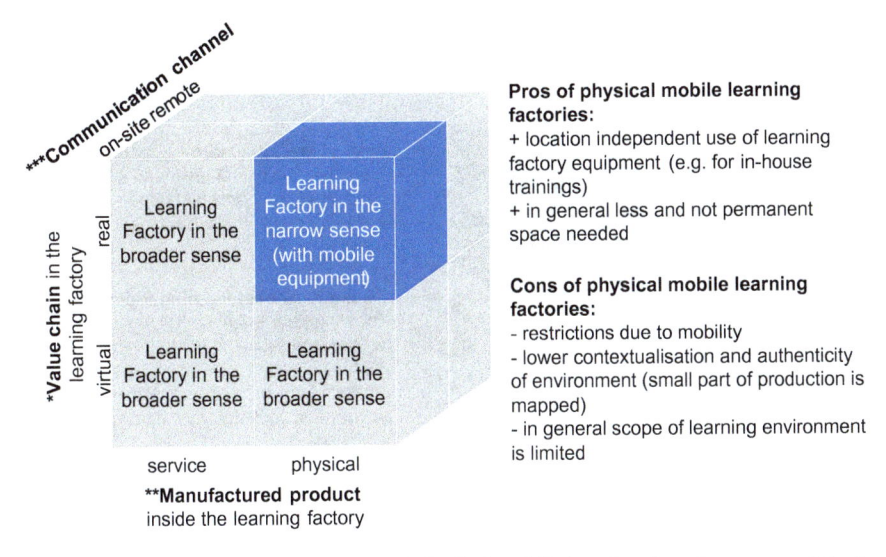

Fig. 9.7 Advantages and disadvantages of physical mobile learning factories compared to the learning factory core concept

9.3.3 *Physical Mobile Learning Factories*

Learning factories are in general built at specific locations and consequently only available in a certain region. One limit of most existing physical learning factories is its immobility, see Sect. 9.2. Thus, trainings at the company's site, with access to the company's own value creation processes and shorter travel times for the employees, are usually not possible. In addition to the use of virtual learning factories, the use of mobile physical learning equipment is a good way to overcome this limitation. The LSP (Lernfabrik für die Schlanke Produktion) at iwb, TU München is a physical, completely mobile learning factory. This learning factory approach is described by Harald Bauer, Felix Brandl, and Prof. Gunther Reinhart in the Best Practice Example 18. For future research on the mobility learning factory approaches, additionally to the physical approach described in this section, virtual[26] and ICT-based remote[27] learning factory concepts can provide added value regarding the mobility of the learning factory concept. Advantages and disadvantages of the use of mobile learning factory environments compared to the learning factory core concept in Sect. 9.3.1 are summarized in Fig. 9.7.[28]

---

[26] See Sect. 9.4.1: Digital, virtual, and hybrid learning factories.

[27] See Sect. 9.4: Teaching factories and remotely accessible learning factories.

[28] See also Tisch and Metternich (2017).

Pros of low cost learning factories:
+ enables the use of the learning factory concept also for smaller budgets
+ starting point for more learning factory activities

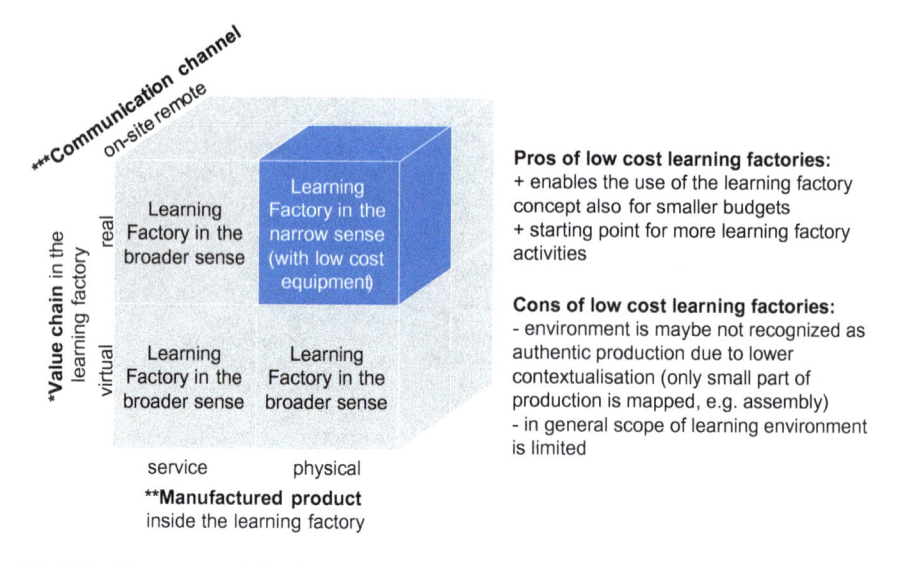

Cons of low cost learning factories:
- environment is maybe not recognized as authentic production due to lower contextualisation (only small part of production is mapped, e.g. assembly)
- in general scope of learning environment is limited

Fig. 9.8 Advantages and disadvantages of low-cost learning factories compared to the learning factory core concept

9.3.4 Low-Cost Learning Factories

Low-cost learning factories enable the use of the learning factory concept with the use of less financial resources. First and foremost, learning factories at low costs are currently made possible primarily by focusing on production processes that can be mapped more cost-effectively. Within these learning factory concepts, assembly and logistics processes in particular are used.[29] The challenge in connection with low-cost learning factories lies particularly in the sufficient contextualization of the learning environment; nonetheless and despite the limited scope of representation, the learning environment should be perceived by the learners as an authentic factory environment.

Learning factories at a lower cost can also be achieved with the integration of so-called learnstruments[30] into the learning factory concept.[31] Exemplarily, the use of learnstruments in a learning factory concept in industry is described in Best Practice Example 22 by Jan Philipp Menn and Dr. Carsten Ulbrich. Figure 9.8 summarizes advantages and disadvantages of low-cost learning factories compared to the learning factory core concept.

---

[29] Examples for those types of learning factories are described in Veza, Gjeldum, and Mladineo (2015), Steffen, May, and Deuse (2012) and the Best Practice Example 19: "Lernfabrik für vernetzte Produktion" at Fraunhofer IGCV, Augsburg, Germany.

[30] The term learnstrument is introduced by Gausemeier, Seidel, Riedelsheimer, and Seliger (2015).

[31] See Muschard and Seliger (2015).

9.3 Learning Factory Concept Variations of Learning Factories … 301

Another alternative for a very small available budget are simulation games using LEGO building blocks or similar components. On the one hand, the activity of learners as an important component of the learning factory concept can be achieved with those games, while on the other hand, a realistic representation of production processes is not taken into account. Since the use of an authentic learning environment is an important part of the learning factory concept, such highly abstract simulation games are not considered a learning factory.[32] Furthermore, simulation games are often integrated into learning factory concepts, see also Sect. 7.1.4 on Game-based learning in learning factories and gamification.

9.3.5 Digitally and Virtually Supported Learning Factories

Regarding the digital and virtual support and extension of physical learning factories, three different types of support can be identified in literature and existing learning factory approaches:

- The use of e-learning in learning factories[33]
- The use of multimedia and ICT support in learning factories[34]
- The use of virtually extended physical learning factories.[35]

e-Learning in learning factories

Since the capacities in learning factories are usually very scarce, the theoretical input can be shifted into the time before or after the actual learning factory training; consequently, no learning factory capacity is occupied to carry out the systematization phases. A good way of implementing this is to integrate e-learning into the learning factory concepts. At home, learners can use their personal computer at a convenient time to prepare via e-learning for the topics dealt with in the learning factory training. A disadvantage, however, is that theory and practice can no longer be alternated short cycles, but rather a longer practical phase follows much theoretical input.[36]

Multimedia and ICT support in learning factories

Hands-on learning in learning factories can be supported using multimedia applications and information and communication technologies (ICT). For example, multimedia support can be used to improve systematization and reflection phases in learning factories where theories related to the topics are to be made accessible to

---

[32] Examples of the use of these simulation games can be found in the literature, see, e.g., Stier (2003) and ElMaraghy and ElMaraghy (2014).

[33] See, e.g., Lanza, Minges, Stoll, Moser, and Haefner (2016), Lanza et al. (2015).

[34] See, e.g., Pittschellis (2015), Tvenge, Martinsen, and Kolla (2016).

[35] See, e.g., Thiede, Posselt, Kauffeld, and Herrmann (2017).

[36] A use case with the integration of e-learning into the learning factory concept is presented by Lanza et al. (2016).

Fig. 9.9 Tec2Screen® enables individual learning paths related to learning speed and preferred media, picture taken from Didactic (2018)

the learners.[37] In this area, tools are needed that take into account an individual learning speed and the learning tendencies in relation to different media. One such tool is Festo Didactic's Tec2Screen® learning device (Fig. 9.9), which is based on an iPad to enable individual learning paths for learners. In addition to simply providing information, the Tec2Screen system can be connected to sensors and other hardware. The device can thus also be used in the context of testing or exploratory learning phases and can be combined with learning factory approaches. The device furthermore enables Augmented Reality applications in learning factory concepts.[38]

Virtually extended physical learning factories

Learning factories with physical learning environments allow learners to act in factory environments close to reality. However, the construction of these physical learning factories is cost-intensive.[39] One way of mapping complete factory structures is to extend the physical learning environment with a supplementary virtual representation, provided that economic or technical limits of the physical representation are reached.[40] In addition, the virtual supplements can also be used in cases

---

[37] Two examples for extension of on-site learning factory trainings with multimedia support are given by Tvenge et al. (2016) and Pittschellis (2015).
[38] For further information regarding the multimedia support in learning factories using Tec2Screen® see also Pittschellis (2015).
[39] See the described space- and cost-related mapping limitations in Sect. 9.2.
[40] See also Thiede et al. (2017) and Tisch and Metternich (2017).

9.3 Learning Factory Concept Variations of Learning Factories … 303

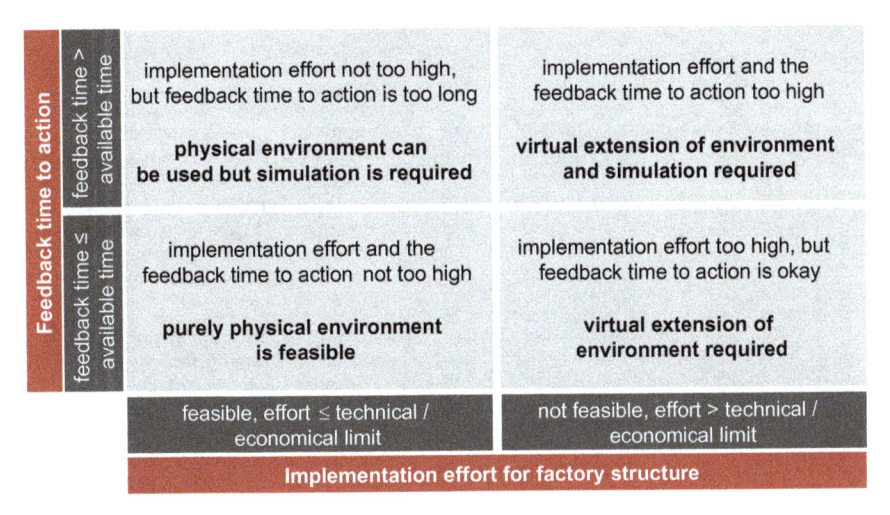

Fig. 9.10 Use of virtual extensions and simulation depending on the implementation effort for a physical environment and the feedback time to actions of the learners based on Thiede et al. (2017)

where feedback from the learning environment would take too long for the learners' actions.[41] Physical but virtually extended learning phases can be used when feedback cycles would take too long compared to the available time in the learning module.[42] Figure 9.10 presents a framework that classifies different actions regarding the feedback duration and the implementation effort for the physical establishment of the learning environment[43] in order to decide

- which actions are planned in the physical learning factory in real time,
- which actions should be addressed in the virtual extension of the learning factory,
- and which actions require simulation in order to speed up the feedback time to actions.

Following those ideas, a purely physical environment can be used in real time when the implementation effort is smaller than the technical/economical limit and the feedback time to action is shorter than the available time in the respective learning module. A virtual extension of the environment is required when the implementation efforts are too high. Additionally, simulation support is required when real-time feedback to actions of the participants takes too long. The resulting portfolio is shown in Fig. 9.10.

Advantages and disadvantages of digitally and virtually supported learning factories using e-learning, ICT, multimedia, and virtual environments are summarized in Fig. 9.11.

---

[41] See the described time-related mapping ability limitation in Sect. 9.2.
[42] See also Thiede et al. (2017) and Tisch and Metternich (2017).
[43] The framework is proposed by Thiede et al. (2017).

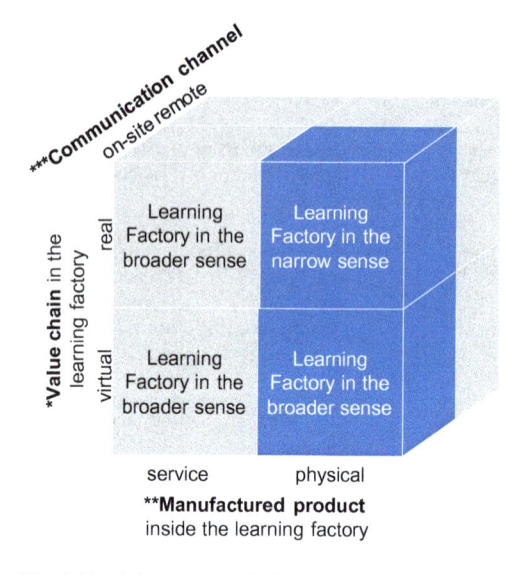

Fig. 9.11 Advantages and disadvantages of digitally and virtually supported learning factories compared to the learning factory core concept

9.3.6 Producing Learning Factories

A small but considerable variation of the learning factory core concept is given with the use of the learning factory-integrated production environment for real production, i.e. the produced goods are manufactured to order or offered on the market. This change in concept leads to some implications:

- The use of the factory environment for real production leads to more realistic framework conditions and finally to more authentic processes inside the learning factory concept.
- The quality of the manufactured products has to be ensured constantly. This is leaving less room for the collection of experiences through free experimentation by the learners. One advantage of the learning factory core concept is the implementation of ideas in an economically risk-free environment.
- Learning processes compete with production processes. In case of doubt, therefore, it must be decided who takes precedence. Supposedly the production has the priority.

An extreme case and a very well-known example of such a producing learning factory is the DFA Demonstration Factory Aachen that contains a small-scale production with a high vertical range of manufacture on 1600 m^2. The electric cars and go-karts that are produced in the DFA are sold on the market with the e.GO Mobile AG as lead customer. In this learning factory concept, the production processes are not only similar to industrial production but also have the same quality

9.3 Learning Factory Concept Variations of Learning Factories ...

Pros of the learning factory core concept:
+ requirements in terms of quality and complexity just like in real production
+ very high motivation and immersion to learn in real production environment
+ income with sold products

Cons of the learning factory core concept:
- learning competes with producing, which has supposedly first priority
- no free experimentation by the learners possible
- factory environment **can't** be modified ad-hoc and in flexible manner

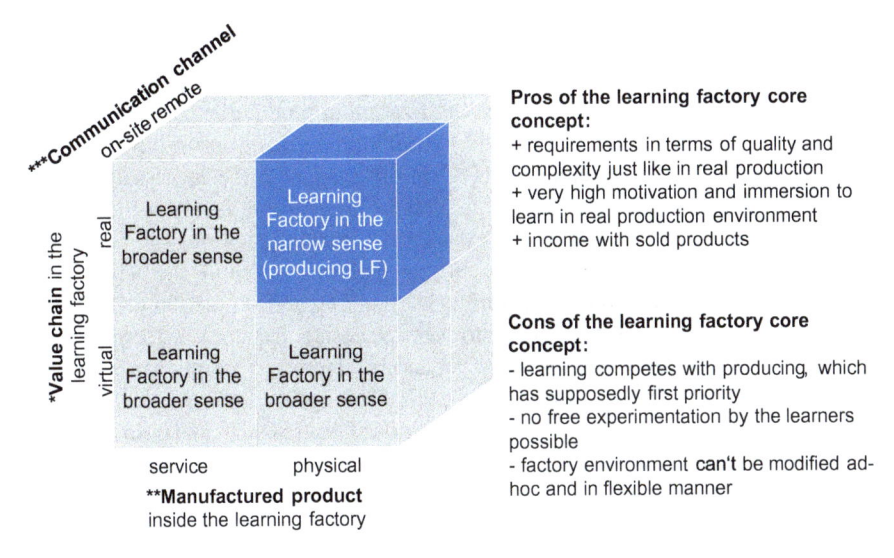

Fig. 9.12 Advantages and disadvantages of producing learning factories compared to the learning factory core concept

and complexity requirements.[44] The Demonstration Factory DFA is presented by Prof. Günther Schuh, Jan-Philipp Prote, Bastian Fränken, and Julian Ays in the Best Practice Example 2. Advantages and disadvantages of producing learning factories are summarized in Fig. 9.12.

9.4 Learning Factory Concept Variations of Learning Factories in the Broader Sense—Advantages and Disadvantages

Additionally to the concept variations of learning factories in the narrow sense,[45] this section provides an overview over the concept variations of learning factories in the broader sense, namely

- digital, virtual, and hybrid learning factories[46] as well as
- remotely accessible learning factories and teaching factories.[47]

---

[44] Schuh, Gartzen, Rodenhauser, and Marks (2015).
[45] See Sect. 9.3.
[46] See Sect. 9.4.1.
[47] See Sect. 9.4.2.

9.4.1 Digital, Virtual, and Hybrid Learning Factories

In the context of the trend toward more digitization in production, greater attention is also being paid to digital, virtual, and hybrid learning factories in the context of production-related teaching and training.[48] In this way, both the addressable topics in learning factories and the application areas of the learning factories themselves can be continuously expanded. Digital and virtual learning factories can be used in similar subject areas as conventional learning factories, whereby the activities of learners can be extended to a broad variety of planning and simulation tasks,[49] like factory planning, layout planning, concurrent engineering, front-loading, ergonomics evaluation, and virtual commissioning.[50] Learning factory concepts can include a digital or virtual learning environment instead[51] or alongside[52] of a physical learning environment to create added value for production-related teaching and training. First, exemplarily different software tools for the facilitation of digital and virtual learning factories are described.

9.4.1.1 Software Tools for Digital and Virtual Learning Factories

For the creation of digital and virtual learning factories, various existing software tools can be identified on the market. Those tools are crucial enablers for digital and virtual learning factories, although the didactic concept that is also an important part of each learning factory concept is no integral part of these software tools themselves.

For example, VisTable[53] and TaraVRbuilder[54] are software tools that allow the visualization, analyzation, and optimization of virtual production environments in 3D without advanced programming or CAD expertise (Fig. 9.13). Both software tools contain a large object library with buildings, machinery, and other manufacturing equipment. On the one hand, VisTable can be used for static factory planning, analysis of material flows, or assembly line optimization. On the other hand, TaraVRbuilder allows the simulation of manufacturing processes that enable the creation of a dynamic virtual production and logistics environment. Furthermore, TaraVRbuilder can also be used for the planning of Industrie 4.0 use cases and applications.[55]

---

[48] See, for example, Weidig, Menck, Winkes, and Aurich (2014), Kesavadas (2013), Haghighi, Shariatzadeh, Sivard, Lundholm, and Eriksson (2014), Hammer (2014), Celar, Turic, Dragicevic, and Veza (2016).
[49] See Weidig et al. (2014).
[50] See Abele et al. (2017a).
[51] The learning factory concepts that use digital and virtual representations instead of a physical environment are named digital and virtual learning factories, see Sect. 9.4.1.2.
[52] Learning factory concepts that use digital and virtual representations additional to physical environments are called hybrid learning factories, see Sect. 9.4.1.3.
[53] See visTABLE (2017a).
[54] See tarakos (2017).
[55] Abele et al. (2017a).

9.4 Learning Factory Concept Variations of Learning Factories … 307

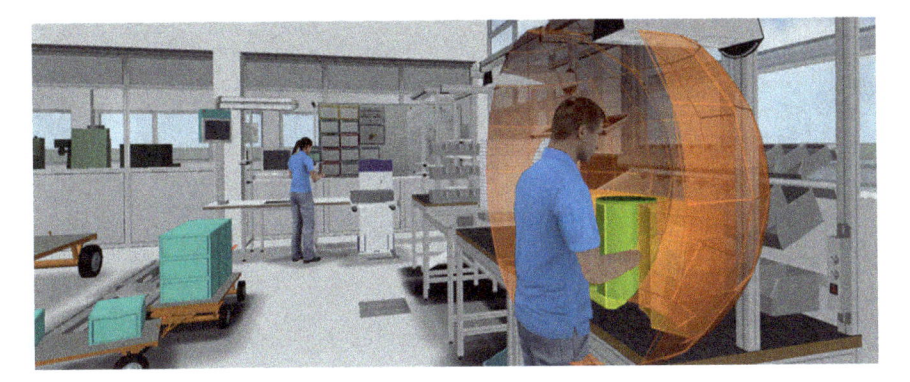

Fig. 9.13 Exemplary virtual factory; visTable® touch software (visTABLE, 2017b)

For the integration of the planning of products, processes, resources, and layouts, formerly separated tools, like Delmia, Catia, Enovia, and Simula, are combined to the software pack 3D experience by Dassault Systèmes.[56] In addition to virtual product development, the software package also supports the creation of a virtual factory environment in which different factory scenarios can be simulated directly. Advanced CAD knowledge is required to use the software package. For the integration of the software into learning factory approaches, an in-depth familiarization with the software components must be provided accordingly.

9.4.1.2 Digital and Virtual Learning Factories

Digital learning factories are mapping processes, resources, products of the (learning) factory environment in digital models making use of information technology and various IT-tools. Beyond that, virtual learning factories also provide a visual representation of digital models through the use of appropriate infrastructure and visual software tools. Virtual learning factories use virtual or augmented reality tools for visualization of digital operations simulations at factory level, this way, virtual process and layout planning, simulation of tasks and the evaluation of alternative factory designs before start of production is enabled.[57] Potential conflicts in the implementation of the factory planning solutions can be simulated and resolved beforehand in order to enable the direct implementation of verified and optimized factory planning solutions.[58]

In the manufacturing domain, virtual learning environments are deployed in order to augment the quality of educational activities.[59] In this regard, virtual learning

---

[56]Dassault Systèms (2017).
[57]See Chryssolouris et al. (2008), Abele et al. (2017a).
[58]See Hummel, Hyra, Ranz, and Schuhmacher (2015).
[59]See Manesh and Schaefer (2010b).

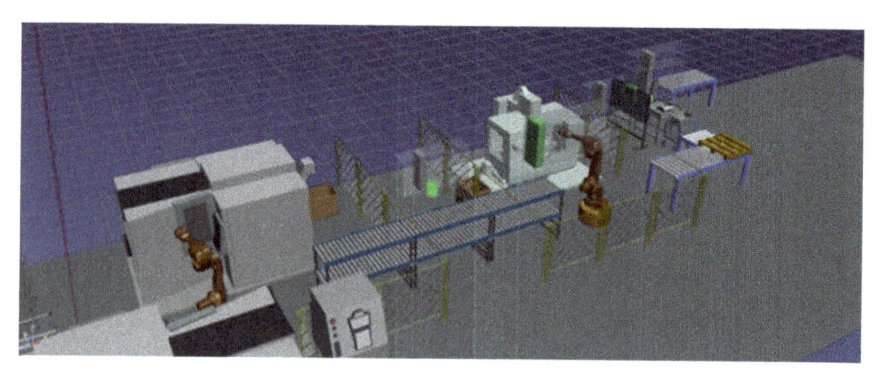

Fig. 9.14 Virtual learning factory XPRES at KTH

environments are seen as a crucial tool for the implementation of high-quality teaching.[60] Consequently, in the manufacturing education and training field, numerous approaches using virtual environments can be noticed.[61] Beyond those purely virtual environments, the use of virtual factory environments in combination with physical environments is well established[62]; those concepts are also referred to as hybrid learning factories, see Sect. 9.4.1.3. In the following, some virtual learning factory examples are described.

The KTH XPRES Lab is a digital and virtual learning factory that is used in the field of manufacturing systems design for the support of cross-disciplinary organizational learning and decision making. The XPRES Lab contains an innovative concept vehicle as well as the digital and virtual factory of the respective manufacturing system. The XPRES Lab intends to exemplify the visualization of digital product, factory, and manufacturing process models in order to identify dependencies and support holistic decision making. Besides these visualizations, learning processes are supported by the active learners' experience through simulations of what-if scenarios evaluating the effects of various changes in the whole manufacturing system. In the virtual learning factory, machining simulations as well as flow simulations are deployed.[63] An impression of the virtual representations of learning factory of the XPRES Lab is shown in Fig. 9.14.

For training purposes regarding various topics, participants from all over the world can use an authentic virtual 3D learning factory by McKinsey & Co.[64] This virtual

---

[60] See Manesh and Schaefer (2010a).

[61] See, for example, Dessouky (1998), Dessouky and Verma (2001), Cassandras et al. (2004), Ong and Mannan (2004), Chi and Spedding (2006), Watanuki and Kojima (2007), Manesh and Schaefer (2010a), Gadre, Cudney, and Corns (2011), Goeser, Johnson, Hamza-Lup, and Schaefer (2011), Abdul-Hadi, Abulrub, Attridge, and Williams (2011).

[62] See, for example, Riffelmacher, Kluge, Kreuzhage, Hummel, and Westkämper (2007), Riffelmacher (2013), Plorin and Müller (2014), Plorin, Jentsch, Hopf, and Müller (2015).

[63] See Sivard and Lundholm (2013).

[64] See Hammer (2014).

learning factory consists of a computer with visualization software for virtual factories, a beamer, and VR glasses for the participants. The virtual learning factory is thus infinitely mobile, but the participants are still in one place and not isolated in front of their own PC. Together, learners can examine, discuss, and optimize the virtual execution of production processes. Within these virtual trainings, experiential learning is realized with the go-see-do-approach. Oriented toward the learning processes in classical learning factories, learners experience the transformation of the virtual factory environment from a non-optimal to a good practice factory through the self-directed identification of potentials and the implementation of improvement ideas. The trainings offered in the virtual learning factory include the application of classic lean methods as well as training on lean management or mindset and behavior, which enables a holistic, company-specific implementation of improvement approaches.

At Reutlingen University, the ESB Logistics Learning Factory is a learning factory used among others for education, training, research, and innovation transfer for the design and optimization of cyber-physical production and logistics systems.[65] The learning factory concept includes a virtual learning environment that is created with the "3DEXPERIENCE" software from Dassault Systèmes' congruent to the physical factory environment (Fig. 9.15).[66] The virtual environment consistently integrates process, product, and resource design including respective simulations at one platform. The virtual models are used complementary to the physical learning factory environment in order to allow preimplementation testing and validation of system changes and system design. The ESB Logistics Learning Factory is described by Prof. Vera Hummel, Fabian Ranz, and Jan Schumacher in the Best Practice Example 5.

The Festo Didactic offered Modular Production System (MPS) learning factory can be used as tailored production system in the academic, industrial, and vocational school sector, see also Best Practice Example 7. These learning factories are supported by CIROS VR,[67] a virtual platform that opens up the opportunity for virtual planning, simulation, and optimization of factory components and interactions.

In order to allow a time- and place-independent learning factory concept, a digital learning game is developed at the IFA, Leibniz Universität Hannover additionally to the physical IFA learning factory.[68] Within the virtual learning environment, the digital learning game maps all company departments involved in order processing, such as storage, assembly, transport, or production controlling of the helicopter also produced in the physical IFA Learning Factory. Additionally, external suppliers and customers are also part of the digital learning game. Each learner takes on one of these roles in the learning game, with a total of 12–15 participants.[69]

On the one hand, the major advantage of these digital and virtual learning factories is that they address various limitations of today's physical learning factories discussed

---

[65] See Hummel et al. (2015).
[66] See Brenner and Hummel (2016).
[67] See Ciros (2016).
[68] The physical IFA learning factory is presented in the Best Practice Example 10.
[69] See Görke, Bellmann, Busch, and Nyhuis (2017).

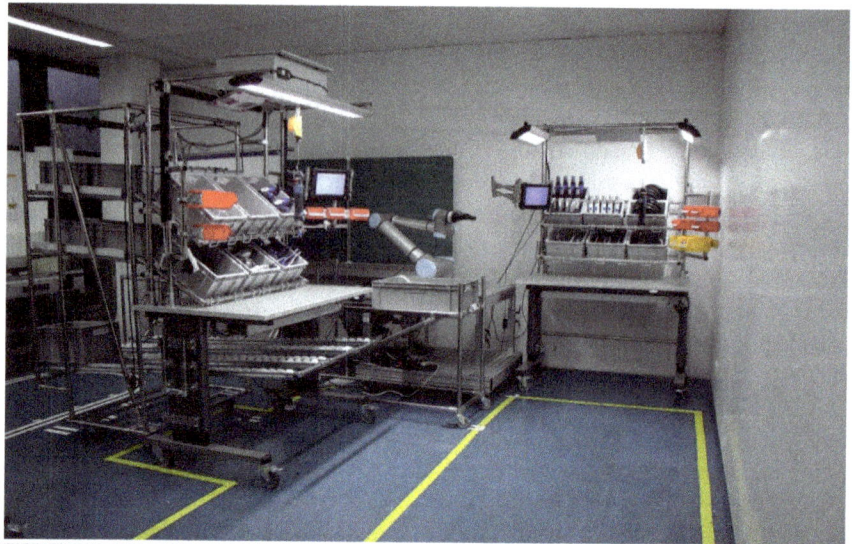

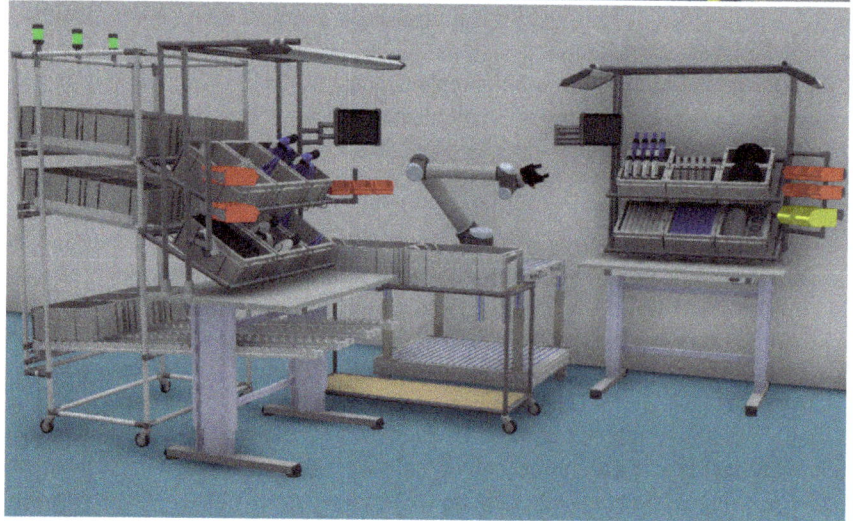

Fig. 9.15 Real assembly environment and its virtual model of the ESB logistics learning factory (LLF) (Abele et al., 2017a)

in Sect. 9.2. For example, virtual learning factories come with comparatively low costs and effort for setup and operations,[70] the low space requirement,[71] and the fact that the learning factory concepts can be used anywhere[72] and at any time.

---

[70] Limited resources issues.

[71] Space- and cost-related mapping ability issues.

[72] Immobility of learning factory approaches.

9.4 Learning Factory Concept Variations of Learning Factories … 311

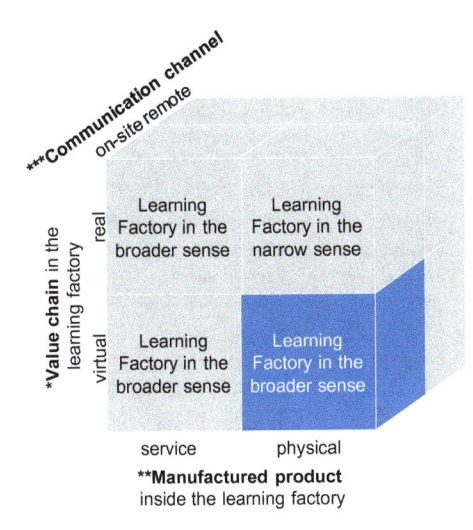

Pros of the virtual learning factory concept:
+ lower resource requirements for set-up and operation
+ mapping of large factory structures is enabled
+ use for various production types possible
+ simulation integration to speed up feedback
+ implementation (preparation) of solution ideas
+ good scalability of learning approach
+ mobility and location-indendent approaches
+ time-independent approaches are enabled

Cons of the virtual learning factory concept:
- learning is less hands-on
- only indirect own experiences and actions
- lower contextualisation and immersion
- activation of learner can be a challenge
- collectivization of learning processes can be a challenge
- complicated integration of thinking and doing
- self-regulation and self-direction limited to the predefined possibilities of virtual environment

Fig. 9.16 Advantages and disadvantages of digital and virtual learning factories compared to the learning factory core concept

Analysis with digital systems offers high speed and the opportunity to simulate long-term periods.[73] All processes along the value or supply chain can be executed using CAD models. Virtual learning factories easily allow the integration of simulation and therefore enable the mapping of topics with long feedback cycles.[74] Digital and virtual learning factories provide the opportunities to simulate an increased number of turbulences and scenarios.[75] Also, implementation ideas of participants can be implemented easier in a virtual compared to a physical factory environment.[76] Furthermore, virtual learning factories can be a suitable approach to integrate learning factory concepts into larger lectures.[77]

On the other hand, when using virtual learning factories, care must be taken to ensure that the success factors of classic learning factories are not neglected, i.e. the hands-on character of learning situations, the integration of learners' own actions, a high degree of contextualization of the factory environment or the possibility of learning processes within groups (collectivization). Furthermore, the creation of a digital or virtual learning factory requires IT infrastructure investment and massive efforts for the implementation, i.e. programming as well as integration efforts. An additional overview of the advantages and disadvantages of virtual learning factories compared to the learning factory core concept can be found in Fig. 9.16.

---

[73] Wiendahl, Harms, and Fiebig (2003).
[74] Time-related mapping ability issues.
[75] See Riffelmacher (2013).
[76] Solution-related mapping ability issues.
[77] Scalability issues of learning factories.

Fig. 9.17 Infrastructure and interfaces of physical, digital, and virtual learning factories (Abele et al., 2017a)

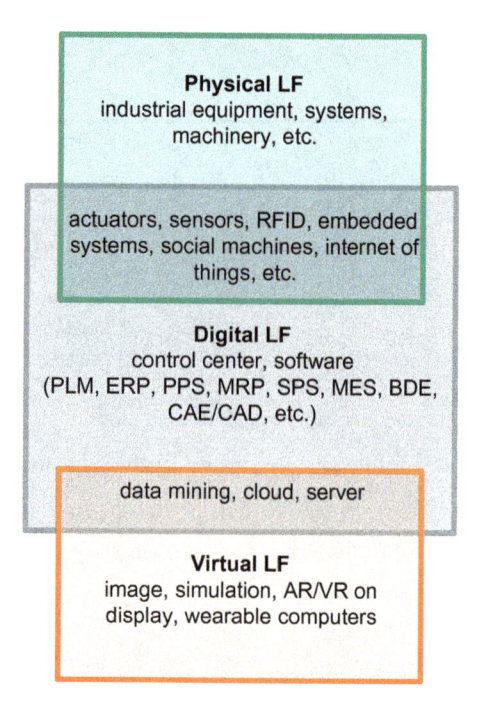

Physical LF
industrial equipment, systems, machinery, etc.

actuators, sensors, RFID, embedded systems, social machines, internet of things, etc.

Digital LF
control center, software (PLM, ERP, PPS, MRP, SPS, MES, BDE, CAE/CAD, etc.)

data mining, cloud, server

Virtual LF
image, simulation, AR/VR on display, wearable computers

9.4.1.3 Hybrid Learning Factories

Digital and virtual learning factories can be seen as a useful complement to learning factories with physical learning environments, as they expand the thematic and application-related scope of the learning systems as explained in the previous section. Digital factories widen the scope to a holistic view on the factory without having strict limitations for the extended field of experimentation.[78] Physical and digital learning environments support the improvement and the changeability of respective factory environment in a different manner.[79] In those cases where physical, digital, and virtual learning factories can be interlinked and the respective strengths of the different environments can be exploited, the so-called hybrid learning factories are created. Hybrid learning factories face the challenge to robustly and reliably bring together various data sources to seamlessly overcome media breaks between the real and the virtual world to create one merged world in this way. Figure 9.17 shows the infrastructure and the interfaces of the physical, the digital, and the virtual learning factories. Figure 9.18 shows the interrelation of the digital, virtual, and physical learning factories' concept. Figure 9.19 summarizes advantages and disadvantages of the holistic learning factory compared to the learning factory core concept.

---

[78] Lu, Shpitalni, and Gadh (1999).

[79] See ElMaraghy, AlGeddawy, Azab, and ElMaraghy (2011), Müller and Horbach (2011).

9.4 Learning Factory Concept Variations of Learning Factories ... 313

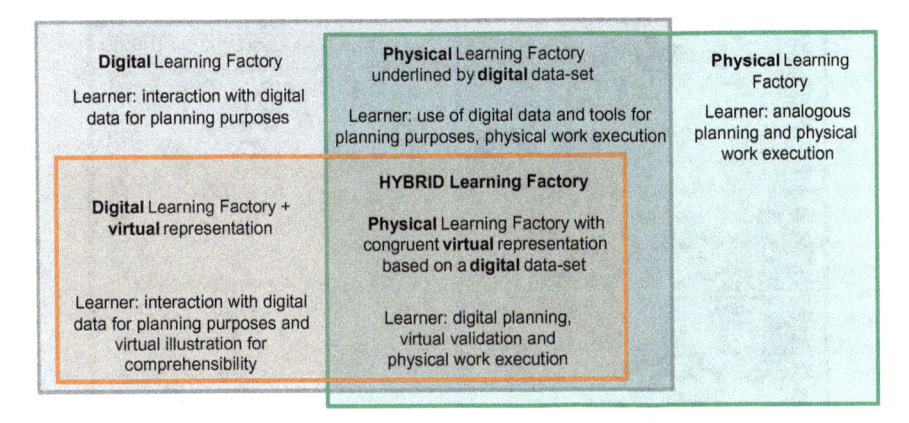

Fig. 9.18 Interrelation of digital, physical, and hybrid learning factories (Abele et al., 2017a)

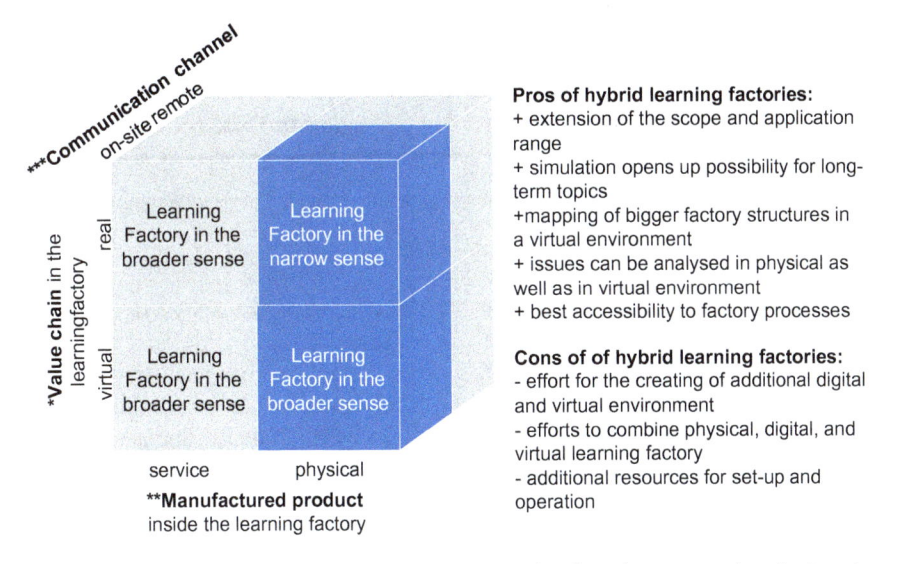

Fig. 9.19 Advantages and disadvantages of hybrid learning factories compared to the learning factory core concept

9.4.2 Teaching Factories and Remotely Accessible Learning Factories

The Teaching Factory is described as a subconcept of the learning factory in this book. The concept intends to bridge the gap between industrial research and the transfer to industrial processes and products.[80] Therefore, the teaching factory is a broad concept that contains several sublearning forms.

---

[80] See Mavrikios, Papakostas, Mourtzis, and Chryssolouris (2013).

factory-to-classroom | **lab-to-factory**

Fig. 9.20 Teaching factory sessions for factory-to-classroom and lab-to-factory knowledge communication, shown in Abele et al. (2017a)

The starting idea for the Teaching Factory concept is the knowledge triangle notion that aims to integrate the three corners research, innovation, and education seamlessly.[81] During the KNOW FACT project, the Teaching Factory as a non-geographically anchored room for learning is implemented. This Teaching Factory concept is enabled by the latest digital technologies and high-quality industrial didactic equipment.[82] Teams of engineers, researchers, and students are able to communicate and interact with the use of equipment at industrial and academic sites in order to work on and find solutions for real-life problems. The Teaching Factory enables a bidirectional knowledge communication channel bringing the real factories to the classroom as well as the academic laboratories to the factories, see Fig. 9.20.[83] The teaching factory concept is also described in the Best Practice Example 30.

At the Cal Poly State University a research and teaching factory is implemented containing a functioning real factory, including a center for production planning and control. The teaching factory uses state-of-the-art networks for communication. In the factory, car parts for different OEMs such as GM and Ford as well as for their suppliers, are produced using high precision machining equipment. This modern factory environment is complemented by the latest learning technologies such as

---

[81] Chryssolouris, Mavrikios, Papakostas, and Mourtzis (2006), Mavrikios et al. (2013).

[82] Chryssolouris, Mavrikios, and Rentzos (2016).

[83] Rentzos, Doukas, Mavrikios, Mourtzis, and Chryssolouris (2014), Rentzos, Mavrikios, and Chryssolouris (2015).

9.4 Learning Factory Concept Variations of Learning Factories …

remote learning technology or online courses. Using this equipment, students gain experiences in conducted industrial projects that provide a highly contextualized setting in the teaching factory. As in most learning factory approaches, the focus here is on effective competency development and application.[84]

Another teaching factory is operated as a complete service engineering and manufacturing center located at an industrial park at the Advanced Manufacturing Institute (AMI), Kansas State University (KSU).[85] In this teaching factory, students are complementing their traditional academic education with hands-on experience designing and developing innovative solutions for industrial partners as a service. This concept aims at the integration of research, education, and innovation in one single industry-academia-partnered activity.[86]

Further teaching factory concepts are presented in the Best Practice Example section:

- Best Practice Example 20: LMS Factory at the Laboratory for Manufacturing Systems and Automation (LMS), University Patras, Greece, by Prof. Dimitris Mourtzis, Dr. Panagiotis Stavropoulos, George Dimitrakopoulos, Harry Bikas, and Stylianos Zygomalas and
- Best Practice Example 30: Teaching Factory: An emerging paradigm for manufacturing education. LMS, University of Patras, Greece, by Dr. Dimitris Mavrikios, Dr. Konstantinos Georgoulias, and Prof. George Chryssolouris.

Since the teaching factory concept in general offers a very broad variety of different implementations,[87] no specific advantages and disadvantages of the general concept compared to the learning factory core concept can be named. Figure 9.21 summarizes pros and cons of remotely accessible teaching and learning factories compared to the learning factory core concept.

Figure 9.22 summarizes the effects on the current limitations of learning factories based on

- concept variations of the learning factory in the narrow sense,[88]
- concept variations of the learning factory in the broader sense,[89] and
- additional methods and approaches along the learning factory lifecycle.[90]

---

[84] See Chryssolouris, Mavrikios, and Mourtzis (2013).
[85] See Azadivar and Kramer (2007).
[86] See Chryssolouris et al. (2006), Tittagala, Bramhall, and Pettigrew (2008).
[87] For example, remote/on-site learning, training/education/research.
[88] See Sect. 9.3.
[89] See Sect. 9.4.
[90] See Chap. 6.

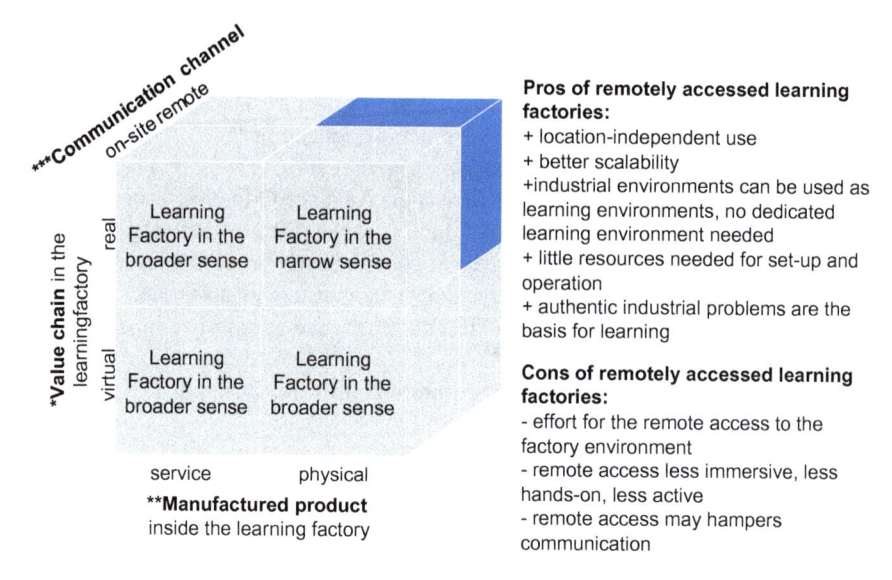

Fig. 9.21 Advantages and disadvantages of remotely accessible learning and teaching factories compared to the learning factory core concept

9.5 Wrap-up of This Chapter

This chapter provides an overview of the different concept variations of learning factories in the narrow and in the broader sense. For all concept variations, advantages and disadvantages in relation to the learning factory core concept are given. For the different learning factory concept variations, various examples from existing learning factories are given.

9.5 Wrap-up of This Chapter

Concepts	Limit 1: Resources	Limit 2: Mapping ability a) content-related	b) space-related	c) time-related	d) solution-related	Limit 3: Scalability	Limit 4: Mobility	Limit 5: Effectiveness
Variations of the LF in the narrow sense								
Model scale learning factories	◐	○	◐	○	○	○	○	○
Physical mobile learning factories	◔	○	●	○	○	○	●	○
Low-cost learning factories	●	○	◐	○	○	○	●	○
Digitally & virtually supported learning factories	◔	◔	◐	◐	◐	◐	◔	◔
Producing learning factories	◔	◔	○	○	○	○	○	○
Variations of the LF in the broad sense								
Digital, virtual, and hybrid learning factories	●	○	●	●	●	◔	●	●
Remotely accessible learning factories	○	○	○	○	○	○	○	◐
Additional methods & approaches								
Systematic learning factory design	◔	○	○	○	○	○	○	●
Turnkey learning factories	○	○	○	○	○	○	○	○
Methods for learning success measurement	○	○	○	○	○	○	○	●
Quality system for learning factories	◔	○	○	○	○	○	○	●

Legend: ○ no effect | ◐ slight effect | ◑ effect | ● major effect | ● focus

Fig. 9.22 Effects of concept variations and methods and approaches along the learning factory life cycle on current limitations

References

Abdul-Hadi, G., Abulrub, A. N., Attridge, A., & Williams, M. A. (2011). Virtual reality in engineering education: The future of creative learning. In *IEEE Global Engineering Education Conference (EDUCON)—Learning Environments and Ecosystems in Engineering Education*, 751–757.

Abele, E., Chryssolouris, G., Sihn, W., Metternich, J., ElMaraghy, H. A., Seliger, G., et al. (2017a). Learning factories for future oriented research and education in manufacturing. *CIRP Annals—Manufacturing Technology, 66*(2), 803–826.

Abele, E., Chryssolouris, G., Sihn, W., Metternich, J., ElMaraghy, H. A., Seliger, G., Sivard, G., ElMaraghy, W., Hummel, V., Tisch, M., & Seifermann, S. (2017b). *Learning factories for future oriented research and education in manufacturing*. Presentation CIRP STC-O keynote-paper, GA 2017. CIRP general asseby 2017, Lugano, Switzerland.

Abele, E., Metternich, J., Tisch, M., Chryssolouris, G., Sihn, W., ElMaraghy, H. A., Hummel, V., & Ranz, F. (2015). Learning factories for research, education, and training. In *5th CIRP-Sponsored Conference on Learning Factories. Procedia CIRP, 32*, 1–6. https://doi.org/10.1016/j.procir.2015.02.187.

Azadivar, F., & Kramer, B. (2007). Rewards and challenges of utilizing university research/economic development centers for enhancing engineering education. *American Society for Engineering Education*, 1212471–12124716.

Brenner, B., & Hummel, V. (2016). A seamless convergence of the digital and physical factory aiming in personalized product emergence process (PPEP) for smart products within ESB logistics learning factory at reutlingen university. *Procedia CIRP, 54*, 227–232. https://doi.org/10.1016/j.procir.2016.06.108.

Cassandras, C., Deng, M., Hu, J.-Q., Panayiotou, C., Vakili, P., & Zhao, C. (2004). Development of a discrete event dynamic systems curriculum using a web-based "Real-Time" simulated factory. In *Proceeding of the 2004 American Control Conference*, Boston, Massachusetts, June 30–July 2, 2004 (p. 1307).

Celar, S., Turic, M., Dragicevic, S., & Veza, I. (2016). Digital learning factory at FESB—University of split. *XXII naučna i biznis konferencija YU INFO, 1*–6.

Chi, X., & Spedding, T. A. (2006). A web-based intelligent virtual learning environment for industrial continuous improvement. In *IEEE International Conference on Industrial Informatics*, August 2006 (pp. 1102–1107). Singapore.

Chryssolouris, G., Mavrikios, D., & Mourtzis, D. (2013). Manufacturing systems—skills and competencies for the future. In *46th CIRP Conference on Manufacturing Systems. Procedia CIRP, 7*, 17–24.

Chryssolouris, G., Mavrikios, D., Papakostas, N., Mourtzis, D., Michalos, M., & Georgoulias, K. (2008). Digital manufacturing: History, perspectives, and outlook. *Journal of Engineering Manufacture, 222*(5), 451–462.

Chryssolouris, G., Mavrikios, D., Papakostas, N., & Mourtzis D. (2006). Education in manufacturing technology and science: A view on future challenges & goals. In *Proceedings of the International Conference on Manufacturing Science and Technology (ICOMAST 2006)* (pp. 1–4). Melaka, Malaysia.

Chryssolouris, G., Mavrikios, D., & Rentzos, L. (2016). The teaching factory: A manufacturing education paradigm. In *49th CIRP Conference on Manufacturing Systems. Procedia CIRP, 57*, 44–48.

Ciros. (2016). *Ciros: Virtual engineering, virtual learning, virtual reality, service*. Retrieved from http://www.ciros-engineering.com/en/home/.

Dassault Systèms. (2017). *3D experience platform*. Retrieved from https://www.3ds.com.

Dessouky, M. M. (1998). A virtual factory teaching system in support of manufacturing education. *Journal of Engineering Education, 87*(4), 459–467.

Dessouky, M. M., & Verma, S. (2001). A methodology for developing a web-based factory simulator for manufacturing education. *IIE Transactions, 33*(3), 167–180.

References

ElMaraghy, H. A., AlGeddawy, T., Azab, A., & ElMaraghy, W. (2011). Change in manufacturing—Research and industrial challenges. In H. A. ElMaraghy (Ed.), *Enabling Manufacturing Competitiveness and Economic Sustainability: Proceedings of the 4th International Conference on Changeable, Agile, Reconfigurable and Virtual production (CARV2011)* (pp. 2–9). Berlin, London: Springer.

ElMaraghy, H. A., & ElMaraghy, W. (2014). Learning factories for manufacturing systems. In *4th Conference on Learning Factories*, Stockholm, Sweden.

ElMaraghy, H. A., & ElMaraghy, W. (2015). Learning integrated product and manufacturing systems. In *5th CIRP-Sponsored Conference on Learning Factories. Procedia CIRP, 32*, 19–24.

Festo Didactic. (2018). Kurse und Simulationen. Retrieved from http://www.festo-didactic.com/de-de/highlights/connected-learning/tec2screen/kurse-und-simulationen/?fbid=ZGUuZGUuNTQ0LjEzLjEwLjc0MjAuNDMyMw.

Gadre, A., Cudney, E., & Corns, S. (2011). Model development of a virtual learning environment to enhance lean education. *Procedia Computer Science, 6*, 100–105. https://doi.org/10.1016/j.procs.2011.08.020.

Gausemeier, P., Seidel, J., Riedelsheimer, T., & Seliger, G. (2015). Pathways for sustainable technology development—The case of bicycle mobility in Berlin. In *12th Global Conference on Sustainable Manufacturing. Procedia CIRP, 26*, 202–207. https://doi.org/10.1016/j.procir.2014.07.164.

Goeser, P. T., Johnson, W. M., Hamza-Lup, F. G., & Schaefer, D. (2011). VIEW-A virtual interactive webbased learning environment for engineering. *Advances in Engineering Education, 2*(3), 1–24.

Görke, M., Bellmann, V., Busch, J., & Nyhuis, P. (2017). Employee qualification by digital learning games. *Procedia Manufacturing, 9*, 229–237. https://doi.org/10.1016/j.promfg.2017.04.040.

Haghighi, A., Shariatzadeh, N., Sivard, G., Lundholm, T., & Eriksson, Y. (2014). Digital learning factories: Conceptualization, review and discussion. In *The 6th Swedish Production Symposium (SPS14)*. Retrieved from http://conferences.chalmers.se/index.php/SPS/SPS14/paper/viewFile/1729/401.

Hammer, M. (2014, August). Making operational transformations successful with experiential learning. In *CIRP Collaborative Working Group—Learning Factories for Future Oriented Research and Education in Manufacturing, CIRP General Assembly*, Nantes, France.

Hummel, V., Hyra, K., Ranz, F., & Schuhmacher, J. (2015). Competence development for the holistic design of collaborative work systems in the logistics learning factory. In *5th CIRP-Sponsored Conference on Learning Factories. Procedia CIRP, 32*, 76–81. https://doi.org/10.1016/j.procir.2015.02.111.

Hummel, V., & Westkämper, E. (2007). *Learning factory for advanced industrial engineering—Integrated approach of the digital learning environment and the physical model factory* (pp. 215–227). Production Engineering, Oficyna Wydawnicza Politechniki Wroctawskiej (needs translation): Krakow, Poland.

Kaluza, A., Juraschek, M., Neef, B., Pittschellis, R., Posselt, G., Thiede, S., & Herrmann, C. (2015). Designing learning environments for energy efficiency through model scale production processes. In *5th CIRP-Sponsored Conference on Learning Factories. Procedia CIRP, 32*, 41–46. https://doi.org/10.1016/j.procir.2015.02.114.

Kesavadas, T. (2013). V-learn-fact: A new approach for teaching manufacturing and design to mechanical engineering students. *ASME 2013 International Mechanical Engineering Congress and Exposition, 5*, 1–6.

Lanza, G., Minges, S., Stoll, J., Moser, E., & Haefner, B. (2016). Integrated and modular didactic and methodological concept for a learning factory. In *6th CIRP-Sponsored Conference on Leanring Factories. Procedia CIRP, 54*, 136–140. https://doi.org/10.1016/j.procir.2016.06.107.

Lanza, G., Moser, E., Stoll, J., & Haefner, B. (2015). Learning factory on global production. In *5th CIRP-Sponsored Conference on Learning Factories. Procedia CIRP, 32*, 120–125. https://doi.org/10.1016/j.procir.2015.02.081.

Lu, S. Y., Shpitalni, M., & Gadh, R. (1999). Virtual and augmented reality technologies for product realization. *CIRP Annals—Manufacturing Technology, 48*(2), 471–495.

Manesh, H. F., & Schaefer, D. (2010a). Virtual learning environments for manufacturing education and training. *Computers in Education Journal*, 77–89.

Manesh, H. F., & Schaefer, D. (2010b). A virtual factory approach for design and implementation of agile manufacturing systems. *American Society for Engineering Education, 15*(111), 1–12.

Mavrikios, D., Papakostas, N., Mourtzis, D., & Chryssolouris, G. (2013). On industrial learning and training for the factories of the future: A conceptual, cognitive and technology framework. *Journal of Intelligent Manufacturing, 24*(3), 473–485. https://doi.org/10.1007/s10845-011-0590-9.

Müller, E., & Horbach, S. (2011). Building blocks in an experimental and digital factory. In H. A. ElMaraghy (Ed.), *Enabling Manufacturing Competitiveness and Economic Sustainability: Proceedings of the 4th International Conference on Changeable, Agile, Reconfigurable and Virtual production (CARV2011)* (pp. 592–597). Berlin, London: Springer.

Muschard, B., & Seliger, G. (2015). Realization of a learning environment to promote sustainable value creation in areas with insufficient infrastructure. In *5th CIRP-Sponsored Conference on Learning Factories. Procedia CIRP, 32*, 70–75. https://doi.org/10.1016/j.procir.2015.04.095.

Ong, S. K., & Mannan, M. A. (2004). Virtual reality simulations and animations in a web-based interactive manufacturing engineering module. *Computers and Education, 43*(4), 361–382. https://doi.org/10.1016/j.compedu.2003.12.001.

Pittschellis, R. (2015). Multimedia support for learning factories. In *5th CIRP-Sponsored Conference on Learning Factories. Procedia CIRP, 32*, 36–40. https://doi.org/10.1016/j.procir.2015.06.001.

Plorin, D., Jentsch, D., Hopf, H., & Müller, E. (2015). Advanced learning factory (aLF)—Method, implementation and evaluation. In *5th CIRP-Sponsored Conference on Learning Factories. Procedia CIRP, 32*, 13–18. https://doi.org/10.1016/j.procir.2015.02.115.

Plorin, D., & Müller, E. (2014). Developing an ambient assisted living environment applying the advanced learning factory (aLF): A conceptual approach for the practical use in the research project A²LICE. *ISAGA, 2013,* 69–76.

Rentzos, L., Doukas, M., Mavrikios, D., Mourtzis, D., & Chryssolouris, G. (2014). Integrating manufacturing education with industrial practice using teaching factory paradigm: A construction equipment application. In *47th CIRP Conference on Manufacturing Systems. Procedia CIRP, 17*, 189–194.

Rentzos, L., Mavrikios, D., & Chryssolouris, G. (2015). A two-way knowledge interaction in manufacturing education: The teaching factory. In *5th CIRP-Sponsored Conference on Learning Factories. Procedia CIRP, 32*, 31–35. https://doi.org/10.1016/j.procir.2015.02.082.

Riffelmacher, P. (2013). Konzeption einer Lernfabrik für die variantenreiche Montage. *Dissertation, Stuttgart. Stuttgarter Beiträge zur Produktionsforschung: Vol. 15*. Stuttgart: Fraunhofer Verlag.

Riffelmacher, P., Kluge, S., Kreuzhage, R., Hummel, V., & Westkämper, E. (2007). Learning factory for the manufacturing industry: Digital Learning shell and a physical model factory – iTRAME for production engineering and improvement. In A. Silva (Ed.), *Proceedings of the 20th International Conference on Computer-Aided Production Engineering CAPE* (pp. 120–131).

Schuh, G., Gartzen, T., Rodenhauser, T., & Marks, A. (2015). Promoting work-based learning through Industry 4.0. In *5th CIRP-Sponsored Conference on Learning Factories. Procedia CIRP, 32*, 82–87. https://doi.org/10.1016/j.procir.2015.02.213.

Sivard, G., & Lundholm, T. (2013). XPRES—A digital learning factory for adaptive and sustainable manufacturing of future products. In G. Reinhart, P. Schnellbach, C. Hilgert, & S. L. Frank (Eds.), *3rd Conference on Learning Factories*, Munich, May 7th, 2013 (pp. 132–154). Augsburg.

Steffen, M., May, D., & Deuse, J. (2012). The industrial engineering laboratory: Problem based learning in industrial engineering education at TU Dortmund University. In *Global Engineering Education Conference (EDUCON), IEEE, —Collaborative Learning and New Pedagogic Approaches in Engineering Education*, Marrakesch, Marokko, April 17–20, 2012 (pp. 1–10).

Stier, K. W. (2003). Teaching lean manufacturing concepts through project-based learning and simulation. *Journal of Industrial Technology, 19*(4), 1–6.

Tarakos. (2017). *Tarakos: Virtual made reality*. Retrieved from http://www.tarakos.de/.

References

Thiede, B., Posselt, G., Kauffeld, S., & Herrmann, C. (2017). Enhancing learning experience in physical action-orientated learning factories using a virtually extended environment and serious gaming approaches. *Procedia Manufacturing, 9*, 238–244. https://doi.org/10.1016/j.promfg.2017.04.042.

Tisch, M. (2018). *Modellbasierte Methodik zur kompetenzorientierten Gestaltung von Lernfabriken für die schlanke Produktion. Dissertation, Darmstadt*. Aachen: Shaker.

Tisch, M., & Metternich, J. (2017). Potentials and limits of learning factories in research, innovation transfer, education, and training. In *7th CIRP-Sponsored Conference on Learning Factories. Procedia Manufacturing* (In Press).

Tittagala, R., Bramhall, M., & Pettigrew, M. (2008). *Teaching Engineering in a Simulated Industrial Learning Environment: A Case Study in Manufacturing Engineering*. Loughborough, England: Engineering Education.

Tvenge, N., Martinsen, K., & Kolla, S. S. V. K. (2016). Combining learning factories and ICT-based situated learning. In *6th CIRP-Sponsored Conference on Leanring Factories. Procedia CIRP, 54*, 101–106.

VDI. (1993). *VDI 2221/Methodik zum Entwickeln und Konstruieren technischer Systeme und Produkte*. Beuth: VDI-Richtlinien. Berlin.

Veza, I., Gjeldum, N., & Mladineo, M. (2015). Lean learning factory at FESB—University of Split. In *5th CIRP-Sponsored Conference on Learning Factories. Procedia CIRP, 32*, 132–137. https://doi.org/10.1016/j.procir.2015.02.223.

VisTABLE. (2017a). *visTABLE: innovative Fabrikplanungswerkzeuge*. Retrieved from http://www.vistable.de/.

VisTABLE. (2017b). *visTABLE® touch Software*. Retrieved from http://www.vistable.de/vistabletouch-software.

Watanuki, K., & Kojima, K. (2007). Knowledge acquisition and job training for advanced technical skills using immersive virtual environment. *Journal of Advanced Mechanical Design, Systems, and Manufacturing, 1*(1), 48–57. https://doi.org/10.1299/jamdsm.1.48.

Weidig, C., Menck, N., Winkes, P. A., & Aurich, J. C. (2014). Virtual learning factory on VR-supported factory planning. In *Collaborative Systems for Smart Networked Environments. 15th IFIP WG 5.5 Working Conferencen Virtual Enterprises*, Amsterdam, Netherlands (pp. 455–462).

Wiendahl, H.-P., ElMaraghy, H. A., Nyhuis, P., Zäh, M. F., Wiendahl, H.-H., Duffie, N., et al. (2007). Changeable manufacturing—Classification, design and operation. *CIRP Annals—Manufacturing Technology, 56*(2), 783–809. https://doi.org/10.1016/j.cirp.2007.10.003.

Wiendahl, H.-P., Harms, T., & Fiebig, C. (2003). Virtual factory design. A new tool for a cooperative planning approach. *International Journal of Computer Integrated Manufacturing, 16*(7–8), 535–540.

Chapter 10
Projects and Groups Related to Learning Factories

As becomes clear, learning factories are an important support for production-related teaching, research, and training. However, which projects or groups are currently dealing with the topic? Who offers forums for exchange and mutual support for the establishment and operation of learning factories? This chapter gives an overview of current projects and groups.

As already described in Sect. 4.1 on the historical development of learning factories, the first isolated, local learning factory approaches have developed since the 1980s. In the following years, however, only a few learning factories were built up and operated. After an accumulation of learning factories in Europe since 2007, these learning factory operators joined forces to form the first network. The Initiative on European Learning Factories[1] and the Conference on Learning Factories[2] was born. On this basis, another European-wide Network was established with the Netzwerk innovativer Lernfabriken.[3] Furthermore, with a CIRP Collaborative Working Group (CWG) on "Learning Factories for Future-Oriented Research and Education in Manufacturing",[4] the topic was elevated to a worldwide level including a scientific consideration of the topic. Finally, in August 2017, a CIRP Keynote Paper on Learning Factories as a result of the intense work in the CWG was presented and published that summarizes the basics, state-of-the-art, learning factory definitions, and variations as well as future challenges of learning factories on 25 pages.[5] In the year 2017, this worldwide working group dealing with learning factory-related topics terminated. As a result of this, the members of the Initiative on European Learning Factories decided to open up to worldwide partners in order to maintain a worldwide platform for exchange, discussion, and implementation of ideas in joint projects. In 2017, the Initiative on European Learning Factories was transformed to the Interna-

---

[1] See Sect. 10.1.
[2] See Sect. 10.2.
[3] Network of innovative Learning Factories, see Sect. 10.3.
[4] See Sect. 10.4.
[5] See Abele et al. (2017a, b).

© Springer Nature Switzerland AG 2019
E. Abele et al., *Learning Factories*, https://doi.org/10.1007/978-3-319-92261-4_10

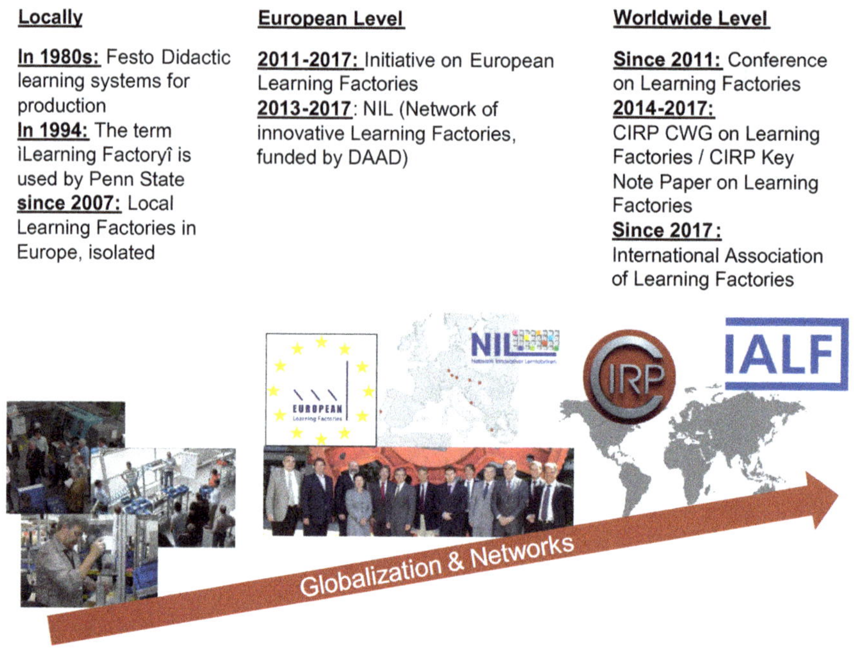
Locally
In 1980s: Festo Didactic learning systems for production
In 1994: The term ìLearning Factoryî is used by Penn State
since 2007: Local Learning Factories in Europe, isolated
European Level
2011-2017: Initiative on European Learning Factories
2013-2017: NIL (Network of innovative Learning Factories, funded by DAAD)
Worldwide Level
Since 2011: Conference on Learning Factories
2014-2017: CIRP CWG on Learning Factories / CIRP Key Note Paper on Learning Factories
Since 2017: International Association of Learning Factories
EUROPEAN
NIL
CIRP
IALF
Globalization & Networks

Fig. 10.1 Learning factory networks, starting from local efforts in 1980s to a worldwide association in 2017

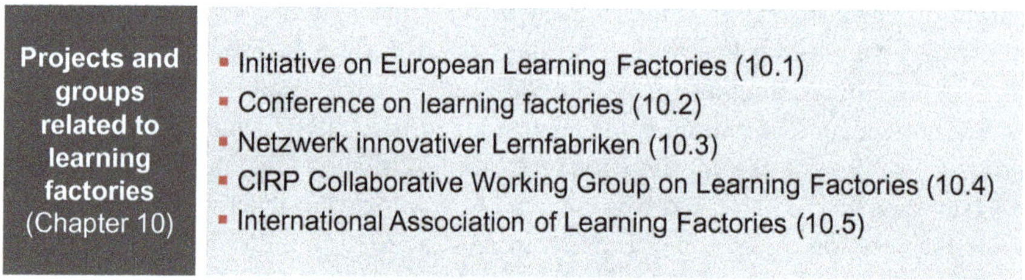
Projects and groups related to learning factories (Chapter 10)
Initiative on European Learning Factories (10.1)
Conference on learning factories (10.2)
Netzwerk innovativer Lernfabriken (10.3)
CIRP Collaborative Working Group on Learning Factories (10.4)
International Association of Learning Factories (10.5)

Fig. 10.2 Structure of projects and groups related to learning factories

tional Association of Learning Factories.[6] The development from isolated learning factories to worldwide learning factory networks is visualized in Fig. 10.1 In the following sections, the mentioned groups, projects, and associations are presented (Fig. 10.2).

[6]See Sect. 10.5.

Fig. 10.3 Founding members of the Initiative on European Learning Factories (from left to right: Professor Laszlo Monostori, Professor Wilfried Sihn, Professor Friedrich Bleicher, Professorin Vera Hummel, Professor Kurt Matyas, Professor Eberhard Abele, Dr. Thomas Lundholm, Dr. Dimitris Mavrikios, Christian Morawetz, Professor Ivica Veza, Professor Toma Udiljak, Jan Cachay, Professor Bengt Lindberg. Not in the picture: Professor Gunther Reinhart, Professor Pedro Cunha)

10.1 Initiative on European Learning Factories

On May, 20, 2011, together with the "1st Conference on Learning Factories" that took place in Darmstadt,[7] the Initiative on European Learning Factories (IELF) was founded as union of several European Learning Factory operators with the goal to start joint projects as well as improve and disseminate the learning factory concept worldwide (Fig. 10.3). The Initiative on European Learning Factories can be understood as an European network of Learning Factory experts with a broad know-how in the field of education, training, and applied research.

First president of the Initiative in the years from 2011 to 2016 was Prof. Eberhard Abele from PTW, TU Darmstadt. Founding members of the Initiative on European Learning Factories are shown in Fig. 10.4.

In summer 2016, Prof. Joachim Metternich, also PTW, TU Darmstadt, took over as president of the Initiative on European Learning Factories. In the following, it was decided that the Initiative on European Learning Factories should be elevated to a worldwide level due to the near end of the CIRP CWG on Learning Factories in 2017. The Initiative was following renamed to "International Association of Learning Factories" with the goal to integrate worldwide learning factory operators and enable a further internationalization of the learning factory topic. For the International Association of Learning Factories, see also Sect. 10.5.

Important dates of the Initiative on European Learning Factories:

- May 20, 2011: Founding of the Initiative on Learning Factories (IELF) in Darmstadt, Germany,

---

[7]See Abele, Cachay, Heb, and Scheibner (2011).

PTW TU DARMSTADT	Institute of Production Management, Technology and Machine Tools (PTW), Technische Universität Darmstadt, Germany	Prof. Eberhard Abele (president 2011-2016)
ESB BUSINESS SCHOOL REUTLINGEN UNIVERSITY	ESB Business School, Reutlingen University, Germany	Prof. Vera Hummel
KTH	Royal Institute of Technology, Stockholm, Sweden	Prof. Bengt Lindberg
LMS Laboratory for Manufacturing Systems & Automation University of Patras	Laboratory for Manufacturing Systems & Automation Department of Mechanical Engineering and Aeronautics, University of Patras, Greece	Prof. George Chryssolouris
FESB	Faculty of Electrical Engineering, Mechanical Engineering and Naval Architecture, University Split, Croatia	Prof. Ivica Veza
imw	Institute of Management Science, Vienna University of Technology, Austria	Prof. Wilfried Sihn (vice president)
IFT	Institute of Production Engineering and Laser Technology, Vienna University of Technology, Austria	Prof. Friedrich Bleicher
iwb	Institute for Machine Tools and Industrial Management, Technische Universität München, Germany	Prof. Gunther Reinhart
ceni	Center for Integration and process Innovation, Setúbal, Portugal	Prof. Pedro Cunha
MTA SZTAKI	The Computer and Automation Research Institute, Hungarian Academy of Sciences, cooperation with Budapest University of Technology & Economics, Hungary	Prof. Laszlo Monostori

Fig. 10.4 Names of the founding members of the initiative on European Learning Factories in 2011 and their institutes

- May 09, 2012: General Assembly of the IELF in Vienna, Austria,
- May 08, 2013: General Assembly of the IELF in Munich, Germany,
- May 27, 2014: General Assembly of the IELF in Stockholm, Sweden,
- July 07, 2015: General Assembly of the IELF in Bochum, Germany; New members: Institute for Machine Tools and Production Technology (IWF), Technical University Braunschweig,
- June 29, 2016: General Assembly of the IELF in Gjøvik, Norway; New members:

Institute of Production Science (wbk), Karlsruhe Institute of technology (KIT), Norwegian University of Science and Technology (NTNU),
- April 04, 2017: General Assembly of the IELF in Darmstadt, Germany; new members: Institut für Textiltechnik, RWTH Aachen.

10.2 Conferences on Learning Factories

The conferences on Learning Factories so far took place in Darmstadt (2011), Vienna (2012), Munich (2013), Stockholm (2014), Bochum (2015), Gjovik (2016), and Darmstadt (2017), and Patras (2018). The conference is growing in popularity and internationality. At the 2017 conference in Darmstadt, 150 participants from 18 countries participated in the two-day conference. Since 2015, the conference has been CIRP-sponsored, which can be seen as an indication of the growing importance of learning factories in manufacturing research.

The dates and topics of the Conferences on Learning Factories so far were

- May 19, 2011: **1st Conference on Learning Factories** in **Darmstadt**, Germany; main topics: learning and competency-building as a competitive factor, learning factories in operational application, leaders as teachers,
- May 10, 2012: **2nd Conference on Learning Factories** in **Vienna**, Austria; main topics: universities, industry, learning, and innovation factory of the Vienna University of Technology,
- May 07, 2013: **3rd Conference on Learning Factories** in **Munich**, Germany; main topics: Learning Factories for optimization of energy efficiency, sustainable efficiency in production and logistics through lean learning factories, creating the future with digital learning factories,
- May 28, 2014: **4th Conference on Learning Factories** in **Stockholm**, Sweden; main topics: learning factories for optimization of resource efficiency, sustainability in production and logistics through lean learning factories, innovation through virtual production in digital learning factories,
- July 07, 2015: **5th CIRP-sponsored Conference on Learning Factories** in **Bochum**, Germany; main topics: didactical approaches, resource efficiency, and sustainability, new learning factory concepts, Industrie 4.0 and cyber-physical systems, productivity management and lean production,
- June 29, 2016: **6th CIRP-sponsored Conference on Learning Factories** in **Gjøvik**, Norway; main topics: learning in Industrie 4.0/cyber physical manufacturing systems, Learning factories, cooperation, flexibility, transparency in manufacturing education an learning, research-based innovation, and learning
- April 04, 2017: **7th CIRP-sponsored Conference on Learning Factories** in Darmstadt, Germany; main topics: Learning factory concepts, "Industrie 4.0" production systems, "Industrie 4.0" use cases, Integration of digital learning, competency development,

- April 12–13, 2018: **8th CIRP-sponsored Conference on Learning Factories** in Patras, Greece; main topics: advanced skills and competencies, learning factory for digitization and Industrie 4.0, learning factories for training and education,

Figure 10.5 gives some impressions of the conferences organized so far.

10.3 Netzwerk Innovativer Lernfabriken

Initiated inside the Initiative on European Learning Factories, the project or network "Netzwerk innovativer Lernfabriken" (NIL, Network of Innovative Learning Factories) was funded by the German Academic Exchange Service (DAAD) and the German ministry for education and research (BMBF). Under the lead of Prof. Vera Hummel from ESB Reutlingen, national and international members mainly of the Initiative participated in the project:

- ESB Business School, Reutlingen-University,
- Institute of Production Management, Technology and Machine Tools (PTW), Technische Universität Darmstadt;
- Royal Institute of Technology, Stockholm;
- Laboratory for Manufacturing Systems and Automation Department of Mechanical Engineering and Aeronautics, University of Patras;
- Faculty of Electrical Engineering, Mechanical Engineering and Naval Architecture, University of Split;
- Institute of Management Science, Vienna University of Technology;
- Institute of Production Engineering and Laser Technology, Vienna University of Technology;
- Institute for Machine Tools and Industrial Management, Technische Universität Munich;
- Center for Integration and process Innovation, The Computer and Automation Research Institute, Hungarian Academy of Sciences, in cooperation with Budapest University of Technology and Economics;
- Chair for Production Systems, Ruhr-Universität Bochum;
- Stellenbosch University, South Africa.

In order to improve and use the learning factory concept for educational purposes, NIL organized exchange of ideas, students and researchers among ten Learning Factories in the field of manufacturing education. However, the NIL network is not considered a closed circle, but rather an open platform that aims to develop cooperation and partnerships.

During the four-year project, a total of about 400 scientists and students from seven countries took part in the workshops, exchanges, summer/winter schools, and other activities of the Network of Innovative Learning Factories. Stays abroad lasted from several days up to six months. In addition, a large number of workshops were

10.3 Netzwerk Innovativer Lernfabriken

organized in the learning factories and the participation at suiting conferences was facilitated. In addition, the project released several publications, including

- three series of papers including articles about the partner learning factories,[8]
- and two high-quality learning videos on line balancing[9] and Industrie 4.0 in learning factories.[10]

The project was successfully concluded with the two-week Winter School at the ESB Business School, where students from Greece, Croatia, and South Africa were able to compete in various challenges, expand their knowledge with practice-oriented educational games, and gain insights into the automotive industry during company visits.[11] Figure 10.6 shows students playing a logistics game.

10.4 CIRP Collaborative Working Group on Learning Factories

Basically, from the members of the Initiative on European Learning Factories with additional partners inside CIRP[12] active in the field of learning factories, a CIRP Collaborative Working Group with the title "Learning factories for future-oriented research and education in manufacturing" was agreed on and consequently initiated in 2014. In connection with the establishment of the CIRP CWG scientific, educational, and industrial objectives have been formulated in order to organize and boost learning factory activities globally[13]:

- Scientific objectives

- Provide a comprehensive overview of the global state of the art of action-oriented learning in Learning and Teaching Factories.
- Identify potentials and limits of Learning and Teaching Factories.
- Secure the knowledge gathered in CIRP CWG on "Learning Factories" as a basis for future research.
- Identify and name future research fields on the topic and potential (inter)national funding programs.

---

[8] See "The Learning Factory an Annual Edition from the Network of Innovative Learning Factories", for example Network of innovative Learning Factories (2015) and Network of innovative Learning Factories (2016).

[9] The video "Line Balancing: Practical example in a Learning Factory" was shot at the Process Learning Factory CiP in Darmstadt, the video can be found at: https://www.youtube.com/watch?v=PJg1DyZElvc&feature=youtu.be.

[10] The video "Industrie 4.0 in Learning Factories" was shot in several learning factories in Reutlingen, Bochum, and Darmstadt, the video can be found at: https://www.youtube.com/watch?v=pg64P0laeTM.

[11] See Bauer (2017).

[12] The International Academy for Production Engineering.

[13] See Abele, Chryssolouris, Sihn, and Seifermann (2014).

- Educational objectives

- Provide a comprehensive overview of education in Learning and Teaching Factories around the globe.
- Simplify the exchange of educational and didactical contents among CIRP members.

- Industrial objectives

- Link CIRP closer with the industry in the Learning and Teaching Factory area by including industrial efforts on this topic.
- Raise the visibility of the topic in the industry by providing scientifically sound data.

All of those objectives were tackled in the three years from 2014 to 2017 among others in well-attended CWG meetings at all General Assemblies as well as all Winter Meetings. Right from the start, the aim was to additionally summarize the results of the CIRP CWG on learning factories in a STC-O Keynote Paper. This Keynote Paper was presented at the General Assembly in Lugano, Switzerland in summer 2017:

Abele. E; Chryssolouris. G; Sihn. W; Metternich. J; ElMaraghy. H; Seliger. G; Sivard. G; ElMaraghy. W; Hummel. V; Tisch. M; Seifermann. S. (2017): Learning Factories for future-oriented research and education in manufacturing. In: *CIRP Annals—Manufacturing Technology* 66 (2), 803–826.

Beyond the contributors named above as authors of the keynote paper, many persons contributed to successful CWG meetings in the three years. Among others, thanks go to the CIRP contributors (in alphabetical order) F. Bleicher, C. Herrmann, A. Jäger, G. Lanza, K. Martinsen, D. Mourtzis, G. Putnik, M. Putz, G. Schuh, K. Schützer, T. Tolio, and J. Vancza. As well as to further contributors: J. Bauer; B. Brenner, Y. Gloy, M. Hammer, D. Mavrikios, J. Menn, G. Michalos, E. Moser, B. Muschard, R. Pittschellis, F. Ranz, C. Reise, and K. Tracht.

At the end of the CIRP CWG, it can be concluded that in recent years there is great progress and a huge development among others regarding learning factory networks and joint research projects in industry and academia. Learning factories have been established as valuable tools for the education of students, the training of employees but also for production-related research and innovation. Still after these developments, a great potential for synergies in the field of learning factories is identified, for example an exchange of learning factory modules and joint courses can be named. For the coming years, especially the challenges of (a) keeping pace with or even forerun industrial innovation and (b) combine digital and virtual learning factories with real learning factories are identified. In order to tackle the problem of scarce resources in connection with the construction and operation of learning factories, it will be crucial in future to work together on the challenges in the field of learning factories. In order to have a platform for these purposes, the globally active Association on Learning Factories[14] was founded.

---

[14]See Sect. 10.5.

10.5 International Association of Learning Factories (IALF)

From 2014 to 2017, the CIRP Collaborative Working Group on learning factories[15] was used as a platform for worldwide exchange; in addition to the Europe-wide exchange inside the IELF.[16] With the ending of the CIRP CWG after three years, the members of the IELF decided in 2017 to open up the Initiative for worldwide members in order to establish an enduring worldwide learning factory community (Fig. 10.7). In the course of this, the "Initiative on European Learning Factories" was renamed because (a) it was not only an European community anymore and (b) over the years, the group has moved beyond the status of an "initiative". The name of the newly formed group is now "International Association of Learning Factories" (IALF).

According to the self-understanding, the International Association of Learning Factories is a circle of research institutions running learning factories that aims to cooperate to reach excellence in education and research in the field of manufacturing engineering. Therefore, the members strive for

- Exchange of knowledge,
- Synergies in the physical establishment of learning factories,
- Leadership through innovative enhancements.

Further information on the International Association of Learning Factories can be found online: http://ialf-online.net/.

10.6 Wrap-up This Chapter

This chapter gives an overview of the different associations, working groups, projects, and initiatives that want to promote, improve, and disseminate learning factories as a practice-oriented form of education, training, and research for manufacturing.

---

[15] See Sect. 10.4.
[16] See Sect. 10.1.

Fig. 10.5 Impressions of the Conferences on Learning Factories from 2011 to 2018

10.6 Wrap-up This Chapter

Fig. 10.6 Students in the NIL winter school playing a logistics game (Bauer, 2017)

Fig. 10.7 Group picture of the members of the International Association of Learning Factories in Darmstadt, 2017

References

Abele, E., Cachay, J., Heb, A., & Scheibner, S., (Eds.). (2011). In *1st Conference on Learning Factories*, Darmstadt. Darmstadt: Institute of Production Management, Technology and Machine Tools (PTW).

Abele, E., Chryssolouris, G., Sihn, W., Metternich, J., ElMaraghy, H. A., Seliger, G., Sivard, G., ElMaraghy, W., Hummel, V., Tisch, M., & Seifermann, S. (2017a). Learning factories for future oriented research and education in manufacturing. *CIRP Annals—Manufacturing Technology, 66*(2), 803–826.

Abele, E., Chryssolouris, G., Sihn, W., Metternich, J., ElMaraghy, H. A., Seliger, G., Sivard, G., ElMaraghy, W., Hummel, V., Tisch, M., & Seifermann, S. (2017b, August). *Learning factories for future oriented research and education in manufacturing*. Presentation CIRP STC-O Keynote-Paper, GA 2017. CIRP General Assembly 2017, Lugano, Switzerland.

Abele, E., Chryssolouris, G., Sihn, W., & Seifermann, S. (2014, January). *CIRP collaborative working group—Learning factories for future oriented research and education in manufacturing*. CIRP. Paris: CIRP Winter Meeting.

Bauer, J. (2017). *Netzwerk innovativer Lernfabriken (NIL): Erfolgreicher Projektabschluss mit NIL-Winter School*. Retrieved from https://www.esb-business-school.de/fakultaet/aktuelles/detail/artikel/netzwerk-innovativer-lernfabriken-nil/.

Network of Innovative Learning Factories (Ed.). (2015). *The learning factory 2015: An annual edition from the network of innovative learning factories*.

Network of Innovative Learning Factories (Ed.). (2016). *The learning factory 2016: An annual edition from the network of innovative learning factories*.

Chapter 11
Best Practice Examples

11.1 Best Practice Example 1: AutFab at the Faculty of Electrical Engineering of the University of Applied Sciences Darmstadt, Germany

Authors: Stephan Simons[a], Stephan Neser[a]
[a]University of Applied Sciences Darmstadt

	Name of learning factory: **AutFab**								
	Operator: **Prof. Dr. S. Simons, Faculty EIT, Hochschule Darmstadt**								
	Year of inauguration: **2012**								
	Floor Space in learning factory: **50 sqm**								
	Manufacture product(s): **Relais**								
	Main topics / learning content: **Industrie 4.0, IIoT**								
	Extract from the morphology								
2.1	main purpose	education		training		research				
2.2	secondary purpose	test environment / pilot environment		industrial production	innovation transfer		public image		
3.1	product life cycle	product planning	product development	prototyping			service	recycling	
3.2	factory life cycle	investment planning	factory concept	process planning	ramp-up	manufacturing	assembly	logistics	main-tenance	recycling
3.3	order life cycle	configuration & order	order sequencing	planning and scheduling			picking, packaging	shipping	
3.4	technology life cycle	planning	development	virtual testing				main-tenance	moderni-zation
3.5	indirect functions	primary activities				secondary activities				
		Inbound & outbound logistics	marketing & sales	ser-vice	firm infra-structure	HR	technology development	Procure-ment	
4.1	learning environment	purely physical (planning + execution)		physical LF supported by digital factory	physical value stream of LF extended virtually		purely virtual (planning + execution)		
4.2	environment scale	scaled down		life-size						

11.1.1 Starting Phase and Purpose

In 2009, Prof. Dr. Stephan Simons came up with the idea to install a fully automated assembly line in the laboratories of the Faculty of Electrical Engineering of the University of Applied Sciences Darmstadt. As it seemed impossible to think about a complete smart factory concerning the limits of a University of Applied Sciences regarding human resources and financial means, the idea at that time was to concentrate on automated assembly. However, it was also planned to create a 3D simulation for virtual commissioning. Fortunately, Prof. Dr. Stephan Neser of the Faculty of Mathematics and Natural Sciences, decided to take care of the machine vision parts of the assembly line. The line was initially funded solely by a special fund for improving the quality of teaching of the State of Hessen. Interestingly, these funds are essentially approved by students, who directly recognized the value of the facility for education. The line, whose basic design and structure were supplied by Köster Systemtechnik GmbH, was inaugurated in February 2012. The first student project, realizing a complete 3D simulation of the line for virtual commissioning using Siemens Tecnomatix Process Simulate, already started in 2010. In 2011, the first thesis dealt with the implementation of control software on the workstations. Students worked on the first laboratory tasks of the visualization laboratory in 2012. Starting in 2011, the system was developed from a purely automated to a full smart factory in a series of projects by students from the Faculty of Electrical Engineering under the guidance of Prof. Simons and from the Faculty of Mathematics and Natural Science under the guidance of Prof. Neser. The dominant topic of these projects was Industrie 4.0, the German high-tech strategy initiative for future production. In 2011, the authors were convinced that small- and medium-size companies needed to see realized use cases to be able to recognize the full potential of Industrie 4.0. At the same time, these companies had a great need for graduates with the necessary skills to implement these new technologies. This led to the idea that students should implement appropriate use cases in the learning factory using new technologies and thus learn the desired skills. These skills include holistic system comprehension, independent problem-solving abilities, interdisciplinary expertise, the knowledge of production processes, of Industrie 4.0 technologies and business projects as well as project management, communication skills, and the ability to abstract. The previous projects have successfully demonstrated that the students acquire these skills in their projects.

In addition to teaching, the learning factory is a valuable research platform and a great means to attract potential students. It is continuously expanded and equipped with new technology, financed by the special fund for improving the quality of teaching of the State of Hessen and component donations from well-known manufactures.

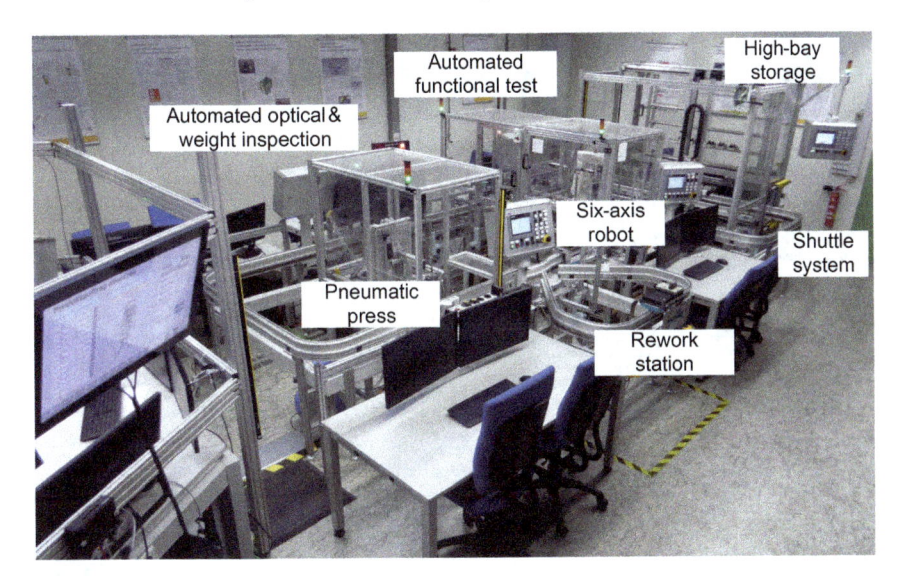

11.1 Best Practice Example 1: AutFab at the Faculty of Electrical … 337

Fig. 11.1 Value Stream of Learning Factory AutFab of h_da

11.1.2 Learning Environment and Products

The Learning Factory AutFab has a size of 50 m^2 and automatically assembles variants of relais. It consists of a high bay storage, two assemblies, and two inspection stations. A six-axis industrial robot and a pneumatic press assemble the relay, which is afterward inspected using an automated optical inspection, a weight inspection, and an automated functional test (Fig. 11.1).

The workstations are controlled by modern programmable logic controllers, which are networked by industrial Ethernet and classical field bus communication. The different products are assembled with lot size 1 using RFID. Additionally, Data Matrix codes are used for the adaption of the inspections. All functional safety components are fully integrated into the network. Energy consumption is monitored, and measures for energy efficiency have been implemented. Human–machine interfaces exist on fixed and mobile touch panels, but also on consumer devices like smartphones and tablets. Digital twins have been created for virtual commissioning, material flow, and energy consumption simulation. Remote control is possible using VPN technology for security reasons. The AutFab has been connected to a manufacturing system in the network of the University and to an enterprise resource planning and manufacturing system running in the cloud. The vertical communication to higher control levels is realized by OPC UA. As a first use case, an assistance of the rework process was implemented using mixed reality technology on the Microsoft HoloLens. In addition, the line has been connected to the MindSphere cloud solution from Siemens AG, which provides worldwide access for extensions such as predictive maintenance and line optimization analysis using cloud computing (Fig. 11.2).

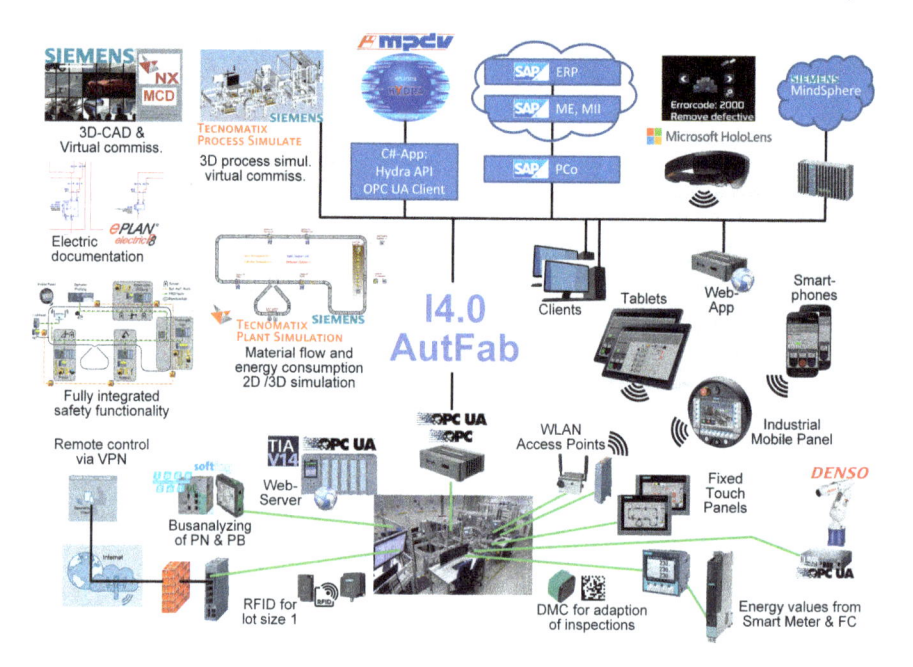

Fig. 11.2 Industrie 4.0 in the AutFab

11.1.3 Operation

Currently, the plant is operated by Prof. Dr. Simons and Prof. Dr. Neser, as supervisors for student projects and laboratory tasks. The vision for future development is a continuous evolution of the AutFab implementing further use cases using new tools and technologies. This is an integrative process, in which teaching and research are combined and in which students acquire key qualifications for starting successful careers in the automation industry.

Up to now, 250 students have already worked in such student projects on the AutFab and 150 students passed the visualization laboratory. Furthermore, 24 videos, partly as tutorials, were created, largely by the students themselves, and uploaded on YouTube. These videos already have in total 51,000 clicks, showing the enormous interest in the technologies implemented in AutFab. Further research projects and collaborations with other colleagues and industry are planned.

11.2 Best Practice Example 2: Demonstration Factory at WZL, RWTH Aachen, Germany

Authors: Günther Schuh[a], Jan-Philipp Prote[a], Bastian Fränken[a], Julian Ays[a]

11.2 Best Practice Example 2: Demonstration Factory at WZL, RWTH …   339

[a]Laboratory for Machine Tools and Production Engineering (WZL), RWTH Aachen University

Name of learning factory: DFA Demonstration Factory
Operator: Smart Logistics Cluster, RWTH Aachen
Year of inauguration: 2013
Floor Space in learning factory: 1600 sqm
Manufacture product(s): e.GO Kart and e.GO Life (E-Mobility vehicles)
Main topics / learning content: Industry 4.0, prototypes and industrialization

Extract from the morphology

2.1	main purpose	education	training		research	
2.2	secondary purpose	test environment / pilot environment	industrial production	innovation transfer	public image	
3.1	product life cycle	product planning	product development	prototyping	service	recycling
3.2	factory life cycle	investment planning	factory concept	process planning	ramp-up	maintenance	recycling
3.3	order life cycle	configuration & order	order sequencing	planning and scheduling	picking, packaging	shipping
3.4	technology life cycle	planning	development	virtual testing	maintenance	modernization
3.5	indirect functions	primary activities: Inbound & outbound logistics, marketing & sales, service	secondary activities: firm infra-structure, HR, technology development, Procurement			
4.1	learning environment	purely physical (planning + execution)	physical LF supported by digital factory	physical value stream of LF extended virtually	purely virtual (planning + execution)	
4.2	environment scale	scaled down	life-size			

11.2.1 Starting Phase and Purpose

Since 2013, the Demonstration Factory DFA in Aachen allows internal researchers and visitors alike to experience and investigate topics of Industrie 4.0 in real life. With a total area of 1600 m², the learning factory is located on the RWTH Campus as part of the Smart Logistics Cluster and was originally founded with internal funds. As a consortium of the university's academic institutions and industry partners, the Smart Logistics Cluster focuses on research combining topics in logistics, productions, and services. The learning factory itself is operating in close cooperation with both the Laboratory for Machine Tools and Production Engineering (WZL) and the Institute for Industrial Management (FIR) of the RWTH Aachen University. Nevertheless, it is organized independently as a "GmbH" and thus has an economical responsibility to produce and sell real products on the market. The factory specializes on prototype construction and industrialization; the main purposes, next to a real production, being experiment based and applied production research as well as training of heterogeneous groups of management, shop floor workers, and students in a production

Fig. 11.3 Inside view of the DFA demonstration factory

environment. In addition to research and trainings, the DFA supports the development of methods and tools for production-ready design and cost-efficient production. A unique characteristic of the DFA is that marketable products are manufactured with the e.GO Mobile AG as lead customer (Fig. 11.3).

11.2.2 Learning Environment and Products

Two products are manufactured in the learning factory: firstly, the e.GO Kart, an electric kart with a 250 W pedelec motor and secondly the body of the e.GO Life, an electric compact car. Both car and kart are fully functional and mostly self-developed. While only the welding of the e.GO Life body takes place in the learning factory, the kart is almost completely manufactured and assembled there and is already available on the market. It is sold in two to four variants and for the latest model up to 100 components are required (Fig. 11.4).

For this purpose, a complete value chain of a small batch production is integrated in the learning factory, including material management, manufacturing, and assembly. With the increasing importance of changeability and thus mobility, modularity, and scalability, all dimensions—layout, product, technologies and quantities—can be adapted in the factory. The digital infrastructure contains IT systems with the integration of different ERP systems and corresponding support software as well as Industrie 4.0 solutions, which cover each step of the production (Fig. 11.5).

Starting in material management, a Pick-by-Voice system helps the employee to find required parts with the support of acoustic instructions and visual aids on smart watches. RTLS tags make it possible to monitor the commissioned parts along the value chain. Location and status of all parts are constantly monitored by this indoor-positioning system. The ERP system uses the current positioning data for generating feedback data (process data, transport, and waiting times), directly in the ERP system, enabling to a production planning and control in real time. Further down the process, the parts are mounted with the support of three-dimensional assembly instructions. Each assembly station is equipped with a touch screen displaying visual

11.2 Best Practice Example 2: Demonstration Factory at WZL, RWTH ...

e.GO Life

Small Batch Production

- Currently in preparation for larger scale production
- Welding of the car body
- Reduction on manufacturing costs by application of cost-effective technologies

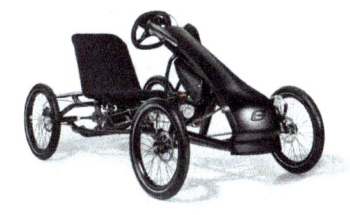

e.GO Kart

Available on the market

- The cart already available on the market and sold in two to four variants
- About 100 components required
- 250 Watt Pedelec electric power
- 3 performance levels

Fig. 11.4 Manufactured products in the Demonstration Factory DFA

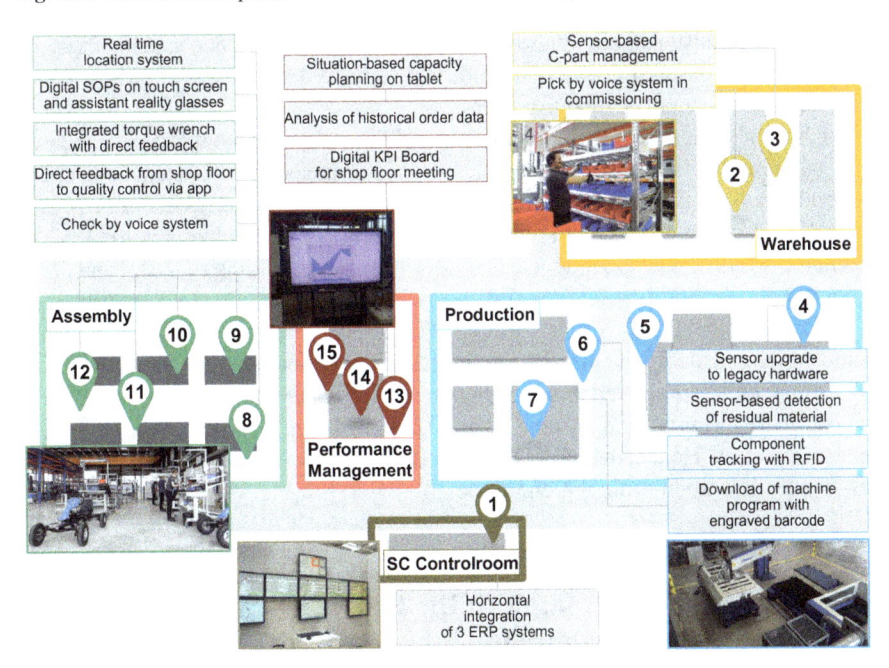

Fig. 11.5 Value chain in the Demonstration Factory DFA

instructions. To efficiently steer and control the production, a live shop floor KPI visualization board is positioned in the factory and supported by data collection. Following the process of change management, an Error-app developed at the WZL enables accelerated uptake of change requests and feed-in into the data-preserving product life cycle management system, by accessing product data and supplementing contextual information (e.g. description of the errors by photo). In the process of deciding on the approval of the change request, a change request cockpit creates transparency about the technical change by using context-dependent. For example, development-relevant information from the ERP systems are summarized, resulting in a sound decision-making basis. All the Industrie 4.0 technologies in the factory are updated regularly as new findings and developments emerge. The value chain as depicted above shows the current configuration in the learning factory.

11.2.3 Operation

The DFA stands open for both external firms and student classes and as a hybrid learning factory serves the dual purpose of training and research parallel to real production. Around 8000 participants have visited the DFA in total until 2017 with capacity utilization near 100%. Five to nine full-time equivalent workers, the majority being student assistants, are employed in the learning factory. The sustainability of the operation is ensured by internal funds and supported by research funds, sales of the produced products, and course fees.

Visitors of the learning factory may choose from more than ten different standardized and individualized trainings, which are all continuously adapted and developed. Depending on the type of training, the number of participants varies between 10 and 30 and the duration from one to five days. Researchers or consultants of the RWTH Aachen University lead the practical laboratory courses providing practical and theoretical guidance. Additionally, cognitive and also psycho-motoric skills are practiced through activities. For this purpose, the trainings include various use cases using demonstration and do-it-yourself approaches to transfer knowledge. While the trainings aim at building up technological and methodological competences, research in the factory revolves around the core topics Industrie 4.0, prototype construction, and industrialization. Hence, the DFA Demonstration Factory unites up-to-date research, real production, and training at one place.

11.3 Best Practice Example 3: Die Lernfabrik at IWF, TU Braunschweig, Germany

Authors: Gerrit Posselt[a]
[a]Institute of Machine Tools and Production Technology, Technische Universität Braunschweig

11.3 Best Practice Example 3: Die Lernfabrik at IWF, TU Braunschweig ... 343

Name of learning factory: **Die Lernfabrik**
Operator: **IWF, Technische Universität Braunschweig**
Year of inauguration: **2012**
Floor Space in learning factory: **450 sqm**
Manufacture product(s): **divers**
Main topics / learning content: **Sustainable Production, Cyber-physical Production systems, Urban Production**
Extract from the morphology

2.1	main purpose	education	training	research				
2.2	secondary purpose	test environment / pilot environment	industrial production	innovation transfer	public image			
3.1	product life cycle	product planning	product development	prototyping		service	recycling	
3.2	factory life cycle	investment planning	factory concept	process planning	ramp-up	manufacturing / assembly / logistics	main-tenance	recycling
3.3	order life cycle	configuration & order	order sequencing	planning and scheduling			picking, packaging	shipping
3.4	technology life cycle	planning	development	virtual testing			main-tenance	moderni-zation
3.5	indirect functions	primary activities				secondary activities		
		Inbound & outbound logistics	marketing & sales	ser-vice	firm infra-structure	HR	technology development	Procure-ment
4.1	learning environment	purely physical (planning + execution)	physical LF supported by digital factory	physical value stream of LF extended virtually			purely virtual (planning + execution)	
4.2	environment scale	scaled down	life-size					

11.3.1 Starting Phase and Purpose

The development of new research questions within current research works and the transfer of knowledge from methods and tools into teaching and training is particularly importance for the Chair of Sustainable Manufacturing and Life Cycle engineering at the Technische Universität Braunschweig. For this reason, it was decided to establish "*Die Lernfabrik*"[1] as an appropriate platform to suit this objective in 2012. The original motivation for the concept of the learning factory came up in 2011 from the question posed by the Ministry of Education and Research as to how very good project results can be made accessible to a broad public, in particular producing SMEs, beyond the duration of common research projects. Thus, the first thematic focus and as well the focus on research and further education emerged from the joint research project "EnHiPro - Energy- and Auxiliary Material-Optimized Production." The focus was further expanded by embedding the learning factory in student teaching, which led to the development of innovative teaching concepts for engineering education integrating the infrastructure of the learning factory. For example, future engineers in production and systems engineering are trained in research-based teaching in the learning factory in the subject areas "energy efficiency in production engineering," "sustainable cyber-physical production systems," and "future production

---
[1] German for: the learning factory.

Fig. 11.6 Die Lernfabrik of the Technische Universität Braunschweig

systems." A scaled production system was developed especially for this purpose in cooperation with Festo Didactic SE, which permits safe, practice-oriented learning, and testing.

At this point, it was clear that the learning factory had to be given a broader organizational structure. Target group-specific and purpose-specific laboratories were defined. The first is the "research lab" with its focus on research activities with the purpose of serving as a test and pilot environment for new technologies. The second laboratory is the "experience lab," which has been addressed earlier, a scaled production environment especially for student teaching and the further training of specialists and managers. The third laboratory focuses on technical-industrial training and skilled workers and is called the "education lab" (Fig. 11.6).

11.3.2 Learning Environment and Products

The research laboratory focuses on the dissemination of research results and the continuous derivation of new research questions. On an area of over 400 m^2, innovative research prototypes and tools are being researched with partners from industry in a real production environment in an infrastructure close to industry. This means that tests can be carried out both at machine and equipment level and at the factory level, taking into account all technical building services. In this way, interactions between machines and the technical building services of the factory and the building shell can be made measurable and evaluable.

The experience laboratory focuses on the transfer of research methods and tools into the teaching of engineering students and the training of experts. For this purpose, a factory was built in the factory. The model factory consists of a real working modular production system in a process chain from additive manufacturing to product assembly up to scrap recycling. Without high electrical voltages and high mechanical forces, learners can define their own research goals and conduct their own experiments to test and deepen their theoretical knowledge in practice. It could be evaluated that through the approach of research-oriented learning, learners reach their admired competences and related knowledge much faster than with conventional teaching methods.

As a central institution of the Technical University of Braunschweig, the education laboratory provides technical and commercial training on more than 50 m^2. The trainees learn the basics of metalworking and electrical, pneumatic, and hydraulic circuit construction in an energetically renovated and technologically reequipped workshop. As a predicate of the training, the curricular prescribed contents are extended by the topics of energy and resource efficiency as well as Industrie 4.0. Energy-efficient circuits, mineral oil-free cooling lubricants, and the effective use of compressed air are part of the education lab's daily business.

11.3.3 Operation

However, the organizational structure shown in Fig. 11.7 also reflects the diversity of user models, business models, and personnel structure. The research focus is mainly represented by research assistants and doctoral candidates in cooperation with colleagues from collaborative projects with industry. The training focus, on the other hand, calls for more solid structures in personnel as well as in the thematic orientation. In most cases, permanent trainers and scientists work here in the field of further education. In addition to the thematic priorities already mentioned above, *"Die Lernfabrik"* is also an expert factory for energy transparency and mixed reality in production at the Competence Centre for Lower Saxony and Bremen in northern Germany, for example. Furthermore, *"Die Lernfabrik"* is active at two international locations. In Singapore, there is a virtual twin concept with the SIMTech from A*STAR. Joint Master Classes for industry in Southeast Asia and Europe are organized there and certified specialists are trained further. The second anchor point is in India, at BITS in Pilani. The stated goal there is to strengthen engineering education in emerging development nations through more practice-oriented training and to make a major contribution to sustainable development.

A selection of representative projects is given below:

- Mit uns digital! Mittelstand 4.0 Kompetenzzentrum—Expert Factory for energy transparency teaches how digitalization is stabilizing energy management (BMWI, 2016–2018).

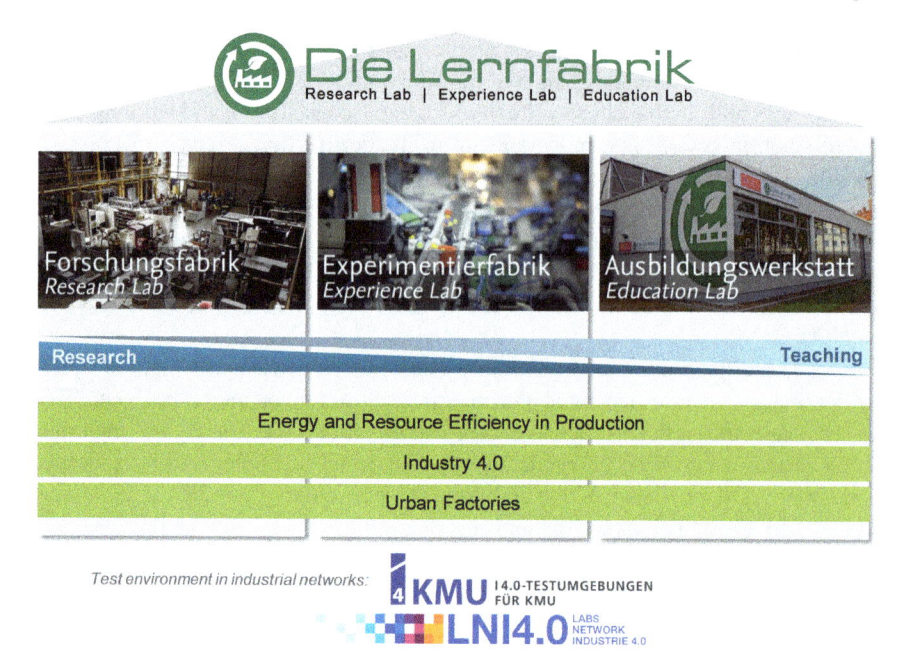

Fig. 11.7 Organizational structure and core topics of Die Lernfabrik and its three laboratories

- ILehLe—The Intelligent Teaching-Learning Factory adapts to the individual learning progress of learners in initial and continuing education and training (BMBF, 2016–2019).
- JoSITURF—Joint SIMTech-TU Braunschweig Research Factory for Urban and Energy Efficient Manufacturing through Information and Communication Technology (BMBF, 2017–2018).
- JInGEL—Joint Indo-German Experience Lab brings practical engineering training into the indian higher education system in cooperation with the german industry (DAAD, 2016–2019).
- PlayING—Game-based learning for mediation of key competences in engineering studies (2013).
- Initiative on European Learning Factories—European network of learning factories in the higher education and university landscape.
- LNI4.0—Test center for drivers and early adopters of I4.0 technologies and services.
- I4.0 Test Environment for SMEs (I4SMEs)—supports small and medium-sized enterprises in the testing and demonstration of I4.0 products and components (BMBF 2017–2019).

11.4 Best Practice Example 4: E|Drive-Center at FAPS, Friedrich-Alexander University Erlangen-Nürnberg, Germany

Authors: Jörg Franke[a], Alexander Kühl[a]
[a]Institute for factory automation and production systems (FAPS), Friedrich-Alexander University Erlangen-Nürnberg

Name of learning factory: **E	Drive-Center**
Operator: **Friedrich-Alexander-University Erlangen-Nürnberg, Institute FAPS**
Year of inauguration: **2011**
Floor Space in learning factory: **867 sqm**
Manufacture product(s): **Electric Motors**
Main topics / learning content: **Production technology**
Extract from the morphology
2.1	main purpose	education		training		research				
2.2	secondary purpose	test environment / pilot environment		industrial production	innovation transfer		public image			
3.1	product life cycle	product planning	product development	prototyping			service	recycling
3.2	factory life cycle	investment planning	factory concept	process planning	ramp-up	manufacturing	assembly	logistics	maintenance	recycling
3.3	order life cycle	configuration & order	order sequencing	planning and scheduling				picking, packaging	shipping
3.4	technology life cycle	planning	development	virtual testing			maintenance	modernization
3.5	indirect functions	primary activities				secondary activities				
		Inbound & outbound logistics	marketing & sales	service	firm infrastructure	HR	technology development	Procurement
4.1	learning environment	purely physical (planning + execution)		physical LF supported by digital factory		physical value stream of LF extended virtually		purely virtual (planning + execution)
4.2	environment scale	scaled down			life-size					

11.4.1 Starting Phase and Purpose

Since the mid-1990s, the Friedrich-Alexander Universität Erlangen-Nürnberg (FAU) at its Chair for Factory Automation and Production Systems (FAPS) is developing production technologies for the assembly of electric drives. Within this framework, various winding applications and magnet mounting technologies were initially developed. The continuous expansion of research activities necessitated an expansion of laboratory capacities. As part of a major research project funded by the Bavarian Ministry of Economic Affairs, in 2011 the E|Drive-Center moved into a former factory building of AEG/Electrolux in Nuremberg. Since then, the learning factory has been filled and expanded with various current technologies for the production of electric drives.

Today, the E|Drive-Center develops innovative drive concepts and associated production technologies with the aim of transferring the knowledge to industrial applications. In addition, production and testing processes for contactless power transmission components in electric vehicles are addressed.

The following three objectives are pursued with the E|Drive research and learning factory:

Firstly, the processes are investigated in greater depth in a large number of publicly funded projects and are also being industrialized in close cooperation with innovative regional, national, and international companies.

Secondly, the production environment and current research results are used to educate FAU students in the production of electric drives within the framework of lectures, exercises, and internships.

In the third step, new knowledge is transferred to industrial users in a practice-oriented way, particularly in the context of seminars.

11.4.2 Learning Environment and Products

In the production environment of the E|Drive-Center, the multistage manufacturing process, the entire value-added chain of an electric motor, is illustrated. More than 25 sophisticated production facilities and a large number of equipment are available within the 800 m^2 factory (Fig. 11.8).

The process chain begins with the production and packaging of electrical steel sheets, e.g. by the use of a laser cutting system. For further processing as a stator, the packages receive their windings. For this process step, various flexible and partly robot-assisted production lines are available, which offer a broad portfolio of different winding topologies and geometries, e.g. single teethes, inner or outer stators in different sizes (diameters up to 500 mm).

Furthermore, various technologies in the field of stripping enameled copper wires and the subsequent contacting process are available, including an 8 kW infrared laser system or ultrasonic machines. For the subsequent insulation of stators, various impregnation and powder coating processes are available, such as different machines for the heating of the stators (e.g. thermal oven or inductors).

The sheet stack of the rotor has to be equipped with magnets. Therefor, assembly technologies for magnetized and non-magnetized magnets are available.

A specific focus of the E|Drive-Center is the laboratory for the quality assurance of magnets (MagLab) in which various measuring methods for magnetic fields, as hall arrays, magneto-optical cameras, and magneto-resistive sensors are applied.

11.4 Best Practice Example 4: E|Drive-Center at FAPS, Friedrich ...

Fig. 11.8 Electric Drives Learning Factory and Process Chains

11.4.3 Operation

The aim of the teaching and training activities is the sustainable development of crucial competences in the field of production of electric drives. The offered courses and seminars can be subdivided according to the process chains for rotors, stators and final assembly. Special emphasis is placed on winding, contacting, insulation, magnetization, and magnet assembly as well as quality assurance technologies. A team of more than 15 scientific staff members teaches the theoretical content. The practical in-depth training then takes place at various plants of national and international partner companies.

The two-day seminars with up to 40 participants are therefore divided into small groups for optimal supervision and knowledge transfer. In addition, various laboratory tours are possible, which should lead to a deeper understanding of the process through a detailed technical discussion directly at the process demonstrator. Various technology experts are available as discussion partners and provide detailed insights into current process research.

11.5 Best Practice Example 5: ESB Logistics Learning Factory at ESB Business School at Reutlingen University, Germany

Authors: Vera Hummel[a], Fabian Ranz[a], Jan Schuhmacher[a]
[a]ESB Business School, Reutlingen University

Name of learning factory: ESB Logistics Learning Factory							
Operator: ESB Business School, Reutlingen University							
Year of inauguration: 2014							
Floor Space in learning factory: 700 sqm							
Manufacture product(s): City scooter and Accessoires							
Main topics / learning content: Design, implementation, optimization and digitalization of partially automated assembly and logistics systems							
Extract from the morphology							
2.1	main purpose	education		training		research	
2.2	secondary purpose	test environment / pilot environment		industrial production	innovation transfer	public image	
3.1	product life cycle	product planning	product development	prototyping	service		recycling
3.2	factory life cycle	investment planning	factory concept	process planning	ramp-up	main-tenance	recycling
3.3	order life cycle	configuration & order	order sequencing	planning and scheduling	picking, packaging		shipping
3.4	technology life cycle	planning	development	virtual testing	main-tenance		moderni-zation
3.5	indirect functions	primary activities			secondary activities		
		Inbound & outbound logistics	marketing & sales	service	firm infra-structure	HR	technology development	Procure-ment
4.1	learning environment	purely physical (planning + execution)	physical LF supported by digital factory	physical value stream of LF extended virtually	purely virtual (planning + execution)		
4.2	environment scale	scaled down			life-size		

11.5.1 Starting Phase and Purpose

After the initiator of the ESB Logistics Learning Factory, Prof. Vera Hummel had made experience in developing and implementing a concept for a Learning Factory for advanced Industrial Engineering (aIE) at the University of Stuttgart, Institute IFF between 2005 and 2008, she was appointed as a full professor at ESB Business School, a faculty of Reutlingen University in March 2010. Lacking a realistic,

11.5 Best Practice Example 5: ESB Logistics Learning Factory at ESB ...

hands-on learning and teaching environment of industrial scale for its industrial engineering students, first ideas for a Learning Factory that would strongly focus on all aspects of production logistics were drafted in 2012. Already back then, a strong integration of virtual and physical factory was desired: While the Learning Factory itself would be physical, the neighboring partners along the supply chain, such as suppliers or distribution warehouses, could be added in a fully virtual way. Considering implementation of the ESB Logistics Learning Factory a strategic initiative of the university, initial funding was provided by the faculty ESB Business School itself. Following its own creed, to provide future-oriented training for the region, also primarily local suppliers and manufacturers were selected as equipment providers to the new Learning Factory. During the initialization phase, 2014, a total of three researchers and nine students worked approximately four months to set up a first assembly line, storage racks, AGVs, or pick-by-light systems in conjunction with the underlying didactical concept. Since then, several hundred of students have participated in trainings and lectures held in the ESB Logistics Learning Factory, several research projects were carried out, and multiple high-level politicians and industry executives have been touring the shop floor. Also, more than EUR 2 million in research and infrastructure funds could be secured for expansion and upgrade—allowing the ESB Logistics Learning Factory today to represent many core aspects of an Industrie 4.0 production environment.

11.5.2 Learning Environment and Products

11.5.2.1 Physical Factory

Starting in an approximately 50 m^2 basement area, the ESB Logistics Learning Factory has been upgraded twice between 2014 and 2018 and now resides in a 700 m^2 stand-alone factory building right on campus that will be expanded further in 2019. The available infrastructure is able to store, assemble, pack, and ship recreational city scooters. The product has been selected due to its modularity, high variance, reusability, and emotionality—after successful production, students can straight go for a test ride. Produced out of approx. 60 single components of different sizes and shapes, the product requires all generic industrial assembly processes and offers various logistics challenges—e.g. for replenishment. Offering all equipment of a state-of-the-art assembly plant, the ESB Logistics Learning Factory is enhanced by lightweight robot systems, autonomously guided vehicles, communication and information technology as well as additive manufacturing technology to be able to represent different technological maturity levels and confront students with complexity induced by high-tech infrastructure.

Fig. 11.9 Impressions of the ESB Logistics Learning Factory

11.5.2.2 Digital/Virtual Factory

In addition, two major tools are in use to digitally and virtually represent the physical factory. For integrated product and factory engineering and life-cycle management, a cloud-hosted instance of the *Dassault Systèmes 3D experience platform* is in use. The platform enables collaborative engineering of several users within the same project at the same time and provides required functionalities, from project management tools to process planning, line balancing or robot simulation through *Apps,* which can be added to or removed from the platform as needed. Thus, the ESB Logistics Learning Factory (Fig. 11.9) and its products can be holistically designed, engineered, visualized, and simulated in one integrated, industry-wide recognized software.

Production planning, scheduling, and execution take place through *BECOS oneiroi*, also a cloud-hosted solution. Available orders and resources are taken into consideration by the tool to simulate various production principles and scenarios and to propose an ideal production setup and layout. Single orders then can be prioritized manually or automatically and released for production. Through its flexible connectivity, *oneiroi* can provide operator interfaces with relevant information, just like it is able to trigger robots, pick-by-light systems, or conveyor systems. The communication between all entities is based on an event-logic, and the analysis of such events enables students to carry out digital shop floor management.

11.5.3 Operation

Trainings are designed for different target groups: For bachelor's degree students, the Learning Factory is used in order to deliver methods competence in essential tools and techniques in the field of work system design, production logistics, and optimization of either. Trainings are held by associates, which take on an instructor role to ensure that specific learning processes are adhered to and a certain application routine in the respective methods or in working with a particular technology is acquired. Bachelor-level trainings take one to two days. Before the students finish their study with the bachelor thesis, they conduct an intensive interdisciplinary seminar which covers the complete product life cycle. In about 2 weeks, spread over one semester, the generation of the product idea, procurement of the parts, conception of the production system up to production of the prototype take place in the ESB Learning Factory. In contrast, on master level, extensive projects are carried out that involve groups of students typically for up to fifteen weeks. Here, students solve comprehensive and interdisciplinary product and production engineering problems that involve ideation and conception phases, take them through design and product/process engineering, prototyping, and end up in the ramp-up of the production system that makes use of the infrastructure of the learning factory. In such projects, associates and professors are involved for regular moderation of the group, while the actual day-to-day project work takes place in a strongly self-instructed manner.

11.5.3.1 Development

The Learning Factory also serves as an industry-related development and validation environment for the further development of innovative solutions such as the cloud-based control system architecture of BECOS oneiroi for changeable factory environments or solutions for orchestrating different components and systems in the Industrie 4.0 context.

Intensive development work to create and link the "digital twins of the entire factory" in the 3D business platform with the physical production environment by means of object-oriented laser scanning takes place. The aim here is to provide an up-to-date digital image of bidirectional integrated engineering at all times in order to shorten design and improvement cycles for production systems and thus increase the attractiveness and cost-effectiveness of digital virtual engineering.

11.5.3.2 Research

To investigate and develop solutions for future-oriented, flexible work and intralogistics systems, several initiatives are underway. A collaborative tugger train system combing the potentials of automation and human–machine collaboration is currently under development in the Learning Factory. The collaborative tugger train system consists of an automated guided vehicle (AGV) with a collaborative robot manipulator on top of it for collaborative bin handling, and the AGV is towing multiple trailers to achieve a high transport capacity. In addition, a flexible transport allocation and control method covering various means of intralogistics transportation resources is investigated using the cloud-based control system of the Learning Factory. To reduce the integration and development effort for service robotics in industrial intralogistics systems, a modular service, tool, and hardware components construction kit for service robotics are being developed in the Learning Factory in close cooperation with other research organizations and industrial partners from different industrial sectors.

Another aspect that is being researched in the ESB Logistics Learning Factory with regard to human–machine collaboration is the interaction between operators and articulated lightweight robots in assembly. Here, labor science aspects such as task allocation between humans and robots for reduced burden and capability-oriented work assignment and workplace design for human–robot collaboration can not only be conceptually conceived, but also put into practice and tested.

In the innovative research master program Digital Industrial Management and Engineering DIME, the ESB Logistics Learning Factory finds its applications in the defined research projects of the students over the entire four semesters for researching and validating innovative solutions and instruments in the context of digital transformation.

11.5.4 Outlook and Future Development

Due to growing digitization, automation within and integration of production systems, cognitive competence requirements, such as contextualization or attention division, will move more into focus in education and training. The ESB Logistics Learning Factory will develop further into an innovation factory in the context of digitalization and *Industrie 4.0* in order to build up and strengthen such cognitive competences in complementation to the competence classes addressed by Learning Factories so far.

11.6 Best Practice Example 6: ETA-Factory at PTW, TU Darmstadt, Germany

Authors: Eberhard Abele[a], Michael Tisch[a], Dominik Flum[a], Mark Helfert[a]
[a]Institute for Production Management, Technology and Machine Tools (PTW), TU Darmstadt

11.6 Best Practice Example 6: ETA-Factory at PTW, TU Darmstadt, Germany

Name of learning factory: **ETA-Factory**
Operator: **PTW, TU Darmstadt**
Year of Opening: **2016**
Floor Space in learning factory: **810 sqm**
Manufactured product(s): **Control Plate for hydraulic pump, Gear-shaft combination**
Topics / Learning Content: **Energy Efficiency, Energy Flexibility**

Extract from the morphology

2.1	main purpose	education	training	research
2.2	secondary purpose	test environment / pilot environment	industrial production	innovation transfer	public image
3.1	product life cycle	product planning	product development	prototyping			service	recycling
3.2	factory life cycle	investment planning	factory concept	process planning	ramp-up	manufacturing	assembly	logistics	maintenance	recycling
3.3	order life cycle	configuration & order	order sequencing	planning and scheduling			picking, packaging	shipping
3.4	technology life cycle	planning	development	virtual testing			maintenance	modernization
3.5	indirect functions	primary activities				secondary activities			
		inbound & outbound logistics	marketing & sales	service	firm infrastructure	HR	technology development	procurement
4.1	learning environment	purely physical (planning + execution)	physical LF supported by digital factory	physical value stream of LF extended virtually	purely virtual (planning + execution)
4.2	environment scale	scaled down	life-size

11.6.1 Starting Phase and Research in the ETA-Factory

A sustainable production that is economically optimized and environmentally friendly is a success factor for manufacturing companies in the future. In this context, the research group "Sustainable Production" was founded in 1996 at the PTW. Starting point of research activities is the question of how energy consumption in production can be reduced and thus the energy efficiency of an entire production system can be improved. Against this background, in the year 2008 the vision was born to transfer the already successful concept of the Process Learning Factory CiP (see Best Practice Example 26) to the field of energy efficiency. An energy efficiency factory was to be built on the campus, where students can get to know industrial processes and, in particular, PhD students are able to conduct practice-oriented research. After the elaboration of the comprehensive concept, construction work of the ETA-Factory started in 2013; just in time for the 120th anniversary of PTW. Through the high level of commitment from all stakeholders and with the support of the BMWi, the state of Hesse and the Technical University of Darmstadt the ETA-Factory inauguration was achieved only three years later. Since then, the ETA-Factory makes production research on energy efficiency tangible and sets new impulses for tomorrow's sustainable production. Notwithstanding the comparatively short history of the ETA-Factory, both the facility and the concept have already received many awards: Deutscher Ingenieurbaupreis (2016, translated: German Engineering Award), ICONIC AWARDS, Architekturpreis Beton (literal translation: Architectural Award for Concrete), DETAIL Produktpreis (literal translation: DETAIL Product Award), Land der Ideen (literal translation: Land of Ideas), German Design Award 2018 (all 2017).

The ETA-Factory is a large-scale research unit offering unique opportunities to explore profoundly the topics energy and resource efficiency and to test the solutions developed in a close-to-reality environment. Furthermore, completely new fields of research are opened up, such as the energy flexibility in production associated with the energy transition toward renewable energies.

11.6.2 Learning Environment and Products

In addition to the excellent opportunities for research, the ETA-Factory also serves as a learning environment in which gained research insights are transferred to industry and education. In particular, in the context of the offered trainings for industry, the complete ETA-Factory learning environment is integrated. On the shop floor of the ETA-Factory, two complete value chains are operated in which market-ready products are produced; see Fig. 11.10.

11.6.2.1 Greenfield Process Chain

The first value chain of the ETA-Factory is shown on the left-hand side of Fig. 11.10. This value chain was created in a greenfield approach in the context of the project "ETA-Factory." A control plate for hydraulic pumps is manufactured here. The hydraulic pump is originally manufactured and distributed by Bosch Rexroth. The production machines integrated in the process chain comply with the latest energy efficiency standards—in course of the ETA-Factory project the machines have been modified as research demonstrators. The process chain essentially comprises: two cutting machine tools made by EMAG, two wet cleaning machines from MAFAC, a gas nitriding retort furnace from IVA. In addition, the research building contains further machines in the factory hall, the technical building equipment of the ETA-Factory as well as various thermal storage units, an absorption chiller, and the thermally activatable building façade. The machine park and the entire technical building services are supplemented by an interactive learning course. Here, not only the basics of energy and energy efficiency are explained, but also detailed descriptions of the innovations in the ETA-Factory process chain as well as in the rest of the factory building. Additionally, a playful introduction to energy efficiency is enabled by a teaser demonstrator (touch screen with integrated software solution) using the example of the greenfield process chain of the ETA-Factory.

11.6 Best Practice Example 6: ETA-Factory at PTW, TU Darmstadt, Germany

Fig. 11.10 ETA-Factory shop floor: Greenfield process chain (left) and Brownfield process chain (right)

11.6.2.2 Brownfield Process Chain

The brownfield value chain of the ETA-Factory is shown on the right-hand side of Fig. 11.10. This is part of the learning factory for energy productivity (LEP), which was installed on the shop floor of the ETA-Factory as part of a cooperation between PTW and McKinsey & Company. In the process chain of the LEP, different gear shaft combinations are manufactured for different gear types. At the end of the process chain, after final assembly with externally purchased housing parts, ready-to-ship gearboxes for various purposes are available. The LEP process chain is representative of discrete mechanical manufacturing operations and consists of a turning machine, a steam cleaning system including steam generator, two robots, a continuous curing oven, a shrink oven, two decentralized air compressors, and several assembly processes.

A special feature of the LEP is that the process chain can be converted from an energy-inefficient to an energy-efficient system design (and vice versa) within a few minutes. This conversion demonstrates the energetic optimization of an existing brownfield system. Based on the inefficient system state, training participants

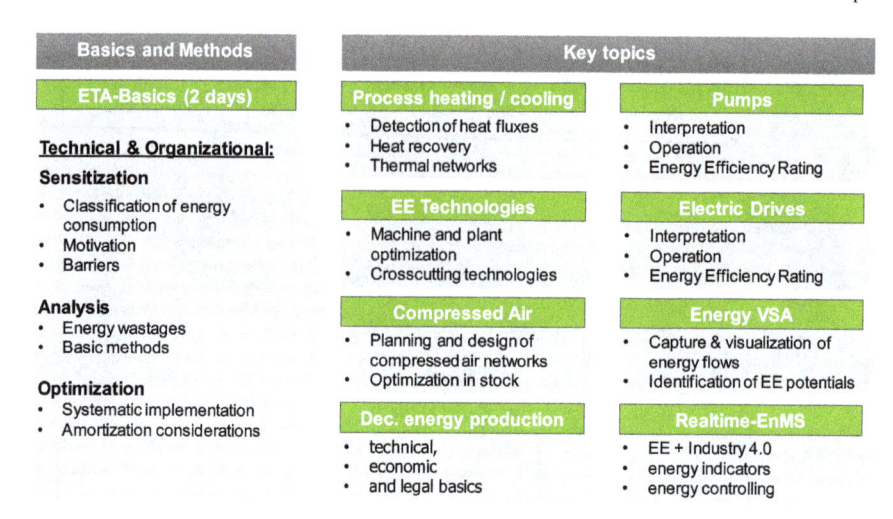

Basics and Methods	Key topics	
ETA-Basics (2 days)	**Process heating / cooling**	**Pumps**
Technical & Organizational: **Sensitization** • Classification of energy consumption • Motivation • Barriers **Analysis** • Energy wastages • Basic methods **Optimization** • Systematic implementation • Amortization considerations	• Detection of heat fluxes • Heat recovery • Thermal networks	• Interpretation • Operation • Energy Efficiency Rating
	EE Technologies	**Electric Drives**
	• Machine and plant optimization • Crosscutting technologies	• Interpretation • Operation • Energy Efficiency Rating
	Compressed Air	**Energy VSA**
	• Planning and design of compressed air networks • Optimization in stock	• Capture & visualization of energy flows • Identification of EE potentials
	Dec. energy production	**Realtime-EnMS**
	• technical, • economic • and legal basics	• EE + Industry 4.0 • energy indicators • energy controlling

Fig. 11.11 ETA-Factory training offers for industry

can learn to understand the energy flows in the processes, identify existing energy wastages, and derive improvement measures. In addition, trainees can immediately implement optimization measures and analyze in detail the potential of individual measures.

11.6.3 Operation

The energy efficiency training sessions in the ETA-Factory are conducted as interactive workshops in which theoretical foundations are linked with practical implementation. While theoretical sessions usually take place in the seminar room, the practical parts use the learning environment of the ETA-Factory. A lively exchange of knowledge and experience between the moderators and participants is intended in order to create a fruitful learning atmosphere. The current training offer consists of a basic module on energy efficiency in industry and several modules addressing specific key topics; see Fig. 11.11.

11.7 Best Practice Example 7: Festo Didactic Learning Factories

Author: Reinhard Pittschellis[a], Dirk Pensky[a]
[a]Festo Didactic

11.7 Best Practice Example 7: Festo Didactic Learning Factories

Name of learning factory: MPS, iCIM, CP Factory
Operator: Customers of Festo Didactic
Year of inauguration: 1989
Floor Space in learning factory: 2 – 50 m2
Manufacture product(s): pneumatic cylinder (model), Deskset, Microcontroller
Main topics / learning content: Automation Technology

Extract from the morphology

2.1	main purpose	education	training	research
2.2	secondary purpose	test environment / pilot environment	industrial production	innovation transfer	public image
3.1	product life cycle	product planning	product development	prototyping			service	recycling
3.2	factory life cycle	investment planning	I factory concept	process planning	ramp-up	manufacturing	assembly	logistics	maintenance	recycling
3.3	order life cycle	configuration & order	order sequencing	planning and scheduling			picking, packaging	shipping
3.4	technology life cycle	planning	development	virtual testing			maintenance	modernization
3.5	indirect functions	primary activities				secondary activities		
		Inbound & outbound logistics	marketing & sales	service	firm infra-structure	HR	technology development	Procurement
4.1	learning environment	purely physical (planning + execution)	physical LF supported by digital factory	physical value stream of LF extended virtually	purely virtual (planning + execution)
4.2	environment scale	scaled down	life-size		

11.7.1 History and Categories

Festo Didactic develops learning factories since 1989. Different to most of the other learning factories described in this volume, the learning factories from Festo Didactic are not operated by Festo itself, but by its customers.

Festo Didactic has been founded as a subsidiary of Festo Automation. Its mission was initially to provide training for products of Festo. Over time Festo Didactic's product range evolved, and today it covers additional fields like production technology, mechatronics, and electronics.

The first learning factory evolved out of training packages for pneumatic and hydraulics (Fig. 11.12). Within the framework of a funded project in partnership with vocational schools, the first so-called "Modular Production System" (MPS) was developed. The intention was to develop an environment to train competencies which could not be trained with the pneumatics trainer, e.g.

- PLC programming,
- fault finding,
- optimization,
- maintenance, and
- setup of industrial production lines.

The characteristics of the first MPS (and still of the actual one) were:

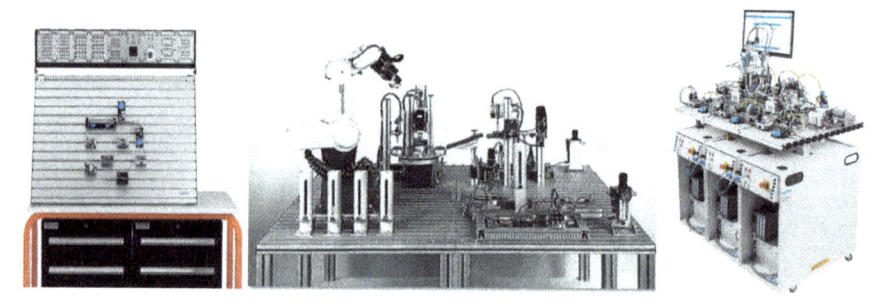

Fig. 11.12 Pneumatic Trainer (1980), MPS first release (1998), MPS fourth release (2017)

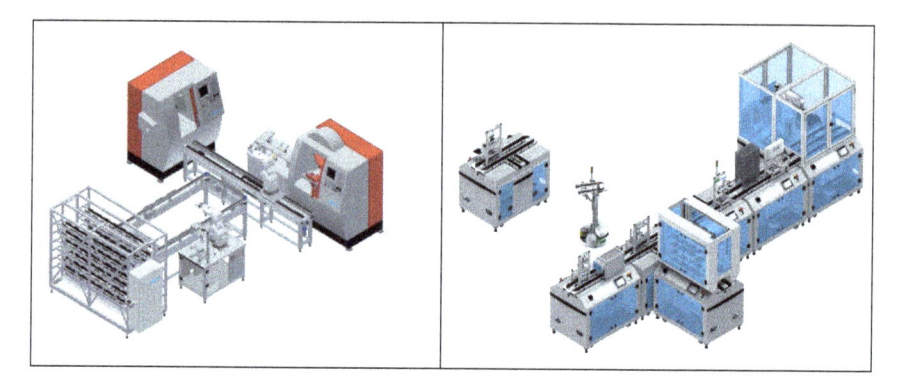

Fig. 11.13 iCIM (2005), CP Factory (2016)

- drastically simplified process,
- compared to real production lines reduced size and simplified, reusable product,
- modular system approach,
- focus on assembly processes,
- designed to train automation technologies,
- usage of industrial components.

Since 1990, MPS is the official equipment for the Worldskills Mechatronics competition.

Soon after the first MPS was launched, customers demanded extended and more complex solutions. Especially the research for "Computer-Integrated Manufacturing" raised demands for installations additionally consisting of,

- CNC machines with robot loading,
- automated warehouses,
- flexible material transport by conveyor systems, and
- cell line controllers.

The result was the product line iCIM (Fig. 11.13), which was sold successfully for many years.

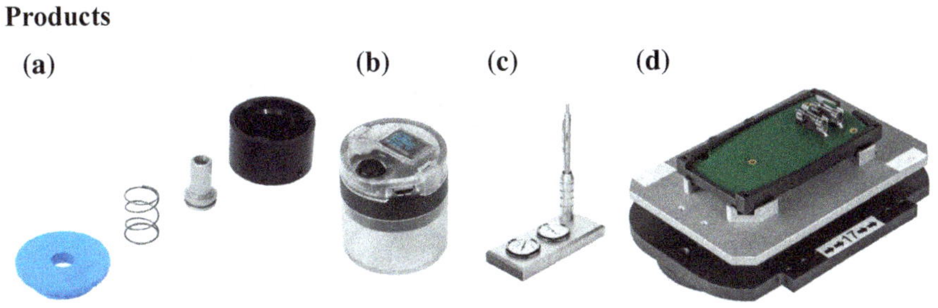

11.7 Best Practice Example 7: Festo Didactic Learning Factories

Products

(a) (b) (c) (d)

Fig. 11.14 Products of the Learning Factories of Festo, **a** single-acting cylinder (MPS A), **b** microcontroller (MPS D), **c** deskset (iCIM), **d** electronic device (CP Factory)

With the upcoming of Industrie 4.0, new solutions became necessary resulting in the development of a new class of learning factories called CP Factory (Cyber-Physical Factory). Main features of this learning factory are:

- refined and more versatile material transport by double conveyor,
- free arrangement of modules without reprogramming,
- intelligent control of material flow by RFID,
- Cyber-physical components like the cyber-physical Stopgate,
- integrated manufacturing execution system (MES), and
- connection to manufacturing cloud services, e.g. for condition monitoring.

11.7.2 Products

Every Learning Factory produces a product (Fig. 11.14). Learning Factories from Festo Didactic in most cases have reusable model products in order to keep operational costs low.

A single-acting pneumatic cylinder was the product of the first MPS. It consisted of a cylinder housing, a piston, a return spring, and a cap. This product could be assembled fully automatically and allowed a simple form of quality check (color and height of the cylinder). It did not have many variants, since the main purpose of MPS was basic and advanced training of PLC programming, so no complex process was required.

With the latest version of MPS, new product was introduced, too. Keeping the old form factor of a cylinder with a diameter of 40 mm, it is now an electronics unit with a display and a joystick. This makes it versatile, since the function of the product can be changed by different programs.

iCIM had a product called deskset, consisting of different aluminum parts that had to be machined in the CNC machines of iCIM. One of the main purposes of iCIM was the training for CNC. This product was not reusable.

The product of CP Factory is a simplified electronic device, consisting of a circuit board and a housing in two parts. The housing can be machined and then assembled with the circuit board. In addition, the production process can be enhanced by other process steps like drilling, gluing, and heating.

11.7.3 Customers and Operation Models

Typical customers for Festo Didactic's learning factories are vocational training institutions and universities. These institutions use the learning factory to teach and train different aspects of automation technology, following a given curriculum.

Some universities use the learning factory for research or research-based learning, too.

The staff to operate such learning factories is usually quite small; very often there is only the teacher.

Many customers use the learning factories together with exercises prepared by Festo Didactic. The students work under supervision of the teacher, who provides the necessary training material and assigns exercises to the students.

Other customers, mainly universities, use the learning factory for project work. Students have to solve problems, e.g. optimizing the output or reducing the energy consumption. They do this by modifying the stations or the layout or by adding stations, which is easy due to the modular design of the systems.

11.7.4 Didactical Environment

Learning factories are didactical products; therefore, they come with extensive Courseware, starting with ready-to-use exercises, multimedia learning programs, and simulation software.

11.8 Best Practice Example 8: Festo Learning Factory in Scharnhausen, Germany

Author: Holger Regber[a]
[a]Festo Didactic SE

11.8 Best Practice Example 8: Festo Learning Factory in Scharnhausen ...

Name of learning factory: **Festo Learning Factory Scharnhausen**
Operator: **Festo AG**
Year of inauguration: **2014**
Floor Space in learning factory: **220m²**
Manufacture product(s): **Pneumatic valves and valve terminals**
Main topics / learning content: **workplace-oriented trainings, Industry 4.0 and lean production**

Extract from the morphology

2.1	main purpose	education	training	research
2.2	secondary purpose	test environment / pilot environment	industrial production	innovation transfer	public image
3.1	product life cycle	product planning	product development	prototyping			service	recycling
3.2	factory life cycle	investment planning	factory concept	process planning	ramp-up	manufacturing / assembly / logistics	maintenance	recycling
3.3	order life cycle	configuration & order	order sequencing	planning and scheduling			picking, packaging	shipping
3.4	technology life cycle	planning	development	virtual testing			maintenance	modernization
3.5	indirect functions	primary activities				secondary activities		
		Inbound & outbound logistics	marketing & sales	service	firm infrastructure	HR	technology development	Procurement
4.1	learning environment	purely physical (planning + execution)	physical LF supported by digital factory	physical value stream of LF extended virtually	purely virtual (planning + execution)
4.2	environment scale	scaled down	life-size

11.8.1 Starting Phase and Purpose

The Festo Learning Factory Scharnhausen was established in 2014 as part of the new Festo Technology Plant what is focused on the production of pneumatic valves and pneumatic valve terminals in a wide range of variants and dimensions. During the planning process of the new technology plant, the project team detected a particular need for small- and demand-oriented learning units. The qualification opportunities for operators, tool setters, and foremen offered by the Festo Academy were too distant from the real work requirements. They were too long and unsuitable for the daily business in production departments.

Otherwise the project team saw no chances to integrate the training directly in the work sequences. They were afraid that the learning sequences could often be disturbed by daily problems and could cause a major setback. They decided in cooperation with the Festo board to establish a learning factory for the technology plant that will be able to provide small-, demand-, and workplace-oriented learning units. Regarding the German working laws, no unit should be longer than two hours. The main focus was on the development of skills; therefore, the real production equipment should be used. A second approach was requested: the development of a serious business game which combines all the learning places to a didactical simplified but real value stream. That should give the operators and team leaders the opportunity to

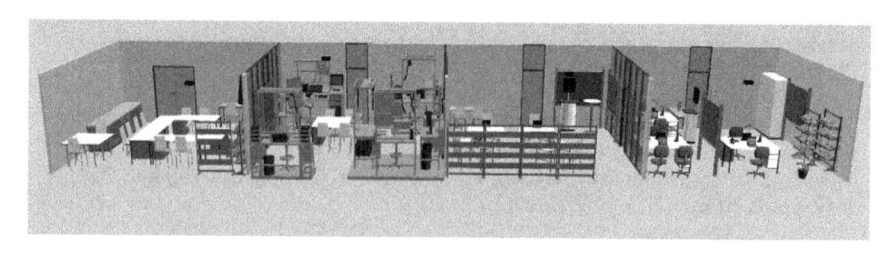

Fig. 11.15 Festo Learning Factory Scharnhausen as cross section

completely understand the concept, test process optimizations, and play an impact of different changes such as higher variance, changed customer requirements, or some disruption by production. This business game is called ProBest.

The learning factory should provide the opportunity to familiarize the production employees with the new developments in digitalization and automation, called Industrie 4.0. In order to fulfill these objectives, the Festo Learning Factory was funded by the Festo board and the Festo owner family.

11.8.2 Learning Environment and Products

The Festo Learning Factory is located on the fourth level of the Festo Technology Factory in Scharnhausen (Fig. 11.15), surrounded by different production departments. The total surface area is 220 m^2, separated into four rooms. Each of the rooms has their own focus. Room 1 is focused on mechanical processing and on general topics, room 2 on valve and valve terminal assembly. Room 3 is focused around automation and process improvements. It contains one of the highlights: the cyber-physical system which is adapted to the used Industrie 4.0 approach in the Festo Technology Plant. There are RFID product identification, different kinds of robots, a fully automated material flow, and networking to the MES.

Finally, room 4 is needed for the administration of the learning factory, which is done by two Festo apprentices: one for the administrative questions and one for all maintenance issues.

The walls between rooms 1 and 3 can be opened to offer a wider space for bigger groups or for learning topics around value stream optimizations.

Currently, the learning factory offers fourteen different learning workplaces with twenty-five learning topics and more than forty learning modules. All learning processes are based on elements and components of VUVG valves and VTUG valve terminals. With this range, it is possible to cover more than three hundred product variants. This is a small number for the Festo production world where over 10^{40} product variants are theoretically covered. However, it is a wide range for the approach of a learning factory, but variety and the handling of complexity are parts of the learning process.

Fig. 11.16 Training session on the cyber-physical system

11.8.3 Operation

The Festo Learning Factory is used exclusively for Festo employees, mainly for the training of new operators, but also for the advanced qualification of incumbent workers. Based on a qualification matrix, the team leaders know the level of skills and knowledge of their team members and take measures for improving the qualification level. A particular solution was founded for the acquisition of trainers. Depending on the small learning units, there was not a chance to contract professional trainers. The cost would be too expensive. So Festo decided the qualification of operators and tool setters should be a leadership task. Each team leader or a qualified team specialist should train the operators by themselves. For their qualification, a special "Train-the-Trainer" program was provided. Based on prepared training modules, the trainers were enabled to train adults regarding the intended training outcomes, handle the learning equipment, and evaluate the learning success. Now more than forty learning units with up to four thousand participants are taught (Figs. 11.16, 11.17, and 11.18).

Furthermore, the development of training is a continued process. Because of the development of new products and production processes, new production equipment, and turbulent market trends, the plant managers use the Festo Learning Factory to join together with their employees and play an impact on the value streams. So, the mentioned ProBest serious game was further developed into the ProBest Advanced. In that format, the real, didactical simplified value stream is confronted

Fig. 11.17 Scene during the optimization phase at the ProBest game

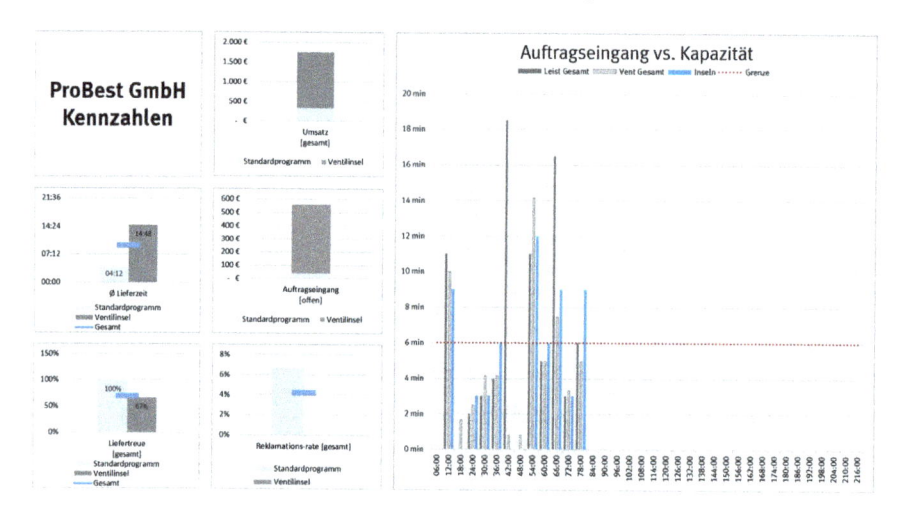

Fig. 11.18 Evaluation of KPI's at the ProBest game by dashboard

with different changes and the participants should develop suitable strategies to manage these influences. Like a live test.

11.9 Best Practice Example 9: IFactory at the Intelligent Manufacturing Systems (IMS) Center, University of Windsor, Canada

Authors: Hoda ElMaraghy[a]
[a]Intelligent Manufacturing Systems (IMS) Center, University of Windsor, Canada

Name of learning factory: **iFactory**
Operator: **Intelligent Manufacturing Systems Center, University of Windsor, Canada**
Year of inauguration: **2011**
Floor Space in learning factory: **200 sqm**
Manufacture product(s): **Family of deskset and Family of belt tensioner**
Main topics / learning content: **Integrated products - systems Learning, Industry 4.0**

Extract from Morphology

2.1	main purpose	education		training		research	
2.2	secondary purpose	test environment / pilot environment	industrial production	innovation transfer		public image	
3.1	product life cycle	product planning	product development	prototyping		service	recycling
3.2	factory life cycle	investment planning	factory concept	process planning	ramp-up	maintenance	recycling
3.3	order life cycle	configuration & order	order sequencing	planning and scheduling		picking, packaging	shipping
3.4	technology life cycle	planning	development	virtual testing		maintenance	modernization
3.5	indirect functions	primary activities				secondary activities	
		Inbound & outbound logistics	marketing & sales	service	firm infrastructure	HR	technology development	Procurement
4.1	learning environment	purely physical (planning + execution)	physical LF supported by digital factory	physical value stream of LF extended virtually		purely virtual (planning + execution)	
4.2	environment scale	scaled down		life-size			

11.9.1 Starting Phase and Purpose

An integrated products/systems Learning Factory, the first of its kind in North America, was set up in 2011 at the Intelligent Manufacturing Systems (IMS) Centre at the University of Windsor in Canada (Fig. 11.19), and is codirected by Professor Hoda ElMaraghy and Professor Waguih ElMaraghy. It was initially funded by research awards including infrastructure grant by Canada Foundation for Innovation (CFI) and Ontario Ministry of Research and Innovation (MRI), along with industrial contributions. On-going operation and research in the Learning Factory is supported by the Canada Research Chairs (CRC), Natural Sciences and Engineering Research Council (NSERC) of Canada and industrial grants and contracts.

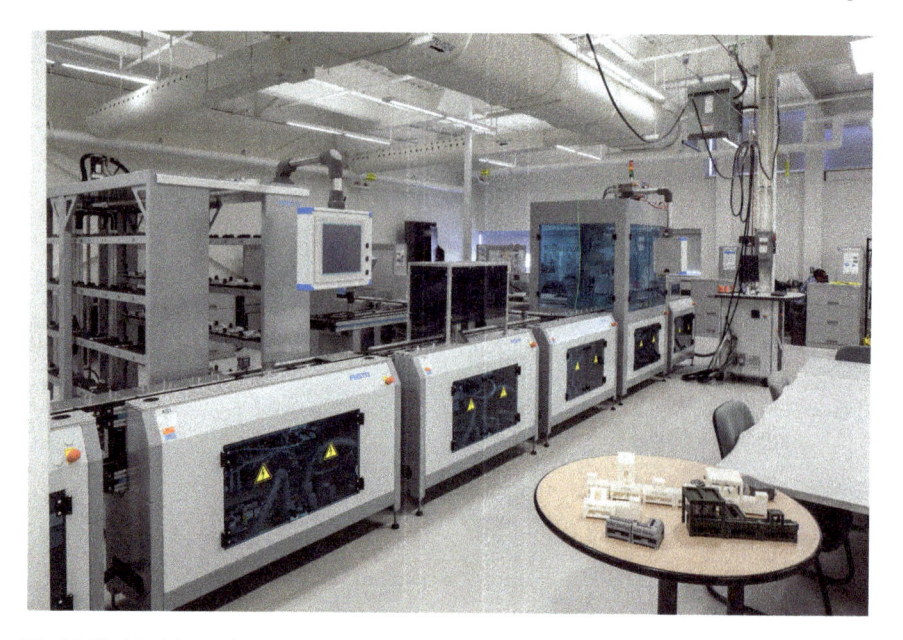

Fig. 11.19 Modular and reconfigurable iFactory at the IMS center

The objective is "Systems Learning," which integrates products design, customization, and personalization, through the *iDesign* and *iOrder* modules, with the design, planning, and control of changeable manufacturing systems, and development of innovative physical and logical enablers of change on the shop floor, e.g. variant-oriented reconfigurable process and production plans through the *iPlan* module (Fig. 11.20). This state-of-the-art Learning Factory provides an experiential design, planning, and realization environment that is conducive to innovation in products, processes, and systems. It is used primarily for research but also for teaching and demonstrations for students and industry.

11.9.2 Learning Environment and Product

The integrated Learning Factory is a complete experiential learning environment from product design in the Design Innovation Studio (iDesign) and design prototyping and metrology equipment to process and production planning tools (*iPlan*) and its manufacture (*iFactory*) supported by systems design synthesis and configuration algorithms and methodologies.

The *iFactory* (by FESTO Didactic) is a truly modular and reconfigurable assembly system with the ability to change both its configuration and layout by modules relocation, addition, and/or removal. This "Factory-in-a-Lab" contains modular Plug'n Produce robotic and manual assembly, computer vision inspection, automated stor-

11.9 Best Practice Example 9: IFactory at the Intelligent …

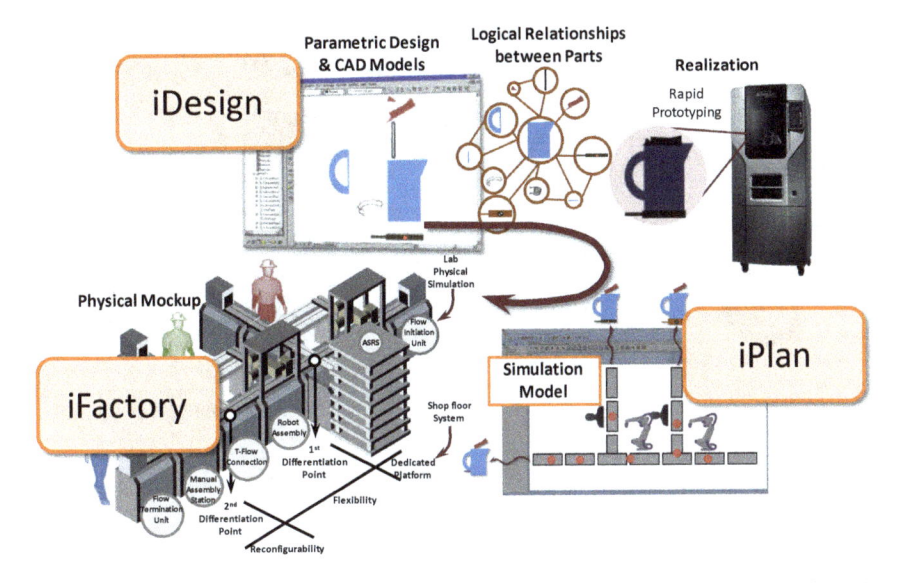

Fig. 11.20 Integrated products and systems design, planning and control demonstrated within the learning Factory environment at the IMS center

age and retrieval system (ASRS), material handling modules, RFIDs communication sensors, and Siemens SCADA control system. Its intelligent control with neighbor awareness capability and modular standardized interfaces do not require reprogramming or change of setup after physical reconfiguration, which is easily done in couple of hours, greatly reduces ramp-up efforts and time.

Innovation Design Studio (*iDesign*) with state-of-the-art equipment to foster innovative design and group interactions includes interactive 3D graphics tablets and displays supported by powerful PC-based applications and a computing environment for design synthesis, configuration, modeling, and analysis that integrates products, processes, and manufacturing systems development.

The *iOrder* (for customized orders entry) module, complemented by the *iDesign, iPlan,* and *iFactory* environment, 3D printing, and coordinate measuring machine facilities constitute the "Learning Factory" (Fig. 11.20) which provides a unique experiential learning, training, and research experience for undergraduate and graduate students, researchers and professional trainees. Knowledge elements covered include products design, prototyping, customization, and personalization; variant-based process planning, order processing for mixed-model production, dynamic production planning and scheduling, and principles and enablers of flexible, reconfigurable, and changeable intelligent manufacturing system.

The current products, assembled in the iFactory, are (i) a family of office desksets and (ii) a family of automobile engine belt tensioners (Fig. 11.21). The carrier base plate, work piece pallet, and product base plate are linked by positioning pins and corresponding holes and is capable of holding members of both products families.

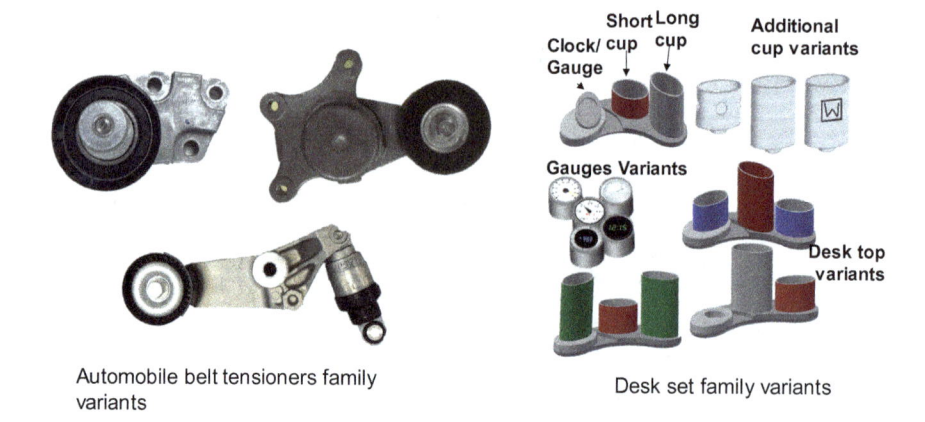

Fig. 11.21 Assembled products families in the Integrated Systems Learning Factory

Additionally, the work piece pallet is equipped with RFID tags, which allow the tracking of processes and production operations planning and scheduling, and principles and enablers of flexible, reconfigurable and changeable intelligent manufacturing systems.

11.9.3 Operation

Since 2011, the integrated learning factory at IMS center has provided an exceptional experiential learning, training, and research experience to undergraduate and graduate students, researchers, and professional trainees. Training is carried out by senior IMS center researchers. Knowledge elements covered include products design, prototyping, customization, and personalization; variant-based process planning, order processing for mixed-model production, dynamic production planning and scheduling, and principles and enablers of flexible, reconfigurable and changeable intelligent manufacturing systems. The Learning Factory facilitates the codesign and codevelopment of products and manufacturing systems for the whole life cycle and developing the enabling technologies needed to increase manufacturing competitiveness, agility, and flexibility.

This environment is also conducive to innovative research such as: (i) product variety management; (ii) manufacturing systems complexity; (iii) coevolution and codevelopment of products and their manufacturing systems inspired by biological evolution; (iv) design synthesis of assembly systems and their optimum granularity; (v) manufacturing systems layout complexity modeling and metrics; (vi) products and systems configuration coplatforming; (vii) development of Digital Twin of the *iFactory*; and (viii) applications of Industrie 4.0.

11.10 Best Practice Example 10: IFA-Learning Factory at IFA, Leibniz University Hannover, Germany

Authors: Melissa Seitz[a], Lars Nielsen[a], Niklas Rochow[a], Peter Nyhuis[a]
[a]Institute of Production Systems and Logistics (IFA), Leibniz Universität Hannover

Name of learning factory: **IFA-Learning Factory**
Operator: **IFA, Leibniz Universität Hannover**
Year of inauguration: **2000**
Floor Space in learning factory: **150 m²**
Manufacture product(s): **Helicopter and components**
Main topics / learning content: **Factory Planning, Lean Production, PPC**

Extract from the morphology

2.1	main purpose	education	training		research	
2.2	secondary purpose	test environment / pilot environment	industrial production	innovation transfer	public image	
3.1	product life cycle	product planning	product development	prototyping	service	recycling
3.2	factory life cycle	investment planning	factory concept	process planning	ramp-up	maintenance	recycling
3.3	order life cycle	configuration & order	order sequencing	planning and scheduling	picking, packaging	shipping
3.4	technology life cycle	planning	development	virtual testing	maintenance	modernization
3.5	indirect functions	primary activities: Inbound & outbound logistics, marketing & sales, service, firm infrastructure			secondary activities: HR, technology development, Procurement	
4.1	learning environment	purely physical (planning + execution)	physical LF supported by digital factory	physical value stream of LF extended virtually	purely virtual (planning + execution)	
4.2	environment scale	scaled down	life-size			

11.10.1 Starting Phase and Purpose

The objective of starting the development of the IFA-Learning Factory (www.ifa-lernfabrik.de) in the year 2000 was the creation of a model factory, in which real situations and conditions out of industry practice are represented in a simplified manner, yet as realistic as possible. Furthermore, it was aimed to implement various trainings and seminars in one single learning environment with connecting elements.

Trainings and seminars nowadays range from designing, planning, controlling, and monitoring production systems. Within the framework of different production scenarios, the participants experience the effects of their own decisions regarding structural adjustments of production setups as well as changes of production planning and control (PPC) parameters without financial risk. The advantages of changes in processes and arrangements can thus be experienced in practice and provide confidence for the knowledge transfer in the participant's companies.

Fig. 11.22 Impressions from the IFA-Learning Factory

11.10.2 Learning Environment and Products

The prerequisite for the realization of various training courses and seminars on different production logistics issues in the same learning factory is its changeability. In the IFA Learning Factory, various existing workstations can be used and arranged in multiple ways. This includes two NC milling machines and six mobile, versatile workstations as well as two administrative workstations and a material supply warehouse. For the provision of materials, semi-finished and finished products, diverse flow racks are also available. Figure 11.22 gives an overview of the IFA Learning Factory and the technology used.

The workstations are equipped with a variety of interchangeable tools and fixtures. Furthermore, they are flexible and easy to move, thanks to equipment with rollers, automatic height adjustment, and wireless network connection. The possibility of repositioning the workstations is also facilitated by a regular ceiling grid of power, compressed air, and data connections. This flexible infrastructure makes it possible to emulate many different setups in a manufacturing company. This not only enables the implementation of scenarios with different focal points, but also the orientation on and implementation of specific ideas and wishes of the training participants.

The product manufactured in the IFA-Learning Factory is a helicopter model consisting of 13 different metal and plastic components including variants. The product complexity is therefore adequately high, in order to teach existing interdependencies and is yet sufficiently manageable to enable learners to familiarize themselves with the analysis and design of a production system in the given time of a training.

11.10 Best Practice Example 10: IFA-Learning Factory at IFA ...

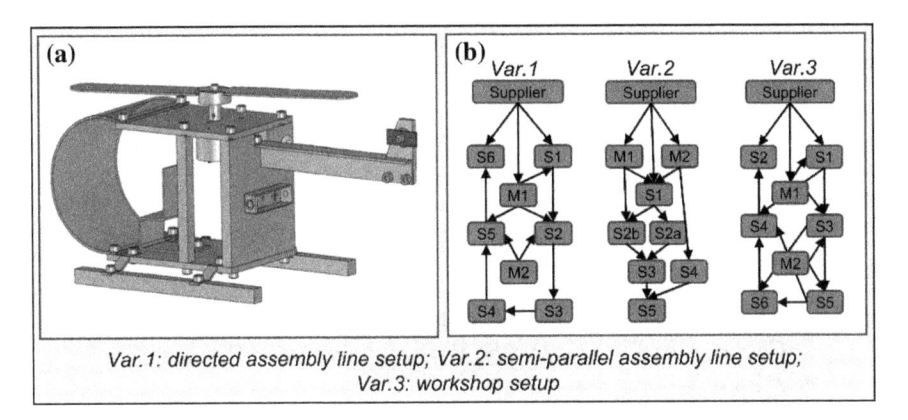

Fig. 11.23 **a** Product model of a helicopter and **b** Selection of possible setups

11.10.3 Operation

In the business simulation game, the participants become employees of a fictitious company that produces the helicopters. Depending on the training scenarios and the chosen setup (Fig. 11.23), different parts of this company are considered. This includes, for example, two scenarios in which the component premanufacturing or the final assembly is represented. The premanufactured components are taken up again in the final assembly scenario and are used to assemble the helicopter. This combination of scenarios and various setups promotes a holistic understanding of production logistics in the internal supply chain.

With regard to the entire company, the training courses address issues of factory planning according to existing standards as well as new methods and tools in addition to the related plant structure development. Using the helicopter's final assembly scenario as an example, the design and evaluation of assembly processes, assembly organization, and layout design with a special focus on the principles of lean production are examined in greater depth. In this context, the focus is set on the effects of harmonization and balancing of work content in the assembly process as well as various material supply options. In addition, at some of the workstations, the emphasis can be set on ergonomic workplace design or human–robot collaboration. Based on the scenario of the component premanufacturing, which is organized according to the workshop principle, questions regarding efficient order processing are addressed. In this context, the areas of PPC as well as production monitoring are amplified. In addition to the production logistics content, specific training courses also integrate leadership aspects with a special focus on the successful management of change processes in the focus of Industrie 4.0.

In view of the challenges and potentials of increasing digitalization, the IFA Learning Factory also serves as a demonstrator for various innovative technologies. For example, order processing is facilitated by a digital order support and dynamic information provision via electronic shelf labels (ESL). This is made possible by the automated collection and evaluation of feedback data in an individually programmed PPC system. In addition, real-time locating systems (RTLS) are used for the analysis of material and personnel movements. Operational processes such as material picking and supply are assisted by the use of augmented reality glasses. With the help of biomechanical measuring systems, the ergonomics of various workplaces can be evaluated and improved.

11.11 Best Practice Example 11: Integrated Learning Factory at LPE & LPS, Ruhr-University Bochum, Germany

Author: Beate Bender[a]
[a]Product Development (LPE), Ruhr-University Bochum

Name of learning factory: **Integrated Learning Factory**
Operator: **LPE & LPS, Ruhr-University Bochum**
Year of inauguration: **2015**
Manufacture product(s): **Percussion drilling machine**
Main topics / learning content: **Collaboration of product development and production**

Extract from the morphology

2.1	main purpose	education		training			research	
2.2	secondary purpose	test environment / pilot environment		industrial production		innovation transfer	public image	
3.1	product life cycle	product planning	product development	prototyping	manufacturing	assembly	logistics	service	recycling
3.2	factory life cycle	investment planning	factory concept	process planning	ramp-up			maintenance	recycling
3.3	order life cycle	configuration & order	order sequencing	planning and scheduling				picking, packaging	shipping
3.4	technology life cycle	planning	development	virtual testing				maintenance	modernization
3.5	indirect functions	primary activities				secondary activities			
		Inbound & outbound logistics	marketing & sales	service	firm infrastructure	HR	technology development	Procurement	
4.1	learning environment	purely physical (planning + execution)		physical LF supported by digital factory		physical value stream of LF extended virtually		purely virtual (planning + execution)	
4.2	environment scale	scaled down				life-size			

11.11.1 Starting Phase and Purpose

This learning factory was tested first in 2015 at Ruhr-University Bochum to prepare future engineers for their job specification. Studies reveal that a significant part of

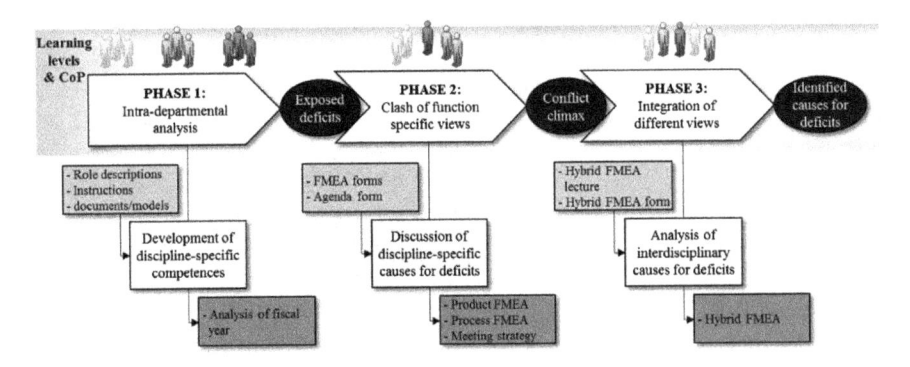

Fig. 11.24 Procedure of the learning factory

university graduates still have a lack of general competences. In addition to domain specific expertise, these are main factors of employability. This applies in particular in the context of simultaneous engineering which is characterized by temporary overlapped and integrated work processes instead of isolated subsequent workflows. Life-cycle spanning disciplines such as production engineering or industrial services must be integrated in the early phases of product development.

In line with the propagated "shift from teaching to learning," this learning factory represents a student-centered approach. Active learning methods facilitate longlasting learning results due to action-orientation and knowledge transfer in an authentic context. The new didactical concept focuses in particular on the improvement of interdisciplinary collaboration between product development and production. Literature review shows that nearly all existing learning factories address the optimization of production processes (e.g. manufacturing or assembly training) whereas the upstream product development processes receives almost no consideration yet.

11.11.2 Learning Environment and Products

The students are made aware of the problems caused by missing collaboration between product development and production by using the conception of negative experience. Based on a role play, the students go through a conflict situation which is representative for a real-life producing company. A company that develops and produces percussion drilling machines constitutes the simulated environment. In the course of this role play (see Fig. 11.24), the students experience issues in the interfunctional contact of product development and production which often arise when these departments work independent from each other. After reaching a conflict climax, the students are supervised in applying an exemplary method that supports interdisciplinary cooperation to overcome the case-specific barriers. Finally, the benefits of a systematic approach are reflected jointly.

Fig. 11.25 Examples of the provided models/simulations; students reassembling a dismantled product

As physical setting for this role play, the existing learning factory at the Chair of Production Systems (LPS) is used which is already covering main aspects of the typical production process and current research topics. In addition, it is not only used for education and research but also for make-to-order production. The machines used in the learning factory have been used for manufacturing of real customer orders before. So, this production area with its heterogeneous machinery of different ages is a particularly realistic image of a manufacturing environment.

11.11.3 Operation

The learning factory is conducted as an optional four-hour exercise for students attending either the course "fundamentals of product development" held by the chair of product development (LPE) or the course "networked production systems" held by the chair of production systems (LPS). In the beginning, the students are divided into small groups, each undergoing the same procedure simultaneously, supported and observed by one lecturer each. In every group, each member takes a specific role: 2–3 students act as engineering product designers, 2–3 as production employees, and the last one takes over the role of the project manager.

The first phase allows for the individual students to identify with their role and to get sufficient insight into the work of "their" department. The product development group is provided with CAD models and manuals to understand the functionality of this product while the production group analyzes the processes of production and assembly on the basis of material flow models and a dismantled product. The project managers receive financial reports and customer feedback to identify major problems, which originate in both departments and are discussed in a joint meeting (Fig. 11.25).

Fig. 11.26 Joint meetings in presence of the management board

The second phase aims to emphasize the difficulties of solving problems when each party sticks solely to their own views and goals. Therefore, the students are instructed to find reasons for the problem in their own department by using the product FMEA in the department of product development or the process FMEA in the production group. The project managers are encouraged to push the workers and emphasize the importance of the meeting by announcing the attendance of the management board. Thereby a conflict between the departments is forced where each party aims to protect their own interests instead of revealing potential own failures (Fig. 11.26).

In a third phase, the students learn how different views on engineering problems can cause conflicts and how interdisciplinary cooperation between different engineering functions can be supported by methods with a mutual approach. Therefor, the conflicting interaction during the previous phases is reflected and discussed in a feedback round. Afterward, the lecturer introduces the hybrid FMEA as an exemplary method to enable better cooperation by integrating the views of both disciplines. The students get the chance to meet again in their interdisciplinary groups to discuss the cost and quality problems once again by using the hybrid FMEA. At the end of this phase, the new experiences are discussed again in the whole group.

11.12 Best Practice Example 12: LEAN-Factory for a Pharmaceutical Company in Berlin, Germany

Authors: Holger Kohl[a,b], Roland Jochem[b], Felix Sieckmann[b], Christoffer Rybski[a]
[a]Fraunhofer IPK, [b]TU Berlin

Name of learning factory: **LEAN-Factory**
Operator: **Fraunhofer IPK, TU Berlin, ITCL GmbH, pharmaceutical company**
Year of inauguration: **2014**
Floor Space in learning factory: **400 sqm**
Manufacture product(s): **Pharmaceutical Tablets (bottled, blistered)**
Main topics / learning content: **Lean Management**

Extract from the morphology

2.1	main purpose	education		training				research	
2.2	secondary purpose	test environment / pilot environment		industrial production		innovation transfer		public image	
3.1	product life cycle	product planning	product development	prototyping				service	recycling
3.2	factory life cycle	investment planning	factory concept	process planning	ramp-up	manufacturing	assembly / logistics	main-tenance	recycling
3.3	order life cycle	configuration & order	order sequencing	planning and scheduling				picking, packaging	shipping
3.4	technology life cycle	planning	development	virtual testing				main-tenance	moderni-zation
3.5	indirect functions	primary activities						secondary activities	
		Inbound & outbound logistics	marketing & sales	ser-vice	firm infra-structure	HR		technology development	Procure-ment
4.1	learning environment	purely physical (planning + execution)		physical LF supported by digital factory		physical value stream of LF extended virtually		purely virtual (planning + execution)	
4.2	environment scale	scaled down		life-size					

11.12.1 Starting Phase and Purpose

According to a "Pharma Operations Benchmarking Study" of McKinsey the pharmaceutical industry faces three major challenges:

- Increase the performance of facilities and plants;
- interconnect and configure facilities and plants;
- increase quality and compliance.

To meet these challenges, a leading German pharmaceutical company established a "LEAN-Factory" together with its partners Fraunhofer IPK TU Berlin and ITCL (International Transfer Center for Logistics) in 2014. A main purpose of this learning factory is to function as a competence center for lean management tools and the lean philosophy within the whole company. Internally, it has become the number one training facility to qualify employees from the shop floor level through to the management level from Germany but also from international sites. The focus is on lean methods and tools, standardization, mind set and behavior as well as performance culture.[2] Since there is mostly a mixture of participants from different sites and functions, the trainings also serve as a knowledge exchange platform. In addition to the employee trainings, there are special trainings for students from universities.

---

[2]For further information, see Rybski and Jochem (2016).

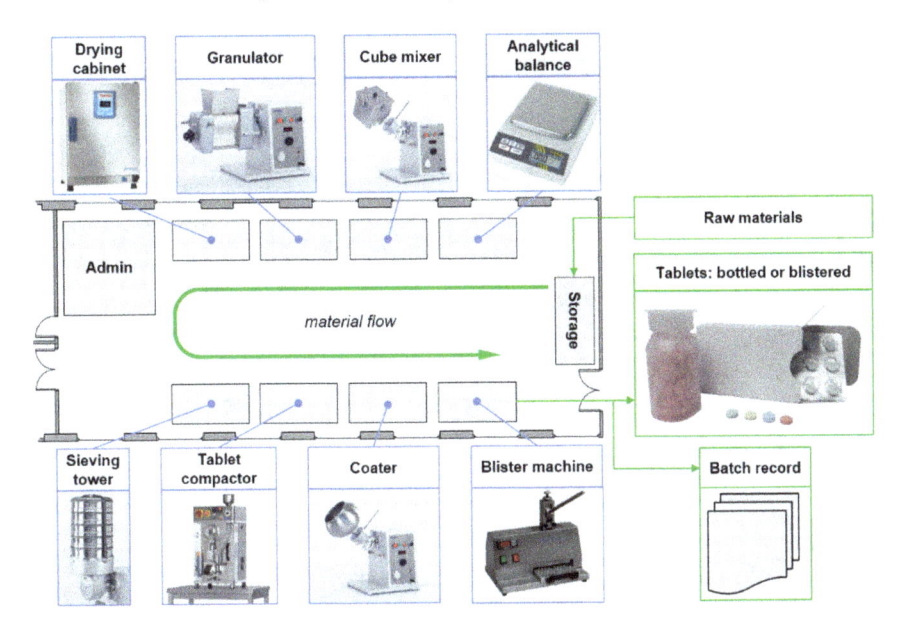

Fig. 11.27 Layout, process, equipment, and products of the production environment

11.12.2 Learning Environment and Products

The core element of the LEAN-Factory is a realistically replicated pharmaceutical production of tablets (Fig. 11.27). Input materials are first weighed and subsequently mixed, granulated, dried, sieved, and compressed into tablets. An additional coating is possible, if a high variant diversity needs to be demonstrated. Afterward, the tablets are manually packaged into blisters or bottles and then into boxes. Several quality tests are performed in between, e.g. testing of tablet hardness or residual moistness. The input materials are substances that are commonly used as excipients in real tablets, with the difference being that food coloring is used instead of an active pharmaceutical ingredient. The production and testing equipment consists of small machines that are normally used for research and development purposes, so that the machine layout can be rearranged easily. To provide a realistic environment of the heavily regulated pharmaceutical production, various activities necessary for the compliance with rules of the "Good Manufacturing Practice" (GMP) are integrated. For each product, the production process is documented in a detailed batch record. All production, testing, and change-over activities are accompanied by thorough cleaning processes.[3]

---

[3] See Seliger, Jochem, Straube, and Kohl (2015).

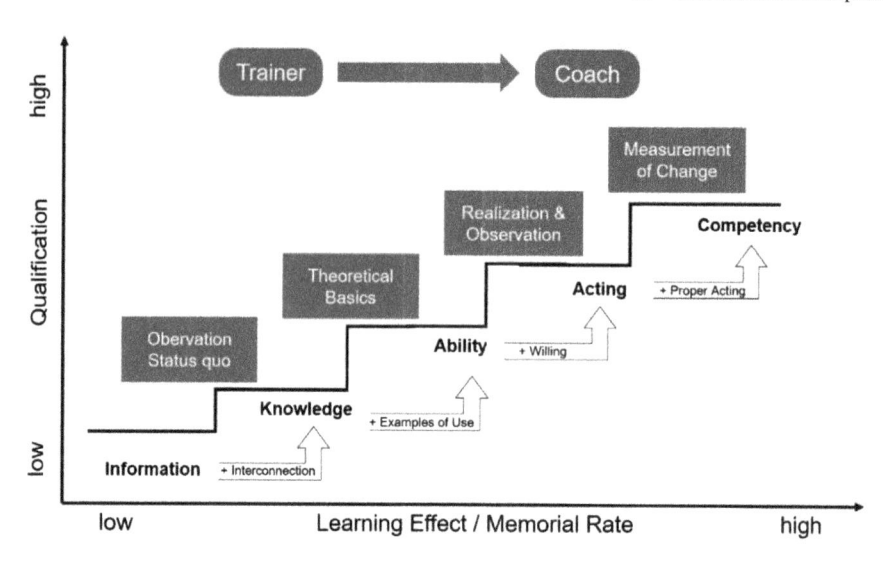

Fig. 11.28 Didactical design of the LEAN-Factory (Rybski & Jochem, 2016)

11.12.3 Operation

When it comes to the conduction of trainings, always two trainers, one internal, and one external (from the research partners) are part of the team. The main topics of the trainings are: seven types of waste, 5S, standard work, key performance indicators (KPIs), performance culture, problem solving, and in some special courses single-minute exchange of die (SMED). So far over 100 trainings with more than 1500 trainees from different hierarchy levels (management and shop floor) were realized. Additionally, round about 150 students attended the special university trainings. The main topics for the management and shop floor trainings are the same but differ regarding the learning objectives. The management trainings focus more on how to lead and support employees to realize the lean philosophy while the shop floor trainings focus on the actual implementation of lean elements. The university trainings combine both trainings and include additional exercises to implement lean elements with a focus on sustainability.[4] For all target groups, a systematic approach (problem pull) is used, shown in Fig. 11.28. It shows the general procedure to teach the different topics. Always starting with an observation of a semi-optimal situation, getting a short theoretical input to realize a better status afterward will be measured in the end. During this process, the trainers function as a typical trainer first to get more and more passive in the role of a coach later. To portray a realistic pharmaceutical production, students are acting in the role of operators to demonstrate problems and improvements. This approach gives the trainees the chance to observe and reflect the situation from a more passive point first.

---

[4] See Stock and Kohl (2018).

Since the learning factory is not only a training facility but also a competence center for different topics, it is an aim of the cooperation partners to improve the whole concept continuously. Therefore, all the taught methods and tools are used to control and live the processes to manage the learning factory itself.

11.13 Best Practice Example 13: Learning and Innovation Factory at IMW, IFT, and IKT, TU Wien, Austria

Authors: Wilfried Sihn[a], Fabian Ranz[a], Philipp Hold[a]
[a]TU Wien, Institute of Management Sciences, Austria

Name of learning factory: **Learning- & Innovation Factory (LIF)** Operator: **TU Wien** Year of inauguration: **2012** Floor Space in learning factory: **200 sqm** Manufacture product(s): **slot car** Main topics / learning content: **Integrated product & process planning, optimization of manufacturing and assembly operations** Extract from the morphology							
2.1	main purpose	education		training		research		
2.2	secondary purpose	test environment / pilot environment		industrial production	innovation transfer	public image		
3.1	product life cycle	product planning	product development	prototyping			service	recycling
3.2	factory life cycle	investment planning	factory concept	process planning	ramp-up	manufacturing / assembly / logistics	maintenance	recycling
3.3	order life cycle	configuration & order	order sequencing	planning and scheduling			picking, packaging	shipping
3.4	technology life cycle	planning	development	virtual testing			maintenance	modernization
3.5	indirect functions	primary activities				secondary activities			
		Inbound & outbound logistics	marketing & sales	service	firm infra-structure	HR	technology development	Procurement
4.1	learning environment	purely physical (planning + execution)		physical LF supported by digital factory	physical value stream of LF extended virtually		purely virtual (planning + execution)	
4.2	environment scale	scaled down			life-size			

11.13.1 Starting Phase and Purpose

The forerunner of the International Association of Learning Factories, the European Initiative on Learning Factories was initiated in 2011 with the TU Wien being one of the founding members. As a reaction, three institutes of the TU Wien, namely the Institute for Management Sciences (IMW), the Institute for Manufacturing Technology (IFT), and the Institute for Engineering Design and Logistics Engineering

(IKL), jointly decided to establish their own learning factory that should integrate contentual aspects from all three partners: The *TU Wien Learning & Innovation Factory (LIF)*. A machine fleet that was already existent at the TU Wien could be complemented by storage racks and assembly workstations from internal funds, while a didactical concept focused on engineering students was developed concurrently. The initial product, a slot car, had limited reusability and was designed to be kept by the students. On this account, an industrial sponsor could be convinced to tribute raw materials and parts and thus allow reducing the running costs of the LIF. The first training was held in 2012 after approximately 6 months of conception and implementation. The LIF has a clear focus on teaching and learning along the integrated product and process creation process.

11.13.2 Learning Environment and Products of the Learning & Innovation Factory

The LIF had formerly been located on a 200 m^2 laboratory floor off campus and was moved to the new TU Science Center in 2017. The LIF infrastructure is able to replicate the production process of the slot car through all stages—from 3D product design through prototyping, manufacturing, assembly, and testing (Fig. 11.29). The application scenario in the LIF starts with a suboptimal product design that needs to be revised in terms of cost, quality, and product performance. Students use 3D CAD tools to develop better designs of the chassis and its production process. The chassis, including an electric motor, includes approx. 30 individual parts (Fig. 11.30). Redesigned parts and components are subsequently manufactured on EMCO educational turning and drilling machines. For tooling, joints, and fixtures, an industrial 3D printer is available. Manual workstations allow for final assembly. Ultimately, students can test their designs and compete against each other on the Carrera car racing track.

11.13.3 Operation

The LIF is an educational facility that offers one single but extensive four-week course for bachelor-level students in mechanical engineering and industrial engineering degree courses. The training is offered once annually and led by researchers and laboratory technicians. Its goal is to deliver comprehensive understanding of and methods competencies along the integrated product and process creation process, from design to customer-ready slot cars. Addressed knowledge areas include product analysis and costing, engineering and design, prototyping, machine programming and manufacturing, assembly planning, and time studies. Students primarily work in self-organized teams of four and are guided mostly with regard to the correct and structured application of the respective methods (Fig. 11.31). Another application of

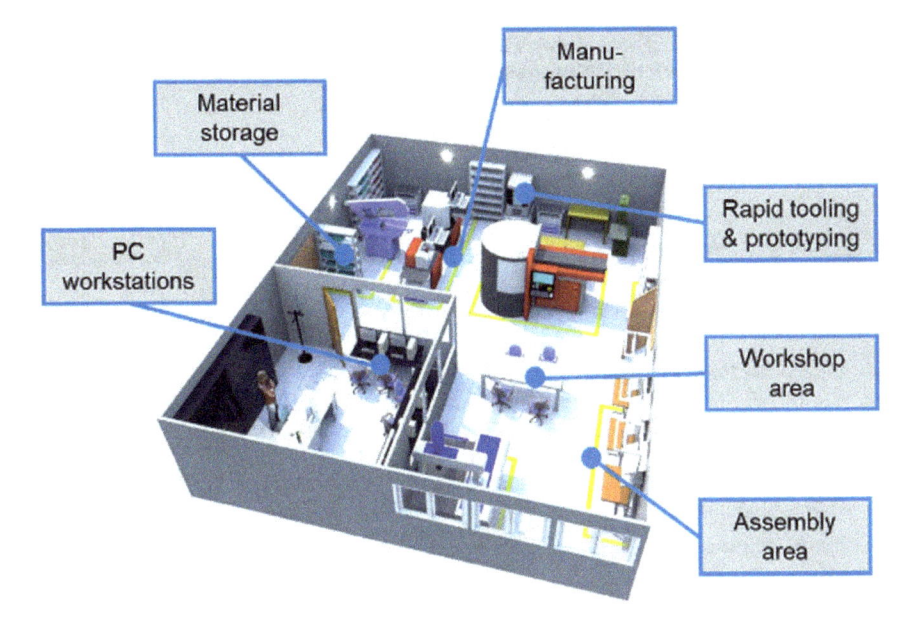

11.13 Best Practice Example 13: Learning and Innovation Factory at IMW ... 383

Fig. 11.29 Virtual representation of LIF

Fig. 11.30 Slot car in exploded view

the learning factory concept at TU Wien, the Pilot Factory Industrie 4.0 for research, testing, and demonstration, is described in Best Practice Example 25.

11.14 Best Practice Example 14: Learning Factory AIE at IFF, University of Stuttgart, Germany

Authors: Thomas Bauernhansl[a], Erwin Gross[a]; Jörg Siegert[a]
[a]Institute of Industrial Manufacturing and Management (IFF), University of Stuttgart

Name of learning factory: **Learning Factory aIE**
Operator: **IFF, University of Stuttgart**
Year of Opening: **2007**
Floor Space in learning factory: **350 sqm**
Manufactured product(s): **desk tool set**
Topics / Learning Content: **Lean Production and Quality Management**

Extract from the morphology

2.1	main purpose	education		training				research	
2.2	secondary purpose	test environment / pilot environment		industrial production		innovation transfer		public image	
3.1	product life cycle	product planning	product development	prototyping				service	recycling
3.2	factory life cycle	investment planning	factory concept	process planning	ramp-up	manufacturing	assembly logistics	maintenance	recycling
3.3	order life cycle	configuration & order	order sequencing	planning and scheduling				picking, packaging	shipping
3.4	technology life cycle	planning	development	virtual testing				maintenance	modernization
3.5	indirect functions	primary activities				secondary activities			
		inbound & outbound logistics	marketing & sales	service	firm infrastructure	HR		technology development	procurement
4.1	learning environment	purely physical (planning + execution)		physical LF supported by digital factory		physical value stream of LF extended virtually		purely virtual (planning + execution)	
4.2	environment scale	scaled down		life-size					

Starting phase and purpose: Opened in 2007, the Learning Factory of advanced Industrial Engineering aIE (LF aIE) in Stuttgart provides education and training for managers, planners, and designers of production processes. For scientists, the LF aIE provides a research facility where new production and organization concepts can be explored. One research focus is competence development during value creation and how it should be designed in the future. Located at the IFF of the University

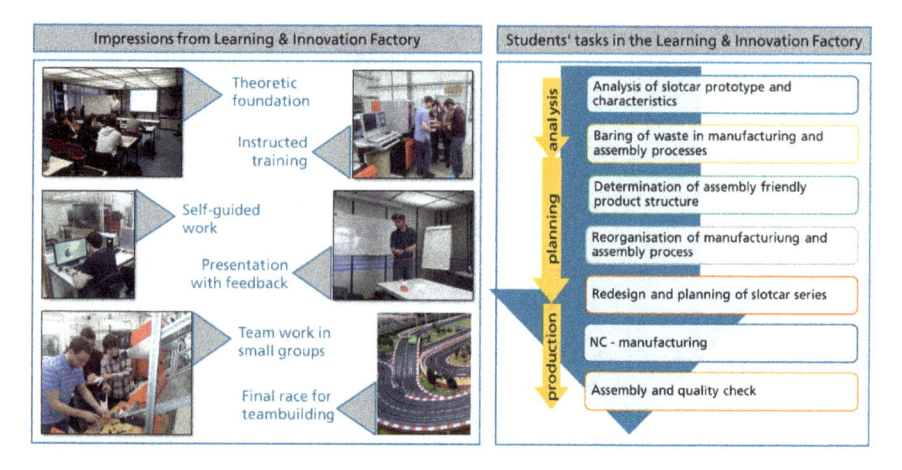

Fig. 11.31 Procedure of the training in the TU Wien Learning & Innovation Factory

11.14 Best Practice Example 14: Learning Factory AIE at IFF …

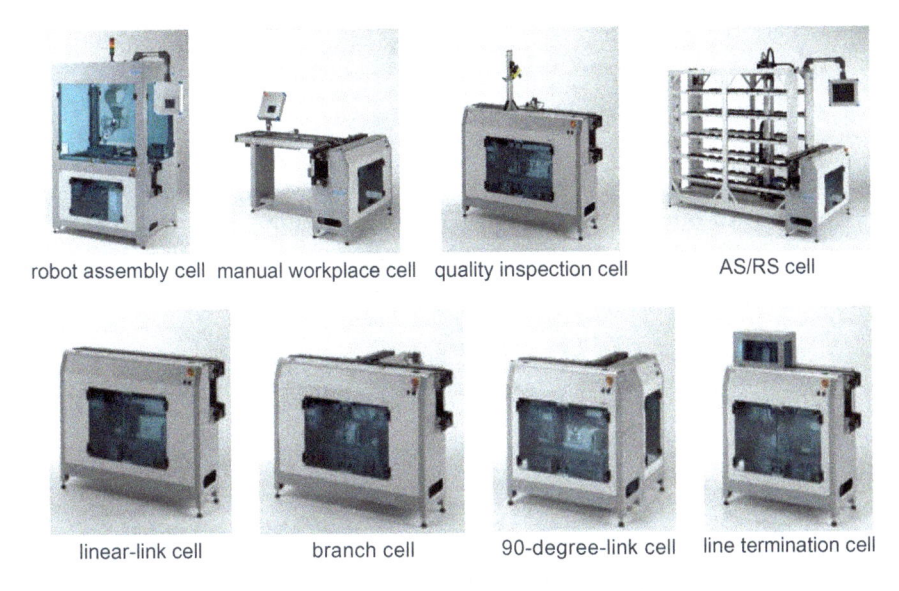

Fig. 11.32 Modules of the Learning Factory aIE, picture from Festo Didactic

of Stuttgart, the LF aIE drives training and further education of industrial engineers in practice, technical managers, planners, and designers of all branches in curricula that are virtual and at the same time practical and real. In the Learning Factory for advanced Industrial Engineering aIE engineers, planners and managers find a unique and innovative combination through a physical model factory, digital learning islands, and theoretical modules. The learning factory of the IFF is also the basic building block of excellent post-university education and training in this field. Methods and tools are applied in a digital learning environment and then tested in reality, the physical model factory. The digital and virtual tools improve the effectiveness, efficiency, safety, and reproducibility of planning. Once recorded or—as a result of a planning step—determined data can be continuously forwarded and reused.

Preliminary work and industrial partners: The foundations have been laid by scientists from Stuttgart in several subprojects of the Collaborative Research Center "Transformable Corporate Structures in Multi-Variant Serial Production"; involved were institutes of business administration, computer science, and mechanical engineering. Project partners of the learning factory are the MTM-Bundesvereinigung, the REFA Bundesverband, and the Festo Didactic GmbH & Co. KG. Other cooperation partners such as Delmia, PTC Parametric Technology GmbH, PSI AG, and Siemens AG support the learning factory by providing systems and products for special conditions.

Learning environment and products: The Learning Factory aIE (iFactory) covers 350 m^2 and includes the modules presented in Fig. 11.32.

different colors of cup insert

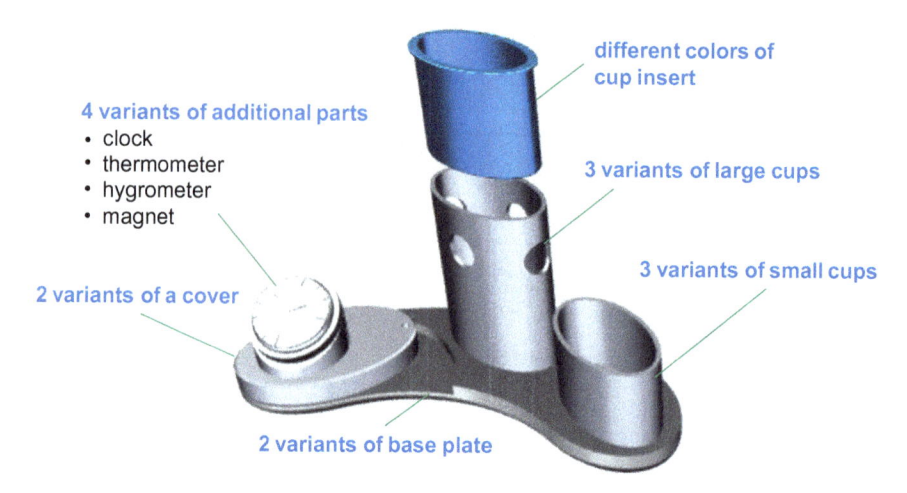

4 variants of additional parts
- clock
- thermometer
- hygrometer
- magnet

3 variants of large cups

3 variants of small cups

2 variants of a cover

2 variants of base plate

Fig. 11.33 Product of the Learning Factory aIE

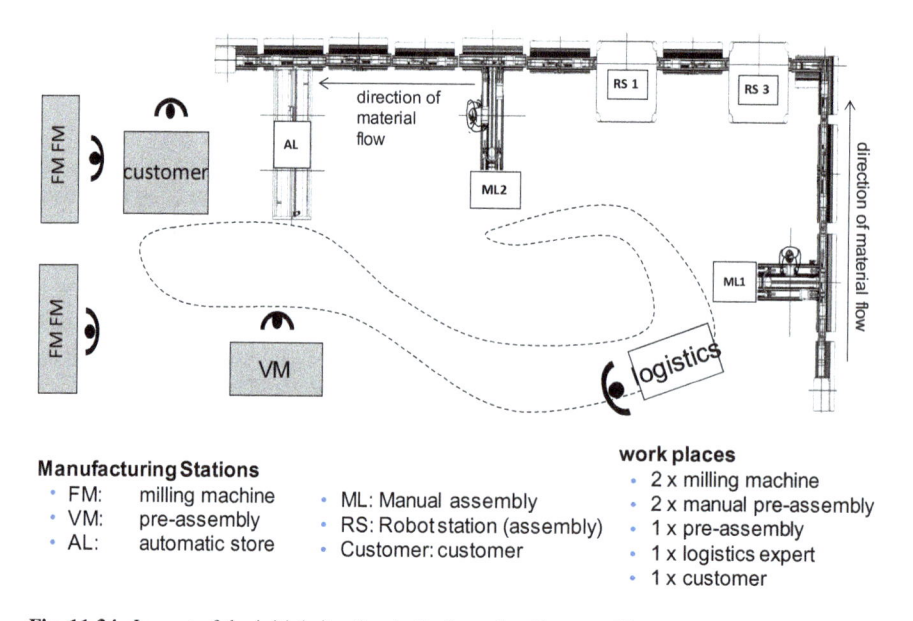

Manufacturing Stations
- FM: milling machine
- VM: pre-assembly
- AL: automatic store
- ML: Manual assembly
- RS: Robot station (assembly)
- Customer: customer

work places
- 2 x milling machine
- 2 x manual pre-assembly
- 1 x pre-assembly
- 1 x logistics expert
- 1 x customer

Fig. 11.34 Layout of the initial situation in the Learning Factory aIE

Depending on the customer's order, different desksets (Fig. 11.33) are produced in the LF aIE. RFID technology enable the complete tracking of the individual production steps.

The initial layout of the factory which is being optimized during the training by the participants in a simulation game is presented in Fig. 11.34.

Operation and didactic concept: The trainers are scientists from the University of Stuttgart and Fraunhofer IPA, working on research projects and industrial projects

11.14 Best Practice Example 14: Learning Factory AIE at IFF ... 387

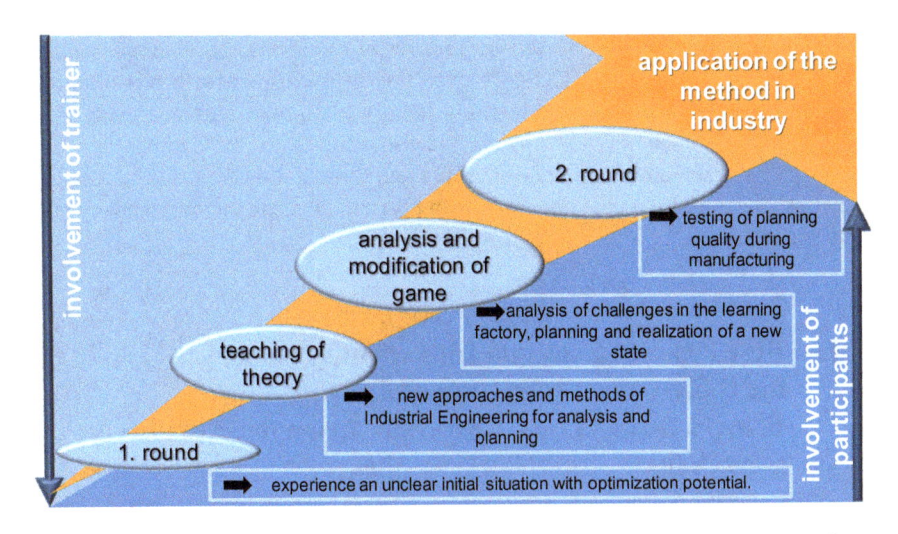

Fig. 11.35 Learning by doing: simulation game in the Learning Factory aIE, concept according to Bonz (2009)

in the field of production optimization. This ensures a comprehensive expertise and the coaches can respond to specific questions of the participants.

The Learning Factory aIE enhances the knowledge transfer from basic research to the application of the methods for industrial engineering. The training modules with a high degree of interaction and application of methods (see Fig. 4) are always ranked higher by the participants than theoretically oriented parts of the curriculum. This proves the effectiveness of the approach to teach these topics in a learning factory where they can be applied.

The relatively wide scope of the curriculum is found helpful by most participants even if they apply only parts of it themselves. The reason for this is that a broad knowledge makes them understand the overall processes in the enterprise and the tasks of their colleagues. The focus of the existing learning concept is the assembly, because this is where most variants are created and the influence from the customers is high. Manufacturing processes are simulated so that they can be included into the value stream. Manufacturing technologies such as injection molding or milling have been integrated during the last years.

The training group usually includes 12 participants, who are supervised by two trainers. A training workshop lasts for two days, covering various topics in the area of lean management and Industrie 4.0. The focus is put on logistics design and Kanban calculation. Other topics include production measurement technology and quality management. Also short training courses are given, in which the most important topics to Lean Management and Industrie 4.0 are mediated. In total, several hundred participants from industry and research are trained each year, but also pupils and students undergo these training courses as a part of their qualification. More than 50% of the participants are international and come mainly from Asia (Fig. 11.35).

Learning factories should always be a mirror of research to provide optimal and credible support for transfer. The pure operation of a learning factory is often not economical. The conversion of a learning factory to an application or application center is very promising. In particular, the use of Industry on Campus concepts opens up new potential in transfer or training and further education. The production of real products is still a challenge, but using the resources to support start-ups can make sense.

11.15 Best Practice Example 15: Learning Factory for Electronics Production at FAPS, Friedrich-Alexander University Erlangen-Nürnberg, Germany

Authors: Thomas Reitberger[a], Jörg Franke[a]
[a]Institute for factory automation and production systems (FAPS), Friedrich-Alexander University Erlangen-Nürnberg

Name of learning factory: Electronics Production
Operator: **FAPS, FAU Erlangen-Nuremberg**
Year of inauguration: **2002**
Floor Space in learning factory: **200m² / 810m²**
Manufacture product(s): **PCB, MID, Power Electronic Component, Opto-MID**
Main topics / learning content: **SMT Production, Printed Electronics and Optics, Power Electronics, 3D-MID Technology**
Extract from the morphology

2.1	main purpose	education		training			research	
2.2	secondary purpose	test environment / pilot environment		industrial production	innovation transfer		public image	
3.1	product life cycle	product planning	product development	prototyping			service	recycling
3.2	factory life cycle	investment planning	factory concept	process planning	ramp-up	manufacturing / assembly / logistics	main-tenance	recycling
3.3	order life cycle	configuration & order	order sequencing	planning and scheduling			picking, packaging	shipping
3.4	technology life cycle	planning	development	virtual testing			main-tenance	moderni-zation
3.5	indirect functions	primary activities				secondary activities		
		Inbound & outbound logistics	marketing & sales	ser-vice	firm infra-structure	HR	technology development	Procure-ment
4.1	learning environment	purely physical (planning + execution)	physical LF supported by digital factory		physical value stream of LF extended virtually		purely virtual (planning + execution)	
4.2	environment scale	scaled down		life-size				

Fig. 11.36 Key topics of the electronics production learning factory since 2011

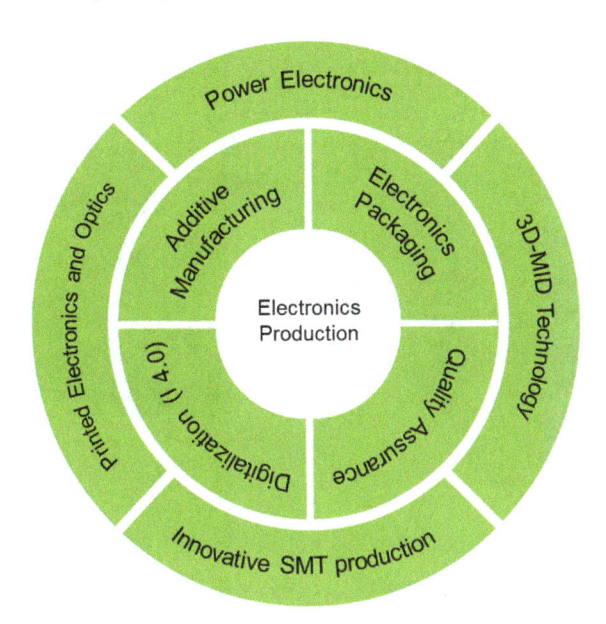

11.15.1 Starting Phase and Purpose

The Institute for Factory Automation and Production Systems (FAPS) started focusing on electronics production in 1992 when the special research field (SFB 356) "production systems in electronics" was launched. In the same year, the research association for molded interconnect devices 3D-MID was founded. These were the two initializing steps toward the electronic production learning factory. In 2002, the learning factory with a size of 200 m^2 in the northeast park Nuremberg was inaugurated. The main purpose consisted of advancing the research on SMT production in a shop floor environment. In the following years, several research activities concerning electronics production started which made it possible to acquire more machines and manpower.

The new and modern learning factory in the former AEG production plant in Nuremberg was inaugurated in 2011. With a floor space of almost 1000 m^2, new possibilities for diversified research on electronics production could be started as displayed in Fig. 11.36. The novel setup made it possible to bring together education, training, and research in a factory-like environment. Today, the focus topics are innovative SMT production, power electronics, 3D-MID technology as well as printed electronics and optics. The goal is to build up a holistic representation to what the future of electronics production will look like. For this, focused investigations on highly relevant themes like digitalization (Industrie 4.0), intralogistics based on autonomous vehicles or possibilities of additive manufacturing of mechatronic functions are also part of research as electronics packaging and quality assurance technologies.

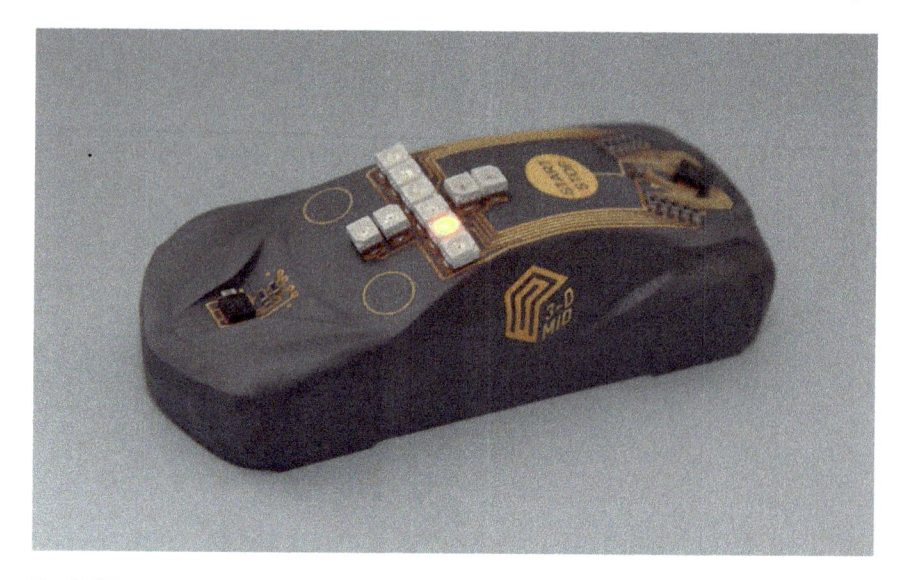

Fig. 11.37 MIDSTER demonstrating the possibilities of integrating electro-mechanical functions onto a 3D surface

11.15.2 Learning Environment and Products

On the modern SMT-factory line, different *printed circuit boards (PCBs)* can be manufactured with high accuracy. The machine repeatability is 12.5 µm at six sigma for the stencil printer and 22 µm at three sigma for the assembly machine. The maximum performance is 67,750 parts per hour. For demonstration and education purposes, a specific PCB with an accelerometer on it was designed. With this, it is possible to measure the inclination and show this by illuminating the LEDs in the tilted direction.

The next step was to bring all functionalities evaluated for the printed circuit board in 2D to a three-dimensional surface. The result was the MID demonstrator which can be seen in Fig. 11.37, the so-called MIDSTER.

The factory line illustrated in Fig. 11.38 starts with the automatic PCB load to the stencil printer where the soldering paste is brought onto the PCB. Thereafter, the miniaturized electronic components are automatically placed with high accuracy and productivity onto the PCB. Inside the reflow-oven, the solder paste is melted according to the defined temperature profile. The last step is quality assurance where all the requirements concerning the PCB are checked. The complete process is interlinked, and the machine data is controlled using Industrie 4.0 tools like analytics and data mining.

11.15 Best Practice Example 15: Learning Factory for Electronics ...

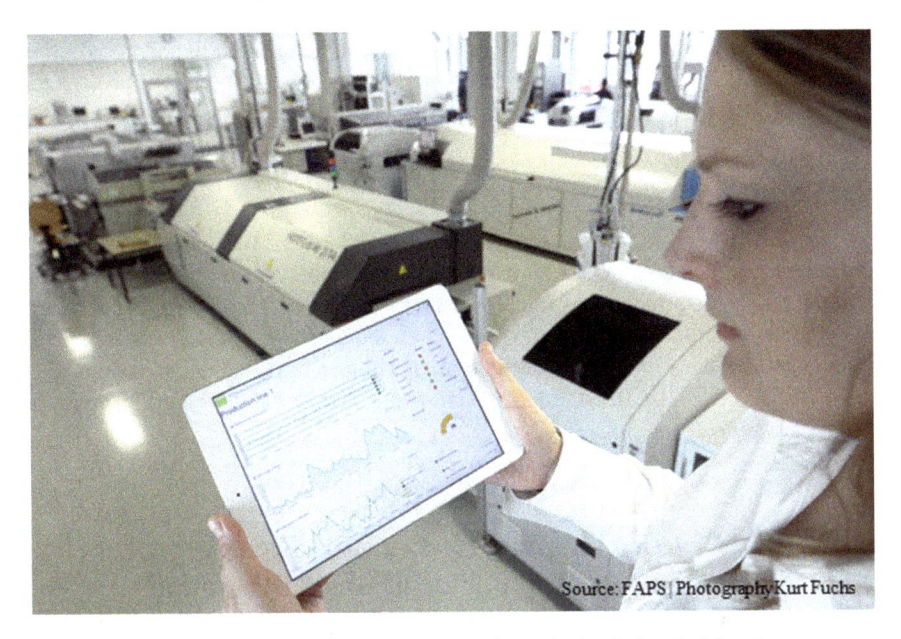

Fig. 11.38 View of the modern SMT-factory line optimized using Industrie 4.0 tools

11.15.3 Operation

The entire electronic productions group consists of about 15 scientists, who cover the complete process chain of electronics production in education and research.

11.15.4 Education

More than 200 students are mostly self-reliant practically educated in the learning factory every year. Beginning with the practical training in manufacturing technology I and II, the students learn to design and built up step by step a specific demonstrator PCB during their bachelor's degree. For the master's degree students, there are two different practical courses. The MID course where the students learn step by step the development and setup of the MIDSTER illustrated in Fig. 11.37 and the power electronics course where a complete inverter is set up. By choosing the theoretical course PRIDE 2 (production processes in electronics production) about 200 students get a complete overview of modern electronics production through 13 intensive lessons every year.

Also a vast number of representatives from industry are guided through the learning factory and about 100 industrial representatives are trained therein every year. With the two-day seminars "Electronics Production" and "Additive Manufacturing," the participants learn all about the different PCB manufacturing steps and the additive mechatronization of three-dimensional components.

11.15.5 Research

The core competencies can be found in Fig. 11.36. With expertise in all these fields, a broad variety of projects has been prepared over the past 26 years. From direct bilateral cooperation with industrial partners on the one hand side to basic scientific projects without and joint research developments funded by different institutions (e.g. DFG, BMBF, AiF/IGF, BFS, EU) with industrial partner on the other, all relevant topics have been covered.

11.16 Best Practice Example 16: Learning Factory for Innovation, Manufacturing and Cooperation at the Faculty of Industrial and Process Engineering of Heilbronn University of Applied Sciences, Germany

Author: Patrick Balve[a]

[a]Heilbronn University of Applied Sciences, Faculty of Industrial and Process Engineering, Study Program Manufacturing and Operations Management (B.Eng.)

11.16 Best Practice Example 16: Learning Factory for Innovation …

Name of learning factory: **L. F. for Innovation, Manufacturing and Cooperation** Operator: **Heilbronn University of Applied Sciences** Year of inauguration: **2011** Floor space in learning factory: **800 sqm** Manufacture product(s): **New products every semester** Main topics/learning content: **Engineering methods along the product creation process, cooperation across departments** Extract from the morphology							
2.1	main purpose	education		training			research	
2.2	secondary purpose	test environment / pilot environment		industrial production		innovation transfer	public image	
3.1	product life cycle	product planning	product development	prototyping	manufacturing	assembly / logistics	service	recycling
3.2	factory life cycle	investment planning	factory concept	process planning	ramp-up		maintenance	recycling
3.3	order life cycle	configuration & order	order sequencing	planning and scheduling			picking, packaging	shipping
3.4	technology life cycle	planning	development	virtual testing			maintenance	modernization
3.5	indirect functions	primary activities				secondary activities			
		inbound & outbound logistics	marketing & sales	service	firm infrastructure	HR	technology development	procurement
4.1	learning environment	purely physical (planning + execution)		physical LF supported by digital factory		physical value stream of LF extended virtually		purely virtual (planning + execution)
4.2	environment scale	scaled down				life size			

11.16.1 Starting Phase and Purpose

It was around 2008 that a team of professors of the Faculty of Industrial and Process Engineering of Heilbronn University started contemplating the formation of a learning factory. The underlying idea was to allow students to develop job-related skills in a much more profound way than any traditional teaching approach or even laboratory experiments could possibly do. It eventually turned out that a combined project-oriented and problem-based concept would best suit the overall mission statement of a university of applied sciences.

After the entire didactical concept was specified in detail, an initial group of 18 students could embark on the first learning factory experience in the winter semester of 2011. However, in the beginning, the tooling machines did not come up to the desired level of versatility and ease-of-use, and meeting places for students were scattered over two different large laboratory halls. Yet, the long-term goal was to pool all equipment in one single building that could also accommodate team rooms and meeting places for students. Along with the Rectorate, a plan was devised on how to upgrade and complete the tooling machines, measuring devices, assembly stations, and alike so that most of the expected manufacturing operations could be carried out on site. That goal was eventually reached in 2016: Five years after the pilot student project was initiated, there is now one single building dedicated to the learning factory. This achievement was made possible thanks to joint funding by

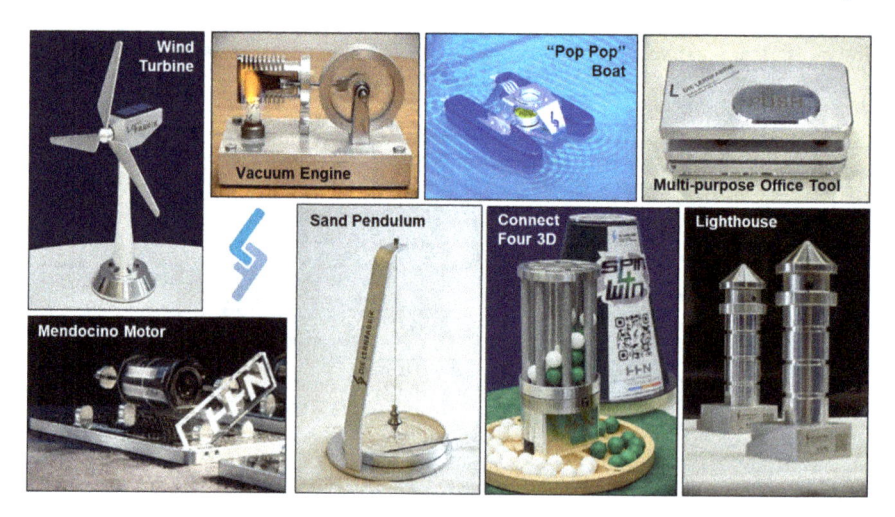

Fig. 11.39 Examples of manufactured products in the Learning Factory

the Faculty, the Rectorate, and the Federal State of Baden-Wuerttemberg in the total amount of EUR 1 million.

The "Learning Factory for Innovation, Manufacturing and Cooperation" is primarily dedicated to hands-on student training in product development, industrial engineering, and management methods toward the end of the "Manufacturing and Operations Management" study program. It is widely acclaimed as a model for large interdisciplinary, project-oriented, and problem-based learning approaches at Heilbronn University and across the southwest of Germany. Besides, it is listed as a best practice example by the *nexus* research project of the German Rectors' Conference (HRK).

11.16.2 Learning Environment and Products

The focus on the creation of new products (Fig. 11.39) along with all necessary manufacturing, logistics, and quality assurance processes needed for a small series production stands for an outstanding feature. For that purpose, the team of supervising professors conceives a unique project assignment consisting of technical requirements, budget, and time constraints each semester. Although it is important to provide technically demanding details as part of the assignment, the students' learning experience coming along with the unfolding problem-solving process and the organizational challenges are valued as high as the technical solution itself.

The Learning Factory thus allows students to fully emulate a real-world start-up company in an authentic industrial environment. Comprising 570 m² of undivided shop floor space with technical equipment and additional 230 m² of student work

11.16 Best Practice Example 16: Learning Factory for Innovation ...

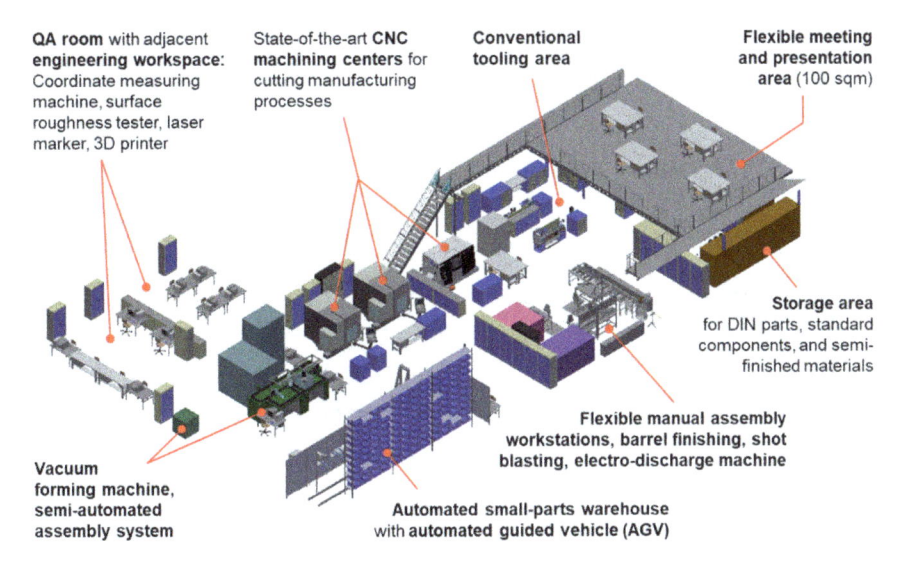

Fig. 11.40 Overview of the Learning Factory facility and operations

space and PC rooms (Fig. 11.40), it is one of the largest single facilities dedicated to student education at Heilbronn University. Careful selection and a vast variety of equipment and computer programs ensure that all direct and indirect key processes are readily accessible to students to perform project tasks.

11.16.3 Operation

Being a mandatory course with 16 credits in the 6th semester, 20–35 students take part in the Learning Factory project twice a year. A team of up to eight professors and additional staff members provide academic supervision, thus assuring appropriate guidance for various functions that need to be covered by students along the product creation process (Fig. 11.41, top right). Although the product varies from semester to semester, the project timeline and its milestone dates are standardized (Fig. 11.41, left). Besides the application and deepening of factual knowledge and working methods, the Learning Factory aims at strengthening students' cross-departmental communication, cooperation, and project management skills as preparation for their upcoming employment in the manufacturing industry.

Applied research is another aspect covered by the Learning Factory at Heilbronn University and pertains, for example, to developing practical Industrie 4.0 applications or agile ways of collaboration. Recently (in 2017), a publicly funded project was launched to research the development of competencies in the learning factory environment as compared to conventional project-oriented learning approaches.

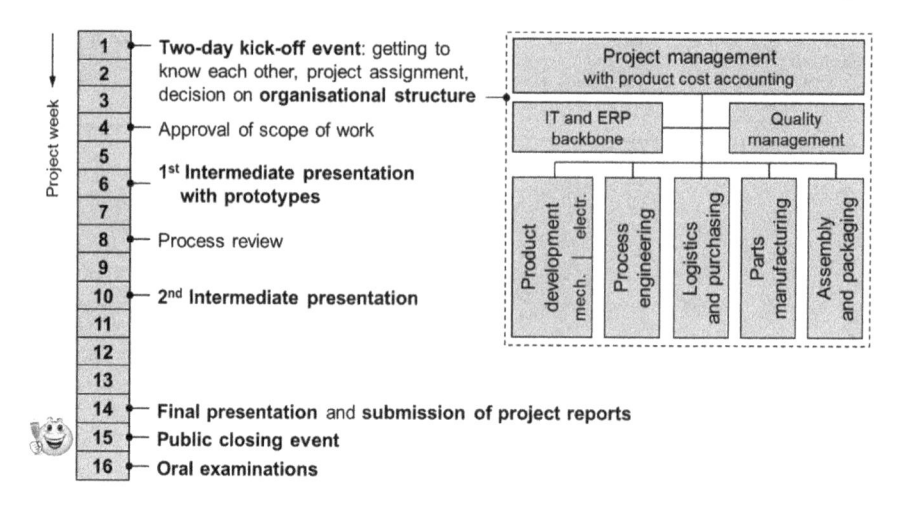

Fig. 11.41 Semester timeline and typical project structure

11.17 Best Practice Example 17: Learning Factory for Global Production at Wbk, KIT Karlsruhe, Germany

Authors: Gisela Lanza[a], Constantin Hofmann[a], Benjamin Haefner[a], Nicole Stricker[a]
[a]wbk, Institute of Production Science, KIT Karlsruhe

11.17 Best Practice Example 17: Learning Factory for Global Production ...

Name of learning factory: **Learning Factory Global Production Karlsruhe**
Operator: wbk Institute of Production Science
Year of inauguration: 2014
Floor Space in learning factory: **200qm**
Manufacture product(s): Electric drive
Main topics/learning content: **Lean Production, Assembly planning, Industry 4.0**

Extract from the morphology

2.1	main purpose	education		training		research	
2.2	secondary purpose	test environment / pilot environment		industrial production	innovation transfer	public image	
3.1	product life cycle	product planning	product development	prototyping		service	recycling
3.2	factory life cycle	investment planning	factory concept	process planning	ramp-up	main-tenance	recycling
3.3	order life cycle	configuration & order	order sequencing	planning and scheduling		picking, packaging	shipping
3.4	technology life cycle	planning	development	virtual testing		main-tenance	moderni-zation
3.5	indirect functions	primary activities				secondary activities	
		Inbound & outbound logistics	marketing & sales	ser-vice	firm infra-structure	HR	technology development	Procure-ment
4.1	learning environment	purely physical (planning + execution)	physical LF supported by digital factory	physical value stream of LF extended virtually		purely virtual (planning + execution)	
4.2	environment scale	scaled down		life-size			

11.17.1 Starting Phase and Purpose

The Learning Factory Global Production Karlsruhe is the emblematic project of Prof. Dr.-Ing. Lanza to materialize the challenges of global production.

In 2012, first preliminary studies regarding the concept were conducted. An early-stage key decision shaping the future learning factory was to implement a real production system for an existing product of a truly globalized company. The electric drives of Robert Bosch GmbH for the automotive market were selected as an ideal, highly diversified product to demonstrate the effects of global production on a cost-sensitive product.

Since flexibility and changeability are key aspects of cutting-edge production systems, the enabler of changeability served as guiding principles during the design phase of the production system. To fully grasp the various faces of global production, all stations have been designed with varying automation degree. Scalable automation allows to seamlessly adjust the degree of automation of each workstation emulating the various production system characteristics that can be found around the globe.

The funding from Prof. Lanza's appointment of professorship initially financed her ambitious project. Thanks to strong industrial support and high personal engagement the learning factory rapidly took shape.

Today the Learning Factory Global Production Karlsruhe (Fig. 11.42) is an established and renowned center for first-class trainings and Industrie 4.0 research in the

Fig. 11.42 Impressions of the learning factory for global production at wbk, KIT Karlsruhe

context of global production. The learning factory demonstrates the effects of the location on the production system and allows the participants to understand how the different components of a production system as well as fundamental principles of waste-free production can be adapted to perfectly fit the surrounding location factors.

The Learning Factory Global Production Karlsruhe is led by the vision to constantly push the boundaries of global production by applying and further developing newest technological solutions and to transmit theses finding in excellent trainings.

The internal development process is driven by new technological options and industrial needs and performed by an agile, Kanban-based development team.

11.17.2 Learning Environment and Products

In the Learning Factory Global Production, fourteen different variants of electric drives are assembled. The ten-stage process chain comprises pressing, screwing, and joining operations as well as a magnetizing process. In addition to the assembly operations, several inline measurement systems as well as an end-of-line functional test system can be added to the process chain to detect and prevent quality issues.

Besides the assembly process, picking and logistics processes have to be designed by the participants. Production control and scheduling as well as customer integration can be varied by the trainees. After assuring the full functionality of the produced drives in a test bed, the drives are reinserted into the system. In 2018, a remanufacturing case will replace the disassembly of the parts.

In the Learning Factory Global Production, numerous Industrie 4.0 solutions have been implemented to improve quality, assist the workers, and aid planning and optimization. Tracking and tracing is a core element of the Learning Factory, not only for material using RFID but also for employees using a real-time high precision indoor location system. The real-time data from the production system is the basis for multiple tools such as digital shop floor management, augmented Go & See, real-time value stream mapping or input to the digital layout planning. To provide seamless interaction with complex systems, several augmented and virtual reality systems have been developed. Since scalable automation is a core pillar of the Learning Factory Global Production, human–robot collaboration is also in the heart of the training center.

11.17.3 Operation

The goal of the learning factory is to materialize global production and to transmit new technological possibilities for production planning in excellent trainings. In 2018, wbk offers seven different trainings addressing Lean Management in the context of Industrie 4.0, Six Sigma-based quality assurance in production networks, Scalable Automation, and Leadership 4.0 (Fig. 11.43). All on-site trainings are accompanied by e-learning such as Supplier Development, Global Site Selection, or Production Network Planning. The trainings are conducted in German, English, and French by wbk and constantly adapted.

11.18 Best Practice Example 18: "Lernfabrik für schlanke Produktion", Learning Factory for Lean Production at iwb, TU München, Germany

Authors: Harald Bauer[a], Felix Brandl[a], Gunther Reinhart[a]

Fig. 11.43 Learning fields of the learning factory for global production at wbk, KIT Karlsruhe

[a]Institut für Werkzeugmaschinen und Betriebswissenschaften (iwb, Institute for Machine Tools and Industrial Management), TU München

11.18 Best Practice Example 18: "Lernfabrik für schlanke Produktion" ...

Name of learning factory: **Lernfabrik für Schlanke Produktion (LSP)**
Operator: **iwb - TUM**
Year of inauguration: **2009**
Floor Space in learning factory: **150m²**
Manufacture product(s): **Gears**
Main topics / learning content: **Lean Philosophy, Lean Assembly**

Extract from the morphology

2.1	main purpose	education	training	research	
2.2	secondary purpose	test environment / pilot environment	industrial production	innovation transfer	public image
3.1	product life cycle	product planning	product development	prototyping	service	recycling
3.2	factory life cycle	investment planning	factory concept	process planning	ramp-up	maintenance	recycling
3.3	order life cycle	configuration & order	order sequencing	planning and scheduling		picking, packaging	shipping
3.4	technology life cycle	planning	development	virtual testing		maintenance	modernization
3.5	indirect functions	primary activities				secondary activities	
		Inbound & outbound logistics	marketing & sales	service	firm infrastructure	HR	technology development	Procurement
4.1	learning environment	purely physical (planning + execution)	physical LF supported by digital factory	physical value stream of LF extended virtually	purely virtual (planning + execution)
4.2	environment scale	scaled down	life-size		

11.18.1 Starting Phase and Purpose

While lean production already has been an essential part in the lectures of the Institute for Machine Tools and Industrial Management (*iwb*) of the Technical University of Munich (TUM), the final kickoff for the establishment of a learning factory took place in 2009. In order to train employees in lean, the company Zeitlauf—a globally active manufacturer of gear technology—worked together with the *iwb*, and the Learning Factory for Lean Production (LSP) was developed. While Zeitlauf provided products and assembly processes, the *iwb* set up the facilities and a modular training program for lean production, always connected to the practical manufacturing of Zeitlauf's gear box units.

In order to achieve a sustainable lean management, focus of the training is the understanding of the lean philosophy including customer orientation, continuous improvement, elimination of waste, and teamwork. Based on lean thinking, methods for lean production are introduced with a problem-based learning approach.

Parallel to the industrial training for Zeitlauf and further organizations, the *iwb* integrated the LSP as a 2-week intensive course into the schedule of the TUM—with great success and enormous demand by students. Today, the LSP is used for both industrial training and educational courses. By inviting guest speakers to the student courses, the LSP provides the opportunity for companies to directly address and recruit lean talents.

11.18.2 Learning Environment and Products

The *iwb*'s LSP provides all necessary material to experience and apply the lean philosophy and continuous improvements within a manual gear box assembly line.

Overall, the facilities are divided into three main areas: the assembly area, the kaizen workshop area, and the theoretical teaching area (Fig. 11.44). Activities within the assembly area include logistics, assembly, quality control, and packaging. An indicator board tracks process performance and improvements, e.g. delivery reliability.

The manufactured product is a real gear box used, e.g. in train doors. It consists of 18 different components and is produced with two or three gear stages in 24 variants. The overall assembly process contains more than 20 steps, which need to be distributed on the assembly stations.

The theoretical teaching area is used to present the lean philosophy and the methods, which are used to implement said philosophy in manufacturing systems. It also is the area for guest presenters to illustrate their lean experience in industry.

In the kaizen workshop area, participants work on their improvements toward their individual lean gear box assembly line. Therefore, whiteboards and rapid prototyping material, e.g. cardboard or 3D-printers, are available.

11.18.3 Operation

The LSP is staffed with internal research associates of the *iwb* with broad experience in lean consulting and coaching. Steady cooperation with industrial partners and clients guarantees a state-of-the-art teaching concept within the learning factory. The training course evolved over the past years continuously adapting to new challenges while the basic idea remains: Lean is a total management philosophy and hence, cannot be mastered by teaching methods only. Instead, a typical lean journey is recreated during the course: An apparently unsatisfactory starting situation causes the participant's need for action to understand and analyze the issues. With the theoretical background including a quick historical recap of the Toyota Production System (TPS), the seven types of waste, and the basic method Value Stream Analysis, the first step toward a lean assembly line is made. The participants assume both roles, the assembly operators and the lean experts. Every improvement step (of in total three) is planned on white boards, implemented in the line, immediately tested in an assembly run, and checked by a set of KPIs that evaluate the product's production costs. This approach is based on the Deming or PDCA-Cycle and represents a continuous improvement in little steps or KAIZEN. Suiting methods for every improvement step are provided in theory lesson slots. The participants are divided into three groups whose assembly lines compete in minimizing production costs. These are determined by the used resources, as well as delivery quality and

11.18 Best Practice Example 18: "Lernfabrik für schlanke Produktion" ...

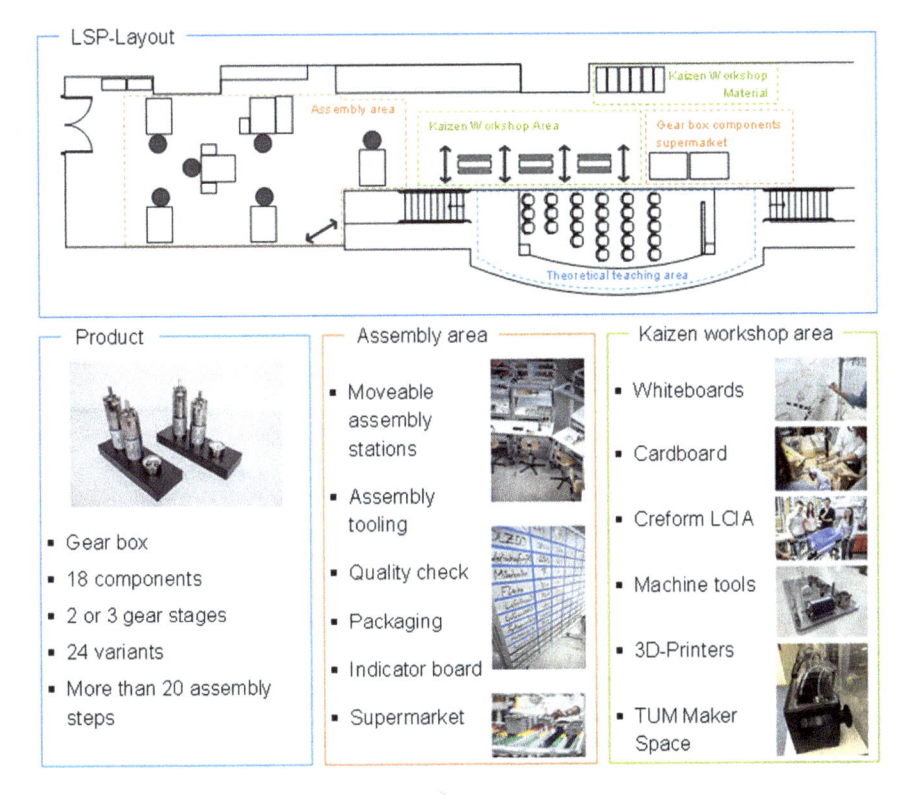

Fig. 11.44 LSP facilities and equipment overview

reliability. While a lot of methods and tools are provided by the trainers, no sample solution exists (Fig. 11.45).

In this setup, the *iwb* offers 4–6 trainings every year, including industrial and educational offers (industrial: WGP, Cluster, individual; educational: summer & winter course, internal trainings). Trainings usually involve 12–21 persons with 2–3 trainers. Individual trainings are elaborated with industrial partners including the ability to transport the 100% mobile equipment to any location.

11.19 Best Practice Example 19: "Lernfabrik für vernetzte Produktion" at Fraunhofer IGCV, Augsburg, Germany

Authors: Lukas Merkel[a], Stefan Braunreuther[a], Gunther Reinhart[a]
[a]Fraunhofer IGCV, Augsburg, Germany

404   11 Best Practice Examples

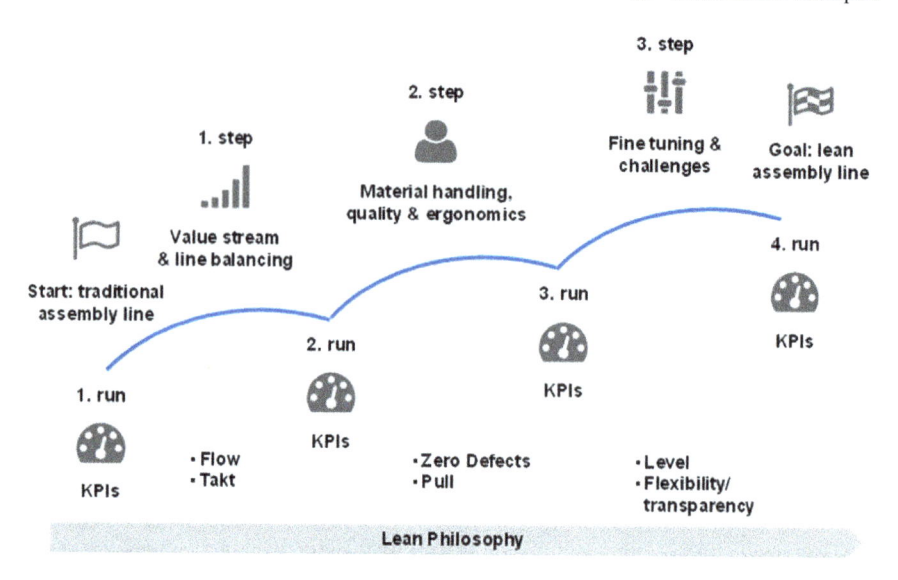

```
                                        3. step
                      2. step
                                    Fine tuning &      Goal: lean
        1. step                      challenges      assembly line
                   Material handling,
     Value stream  quality & ergonomics
     & line balancing                                     4. run
Start: traditional                    3. run
assembly line
                      2. run                              KPIs
   1. run                             KPIs
                      KPIs
   KPIs      • Flow            • Zero Defects    • Level
             • Takt            • Pull            • Flexibility/
                                                 transparency

                          Lean Philosophy
```

Fig. 11.45 LSP training concept

Name of learning factory: **Lernfabrik für vernetzte Produktion (LVP)**
Operator: **Fraunhofer IGCV**
Year of inauguration: **2016**
Floor Space in learning factory: **91 m²**
Manufacture product(s): **Remote-controlled cars**
Main topics / learning content: **Digitization, paperless production**

Extract from the morphology

2.1	main purpose	education	training		research
2.2	secondary purpose	test environment / pilot environment	industrial production	innovation transfer	public image
3.1	product life cycle	product planning	product development	prototyping		service	recycling
3.2	factory life cycle	investment planning	factory concept	process planning	ramp-up	manufacturing / assembly / logistics	maintenance	recycling
3.3	order life cycle	configuration & order	order sequencing	planning and scheduling		picking, packaging	shipping
3.4	technology life cycle	planning	development	virtual testing		maintenance	modernization
3.5	indirect functions	primary activities			secondary activities		
		Inbound & outbound logistics	marketing & sales	service	firm infrastructure	HR	technology development	Procurement
4.1	learning environment	purely physical (planning + execution)	physical LF supported by digital factory	physical value stream of LF extended virtually	purely virtual (planning + execution)
4.2	environment scale	scaled down	life-size		

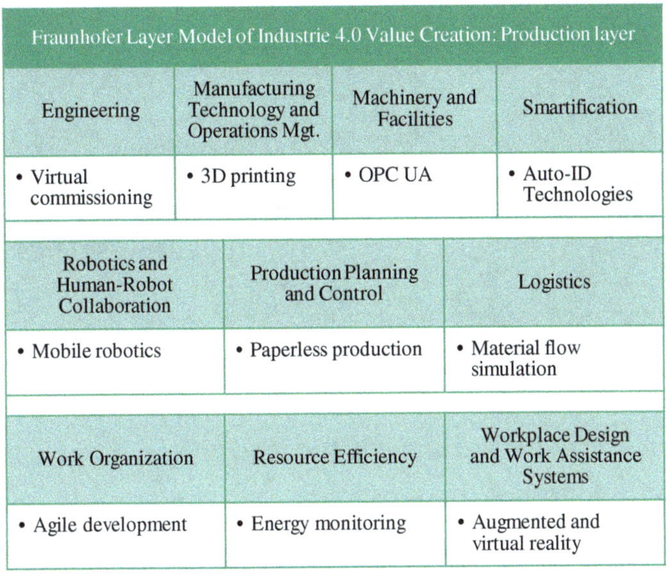

11.19 Best Practice Example 19: "Lernfabrik für vernetzte Produktion" ...

Fraunhofer Layer Model of Industrie 4.0 Value Creation: Production layer			
Engineering	Manufacturing Technology and Operations Mgt.	Machinery and Facilities	Smartification
• Virtual commissioning	• 3D printing	• OPC UA	• Auto-ID Technologies
Robotics and Human-Robot Collaboration	Production Planning and Control	Logistics	
• Mobile robotics	• Paperless production	• Material flow simulation	
Work Organization	Resource Efficiency	Workplace Design and Work Assistance Systems	
• Agile development	• Energy monitoring	• Augmented and virtual reality	

Fig. 11.46 Classification of learning modules in the learning factory

11.19.1 Starting Phase and Purpose

The learning factory for cyber-physical production systems (Lernfabrik für vernetzte Produktion) was established in 2016 at the Fraunhofer Research Institution for Casting, Composite and Processing Technology IGCV. It was initially funded by the Bavarian Ministry of Economic Affairs and Media, Energy and Technology. The main purpose of the learning factory is offering trainings in Industrie 4.0-related topics, such as assistance systems and digital engineering methods. Besides trainings, there are several secondary purposes of the learning factory. Data gained during trainings can be used for research in order to evaluate potentials of digital technologies. The learning factory is also used as a test bed for hardware and software partners from the industry as well as for demonstrators of ongoing research projects. Furthermore, the learning factory is used as a meeting space to get in touch with potential customers in an industrial environment.

The strategic goal of the learning factory is offering trainings in every segment of the production layer of the "Fraunhofer Layer Model of Industrie 4.0 Value Creation." Figure 11.46 shows the target set of learning modules offered in the learning factory.

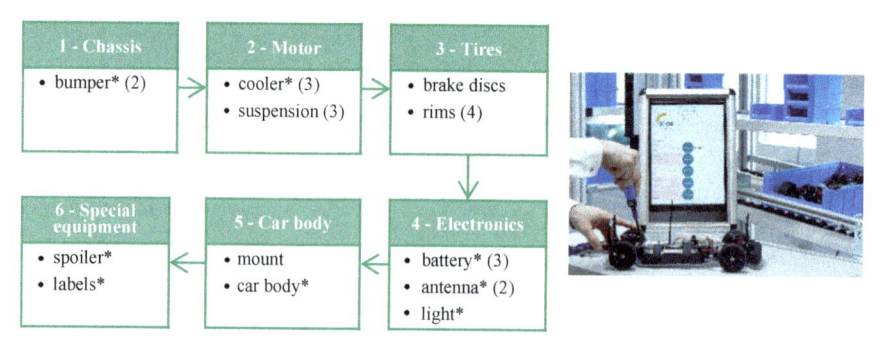

Fig. 11.47 Manual assembly process for remote-controlled cars

11.19.2 Learning Environment and Products

One scenario in the learning factory is the assembly process of customized products. As an exemplary product, remote-controlled cars were chosen since they can be configured in several variants by changing motor cooler, suspension, bumper, and other elements. Figure 11.47 shows an overview of the manual assembly process of the remote-controlled cars in the learning factory. Each box represents one assembly workstation with dedicated accessories.

Figure 11.47 also shows the number of alternative options per accessory in brackets. Optional accessories are marked with a star. In total, there are 27,648 possible product variants. The used assembly system is not a linked assembly, but a flexible assembly system. The number of assembly station a product needs to pass is depending on the product configuration. If more accessories are chosen, the remote-controlled car is therefore assigned to more assembly stations. In addition to the assembly workstations, there are two warehouse shelves for base chassis and finished products.

The participants can fulfill one of three roles in the learning factory: shop floor manager, logistics worker, and assembly worker. The shop floor manager controls the assembly systems and is responsible for communication with the customer. If assembly workers face problems, the shop floor manager is their first point of contact. His workplace is located next to the assembly workstations. Logistics workers have the responsibility to transfer unfinished products to the next assembly station. Since every order takes a different way through the flexible assembly system, logistics need to pay attention to the correct assembly station. Assembly workers stay at one dedicated workstation and are responsible for a high-quality assembly. Figure 11.48 shows two assembly workers in the learning factory.

11.19 Best Practice Example 19: "Lernfabrik für vernetzte Produktion" ...

Fig. 11.48 Assembly of remote-controlled cars in the learning factory

11.19.3 Operation

The interactive educational game "From paper-based to paperless assembly" is a one-day training where three 40-min rounds of assembly take place. It starts with a general introduction about digitization and the learning game and ends with a moderated workshop to discuss the impact of digitization on the participants' company. The level of digitization in the learning factory increases during the three rounds. In each of the three assembly rounds, the same orders are used in order to compare output numbers.

The described training can be offered for groups from 7 to 13 participants and trainings are currently held by two researchers of the Fraunhofer IGCV. In order to continue the operation of the learning factory after government funding ends, new business models are currently developed.

11.20 Best Practice Example 20: LMS Factory at the Laboratory for Manufacturing Systems & Automation (LMS), University Patras, Greece

Authors: Dimitris Mourtzis[a], Panagiotis Stavropoulos[a], George Dimitrakopoulos[a], Harry Bikas[a], Stylianos Zygomalas[a]

[a]Laboratory for Manufacturing Systems & Automation (LMS)

Name of learning factory: LMS Factory
Operator: Laboratory for Manufacturing Systems & Automation (LMS)
Year of inauguration: 2012
Floor Space in learning factory: 300 sqm
Manufacture product(s): N/A
Main topics / learning content: Training, Education

Extract from the morphology

2.1	main purpose	education	training		research
2.2	secondary purpose	test environment / pilot environment	industrial production	innovation transfer	public image
3.1	product life cycle	product planning	product development	prototyping		service	recycling
3.2	factory life cycle	investment planning	factory concept	process planning	ramp-up	main-tenance	recycling
3.3	order life cycle	configuration & order	order sequencing	planning and scheduling		picking, packaging	shipping
3.4	technology life cycle	planning	development	virtual testing		main-tenance	moderni-zation
3.5	indirect functions	primary activities				secondary activities	
		Inbound & outbound logistics	marketing & sales	service	firm infra-structure	HR	technology development	Procure-ment
4.1	learning environment	purely physical (planning + execution)	physical LF supported by digital factory	physical value stream of LF extended virtually	purely virtual (planning + execution)
4.2	environment scale	scaled down	life-size		

11.20.1 Starting Phase and Purpose

Laboratory for Manufacturing Systems & Automation (LMS) of the University of Patras has established multiple Teaching Factories (TF) over the years, recognizing the potential for conforming the theoretical knowledge, research, and innovation into industrial practice.

The Teaching Factory is the concept where manufacturing practitioners "teach" students of Engineering Schools about manufacturing problems, manufacturing issues and manufacturing practices, aiming at gaining experience and skills tailored to the industrial requirements, toward their gradual integration in the industrial field. The concept also involves information flow from a classroom to a factory, where students and faculty "teach" manufacturing practitioners about advances made in manufacturing technology, new trends, and results of research and development work. The Teaching Factory is a non-geographically anchored learning "space"; instead it is realized via internet and facilitated by advanced ICT tools and high-grade industrial didactic equipment, operating as a bidirectional knowledge communication channel, "bringing" the real factory to the classroom and the academic laboratory to the

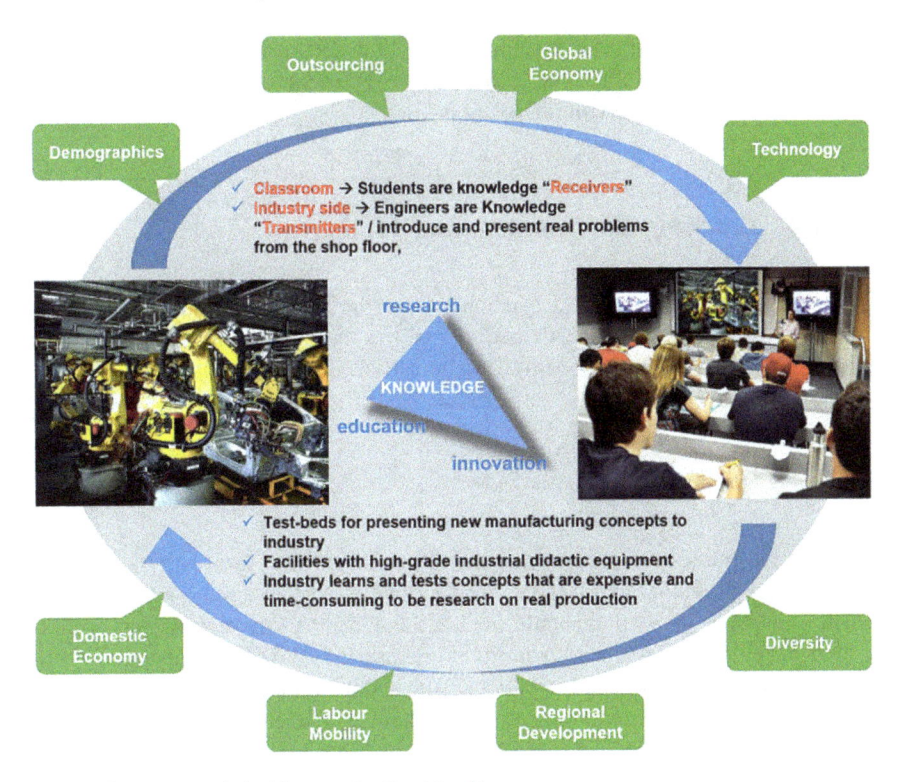

11.20 Best Practice Example 20: LMS Factory at the Laboratory for …

Fig. 11.49 Framework Architecture for Teaching Factory

factory.[5] Moreover, it is a continuous process over a lengthier period of time, with regular sessions and continuous interaction between the factory and the classroom.[6]

Context and content modular configurations allow learning and training on multiple study contents, engaging different factory facilities, engineering activities, delivery mechanisms, and academic practices. The overall framework architecture for a Teaching Factory is shown in Fig. 11.49.

On the other hand, the Learning Factory is a "learning environment specified by processes that are authentic, include multiple stations, and comprise technical, as well as, organizational aspects, a setting that is changeable and resembles a real value chain, a physical product being manufactured, and a didactical concept that comprises formal, informal, and non-formal learning, enabled by own actions of the trainees in an on-site learning approach."[7] The concept refers to interdisciplinary hands-on engineering projects closely linked to the industry, whereby in the facilities of the University, equipment that resembles manufacturing facilities is installed. People

---
[5]See Rentzos, Mavrikios, and Chryssolouris (2015), Rentzos, Doukas, Mavrikios, Mourtzis, and Chryssolouris (2014), Mavrikios, Papakostas, Mourtzis, and Chryssolouris (2013).
[6]See Chryssolouris, Mavrikios, and Mourtzis (2013).
[7]Abele et al. (2017), Abele (2016).

from both academia and industry can participate in specified courses that have as a purpose, to acquire manufacturing concepts, trends, and knowledge in the academic environment, however with an industrial flavor.[8]

11.20.2 Learning Environment and Products

The learning environment of a TF is unique, due to the fact that the concept is based on "remote interconnection" of participating facilities. A number of ICT tools available to all participating parties (academic and industrial) must be established.

11.20.3 Operation

Based on the framework presented, a number of LF pilots have been run in LMS. The first case presented involves the collaborative design of a radio-controlled car, which introduces the participants of the LMS Factory, to a more advanced cycle of product development and manufacturing. This case enables the participants to obtain a hands-on experience with the product design and manufacturing "chain," by using conventional and modern methods and technologies, in a team-working and collaborative environment.[9] The development starts with the design of the product using a CAD software. By introducing the concept of Design for Assembly (DFA), the participants are familiarized with the importance of design for the efficiency of the product manufacturing. Virtual reality technology is used for the virtual validation of the design and possible refinement of the product specifications. Simulation and scheduling topics are designed for introducing the participants on how to program the production and manufacturing of the product. Finally, the participants move to the production and manufacturing phase, where, by using laser and CNC machines they prepare the components of the product (Fig. 11.50). At the last stages of the production process, an introduction to statistical process control, brings the participants close to the concept of quality control at a production level. The RC car performs a set of dynamic tests, such as steering capability and velocity. Furthermore, the overall quality of its design and manufacture is examined.

The second case revolves around a project of designing and manufacturing a two-speed gearbox. This concept aims to impart real-life industrial practices to the participants. The gearbox consists mainly of a steel casing, four gears, and two shafts, and is aiming to introduce to the participants of the LMS Factory the basic principles of manufacturing, having as a topic the main machining processes. The participants of the LMS Factory manufacture the gearbox using the processes of turning, milling, drilling, and grinding (Fig. 11.51). Afterward, the parts are welded and assembled

---

[8]See Abele et al. (2015).
[9]See Chryssolouris and Mourtzis (2008).

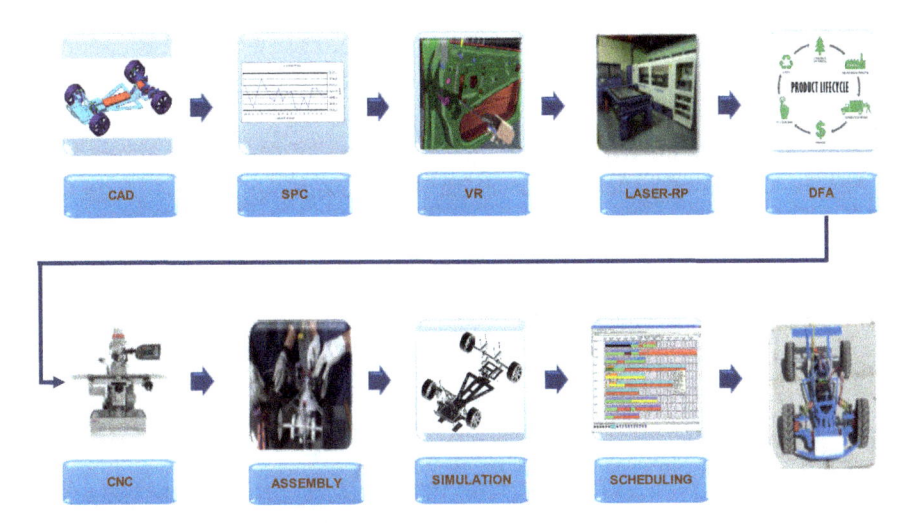

11.20 Best Practice Example 20: LMS Factory at the Laboratory for ... 411

Fig. 11.50 Production stages of the RC car

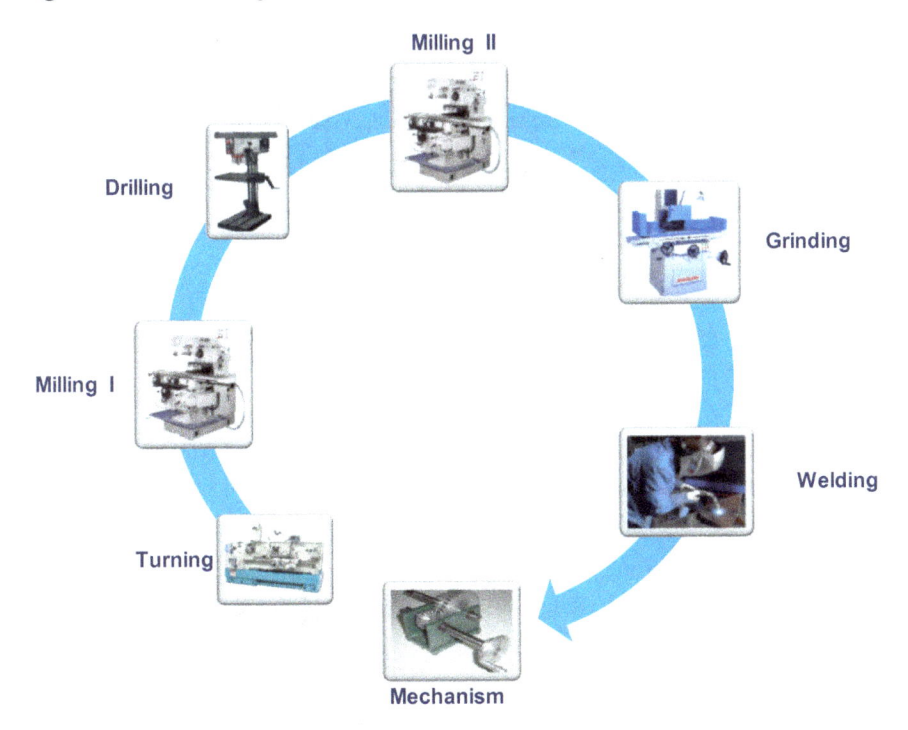

Fig. 11.51 Various processes used for manufacturing the gearbox

together in order to form the final product. Subsequently, the evaluation procedure for the mechanism takes place, with rpm and noise measurements.

11.21 Best Practice Example 21: LPS Learning Factory at LPS, Ruhr-Universität Bochum, Germany

Authors: Christopher Prinz[a], Dieter Kreimeier[a]
[a]Chair for Production Systems, Ruhr-Universität Bochum

Name of learning factory: **LPS Learning Factory**
FactoryOperator: **LPS, Ruhr-Universität Bochum**
Year of inauguration: **2009**
Floor Space in learning factory: **1800 sqm**
Manufacture product(s): **bottle cap, bottle cap holder, various make-to-order products**
Main topics / learning content: **lean production, Industrie 4.0, Ressource Efficiency, workers' participation, labour 4.0**

Extract from the morphology

2.1	main purpose	education	training			Research	
2.2	secondary purpose	test environment / pilot environment	industrial production	innovation transfer		public image	
3.1	product life cycle	product planning	product development	prototyping		service	recycling
3.2	factory life cycle	investment planning	factory concept	process planning	ramp-up	main-tenance	recycling
3.3	order life cycle	configuration & order	order sequencing	planning and scheduling		picking, packaging	shipping
3.4	technology life cycle	planning	development	virtual testing		main-tenance	moderni-zation
3.5	indirect functions	primary activities				secondary activities	
		Inbound & outbound logistics	marketing & sales	ser-vice	firm infra-structure	HR	technology development	Procure-ment
4.1	learning environment	purely physical (planning + execution)	physical LF supported by digital factory	physical value stream of LF extended virtually		purely virtual (planning + execution)	
4.2	environment scale	scaled down		life-size			

11.21.1 Starting Phase and Purpose

The learning factory of the Chair of Production Systems (LPS) has been a central component of research and teaching at the LPS since 2009. The idea of developing its own learning factory was born through a visit to CiP in Darmstadt. As it became difficult to find sponsors for the development, the LPS developed the learning factory independently and with its own resources and ideas. Initially, trainings were developed and offered only for students. Due to the well-equipped research facility of the LPS, these trainings could take place in an already very realistic factory area (machinery in operation with make-to-order production). In 2011, the joint "Institut für WertschöpfungsExzellenz" (IWEX; engl. institut for value creation excellence) was founded through the cooperation with the consulting firm LMX Business Consulting GmbH in order to also offer training courses for industry. Through further cooperations, for example with the Joint Employment Office RUB/IG Metall or the Effizienz-Agentur NRW, more and more topics, such as codetermination and resource efficiency, were included in the training assortment.

Since then, the Learning Factory has been serving both for the university education of students (e.g. from the fields of mechanical engineering, social sciences) and for the qualification and advanced training (lean management, Industrie 4.0, employee participation) of industrial participants (e.g. operators, divisional managers, work councils, middle and top managements). The Learning Factory is also a place of research for the numerous research projects of the LPS (e.g. Industrie 4.0 maturity models, assistance and learning systems, cyber-physical production systems, digitization, robo-forming, shape memory actuators, industrial robotics, human–robotics collaboration). Hundreds of events have already been organized in the LPS Learning Factory. Every year, exercises are carried out with about 900 students in the Learning Factory. Since the beginning of IWEX, several hundred industry participants have been further qualified. The annual visitor numbers vary between over 800 and 1000.

Since 2018, the Learning Factory is part of the SME 4.0 competence center Siegen (Mittelstand 4.0 Kompetenzzentrum Siegen). As one of the initiators of this competence center, the LPS focuses on the subjects "work and organization 4.0" and gives SMEs the opportunity to see and learn about Industrie 4.0 applications within the LPS Learning Factory.

11.21.2 Learning Environment and Products

The technical infrastructure reflects the production environment of small and medium-sized enterprises, thus representing a "realistic factory." This includes various machine tools for machining (CNC milling and lathes), load transports (pallet truck, crane), manual assembly stations, and various industrial robots (Kuka, ABB, Universal Robot, etc.) (Fig. 11.52).

Since the relocation in 2017/18, the learning factory now covers more than 1800 m^2 of floor space. On this surface, the topics already discussed are considered on different machines and assembly and robot cells. Participants have the opportunity to observe both production processes and to immerse themselves in them in order to become part of the value-added process. This allows the participants a completely new and better learning experience, as it relies not only on formal, but also on informal learning, thus enabling action-oriented learning.

In order to ensure an excellent university teaching and realistic factory environment for the qualification offers of the LPS learning factory, the learning factory also produces real products for industry. These are make-to-order products. The offer varies from small series over prototypes to individual products (Fig. 11.53). By manufacturing these components, the learning factory can draw on real experience based on real products and manufacturing processes.

Fig. 11.52 LPS Learning Factory at LPS, Ruhr-Universität Bochum

Bottle cap

UniLokk
- Complete value stream (manufacturing, assembly and logistic)
- Individualization of products possible

Bottle cap mount

UniLokk-Holder
- Complese ressource-value stream (manufacturing)
- Individualization of products possible

Fig. 11.53 Manufactured products in the LPS Learning Factory

11.21.3 Operation

The goal of learning factory training is always to enable problem- and action-oriented learning, which promotes the combination of formal (appropriation of theoretical knowledge) and informal learning (acquisition of practical knowledge) in a realistic production environment in order to increase learning and competence increase.

The qualification opportunities for industry participants are particularly suited to employees from companies (such as managers, department heads, coordinators, works councils), who deal with lean management or resource-efficient production.

11.21 Best Practice Example 21: LPS Learning Factory at LPS …

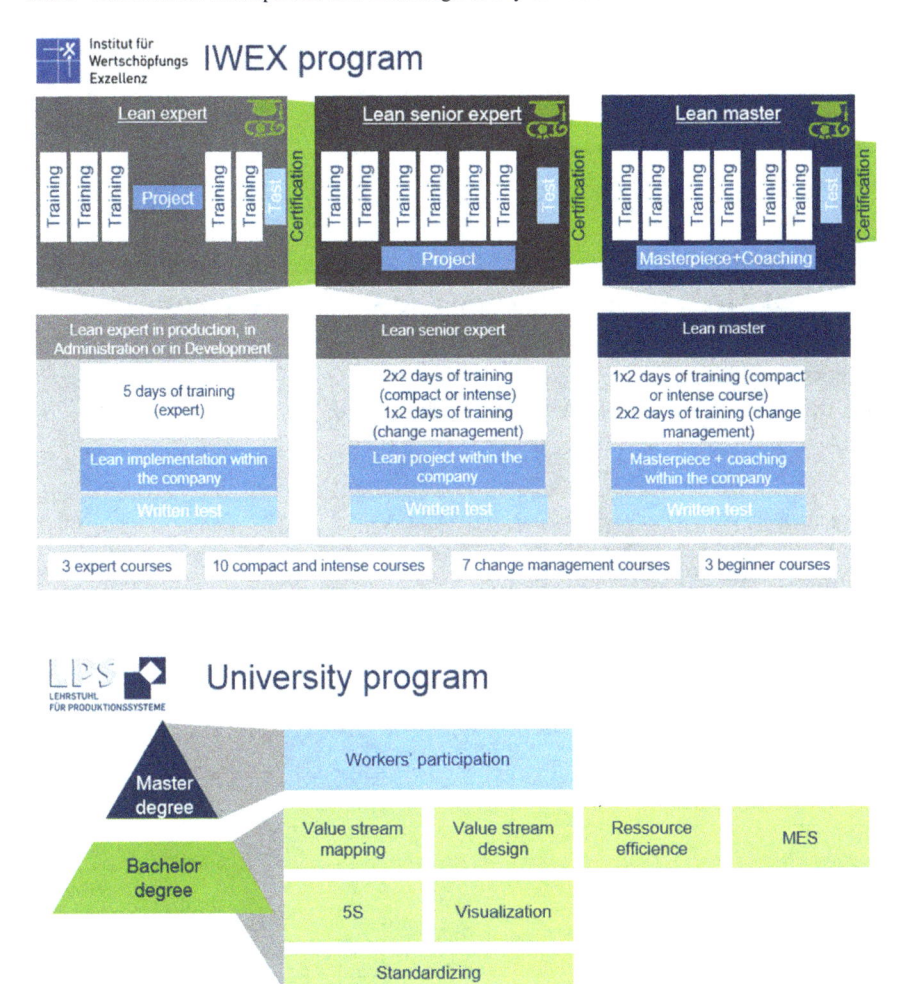

Fig. 11.54 IWEX training program with certification and university education

Lean management seminars as well as a Lean meets Industrie 4.0 seminar are offered by the Institute for Value Creation Excellence (IWEX), run by the LPS and LMX Business Consulting, and include simulations in the Learning Factory (Fig. 11.54). To train a resource-efficient production, the LPS conducts seminars together with experts from the Effizienz-Agentur NRW using the HYDRA MES from MPDV.

Many years of cooperation with the RUB/IG Metall Joint Research Center has been intensified in 2012 through a joint series of lectures aimed at students of engineering as well as sociology. The event "Management and Organization of Work" (MAO) has been carried out for 5 years now and provides the students of both disciplines with theoretical input about the technical (e.g. lean management, mutability, industrie 4.0) as well as the organizational-personnel aspects (e.g. employee participation, industrial relations, employment, and employment conditions). Practical application possibilities through the Learning Factory exercises and practical semester-related elaborations (in companies) in cooperation with works councils supplement the theoretical aspects.

11.22 Best Practice Example 22: MAN Learning Factory at MAN Diesel & Turbo SE in Berlin, Germany

Authors: Jan Philipp Menn[a], Carsten Ulbrich[b]
[a]TU Berlin, [b]MAN Diesel & Turbo SE, Berlin

Name of learning factory: **MAN Learning Factory**
Operator: **MAN Diesel & Turbo SE, Berlin**
Year of inauguration: **2018**
Floor space in learning factory: **45 sqm (workshop) + 50 sqm (assembly area)**
Manufactured product: **integrally geared compressor**
Main topics / learning content: **assembly and maintenance of compressors**

Extract from the morphology

2.1	main purpose	education	training		research	
2.2	secondary purpose	test environment / pilot environment	industrial production	innovation transfer	public image	
3.1	product life cycle	product planning	product development	prototyping		service	recycling
3.2	factory life cycle	investment planning	factory concept	process planning	ramp-up	maintenance	recycling
3.3	order life cycle	configuration & order	order sequencing	planning and scheduling		picking, packaging	shipping
3.4	technology life cycle	planning	development	virtual testing		maintenance	modernization
3.5	indirect functions	primary activities			secondary activities		
		Inbound & outbound logistics	marketing & sales	service	firm infra-structure	HR	technology development	Procurement
4.1	learning environment	purely physical (planning + execution)	physical LF supported by digital factory	physical value stream of LF extended virtually	purely virtual (planning + execution)	
4.2	environment scale	scaled down	life-size			

11.22 Best Practice Example 22: MAN Learning Factory at MAN Diesel … 417

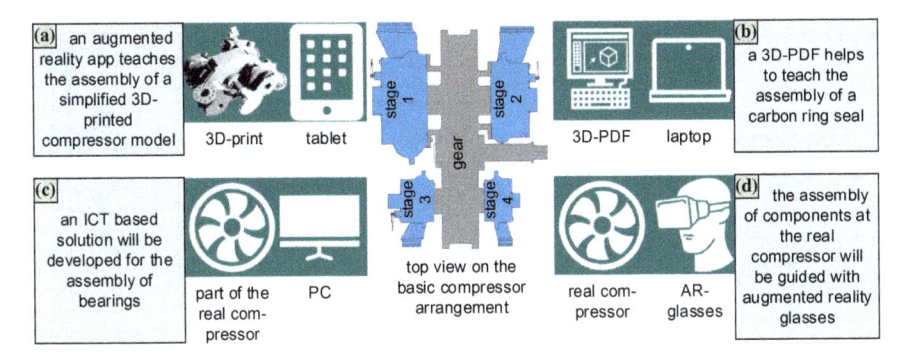

Fig. 11.55 Four exemplary Learnstruments to teach basic (**a**) and variant-specific (**b–d**) assembly knowledge (Menn, Sieckmann, Kohl, & Seliger, 2018)

11.22.1 Starting Phase and Purpose

The idea of a Learning Factory for service related assembly work at MAN Diesel & Turbo SE, Berlin came up in 2017 in collaboration between MAN and TU Berlin. MAN Diesel & Turbo SE produces integrally geared compressors with a high efficiency, low operating cost, a low investment outlay, and an excellent operating range at the German capital Berlin. Specific applications for these compressors include Industrial Gases, Oil & Gas, CO_2, Urea, Purified Terephthalic Acid or Nitric Acid processes. In a sustainable manner, an old MAN compressor, running for nearly 25 years in the field, was bought back from a customer and refurbished for teaching purposes. In 2018, the required adjustments to the compressor were finalized. The compressor builds the center of the Learning Factory with fixed-site production in this one-of-a-kind producing business. As every compressor is specifically made to customer requirements, learning the assembly of such a compressor inhabits a great challenge for international Field Service Engineers (FSEs), who are responsible for the erection and maintenance at the customer site. To cope with this challenge, the compressor was modified to represent the least common denominator regarding product variety of integrally geared compressors. A basic understanding can be easily taught by this arrangement. As the Learning Factory is developed further in the future, additional learning stations will be arranged around the compressor itself, as displayed in Figs. 11.55 and 11.56. They represent compressor subassemblies for different product variants. These learning stations are designed as so-called Learnstruments. The neologism Learnstrument consists of the words *learning* and *instrument* and represents technologic artefacts, both tangible and intangible, automatically demonstrating their functionality to users. The Learnstruments a-d in Fig. 11.55 represent different Information and Communication Technology (ICT)-based artifacts to teach basic (a) and variant-specific (b–d) assembly knowledge.

Fig. 11.56 MAN Learning Factory, workshop area layout, including a basic compressor and Learnstruments **a–d** as described in Fig. 11.55

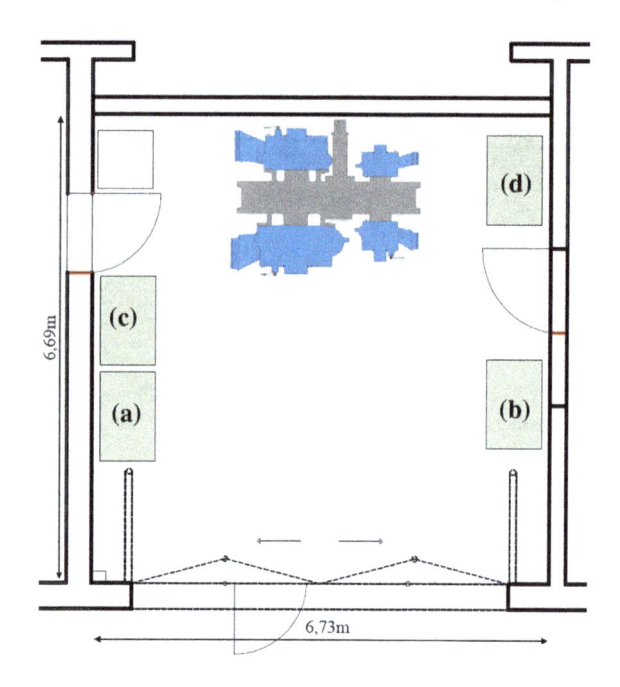

11.22.2 Learning Environment and Products

The MAN Learning Factory is split into two areas. The workshop area (45 m^2) is used for presentations and accommodates several Learnstruments and the basic compressor, see Fig. 11.56. For assembly training, the compressor is moved to an assembly hall nearby, which is equipped with cranes and additional equipment. As the real production of other compressors takes place in this assembly hall as well, depending on the availability about 50 m^2 are used here for the assembly training. Basic (dis-)assembly, component exchange and adjustments, quality checks and service activities are trained here. After assembly training, the compressor is moved back into the workshop area as an exhibit for customers. The integrally geared training compressor consists of six compression stages, see Fig. 11.57. Each stage differs in size and number of parts, but consists of the same main elements. The modular design of the compressor allows in combination with the used ICT a huge variety of different scenarios for machine failures and service activities. Each basic training session starts with the disassembly of a compressor stage, comparable to a real maintenance or service case at a customer site. Finally, the compressor is assembled again. During this process, old or broken parts can be exchanged with new and improved parts.

11.22 Best Practice Example 22: MAN Learning Factory at MAN Diesel … 419

Fig. 11.57 Assembly of an integrally geared compressor at MAN Diesel & Turbo SE, Berlin

11.22.3 Operation

The goal of the MAN learning factory is to teach Field Service Engineers maintenance operations for integrally geared compressors as well as the required assembly knowledge and skills. One training program for the basic assembly of integrally geared compressors has been developed so far. It is used to teach between 10 and 15 FSEs per training. About twelve different experts from the MAN engineering and service department are working as trainers for the individual learning modules. An additional instructor accompanies the whole training program. Further training programs for the assembly of special components are under development. In the future, the basic compressor assembly training will also be offered to apprentices, new employees, and internal assembly workers.

11.23 Best Practice Example 23: MPS Lernplattform at Daimler AG in Sindelfingen, Germany

Author: Michael Schwarz[a]
[a]MPS Lernplattform Sindelfingen, Daimler AG

Name of learning factory: **MPS Lernplattform** Operator: MPS Sindelfingen, **Daimler AG** Year of inauguration: **2011** Floor Space in learning factory: **3.000 qm** Manufacture product(s): **different products** Main topics / learning content: **Lean**								
Extract from the morphology								
2.1	main purpose	education			training			research	
2.2	secondary purpose	test environment / pilot environment		industrial production		innovation transfer		public image
3.1	product life cycle	product planning	product development	prototyping			service	recycling
3.2	factory life cycle	investment planning	factory concept	process planning	ramp-up		maintenance	recycling
3.3	order life cycle	configuration & order	order sequencing	planning and scheduling	manufacturing	assembly / logistics	picking, packaging	shipping
3.4	technology life cycle	planning	development	virtual testing			maintenance	modernization
3.5	indirect functions	primary activities				secondary activities			
		Inbound & outbound logistics	marketing & sales	service	firm infra-structure	HR	technology development	Procurement
4.1	learning environment	purely physical (planning + execution)		physical LF supported by digital factory		physical value stream of LF extended virtually		purely virtual (planning + execution)
4.2	environment scale	scaled down			life-size				

11.23.1 Starting Phase and Purpose

In the middle of the twentieth century, methods for lean management were emerged in Japan for the first time. Over the years, "lean" has become a trend word, and so the Mercedes-Benz Production System (MPS) of the Daimler AG considered how lean production processes could be used to contribute to the CIP (continuous improvement process) of the company. For this purpose, lean specialists are required, who teach this increasingly important topic in a practical way to all other employees of the company. In 2011, the decision was made to create a learning center around the topic "lean"—the birth of the MPS lernplattform. For this purpose, specialists from various fields were brought together, who were trained to become qualified business trainers. Over the years, the team of trainers has grown to over 10 employees. Up to now, almost 10,000 Daimler employees have been trained at various seminars on the MPS lernplattform. These are in particular executives, production planners, plant engineers, and improvement managers. The training offer is diversified and covers various contents of all areas of the entire production process. The training contents are specially adapted to the customer needs of the respective target group as well as to current trends in the production area. Thereby, the didactic focus is always on the imparting of lean knowledge as practically as possible, which is why the training courses are strongly geared toward implementation in the individual areas and work with real components. Therefore, a better understanding and ideal solutions should be developed. The successful concept of the MPS lernplattform has meanwhile been extended to more than four worldwide production sites of Daimler AG.

11.23.2 Learning Environment and Products

The MPS lernplattform (Figs. 11.58 and 11.59) offers the participants the advantage that the training takes place in a production-oriented teaching area. For this, original components of the production as well as additional 1:10 models and various simulations are used on an area of almost 3000 m^2. During the various training courses and simulation rounds, the complete production area with all shops (press shop, body shop, paint shop, assembly, and logistics) is shown intensively. In particular, the focus is not only the value-added process, but also the planning and understanding of lean management. For the important practice units, various products are used which can be reused after the training. These are, for example, real components such as roof control units, sun visors, covers, floor mats, room tears, or the assembly of smaller model cars.

11.23.3 Operation

The training courses are carried out by in-house employees from MPS. They were qualified as trainers by means of special training and, through the successful combination of didactic background knowledge and their own long-term experience in the production area, they can provide the participants with an optimal understanding of the content of course. The MPS lernplattform increasingly relies on cooperation with external partners such as the TU Darmstadt. The didactic concept of the training consists of 20% theory and 80% practice. This division is the secret of success of the MPS lernplattform because the participants can take important insights from the training for their own daily work. In addition to lean basic training, the MPS lernplattform offers more than ten different advanced training courses. Small groups of 10–20 people allow for a pleasant working atmosphere as well as the possibility of joint learning in a team. This interaction between trainers and participants from different areas is very important to open new ways of thinking and to develop a better understanding of different topics. In order to achieve a longlasting effect in the

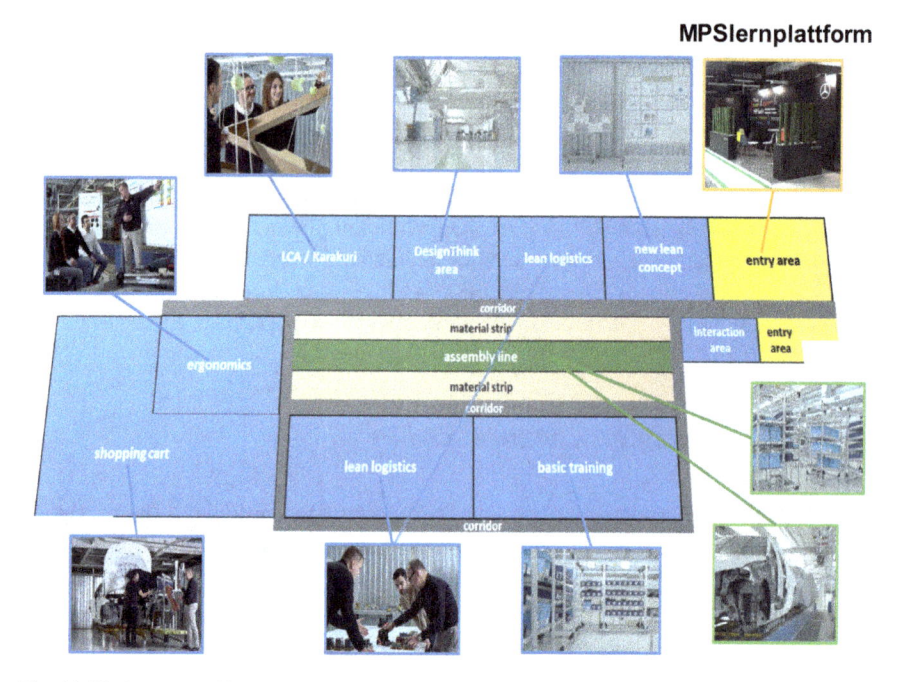

Fig. 11.58 Layout and impressions of the MPS Lernplattform in Sindelfingen at Daimler AG

qualification, theory training is combined with practical training as well as with the implementation in the daily work of the participants, since the learning and memory effect through the personal experience and comprehension of the contents is significantly higher. Consequently, the MPS lernplattform not only creates a lasting added value for the participants, but also contributes to the CIP (continuous improvement process) of the Daimler AG.

11.24 Best Practice Example 24: MTA SZTAKI Learning Factory at the Research Laboratory on Engineering and Management Intelligence, MTA SZTAKI in Győr, Hungary

Authors: Zsolt Kemény[a], Richárd Beregi[a], Gábor Erdős[a], János Nacsa[a]
[a]Institute for Computer Science and Control, Hungarian Academy of Sciences (MTA SZTAKI)

11.24 Best Practice Example 24: MTA SZTAKI Learning Factory …

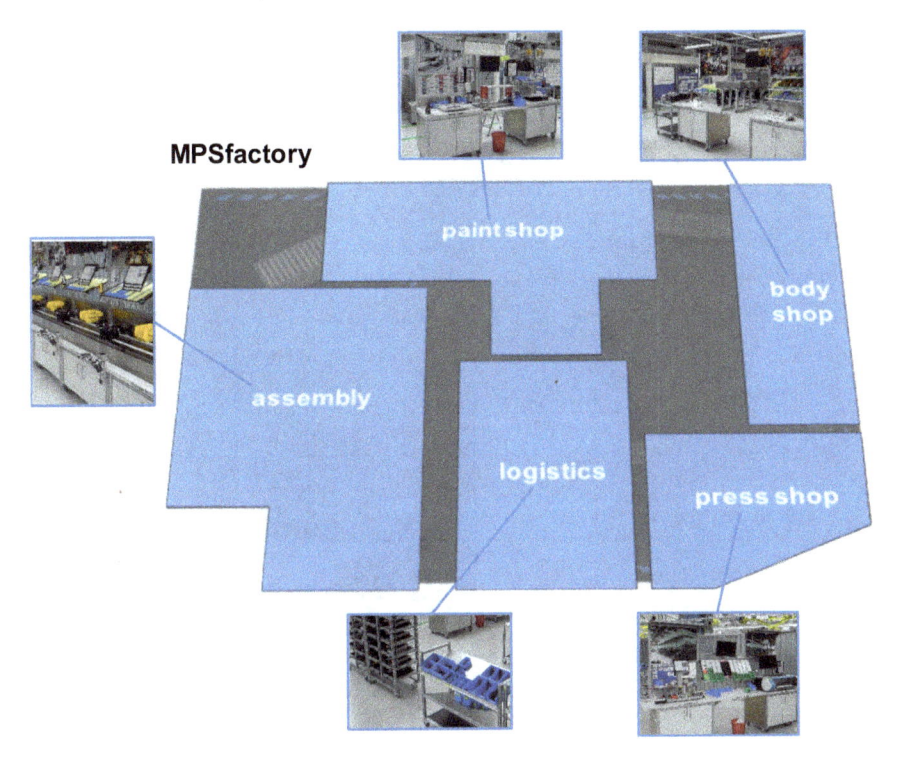

Fig. 11.59 MPSfactory at Daimler AG in Sindelfingen

Name of learning factory: **MTA SZTAKI Learning Factory Győr** Operator: **MTA SZTAKI, dept. EMI (Research Laboratory on Engineering and Management Intelligence)** Year of inauguration: **2017** Floor Space in learning factory: **150m²** Manufacture product(s): **recyclable dummy workpieces** Main topics / learning content: **CPPS aspects, primarily: (1) human–robot collaboration and (2) production planning, scheduling & execution in CPPS**									
Extract from the morphology									
2.1	main purpose	education			training			research	
2.2	secondary purpose	test environment / pilot environment		industrial production		innovation transfer		public image	
3.1	product life cycle	product planning	product development	prototyping			service		recycling
3.2	factory life cycle	investment planning	factory concept	process planning	ramp-up	manufacturing	main-tenance		recycling
3.3	order life cycle	configuration & order	order sequencing	planning and scheduling		assembly / logistics	picking, packaging		shipping
3.4	technology life cycle	planning	development	virtual testing			main-tenance		moderni-zation
3.5	indirect functions	primary activities				secondary activities				
		Inbound & outbound logistics	marketing & sales	ser-vice	firm infra-structure	HR	technology development		Procure-ment
4.1	learning environment	purely physical (planning + execution)		physical LF supported by digital factory		physical value stream of LF extended virtually		purely virtual (planning + execution)	
4.2	environment scale	scaled down			life-size					

11.24.1 Starting Phase and Purpose

Design and construction of the MTA SZTAKI Learning Factory in Győr commenced as part of an R&D&I project supported by national funds, and is now among the contributed assets in the follow-up of the EU-wide EPIC (Centre of Excellence in Production Informatics and Control) as well. Most of the infrastructure and equipment in the facility was built up during 2017, with some activities continuing in 2018. Full functionality and capacity for higher education are planned to be available by the beginning of the 2018–19 academic year in September 2018. The facility itself is located at the premises of Széchenyi University in Győr; however, design and construction of the learning factory are, for most parts, conducted by the Research Laboratory on Engineering and Management Intelligence (EMI), a department of the Institute for Computer Science and Control of the Hungarian Academy of Sciences (MTA SZTAKI). EMI will also maintain and operate the facility in the coming years, with some of EMI staff members also involved in other courses at Széchenyi University.

The Learning Factory in Győr is the second of such facilities constructed by MTA SZTAKI. As opposed to the "Smart Factory" in Budapest, the site in Győr serves education as its primary purpose, being able to support one-shot laboratory exercises as well as complete courses where student groups will receive hands-on experience in an entire series of life-cycle stages of production assets and products in a production environment emphasizing aspects of cyber-physical production systems (CPPS) and human–robot collaboration (HRC). To this end, the equipment is laid out in the form of reconfigurable workstations with collaborative robots, surrounded by open floor space shared by humans and automated guided vehicles (AGVs) performing intralogistics.

11.24.2 Learning Environment and Products

The MTA SZTAKI Learning Factory in Győr has a shop floor area of ca. 150 m^2 which is shared in part with conventional automation equipment owned and operated by Széchenyi University. The Learning Factory consists of four flexibly configurable robotic workstations equipped with UR5/10 robots. Each station is organized around a central support for one (or two) robot arm(s), while the surrounding space is configurable using prefabricated frames and work surfaces. Also part of the infrastructure are indoor positioning, devices for acquiring optical images and 3D point clouds, and reconfigurable human–machine interface components. Intralogistics is performed by AGVs, one of them equipped with a UR3 robot for loading/unloading operations performed at the workstations. Augmenting the facility is a room of 75 m^2 for lectures, product design, and procurement, the latter being supported by an FDM 3D printer, also accessible by remote users.

Material passing through the production equipment will consist of workpieces emphasizing mechanical/geometrical aspects of assembly, preferably laid out to allow handling by humans and robots likewise, thereby allowing variable assignment of production resources to the individual tasks. The assembly process is planned to be fully reversible, i.e. the workpieces can be returned to their initial state after a production run. Due to the availability of a 3D printer and resources for product design, a new set of recyclable products can easily be designed and procured, even as part of a comprehensive curriculum giving insight into the interdependencies of product and production equipment.

11.24.3 Operation

The MTA SZTAKI Learning Factory in Győr is primarily designed to support longer, possibly connected, courses that closely follow life-cycle stages of production assets and products throughout a longer curriculum. The assignment of life-cycle stages to subsequent courses is still under elaboration and is likely to undergo much refinement in the first years of operation, nonetheless, the available spectrum is already outlined. Regarding production equipment, layout design, process planning and production ramp-up will be in the focus, with emphasis on new aspects that are introduced by

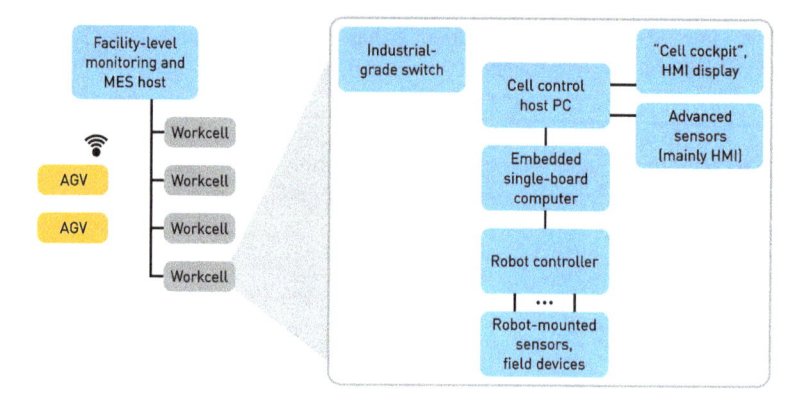

Fig. 11.60 Architectural overview of the MTA SZTAKI Learning Factory in Győr

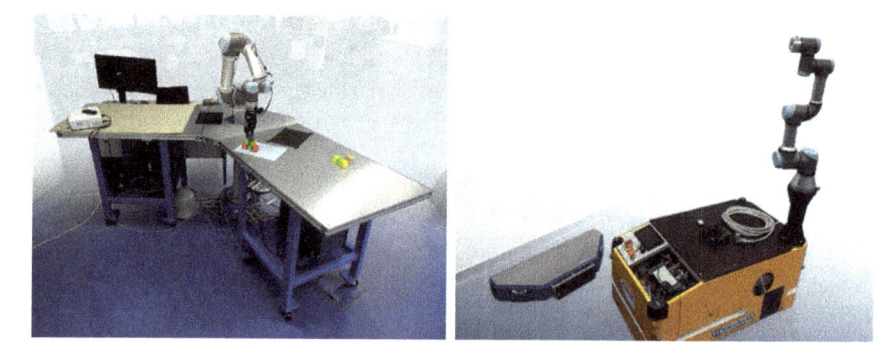

Fig. 11.61 Reconfigurable workstation with collaborative robot (left), and manipulator-equipped AGV with docking station (right)

CPPS and HRC (Figs. 11.60 and 11.61). Students will also have the opportunity of a direct comparison with conventional automation equipment available on site. While currently less in the focus, a series of product life-cycle stages can also be subject to courses, spanning product design, prototype procurement and production.

In addition, stand-alone laboratory exercises and demonstrations can also be hosted by the facility, as well as independent work in preparation of BSc and MSc theses, or research activities of PhD students.

The strong presence of the automotive industry in the area of Győr makes it a natural choice to offer the facility for training, demonstration and performing tests of automation solutions by contract with industrial enterprises. In addition, MTA SZTAKI will also conduct some of its research activities at the site, partly under involvement of remotely connected resources residing at the institute's Budapest headquarters.

11.25 Best Practice Example 25: Pilot Factory Industrie 4.0 at IMW, IFT, and IKT, TU Wien, Austria

Authors: Wilfried Sihn[a], Fabian Ranz[a], Philipp Hold[a]
[a]TU Wien, Institute of Management Sciences, Austria

Name of learning factory: **Pilot Factory Industrie 4.0**
Operator: **TU Wien**
Year of inauguration: **2015**
Floor Space in learning factory: **900 sqm**
Manufacture product(s): **3D printer**
Main topics / learning content: **Factory virtualization, adaptive manufacturing, cyber-physical assembly & logistics**
Extract from the morphology

2.1	main purpose	education	training		research	
2.2	secondary purpose	test environment / pilot environment	industrial production	innovation transfer	public image	
3.1	product life cycle	product planning	product development	prototyping	service	recycling
3.2	factory life cycle	investment planning	factory concept	process planning	ramp-up	maintenance	recycling
3.3	order life cycle	configuration & order	order sequencing	planning and scheduling	picking, packaging	shipping
3.4	technology life cycle	planning	development	virtual testing	maintenance	modernization
3.5	indirect functions	primary activities: Inbound & outbound logistics / marketing & sales / service / firm infrastructure			secondary activities: HR / technology development / Procurement	
4.1	learning environment	purely physical (planning + execution)	physical LF supported by digital factory	physical value stream of LF extended virtually	purely virtual (planning + execution)	
4.2	environment scale	scaled down		life-size		

11.25.1 Starting Phase and Purpose

After three years of successful training operation in the Learning & Innovation Factory (see Best Practice Example 13), the Austrian Ministry for Traffic, Innovation and Technology issued a call for proposals for the establishment of Industrie 4.0 pilot factories. Such pilot factories were intended to strengthen innovation transfer between research and industry in the context of manufacturing digitalization. The TU Wien seized the opportunity and was awarded with significant funds to implement its second factory facility, the *TU Wien Pilot Factory Industrie 4.0*—again under involvement of the same three institutes that already established the LIF: the Institute for Management Sciences (IMW), the Institute for Manufacturing Technology (IFT) and the Institute for Engineering Design and Logistics Engineering (IKL).

While the LIF has a clear focus on teaching and learning, the Pilot Factory Industrie 4.0 is primarily a research, testing/evaluation and demonstration environment. It was inaugurated in 2015. More than 20 companies are involved in equipping the Pilot Factory Industrie 4.0, while researchers from the TU Wien integrate the different infrastructure components into a highly interconnected production environment. Every week, the Pilot Factory Industrie 4.0 opens doors to the general public for demonstration runs—reflecting its character as an open space for innovation and technology transfer.

11.25.2 Learning Environment and Products TU Wien Pilot Factory Industrie 4.0 (PF)

In order to develop, test, and improve new production strategies for the industry, a realistic test environment is needed—real machines, real production chains, a real product (Fig. 11.62). Fundamental ideas that are reflected in the implemented technology in the Pilot Factory Industrie 4.0 on 900 m^2 shop floor include process and layout adaptivity, high degree of human–machine interaction and use of data analytics for transparency and optimization. The product in use is a 3D printer (Fig. 11.63) that uses the principle of fused deposition modeling. It is composed by approx. 90 individual components, including drives, the extruder, the kinematics, and electronics. Core components can be manufactured in-house on machining centers, others are purchased. The printer dimensions can be configured on customer request, resulting in a batch size 1 process. Manufacturing and assembly areas are connected through AGV transportation. AGVs are also used for transportation and as assembly benches along the completion process, yielding high flexibility for reconfiguration of the system. Subsystems in use, among others, include collaborative robot systems, operator assistance systems, sensor technology, image processing, and automatic replenishment. The production system in the Pilot Factory Industrie 4.0 is fully interlinked to various computer system, enabling a real-time digital process illustration, tracing, and optimization.

11.25.3 Operation

The two facilities operated by the TU Wien follow entirely different operating models, for the operating model of the LIF see Best Practice Example 13. The TU Wien Pilot Factory Industrie 4.0 is a primary research, testing, and demonstration environment, which is in continuous operation. The demonstration factory is operated by approximately six researchers who coordinate the various activities that take place at the same time. Those include, but are not limited to, demonstration events, trial and test production for new products of industrial partners, setup of work systems

11.25 Best Practice Example 25: Pilot Factory Industrie 4.0 at IMW … 429

Fig. 11.62 Cyber-physical assembly system in Pilot Factory Industrie 4.0

Fig. 11.63 3D printer in the Pilot Factory Industrie 4.0

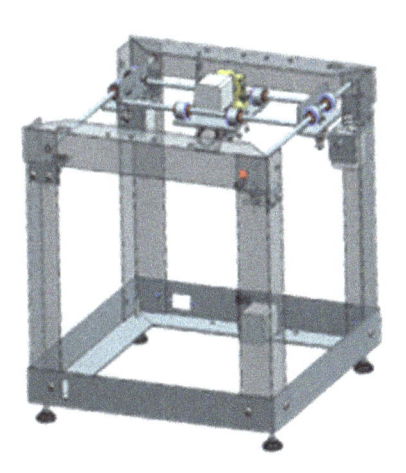

in accordance with new production principles or trainings for industry participants. Research foci of the Pilot Factory Industrie 4.0 are distributed between the involved institutes in conformity with their respective fields of competence (Fig. 11.64).

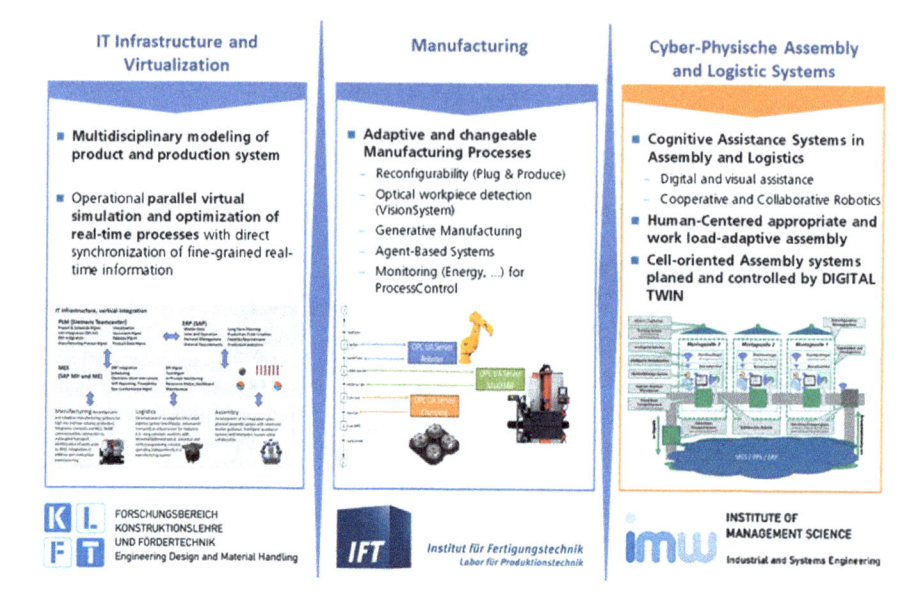

Fig. 11.64 Research foci within the TU Wien Pilot Factory Industrie 4.0

11.26 Best Practice Example 26: Process Learning Factory CiP at PTW, TU Darmstadt, Germany

Authors: Joachim Metternich[a], Eberhard Abele[a], Michael Tisch[a], Jens Hambach[a]
[a]Institute for Production Management, Technology and Machine Tools (PTW), TU Darmstadt

11.26 Best Practice Example 26: Process Learning Factory CiP at PTW ...

Name of learning factory: **Process Learning Factory CiP**
Operator: **PTW, TU Darmstadt**
Year of inauguration: **2007**
Floor Space in learning factory: **500 sqm**
Manufacture product(s): **Pneumatic Cylinder, Electric Gear Drive**
Main topics / learning content: **Lean Production, Industrie 4.0**

Extract from the morphology

2.1	main purpose	education	training	research
2.2	secondary purpose	test environment / pilot environment	industrial production	innovation transfer	public image
3.1	product life cycle	product planning	product development	prototyping			service	recycling
3.2	factory life cycle	investment planning	factory concept	process planning	ramp-up	manufacturing / assembly / logistics	maintenance	recycling
3.3	order life cycle	configuration & order	order sequencing	planning and scheduling			picking, packaging	shipping
3.4	technology life cycle	planning	development	virtual testing			maintenance	modernization
3.5	indirect functions	primary activities				secondary activities		
		Inbound & outbound logistics	marketing & sales	service	firm infrastructure	HR	technology development	Procurement
4.1	learning environment	purely physical (planning + execution)	physical LF supported by digital factory	physical value stream of LF extended virtually	purely virtual (planning + execution)
4.2	environment scale	scaled down	life-size

11.26.1 Starting Phase and Purpose

In 2004, first thoughts about building an almost real factory on the campus came up at PTW, TU Darmstadt. Initially, the university and potential donors were skeptical about the idea to sustainably run a factory on the campus; not to mention to finance it. When in 2006 a factory hall was built as part of a big research project, the former university management decided to build the facility larger as it needed to be in order to accommodate a possible learning factory. Visionary business leaders, for example, from Bosch or SEW, recognized the benefits of this learning factory idea and signed a cooperation agreement that provided the initial foundation of the Process Learning Factory CiP (Center for industrial Productivity). A special stroke of luck at this point was the long-standing cooperation with McKinsey, in particular with the former head of McKinsey Germany, Professor Jürgen Kluge. In close cooperation, a curriculum was established and first pilot trainings were held. Through continuous and dedicated work of the team, the learning factory has been continuously expanded since then.

Today, the Process Learning Factory CiP is an innovative training and research center that enables a comparison of different production systems and organizations. For research in the field of production organization, such a possibility of direct comparison is particularly valuable. The learning factory is therefore constantly utilized in evaluation of newly developed approaches. Additionally, the industry's interest in adult education activities has grown steadily, with over 3000 experts,

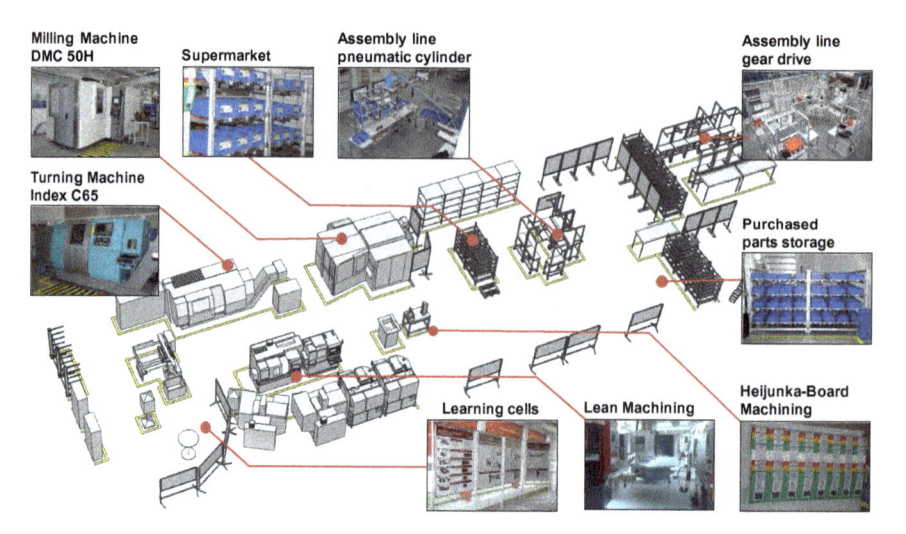

Fig. 11.65 Value stream of the Process Learning Factory CiP

production planners, but also plant managers, and managing directors being trained in lean topics in 10 years of operation (status 2017). Furthermore, the PTW won the Hessian Excellence Award for higher education for the establishment and use of the learning factory in student education.

11.26.2 Learning Environment and Products

In the simulated, but authentic production environment of the Process Learning Factory CiP, the complete value chain of the production of a compact pneumatic cylinder is mapped in a multistage manufacturing process (Fig. 11.65); it covers the delivery of raw material, machining, quality control, assembly, and packaging/shipping. In addition, indirect processes are mapped around the internal logistics and the management of the factory. The pneumatic cylinder consists of a cylinder, the cylinder base and top, piston and piston rod, and mounting elements. The cylinder base is manufactured in two variants, the piston rod in eight variants; the remaining parts are purchased. In general, products are not sold, but dismantled and returned again to the materials cycle. To demonstrate the impact of a large number of variants in assembly, an additional flexible assembly cell which produces electric gear drives is integrated (Fig. 11.66). The gear drive is assembled in more than 8000 variants. The modular design of the assembly cell allows fast changes in organization and line design by the participants. Learners can understand concepts based on own experiences, solve problems, and apply lean production methods on about 500 m^2.

Since 2014, a number of research projects helped to build up a digital value stream in the Process Learning Factory CiP, which can now be modified to address topics

11.26 Best Practice Example 26: Process Learning Factory CiP at PTW ... 433

Compact Pneumatic Cylinder

High volume production
- Complete value Stream including machining, assembly and indirect areas
- Product is not soled, but disassembled

Gear drive

High mix assembly
- More than 8000 product variants
- Simplified product model
- All parts can be disassembled & reused

Fig. 11.66 Manufactured products in the Process Learning Factory CiP

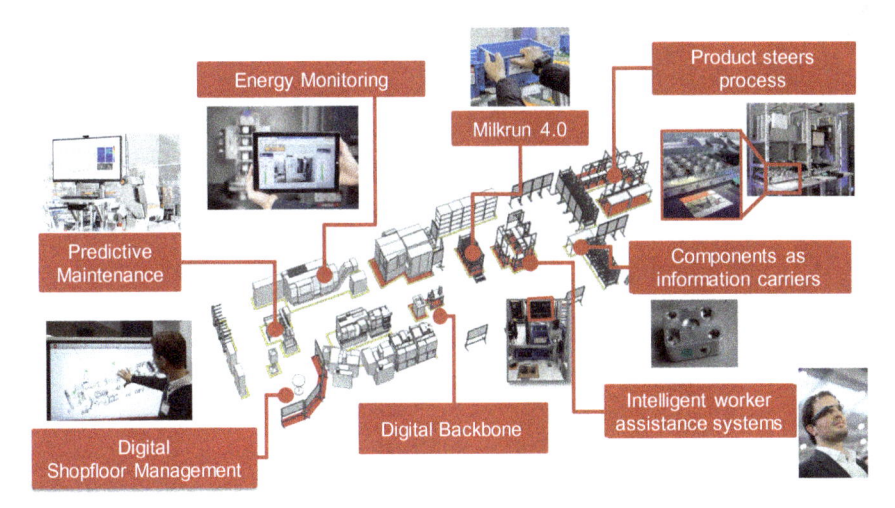

Energy Monitoring
Product steers process
Milkrun 4.0
Predictive Maintenance
Components as information carriers
Digital Backbone
Intelligent worker assistance systems
Digital Shopfloor Management

Fig. 11.67 Alternative value stream for trainings in the field of Industrie 4.0

in the field of Industrie 4.0 as well (Fig. 11.67). The underlying principle consists of the concept to start from where a company and its personnel is situated on the journey to the reduction of waste, meeting customer demands and increasing quality to name a few targets. There trainings can start in a wasteful und unbalanced production environment where the basics of lean need to be taught. However, for example trainings can also start in a lean, yet digitalized shop floor where the introduction of digital shop floor management is planned and executed.

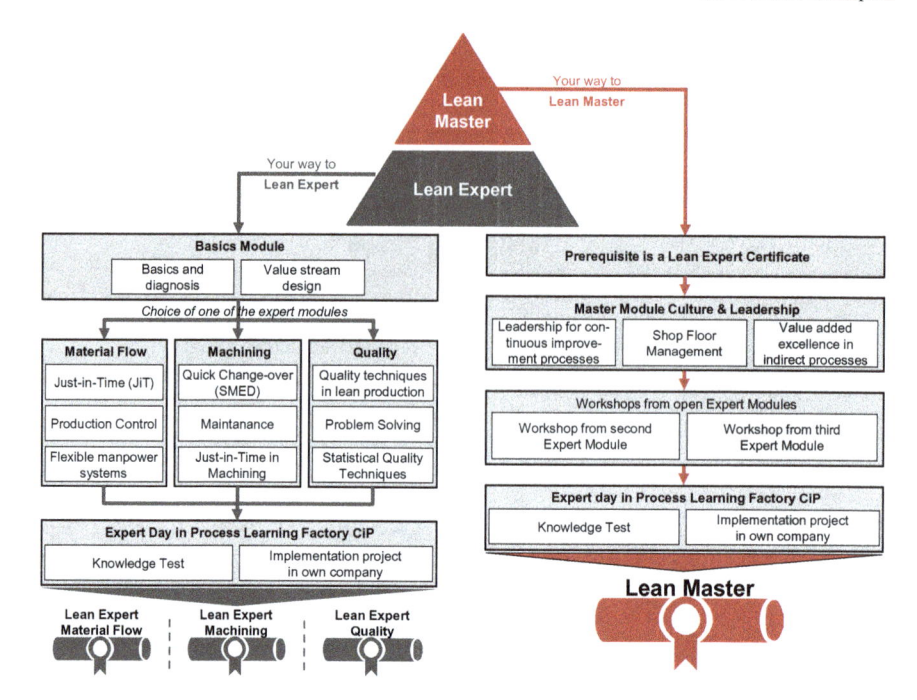

Fig. 11.68 Lean and Industrie 4.0 training program certificates of the Process Learning Factory CiP

11.26.3 Operation

Goal of teaching and training activities is the sustainable development of crucial competencies in the fields of lean management and Industrie 4.0. The lean management learning modules offered to the approximately 20 partner companies, who have a particular contingent on training days, can be divided in the fields lean understanding (basics), lean core elements (material flow, quality, machining), and lean thinking (culture and leadership). All training modules are led by two scientific staff members of the research group CiP supported by students and a technician. Figure 11.68 gives an overview over the prerequisites of the two lean certificates "Lean Expert" and "Lean Master" from the Process Learning Factory CiP.

Additionally, the Process Learning Factory CiP is a SME competency center for the region of Rhine-Main funded by the Federal Ministry for Economic Affairs and Energy. Within the center, trainings in the topic of Industrie 4.0 are developed and performed, especially for SME to promote the concept of the digitalization of production environments.

11.27 Best Practice Example 27: Smart Factory at the Research Laboratory on Engineering & Management, MTA SZTAKI in Budapest, Hungary

Authors: Zsolt Kemény[a], Richárd Beregi[a], Gábor Erdős[a], János Nacsa[a]
[a]Institute for Computer Science and Control, Hungarian Academy of Sciences (MTA SZTAKI)

\multicolumn{9}{l}{Name of learning factory: **Smart Factory**}
\multicolumn{9}{l}{Operator: **MTA SZTAKI, dept. EMI (Research Laboratory on Engineering and Management Intelligence)**}
\multicolumn{9}{l}{Year of inauguration: **2013**}
\multicolumn{9}{l}{Floor Space in learning factory: **30m²**}
\multicolumn{9}{l}{Manufacture product(s): **recyclable dummy workpieces**}
\multicolumn{9}{l}{Main topics / learning content: **CPPS aspects, primarily: (1) production planning, scheduling and execution in CPPS, (2) mechatronics and automation in CPPS**}
\multicolumn{9}{l}{Extract from the morphology}

2.1	main purpose	education		training			research	
2.2	secondary purpose	test environment / pilot environment		industrial production		innovation transfer	public image	
3.1	product life cycle	product planning	product development	prototyping	manufacturing		service	recycling
3.2	factory life cycle	investment planning	factory concept	process planning	ramp-up	assembly / logistics	main- tenance	recycling
3.3	order life cycle	configuration & order	order sequencing	planning and scheduling			picking, packaging	shipping
3.4	technology life cycle	planning	development	virtual testing			main- tenance	moderni- zation
3.5	indirect functions	\multicolumn{4}{l	}{primary activities}	\multicolumn{4}{l}{secondary activities}
		Inbound & outbound logistics	marketing & sales	ser- vice	firm infra- structure	HR	technology development	Procure- ment
4.1	learning environment	purely physical (planning + execution)		physical LF supported by digital factory		physical value stream of LF extended virtually	purely virtual (planning + execution)	
4.2	environment scale	\multicolumn{3}{l	}{scaled down}	\multicolumn{5}{l}{life-size}

11.27.1 Starting Phase and Purpose

Relying on initial funds from national sources and internal budget of the home institute, design and construction of the Smart Factory laboratory at the Institute for Computer Science and Control, Hungarian Academy of Sciences (MTA SZTAKI) started in 2011. In its current form, the facility operates since 2013 and keeps receiving regular hardware and control infrastructure updates as outcomes of research assignments, scheduled development, and student projects.

The initial motivation for the design and construction of the facility was the need for a tangible demonstration platform for the core competencies of the Research Laboratory on Engineering and Management Intelligence (EMI), the department in charge of the Smart Factory. The fundamental setting of the facility is a simplified and scaled-down model of production and intralogistics within a single facility which still retains sufficient functionalities to examine production planning and control, process transparency, and the handling of disturbances, inaccurate information and changes of external origin. Later additions to the hardware and control infrastructure extended the scope of demonstration and research capacities by human–robot collaboration, various means of interaction with personnel at different levels of production hierarchy, extension of "digital twin" models, virtual components, and remote connectivity.

Deployment in higher education has been a secondary purpose of the facility since its beginnings, and the size, mode of operation and regular use of the equipment do not make the facility eligible for being a full-fledged learning factory in the strict sense. Nevertheless, several laboratory exercises (as part of a course), and individual student projects (in support of a complete project-oriented mechatronics course, or in preparation for a BSc or MSc thesis) are hosted by the Smart Factory every year.

11.27.2 Learning Environment and Products

The Smart Factory is a compact (30 m^2), high-level representation of a manufacturing setting with four automated workstations, a warehouse, a loading/unloading station, a collaborative cell with two robot arms, a closed-path conveyor system, and floor space for two mobile robots (Fig. 11.69). Workpieces are of identical geometry, carrying cardboard inserts processed at the workstations. Each workpiece has a 1K NFC tag with a unique identifier and storage for additional data traveling with the product.

In the full production scenario, the workpieces arrive at a collaborative processing station where cardboard pieces are inserted either manually or using a robot. Once fitted with the "blanks," the workpieces are placed on the conveyor and brought to the warehouse (Fig. 11.70). Upon receiving production orders, the prepared workpieces are removed from the warehouse and brought to the processing stations where the cardboard inserts undergo a stamping and drilling sequence in accordance with the production order, along with an optional deposition of production data in the NFC memory of the workpieces. Transfer of workpieces during production relies on both the conveyor and two mobile robots. Depending on the production scenario, finished products are returned to the warehouse or removed from the intralogistic stream at the collaborative workstation. After production, the outcomes can be tested by visual inspection and by reading the NFC tags embedded in the workpieces. The workpieces can then be returned to their initial state by removing the cardboard inserts and erasing the contents of the embedded NFC tags.

11.27 Best Practice Example 27: Smart Factory at the Research …

Fig. 11.69 View of the Smart Factory with 3 of the 4 processing stations (left), two mobile robots (right), and the collaborative handling station equipped with two manipulators (top)

Fig. 11.70 Warehouse loading/unloading mechanism with four palletized workpieces

11.27.3 *Operation*

In the research and demonstration context, the Smart Factory is a tangible test bed for process planning, scheduling and execution in a cyber-physical environment, as well as human–robot collaboration and interaction with human personnel at various decision points of the production hierarchy. Connection to resources maintained by other departments of the hosting institute (immersive virtual reality, cloud services) are also involved in a considerable part of the research and demonstration activities in the Smart Factory.

The Smart Factory has its share in higher education taking place at the Budapest University of Technology and Economics (Department of Manufacturing Science and Engineering), in several forms:

- Laboratory exercises in scheduling and execution in cyber-physical production systems are hosted by the Smart Factory. On average, ca. 50 students are involved in the spring semester of each year, forming several groups of ideally 5–6 students.
- The Smart Factory provides a design and test environment for selected student groups in the "Mechatronics Project" course. In the one-semester course, groups of 3–4 students collaborate as a team in the design and construction automation solutions for an existing deployment environment.
- The Smart Factory hosts individual student projects leading up to BSc and MSc theses and research contest entries.

The outcomes of project-based student work are hardware or software components. If they are deemed fit for reliable live operation, they typically remain part of the equipment of the Smart Factory until replacement by subsequent solutions, similarly to incremental changes in live production environments.

11.28 Best Practice Example 28: Smart Factory at DFKI, TU Kaiserslautern, Germany

Authors: Haike Frank[a], Dennis Kolberg[a], Detlef Zühlke[a]
[a]Deutsche Forschungszentrum für Künstliche Intelligenz GmbH (DFKI), TU Kaiserslautern

11.28 Best Practice Example 28: Smart Factory at DFKI, TU ...

Name of learning factory: **SmartFactory-KL Industrie 4.0 production plant**
Operator: **Technologie-Initiative SmartFactory KL e.V.**
Year of inauguration: **2014**
Floor Space in learning factory: 200 sqm
Manufacture product(s): **Various components (vendor independent approach)**
Main topics / learning content: **Industrie 4.0; Digitalization; Factory Automation**

Extract from the morphology

2.1	main purpose	education	training	research
2.2	secondary purpose	test environment / pilot environment	industrial production	innovation transfer	public image
3.1	product life cycle	product planning	product development	prototyping			service	recycling
3.2	factory life cycle	investment planning	factory concept	process planning	ramp-up	manufacturing assembly	logistics	main-tenance	recycling
3.3	order life cycle	configuration & order	order sequencing	planning and scheduling			picking, packaging	shipping
3.4	technology life cycle	planning	development	virtual testing			main-tenance	moderni-zation
3.5	indirect functions	primary activities					secondary activities	
		Inbound & outbound logistics	marketing & sales	ser-vice	firm infra-structure	HR	technology development	Procure-ment
4.1	learning environment	purely physical (planning + execution)	physical LF supported by digital factory	physical value stream of LF extended virtually	purely virtual (planning + execution)
4.2	environment scale	scaled down	life-size		

11.28.1 The Aim of SmartFactory^{KL} Is to Pave the Way for the Intelligent Factory of Tomorrow

The technologies required for the implementation of a smart factory are already available in the consumer electronics sector. The next logical step is their industrial application. Modern information and communication technologies such as wireless sensor networks, semantic product memories, mobile interaction or ubiquitous access to the web, are enabling the merger of production technologies with information technologies.

With the help of today's omnipresent internet, we can integrate individual elements of a production process into a flexible, self-organizing network of actuators, sensors, and complete modules. Paradigms such as ubiquitous computing and the "Internet of Things" are facilitating the creation of intelligent environments in modern factories where the gulf between the real and the digital worlds is growing ever smaller. The "Internet of Things" in everyday life has delivered the vision for a production environment in the "Factory of Things"—to wit, the birth of the Smart Factory.

11.28.2 SmartFactory^{KL} is Industrie 4.0 You Can Touch—And Take Part In

As a manufacturer-independent technology platform, together with its partners, the *SmartFactory*^{KL} Technology Initiative creates and implements innovative factory systems following the concept of Industrie 4.0. A substantial part of the team's work is application-based research on current questions raised by industry. For the Technology Initiative, it is very important to illustrate findings and results. That is why the team develops and builds demonstrators—to put research into action.

Besides the well-known Industrie 4.0 production plant (Fig. 11.71), *SmartFactory*^{KL} offers to experience and test its demonstrators in the demo center. Topics such as scalable automation, cyber-physical systems, TSN, augmented reality, and virtual training in assembly environments can be experienced in the living laboratory with the help of demonstration modules.

The biggest project to date is the world's first manufacturer-independent Industrie 4.0 production plant—designed, engineered, built and implemented together with partners from industry and research who are all members of the Technology Initiative. The following virtual tour gives a direct impression: http://smartfactory.de/en/industrie-4-0-demonstration/demonstratoren-2/

This production plant shows just how high-quality, flexible manufacturing can be efficiently implemented even for a batch size of one—regardless of whether in an existing production operation or a greenfield. Uniform interface standards enable a manufacturer-independent link to the production units, logistic systems, supply infrastructure, and IT systems. Challenging requirements already affecting production such as custom products, shorter innovation cycles, and more efficient on-site

Fig. 11.71 Modules of the Industrie 4.0 production plant of the *SmartFactory*^{KL} partner consortium comprise various production cells, a manual workstation and a flexible transport system, making flexible production possible. ©*SmartFactory*^{KL}/C. Arnoldi

production can now be met. The production plant functions as a **demonstration unit, a learning environment, and a test bed**, all in one.

A sample product is actually manufactured in the facility: A customized business card holder, where the color, laser engravings, and optional inlays can be ordered online by the customer. The product itself stores all information in a RFID tag and gives guidance to the production modules. The production process proceeds in different ways, depending on the design and availability of the production plant's modules. The flexible transport system dynamically connects with the various production cells and manual assembly stations.

As a prerequisite, the team defined manufacturer-independent standards to enable a flexible system expansion. They include the RFID tag description that is defined from the data structure and data coding of the product memory. The second standard is the OPC UA communication. The third point is the standardization of hardware which means that the docking station as well as the modules have the identical mechanical basic functions. Manufacturer-independent IT systems can be integrated into the production process thanks to these uniform standards.

11.28.3 A Plant that Writes History

The ideas for the continuous advancement of the Industrie 4.0 production plant are developed in the *SmartFactory*^{KL} working groups, following the strategy: "Progress within the network".

2014—First concept of a completely modular Industrie 4.0 production plant put into practice for the first time. At Hannover Messe, *SmartFactory*^{KL} presented the world's first manufacturer-independent production plant. It consisted of five modules and was realized in this first version together with 10 partners from industry and research.

2015—A consortium of 16 *SmartFactory*^{KL} partners presented the expanded, modular Industrie 4.0 plant comprising 7 modules at Hannover Messe. In addition, modularization was not only limited to one dimension, meaning one production line, but now possible for two dimensions or all directions. The plant also featured the manual workstation. By presenting elements of human–machine interaction, the team could now integrate the important topic of "future labor" into the demonstration.

2016—The exhibition of the Industrie 4.0 production plant became proof of concept: Central questions on the factory of the future had been solved: Plug & produce, predictive maintenance, scalable automation, failure-proof maintenance, and worker assistance. The strict modularization of the plant led to new potential for the IT system partners such as the topic of predictive maintenance. This made it possible to integrate the IT partners more into the further development of the plant. The cooperation of the partners had led to market-ready technologies and first products. The consortium consisted of 18 partners.

2017—The expanded Industrie 4.0 production plant appeared in a completely new layout: distributed over two production cells and a manual workstation. The cells are connected thanks to a flexible transport system which distributes the products from station to station. This improvement showed that *SmartFactory*[KL]'s approach was not only valid for one production line, but for various lines—or even various sites, thanks to the now decoupled logistics. Again, 18 partners participated in the consortium.

2018—Together with 19 partners in the consortium, *SmartFactory*[KL] is currently working on topics such as vertical integration (implementation of Edge Devices to prepare for Cloud Computing), the use of 5G applications in an industrial setting and an advanced modular safety concept.

11.29 Best Practice Example 29: Smart Mini-Factory at IEA, Free University of Bolzano, Italy

Authors: Dominik Matt[a,b], Erwin Rauch[a]
[a]Free University of Bozen-Bolzano, Faculty of Science and Technology, Research Area "Industrial Engineering and Automation (IEA)"
[b]Fraunhofer Italia

Name of learning factory: **Smart Mini Factory – Laboratory for Industry 4.0**
Operator: **Ind. Eng. and Autom. (IEA) – Free University of Bolzano (Italy)**
Year of inauguration: **2012 Lean Mini Factory and 2017 Smart Mini Factory**
Floor Space in learning factory: **250 sqm**
Manufacture product(s): **Pneumatic Cylinder, Pneumatic Impact Wrench**
Main topics / learning content: **Smart Manufacturing Systems, Automation**

Extract from the morphology

2.1	main purpose	education		training		research	
2.2	secondary purpose	test environment / pilot environment		industrial production		innovation transfer	public image
3.1	product life cycle	product planning	product development	prototyping		service	recycling
3.2	factory life cycle	investment planning	factory concept	process planning	ramp-up	maintenance	recycling
3.3	order life cycle	configuration & order	order sequencing	planning and scheduling		picking, packaging	shipping
3.4	technology life cycle	planning	development	virtual testing		maintenance	modernization
3.5	indirect functions	primary activities				secondary activities	
		Inbound & outbound logistics	marketing & sales	service	firm infrastructure	HR	technology development	Procurement
4.1	learning environment	purely physical (planning + execution)		physical LF supported by digital factory	physical value stream of LF extended virtually		purely virtual (planning + execution)
4.2	environment scale	scaled down		life-size			

11.29.1 Starting Phase and Purpose

The Learning Factory Laboratory at the Free University of Bolzano was founded in 2012 as a "Mini-Factory" with start-up funds from the Chair of Production Systems and Technologies in the research area Industrial Engineering and Automation (IEA). The name "Mini-Factory" was chosen as name for the learning factory laboratory because it should reflect the principles of lean and agile production in a small and realistic scale. Furthermore, the concept of small and distributed manufacturing systems ("mini-factories") pursues the goal of producing mass-customized products on demand and in close proximity to the customer (Fig. 11.72).

In 2015, the Autonomous Province of Bolzano allocated a budget of 2.3 million euros to the Free University of Bolzano due to IEA's efforts to establish research and teaching competence in the field of Industrie 4.0 in Italy. These funds are planned for capacity building in the sense of suitable laboratory space, research personnel and investments in machinery and equipment. Due to the tight spatial situation in Bolzano, it took until 2017 to find a temporary solution to accommodate the learning factory in the city center. In the medium term, the learning factory together with a part of the university campus is to be relocated to a new building in the brand-new NOI Technology Park in the south of Bolzano.

11.29.2 Learning Environment and Products

The learning factory is divided into two different areas. One area comprises 3D printers for additive manufacturing and several numerically controlled machine tools in the form of a CNC machining workshop, which can be loaded and unloaded via a collaborative robot. This allows parts to be manufactured and then brought into the second area for assembly. There is a manual and hybrid assembly area with assembly stations from Bosch Rexroth, an assembly assistance system and a collaborative robot from UR. A mobile platform from Kuka equipped with a Kuka iiwa lightweight robot takes the semi-finished part to another automated area. There, automated assembly and packaging processes are carried out using a six-axis robot, a SCARA robot, and a delta robot, which will in future be connected via a transfer system. The various systems are connected as CPSs via a horizontal proximity and a vertical access network in a cyber-physical production system environment. Using VR and AR headsets and a virtual 3D data model, a digital mock-up of the production system can be generated and adapted in advance.

Currently, two different products are produced on the assembly line. On the one hand, a pneumatic cylinder and on the other hand a pneumatic impact wrench (Fig. 11.73). A product analysis is carried out at the beginning defining the process steps with an assembly precedence graph. After preparation of the inventory of individual parts, the practical implementation in the line, the balancing of the assembly

Fig. 11.72 Exemplary pictures and 3D virtual data model of the lab

tasks as well as several loops to increase efficiency in the production system are carried out.

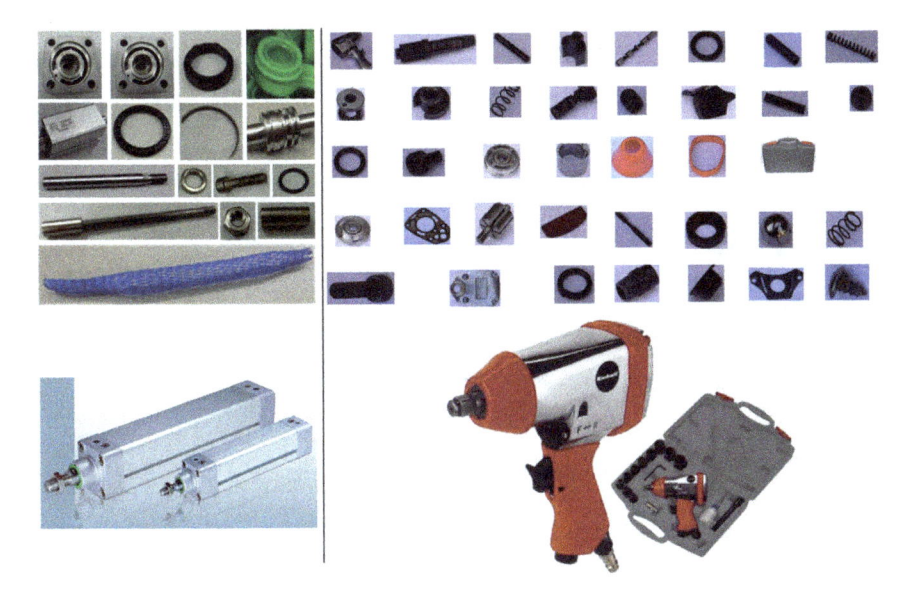

Fig. 11.73 Produced pneumatic cylinder and pneumatic impact wrench in the Smart Mini-Factory

11.29.3 Operation

The use of the learning factory laboratory is divided into three parts. On the one hand, it is used for applied research and therefore serves for building up individual test setups. Secondly, it is used as part of the bachelor's and master's programs for industrial and mechanical engineering (mainly the lectures "Production Systems and Industrial Logistics" as well as "Industrial Automation and Mechatronics"). Thirdly, seminars on all aspects of Industrie 4.0 for industrial companies are offered (Fig. 11.74). The exercises and seminars are designed and conducted by researchers, postdocs, and PhD students. The industrial seminars allow a close contact with local industry and thus facilitate the transfer of knowledge from research to industry.

11.30 Best Practice Example 30: Teaching Factory: An Emerging Paradigm for Manufacturing Education. LMS, University of Patras, Greece

Authors: Dimitris Mavrikios[a], Konstantinos Georgoulias[a], George Chryssolouris[a]
[a]Laboratory for Manufacturing Systems and Automation, University of Patras

Smart Manufacturing and Assembly Systems	1. Human-Centred Design of Cyber-Physical Production Systems
	2. Eye Tracking Technology in Manufacturing
	3. Industrial Data Analytics for Production 4.0
	4. Computational Design for Mass Customization
Industrial Automation and Robotics	5. Collaborative Robotics
	6. Robotic Grasping Devices
	7. Smart Electric Drives
	8. Embedded and Cyber-Physical Systems
SCM-Logistics	9. Augmented Reality in Logistics
	10. Construction 4.0 – Digital Tools for the Construction Site

Fig. 11.74 Seminar offer in the Smart Mini-Factory

Name of learning factory: **Teaching Factory**
Operator: **Laboratory for Manufacturing Systems & Automation, University of Patras**
Year of Opening: **N/A**
Floor Space in learning factory: **Non-geographically anchored learning space**
Manufactured product(s): **N/A**
Topics / Learning Content: **Industrial problems, real-life engineering practices, R&D outputs**

Extract from the morphology

2.1	main purpose	education		training			research	
2.2	secondary purpose	test environment / pilot environment		industrial production	innovation transfer		public image	
3.1	product life cycle	product planning	product development	prototyping			service	recycling
3.2	factory life cycle	investment planning	factory concept	process planning	ramp-up	manufacturing assembly logistics	maintenance	recycling
3.3	order life cycle	configuration & order	order sequencing	planning and scheduling			picking, packaging	shipping
3.4	technology life cycle	planning	development	virtual testing			maintenance	modernization
3.5	indirect functions	primary activities				secondary activities		
		inbound & outbound logistics	marketing & sales	service	firm infrastructure	HR	technology development	procurement
4.1	learning environment	purely physical (planning + execution)	physical LF supported by digital factory	physical value stream of LF extended virtually			purely virtual (planning + execution)	
4.2	environment scale	scaled down		life-size				

11.30.1 Purpose

Manufacturing enters a new era, where novel life-long learning schemes need to keep up with the rapid advances in production-related technologies, tools, and techniques.[10] In order to effectively address the emerging challenges for manufacturing education and skills delivery, the educational paradigm in manufacturing needs to be revised.[11]

Within this context, an extended Teaching Factory paradigm, based on the knowledge triangle notion, has been suggested.[12] The aim is to effectively integrate education, research and innovation activities into a single initiative, involving industry and academia. The concept of the **Teaching Factory** (TF) has its origins in the medical sciences discipline and specifically in the paradigm of the Teaching Hospital that aim at connecting education/training with real life. Real-life changes rapidly and so does industrial practice, i.e. manufacturing technology, industrial settings, engineering problems. The TF paradigm follows these developments by "bringing" the **real factory** to the **classroom** and the classroom to the real factory. It uses advanced information and communication technologies (ICTs) and high-grade industrial didactic equipment in order to facilitate a 2-ways learning channel communicating industrial practices to the classroom and new knowledge to the factory.

11.30.2 Learning Environment

The TF operates as a non-geographically anchored learning environment. Modern ICTs are used in order to allow remotely located teams of manufacturing practitioners and students to communicate and interact through a 2-ways learning channel (Fig. 11.75).

In the "factory-to-classroom" learning mode, manufacturing practitioners "teach" students in engineering schools about manufacturing problems, manufacturing issues, and manufacturing practice. The real production site is used for teaching purposes in order to enhance the teaching activity with the knowledge and experience existing in the processes of every day industrial practice. Delivery mechanisms that enable classroom students to apprehend the production environment in full context and address real-life engineering problems are used. Using different modules, a "factory-to-classroom" training session may be configured according to the training needs and available resources (Fig. 11.76).

In the "lab-to-factory" learning mode, students and faculty "teach" manufacturing practitioners about advances in manufacturing technology, new trends, and results of research and development projects. Engineering or management teams are intro-

---

[10] See International Monetary Fund [IMF] (2013).

[11] See Hanushek and Woessmann (2007), Eurostat (2011), Vieweg et al. (2012).

[12] See Chryssolouris, Mavrikios, Papakostas, and Mourtzis (2006), Mavrikios et al. (2013), Chryssolouris, Mavrikios, and Rentzos (2016).

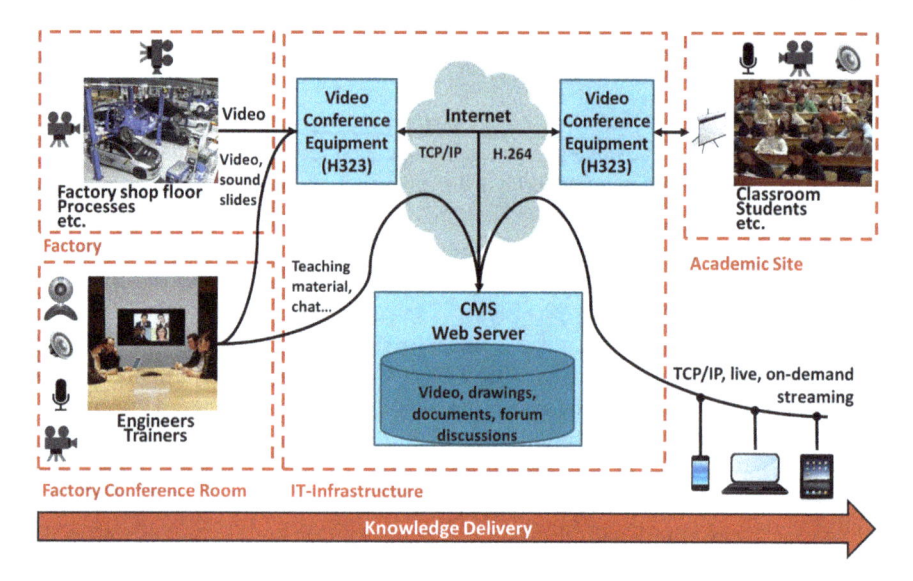

Fig. 11.75 A TF setting for "bringing" the real factory to the classroom

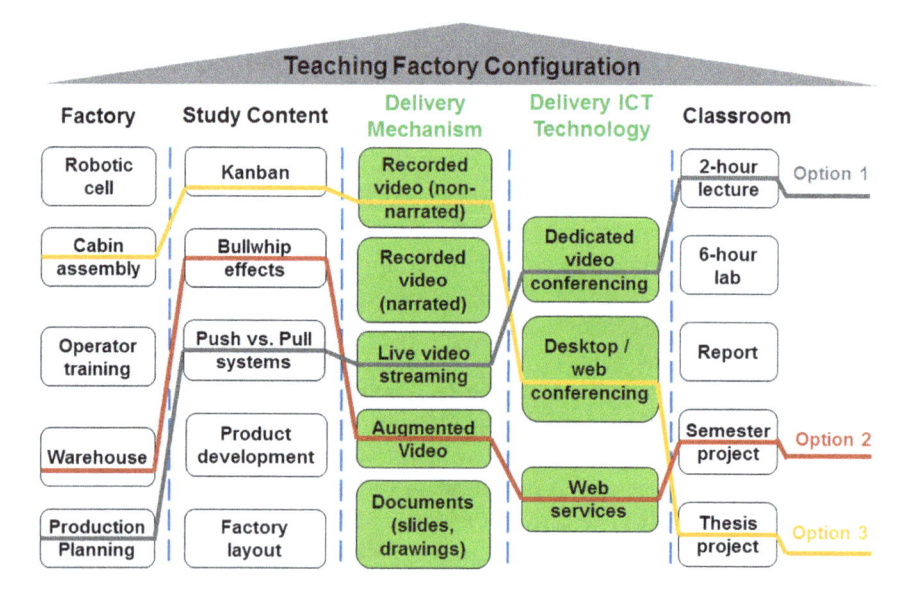

Fig. 11.76 Configuration of a TF training scenario

duced to new concepts or solutions, and operators are trained on new manufacturing technologies and concepts. Industrial-grade or didactic equipment installed into the academic facilities is used as test-beds and demonstrators for new technological concepts.

11.30 Best Practice Example 30: Teaching Factory: An Emerging Paradigm … 449

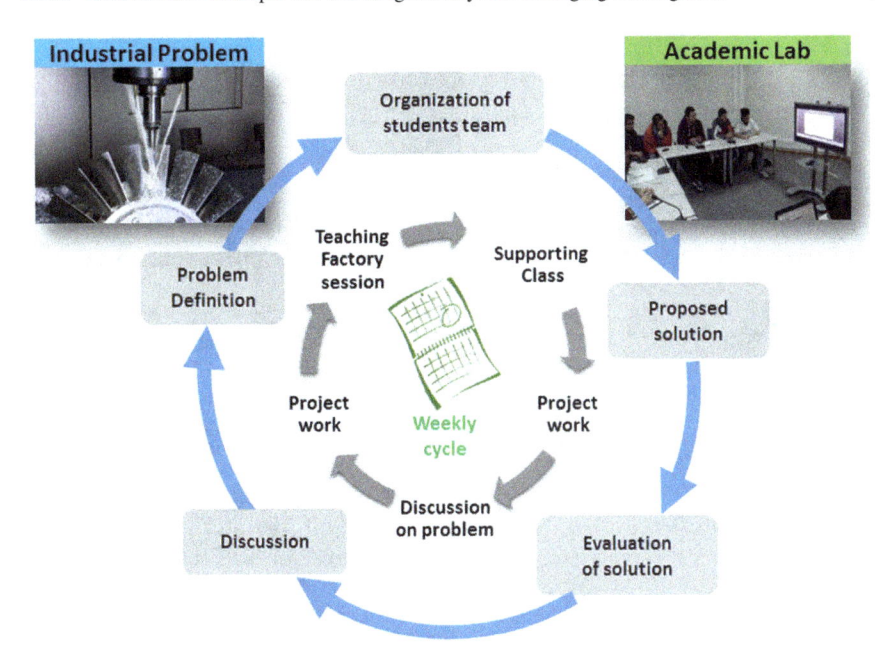

Fig. 11.77 A typical TF cycle for a "factory-to-classroom" knowledge transfer

11.30.3 Operation

The implementation of the "factory-to-classroom" operating scheme is carried out through the launch of an industrial training project by an academic institution and a manufacturing company. Manufacturing practitioners introduce a real-life industrial problem, e.g. the line balancing of a new production area in the factory of a construction equipment company, to student teams that are requested to come up with some engineering solutions (Fig. 11.77). Interaction includes discussions, sharing of presentations, live videos from the production and other knowledge delivery mechanisms, depending on the content of the problem.

In the "lab-to-factory" operating scheme, interactive training sessions aim at knowledge transfer from an academic to an industrial environment, e.g. advances in flexible robotic cells and how they could apply to real-life problems of a company producing automation and control equipment. Delivery mechanisms include live videos and audio interactions, real-time presentations, and demonstrators, through advanced ICT configurations (Fig. 11.78).

Future work on further elaborating the TF operations includes the definition of a new business model facilitating the knowledge communication and training delivery through Teaching Factories network. This would happen by establishing learning and training channels for the communication of manufacturing knowledge among multiple remotely located industrial and academic sites. The structuring and launch of

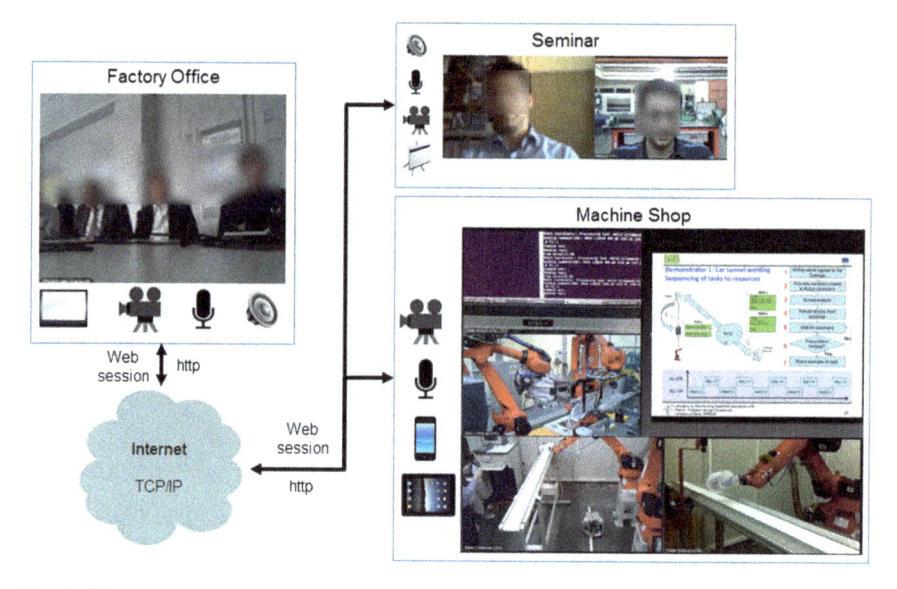

Fig. 11.78 Interaction during a "lab-to-factory" knowledge transfer

a TF network will enable the efficient allocation of learning content and engagement of resources and delivery mechanisms.

11.31 Best Practice Example 31: VPS Center of the Production Academy at BMW in Munich, Germany

Authors: Carolin Lorber[a], Tobias Stäudel[a]
[a]BMW Group

11.31 Best Practice Example 31: VPS Center of the Production Academy ...

Name of learning factory: VPS Center of the Production Academy
Operator: BMW Group
Year of inauguration: 2012
Floor Space in learning factory: 600 sqm
Manufacture product(s): Engines (3/4 cylinder, petrol and diesel)
Main topics / learning content: Lean Production

Extract from the morphology

2.1	main purpose	education	training		research	
2.2	secondary purpose	test environment / pilot environment	industrial production	innovation transfer	public image	
3.1	product life cycle	product planning	product development	prototyping	service	recycling
3.2	factory life cycle	investment planning	factory concept	process planning	ramp-up	maintenance	recycling
3.3	order life cycle	configuration & order	order sequencing	planning and scheduling	picking, packaging	shipping
3.4	technology life cycle	planning	development	virtual testing	maintenance	modernization
3.5	indirect functions	primary activities			secondary activities	
		Inbound & outbound logistics	marketing & sales	service	firm infrastructure	HR	technology development	Procurement
4.1	learning environment	purely physical (planning + execution)	physical LF supported by digital factory	physical value stream of LF extended virtually	purely virtual (planning + execution)
4.2	environment scale	scaled down		life-size	

(assembly / manufacturing / logistics span rows 3.1–3.4)

11.31.1 Starting Phase and Purpose

The value-added production system (VPS) is the BMW application of lean management principles, methods, and mindset. It is furthermore the foundation of the BMW Production System. To qualify employees within the theme of lean and offer a place where participants can get into contact with the matter, the BMW Group decided to found a central VPS Center for the production network in 2011. Contemporary existing solutions were examined and a prototype was created. After some circuits of improving concept and content, the final center was set up and opened in 2012 at the BMW Group Academy in Munich. The main focus of the installation is the qualification of management and lean experts within the company. Since 2015 trainings are also offered to suppliers. To fulfill the kaizen mindset, the VPS Learning factory pursues not only qualification targets but also research targets concerning the continuous improvement of lean tools and methods, which will be amplified later on. Since 2018, the VPS Center is a central part of the Production Academy.

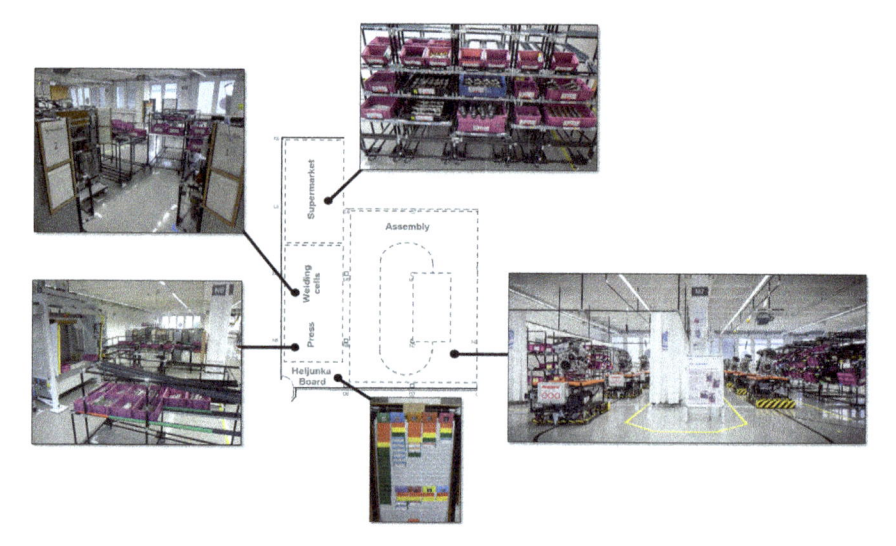

Fig. 11.79 Value stream heat protection plate

11.31.2 Learning Environment and Products

On the 600 m² of the VPS learning factory the value stream of a heat protection plate from the receipt of raw materials over pressing, welding to assembling on the engine is mapped. Because the Engines are not sold, the parts are disassembled and returned to the life cycle of the VPS learning factory (Fig. 11.79).

11.31.3 Operation

The goal of the teaching and training activities is to convey the VPS principles, methods and mindset to ensure the sustainable development of lean management skills within the BMW Group. The VPS learning modules can be divided into three core fields: VPS standard training for employees and managers (knowledge transfer from beginner to expert level), VPS-specific trainings (knowledge transfer in the use of specific lean methods), and customized family trainings (knowledge transfer in a mix of basics and methods customized for homogeneous training groups).

All training modules are conceptualized and conducted by intern BMW Group trainers. Only the standard trainings are conducted in collaboration with trainers of external training institutes.

Figure 11.80 gives an overview over the qualification map of the learning factory including the qualification topics for customized trainings.

Beside the already above described use for trainings within the internal lean qualification of the BMW Group the VPS learning factory is used for production research.

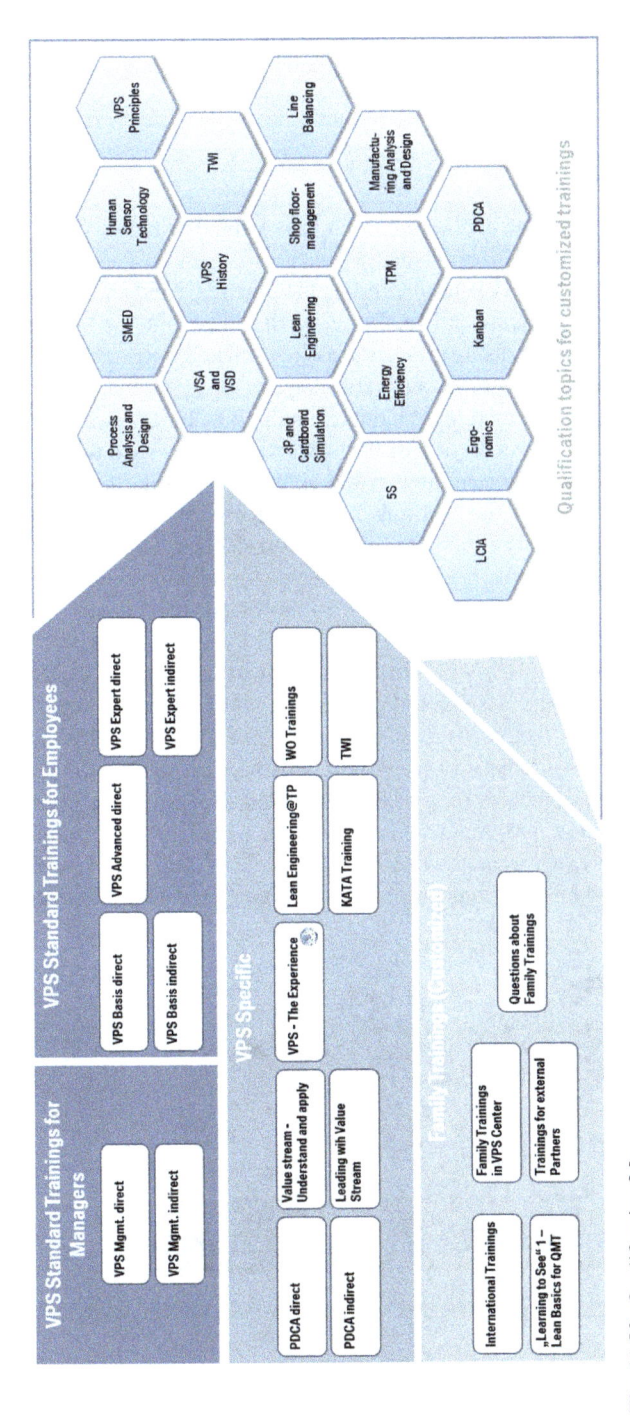

11.31 Best Practice Example 31: VPS Center of the Production Academy ...

Fig. 11.80 Qualification Map

Two of the current research projects are the integration of Low Cost Intelligent Automation into production on the one hand and the implementation of Augmented Reality tools in the employee qualification on the other hand.

11.31.3.1 Low Cost Intelligent Automation (LCIA)

Research concerning the method of LCIA[13] is performed in two dimensions within the learning factory.

First, the whole facility is used to prove the potentials and abilities of the method focusing on logistic topics and finally retransfer these into a generalized solution catalog. Therefore, the factory is used as a simulated production area and tasks that—based on a value stream analysis—appear as a bottleneck are continuously improved by using the method.[14] Concrete solutions are afterward generalized by describing the motion vector and further needed elements such as separation of the handled material and presentation of the single parts to the assembly associates. Following this process, the learning factory secures the developed elements for usage in real production environment and supports Kaizen Teams by offering already approved modules for material handling. Beside the solution catalog as a theoretical part, physical demonstrators are available to show basic movements and research the long-term reliability (Fig. 11.81).

The second dimension is represented by offering the method of LCIA as a solution to the trainees as they are asked for improvements within their training in the assembly line of the learning factory using cardboard engineering to implement the created solutions for testing. This way provides an immediate contact to the method and therefore a deeper understanding of the benefits and makes LCIA more visible.

Combining these two dimensions also offers a possibility to reduce one of the biggest problems of the method: The conflict that LCIA is meant to be realized within the line while the industrialization process has already decided traditional

Fig. 11.81 Low-Cost Automated Logistics (left) and LCIA demonstrator (right)

---

[13]See Seifermann, Stäudel, Abele, and Metternich (2014).
[14]See Stäudel and Ahrens (2017).

11.31 Best Practice Example 31: VPS Center of the Production Academy ...   455

automation solutions and therefore LCIA is only a way of improving after the line design has already taken place by getting the process engineers in touch with the method.

11.31.3.2 Augmented Reality (AR)

The goal of the research concerning AR is to get an insight how AR can be used in assembly trainings. To train motor skills as well as procedural knowledge of employees working in assembly lines is necessary to achieve the highest quality in products and processes. As a head-mounted display solution, a self-programmed application on the Microsoft HoloLens is used to display the assembly training content. A com-

Fig. 11.82 Instructions by trainer (above), instruction by HoloLens (below)

parison of face-to-face training with augmented reality training measuring effects on time, quality, picking mistakes, assembly order mistakes, rework and knowledge retention showed that both training methods generate the same assembly quality (Fig. 11.82). Taking ergonomic aspects into consideration a use of the AR application no longer than one hour is suggested. Despite the potential was proven and the advantages of AR are clear, it isn't in daily use in the automotive industry until today. Therefore, further work will be done in the VPS learning factory in using AR to display lean methods like 5S and value stream analysis.

List of Contributors

Julian Ays, Laboratory for Machine Tools and Production Engineering (WZL), RWTH Aachen University

Patrick Balve, Heilbronn University of Applied Sciences, Faculty of Industrial and Process Engineering, Study Program Manufacturing and Operations Management (B.Eng.)

Harald Bauer, Institut für Werkzeugmaschinen und Betriebswissenschaften (iwb, Institute for Machine Tools and Industrial Management), TU München

Thomas Bauernhansl, Institute of Industrial Manufacturing and Management (IFF), University of Stuttgart

Beate Bender, Product Development (LPE), Ruhr-University Bochum

Richárd Beregi, Institute for Computer Science and Control, Hungarian Academy of Sciences (MTA SZTAKI)

Harry Bikas, Laboratory for Manufacturing Systems & Automation (LMS)

Felix Brandl, Institut für Werkzeugmaschinen und Betriebswissenschaften (iwb, Institute for Machine Tools and Industrial Management), TU München

Stefan Braunreuther, Fraunhofer IGCV, Augsburg, Germany

George Chryssolouris, Laboratory for Manufacturing Systems and Automation, University of Patras

George Dimitrakopoulos, Laboratory for Manufacturing Systems & Automation (LMS)

Hoda ElMaraghy, Intelligent Manufacturing Systems (IMS) Center, University of Windsor, Canada

Gábor Erdős, Institute for Computer Science and Control, Hungarian Academy of Sciences (MTA SZTAKI)

Dominik Flum, Institute for Production Management, Technology and Machine Tools (PTW), TU Darmstadt

Haike Frank, Deutsche Forschungszentrum für Künstliche Intelligenz GmbH (DFKI), TU Kaiserslautern

Jörg Franke, Institute for factory automation and production systems (FAPS), Friedrich-Alexander University Erlangen-Nürnberg

Bastian Fränken, Laboratory for Machine Tools and Production Engineering (WZL), RWTH Aachen University

List of Contributors

Konstantinos Georgoulias, Laboratory for Manufacturing Systems and Automation, University of Patras
Erwin Gross, Institute of Industrial Manufacturing and Management (IFF), University of Stuttgart
Benjamin Haefner, wbk, Institute of Production Science, KIT Karlsruhe
Jens Hambach, Institute for Production Management, Technology and Machine Tools (PTW), TU Darmstadt
Mark Helfert, Institute for Production Management, Technology and Machine Tools (PTW), TU Darmstadt
Constantin Hofmann, wbk, Institute of Production Science, KIT Karlsruhe
Philipp Hold, TU Wien, Institute of Management Sciences, Austria
Vera Hummel, ESB Business School, Reutlingen University
Roland Jochem, TU Berlin
Zsolt Kemény, Institute for Computer Science and Control, Hungarian Academy of Sciences (MTA SZTAKI)
Holger Kohl, Fraunhofer IPK & TU Berlin
Dennis Kolberg, Deutsche Forschungszentrum für Künstliche Intelligenz GmbH (DFKI), TU Kaiserslautern
Dieter Kreimeier, Chair for Production Systems, Ruhr-Universität Bochum
Alexander Kühl, Institute for factory automation and production systems (FAPS), Friedrich-Alexander University Erlangen-Nürnberg
Gisela Lanza, wbk, Institute of Production Science, KIT Karlsruhe
Carolin Lorber, BMW Group
Dominik Matt, Free University of Bozen-Bolzano, Faculty of Science and Technology, Research Area "Industrial Engineering and Automation (IEA)" & Fraunhofer Italia
Dimitris Mavrikios, Laboratory for Manufacturing Systems and Automation, University of Patras
Jan Philipp Menn, TU Berlin
Lukas Merkel, Fraunhofer IGCV, Augsburg, Germany
Dimitris Mourtzis, Laboratory for Manufacturing Systems & Automation (LMS)
János Nacsa, Institute for Computer Science and Control, Hungarian Academy of Sciences (MTA SZTAKI)
Stephan Neser, University of Applied Sciences Darmstadt
Lars Nielsen, Institute of Production Systems and Logistics (IFA), Leibniz Universität Hannover
Peter Nyhuis, Institute of Production Systems and Logistics (IFA), Leibniz Universität Hannover
Dirk Pensky, Festo Didactic
Reinhard Pittschellis, Festo Didactic
Gerrit Posselt, Institute of Machine Tools and Production Technology, Technische Universität Braunschweig
Christopher Prinz, Chair for Production Systems, Ruhr-Universität Bochum
Jan-Philipp Prote, Laboratory for Machine Tools and Production Engineering (WZL), RWTH Aachen University

Fabian Ranz, ESB Business School, Reutlingen University & TU Wien, Institute of Management Sciences, Austria

Erwin Rauch, Free University of Bozen-Bolzano, Faculty of Science and Technology, Research Area "Industrial Engineering and Automation (IEA)"

Holger Regber, Festo Didactic

Gunther Reinhart, Institut für Werkzeugmaschinen und Betriebswissenschaften (iwb, Institute for Machine Tools and Industrial Management), TU München & Fraunhofer IGCV, Augsburg, Germany

Thomas Reitberger, Institute for factory automation and production systems (FAPS), Friedrich-Alexander University Erlangen-Nürnberg

Niklas Rochow, Institute of Production Systems and Logistics (IFA), Leibniz Universität Hannover

Christoffer Rybski, Fraunhofer IPK

Günther Schuh, Laboratory for Machine Tools and Production Engineering (WZL), RWTH Aachen University

Jan Schuhmacher, ESB Business School, Reutlingen University

Michael Schwarz, MPS Lernplattform Sindelfingen, Daimler AG

Melissa Seitz, Institute of Production Systems and Logistics (IFA), Leibniz Universität Hannover

Felix Sieckmann, TU Berlin

Jörg Siegert, Institute of Industrial Manufacturing and Management (IFF), University of Stuttgart

Wilfried Sihn, TU Wien, Institute of Management Sciences, Austria

Stephan Simons, University of Applied Sciences Darmstadt

Panagiotis Stavropoulos, Laboratory for Manufacturing Systems & Automation (LMS)

Tobias Stäudel, BMW Group

Nicole Stricker, wbk, Institute of Production Science, KIT Karlsruhe

Carsten Ulbrich, MAN Diesel & Turbo SE, Berlin

Detlef Zühlke, Deutsche Forschungszentrum für Künstliche Intelligenz GmbH (DFKI), TU Kaiserslautern

Stylianos Zygomalas, Laboratory for Manufacturing Systems & Automation (LMS)

References

Abele, E. (2016). Learning factory. *CIRP Encyclopedia of Production Engineering*.
Abele, E., Chryssolouris, G., Sihn, W., Metternich, J., ElMaraghy, H. A., Seliger, G., et al. (2017). Learning Factories for future oriented research and education in manufacturing. *CIRP Annals—Manufacturing Technology, 66*(2), 803–826.
Abele, E., Metternich, J., Tisch, M., Chryssolouris, G., Sihn, W., ElMaraghy, H. A., et al. (2015). Learning Factories for research, education, and training. In *5th CIRP-sponsored Conference on Learning Factories, Procedia CIRP, 32*, 1–6. https://doi.org/10.1016/j.procir.2015.02.187.
Bonz, B. (2009). Methoden der Berufsbildung. *Ein Lehrbuch* (2., neubearb. und erg. Aufl.). *weiter @ lernen*. Stuttgart: Hirzel.

References

Chryssolouris, G., Mavrikios, D., & Mourtzis, D. (2013). Manufacturing systems—Skills & competencies for the future. In *46th CIRP Conference on Manufacturing Systems. Procedia CIRP, 7,* 17–24).

Chryssolouris, G., Mavrikios, D., Papakostas, N., & Mourtzis D. (2006). Education in manufacturing technology & science: A view on future challenges & goals. In *Proceedings of the International Conference on Manufacturing Science and Technology (ICOMAST 2006)* (pp. 1–4), Melaka, Malaysia.

Chryssolouris, G., Mavrikios, D., & Rentzos, L. (2016). The teaching factory: A manufacturing education paradigm. In *49th CIRP Conference on Manufacturing Systems. Procedia CIRP, 57,* 44–48).

Chryssolouris, G., & Mourtzis, D. (2008). Challenges for manufacturing education. In *CIRP International Conference on Manufacturing Engineering Education (CIMEC 08)*, Nantes, France.

Eurostat. (2011). *Science, technology and innovation in Europe, 2011 Edition. Theme science and technology*. Luxembourg: Publications Office of the European Union.

Hanushek, E. A., & Woessmann, L. (2007). *The role of education quality in economic growth. Policy research working paper: Vol. 4122*. Washington, DC: World Bank Human Development Network Education Team.

International Monetary Fund (IMF). (2013). *World economic outlook—Hopes, realities, risks, world economic and financial survey*. Retrieved from http://www.imf.org/external/pubs/ft/weo/2013/01/pdf/text.pdf.

Mavrikios, D., Papakostas, N., Mourtzis, D., & Chryssolouris, G. (2013). On industrial learning and training for the factories of the future: A conceptual, cognitive and technology framework. *Journal of Intelligent Manufacturing, 24*(3), 473–485. https://doi.org/10.1007/s10845-011-0590-9.

Menn, J. P., Sieckmann, F., Kohl, H., & Seliger, G. (2018). Learning process planning for special machinery assembly. In *8th CIRP Conference on Learning Factories—Advanced Engineering Education & Training for Manufacturing Innovation* (Vol. 23, pp. 75–80).

Rentzos, L., Doukas, M., Mavrikios, D., Mourtzis, D., & Chryssolouris, G. (2014). Integrating manufacturing education with industrial practice using teaching factory paradigm: A construction equipment application. In *47th CIRP Conference on Manufacturing Systems. Procedia CIRP, 17,* 189–194).

Rentzos, L., Mavrikios, D., & Chryssolouris, G. (2015). A two-way knowledge interaction in manufacturing education: The teaching factory. In *5th CIRP-sponsored Conference on Learning Factories, Procedia CIRP, 32,* 31–35). https://doi.org/10.1016/j.procir.2015.02.082.

Rybski, C., & Jochem, R. (2016). Benefits of a learning factory in the context of lean management for the pharmaceutical industry. *Procedia CIRP, 54,* 31–34. https://doi.org/10.1016/j.procir.2016.05.106.

Seifermann, S., Stäudel, T., Abele, E., & Metternich, J. (2014). Schlanke Produktion und Automatisierung. Identifikation und Systematisierung der Begrifflichkeiten im Themenfeld der Low-Cost-Automatisierung. *Werkstattstechnik online: wt, 104*(9), 546–551.

Seliger, G., Jochem, R., Straube, F., & Kohl, H. (2015). Die Lernfabrik: Interdisziplinäre Zusammenarbeit von Praxis und Wissenschaft in der Prozessindustrie. *Management und Qualität, 4,* 14–16.

Stäudel, T., & Ahrens, T. (2017, April). *Gemba Engineering – Einfachautomatisierung auf dem Shopfloor*. OpexCon 2017, Stuttgart. Retrieved from http://opexcon.de/wp/session/gemba-engineering-einfachautomatisierung-auf-dem-shopfloor/.

Stock, T., & Kohl, H. (2018). Perspectives for international engineering education: Sustainable-oriented and transnational teaching and learning. In *Keynote at the 15th Global Conference on Sustainable Manufacturing; Haifa, Israel. Procedia CIRP, 15*.

Vieweg, H.-G., Claussen, J., Essling, C., Reinhard, M., Alexandri, E., Hay, G., ... Frøhlich Hougaard, K. (2012). *An introduction to mechanical engineering: Study on the competitiveness of the EU mechanical engineering industry: Within the framework contract of sectoral competitiveness studies ENTR/06/054*. Final Report. Retrieved from https://ec.europa.eu/docsroom/documents/12329/attachments/1/translations/en/renditions/native.

Chapter 12
Conclusion and Outlook

The book shows how learning factories have been established in recent years as a platform for production-oriented education, training, research, and innovation transfer in academia, industry, and also vocational schools. This book gives a comprehensive overview of

- the challenges to be addressed for future production with the help of learning factories,[1]
- the most important fields of competence in connection with learning factories,[2]
- the historical development, the forms and types of work-related learning for and in production,[3]
- the historical development, the terminology, and the definitions around the topic learning factory,[4]
- the broad variety of existing learning factory concepts in the learning factory morphology,[5]
- the methods, tools, and guidelines along the learning factory life cycle,[6]
- a multitude of learning factory concepts and practical examples regarding different fields of application,[7] contents[8] and concept variations,[9]
- the projects, groups, and associations connected with the learning factory topic[10] as well as,

---

[1] See Chap. 1.
[2] See Chap. 2.
[3] See Chap. 3.
[4] See Chap. 4.
[5] See Chap. 5.
[6] See Chap. 6.
[7] See Chap. 7.
[8] See Chap. 8.
[9] See Chap. 9.
[10] See Chap. 10.

© Springer Nature Switzerland AG 2019
E. Abele et al., *Learning Factories*, https://doi.org/10.1007/978-3-319-92261-4_12

- many best practice examples covering the complete range of the learning factory concept.[11]

In many places, it was difficult to give this comprehensive overview on the subject of learning factories, which has now emerged in the end. This is partly because the topic is comparatively new and partly because many different scientific disciplines have to be reconciled. In addition, the players active in the field of learning factories and learning factory-like systems represent a partly confusing community, whose mental order and structuring required many collaborative efforts. The learning factory concept itself can be used in many ways, for teaching as well as for further training or research and innovation transfer, and for various challenges, for the qualification of personnel to be more competitive or to bring the latest ideas in industry as well as for the stimulation of innovation activities.

Nevertheless, a number of significant challenges await to be faced in context of the use of learning factory concepts in the coming years. One of the biggest challenges is the technological and organizational development of real factories. Technologies that play a role in manufacturing, industrial environments, and processes, and related engineering problems are rapidly changing in industrial practice. With this speed at which production systems change, for example, through the introduction of new technologies and approaches, learning factories must be able to keep up or even develop faster in order to be able to access current or even future production environments within existing learning factory concepts even after a few years.

Furthermore, it was shown that production environments are very different from each other. Individual learning factories usually only pick out a small part of industrial reality, related to technologies, process chains, products to be manufactured and so on. In order to be able to depict the complete industrial reality in learning factories, or at least as much of it as possible, different approaches to solutions are pursued:

- The use of networked learning factories on different contents and foci in order to gain synergies for the single facilities,[12]
- The integration of ICT-equipment, simulation, and virtual environments into the learning factory concept in order to extend ranges of single learning factories.[13]
- Non-geographically anchored leaning factories through the use of advanced ICT-technology and industrial didactic equipment.[14]

Although those digital and virtual learning factories are considered very interesting in terms of widening the scope of learning factories, they somehow lack the hands-on learning characteristics, sometimes the teamwork qualities as well as the physical interaction that are facilitated in physical learning factories. The different types of learning factories have different advantages. In digital and virtual approaches, external factors that are not interesting for the current learning situation, such as machine noise or high temperatures, can be hidden in order to get a

---

[11] See Chap. 11.
[12] See Weeber et al. (2016).
[13] See Sects. 9.3.5 and 9.4.1.
[14] See Sect. 9.4.2.

clearer view of the currently interesting facets of the factory environment. In physical learning factories, this is not possible or only possible to a very limited extent. However, the physical approaches provide an authentic experience that also includes factory-specific uncertainties such as human or machine-related errors. Finally, it can be stated that physical and virtual learning factory approaches are complementary. With the individual advantages of the concepts, an integration and fusion to a hybrid learning factory environment show large potential.[15] In order to fully establish, those hybrid learning factories in education and training as well as in research further research are required regarding the interfaces and the general cooperation between the physical, the digital, and the virtual environment. Further, innovations are desirable regarding individualized learning paths for all participants.

In addition to a worldwide exchange and systematization of the learning factory topic, it is also interesting to network the range of learning factory training courses offered worldwide. This requires new business models and platforms that aim to create learning factory networks. This would bring many advantages:

- Primarily, the worldwide exchange of learning factory modules or environments could greatly reduce the resource intensity of building and operating learning factories and design learning factory modules. Many different topics and challenges could be taken up accordingly by the individual learning factories. Decisive questions to be clarified here are how incentives can be created to make learning factory concepts, but also individual training courses, accessible to other operators and to what extent the learning factory environments and training courses can be passed on in a standardized form.
- In addition, a network of learning factories could ensure good regional coverage with hands-on training regarding various production-relevant topics. A platform as a central point of contact for people and companies interested in learning factory trainings could also broker demand to the individual regional learning factories; the platform as the central point of contact to the network could efficiently link interregional supply and demand.

Furthermore, learning factories are usually cost-intensive, especially in the start-up phase but also for further operation. A challenge in the coming years will be to enable learning factories for the smaller budget, in which simpler but still industry-oriented equipment is used as well as software and virtual environments are used intelligently to expand the possibilities for training and teaching. Here, game-based learning is an interesting approach that may enable simple but effective learning factory concepts. Low-cost learning factories could contribute to an even wider dissemination of the learning system.

In order to make optimum use of the opportunities offered by the learning factories in the coming years, ideas for the good implementation of the learning factory concept should be shared in networks and at conferences. All the academic and industrial learning factory operators are experts in different areas, through exchange

---

[15] See Sect. 9.4.1.3.

and joint coordination, an international learning factory curriculum can emerge that completely attacks the pressing challenges of today's and tomorrow's production.

Reference

Weeber, M., Gebbe, C., Lutter-Günther, M., Böhner, J., Glasschröder, J., Steinhilper, R., et al. (2016). Extending the scope of future learning factories by using synergies through an interconnection of sites and process chains. In *6th CIRP-Sponsored Conference on Learning Factories. Procedia CIRP, 54,* 124–129. https://doi.org/10.1016/j.procir.2016.04.102.